Springer Monographs in Mathematics

Springer
Berlin
Heidelberg
New York
Barcelona
Hong Kong
London
Milan
Paris
Singapore
Tokyo

Władysław Narkiewicz

The Development of Prime Number Theory

From Euclid to Hardy and Littlewood

Springer

Władysław Narkiewicz
Institute of Mathematics
Wrocław University
Plac Grunwaldzki 2–4
50384 Wrocław, Poland
e-mail: narkiew@math.uni.wroc.pl

Library of Congress Cataloging-in-Publication Data applied for

Die Deutsche Bibliothek - CIP-Einheitsaufnahme

Narkiewicz, Wladyslaw:
The development of prime number theory : from Euclid to Hardy and Littlewood / Wladyslaw Narkiewicz. - Berlin ; Heidelberg ; New York ; Barcelona ; Hong Kong ; London ; Milan ; Paris ; Singapore ; Tokyo : Springer, 2000
(Springer Monographs in mathematics)
ISBN 3-540-66289-8

Mathematics Subject Classification (1991):
11-03, 11A41, 11Mxx, 11N05, 11N13, 11N25

ISSN 1439-7382
ISBN 3-540-66289-8 Springer-Verlag Berlin Heidelberg New York

Springer-Verlag Berlin Heidelberg New York
a member of BertelsmannSpringer Science+Business Media GmbH

Printed on acid-free paper SPIN 10886979 44/3111 - 54321

Preface

1. People were already interested in prime numbers in ancient times, and the first result concerning the distribution of primes appears in Euclid's *Elementa*, where we find a proof of their infinitude, now regarded as canonical. One feels that Euclid's argument has its place in *The Book*, often quoted by the late Paul Erdős, where the ultimate forms of mathematical arguments are preserved. Proofs of most other results on prime number distribution seem to be still far away from their optimal form and the aim of this book is to present the development of methods with which such problems were attacked in the course of time.

This is not a historical book since we refrain from giving biographical details of the people who have played a role in this development and we do not discuss the questions concerning why each particular person became interested in primes, because, usually, exact answers to them are impossible to obtain. Our idea is to present the development of the theory of the distribution of prime numbers in the period starting in antiquity and concluding at the end of the first decade of the 20th century. We shall also present some later developments, mostly in short comments, although the reader will find certain exceptions to that rule. The period of the last 80 years was full of new ideas (we mention only the applications of trigonometrical sums or the advent of various sieve methods) and certainly demands a separate book.

2. There are many treatises from which one can learn prime number theory and any list of them should start with E.Landau's monumental *Handbuch der Lehre von der Verteilung der Primzahlen* (Landau 1909a). The next texts treating our topic were published after a long interval: E.C.Titchmarsh's *The Zeta-Function of Riemann* (Titchmarsh 1930c) and *The Distribution of Prime Numbers* by A.E.Ingham (1932). The first (an extended version of which appeared after twenty years (Titchmarsh 1951)) was a monograph, devoted to analytical properties of the zeta-function, which forms an indispensable tool for dealing with various aspects of prime number theory, while the second was a rather concise exposition of the status of that theory in the 1930s. The next important exposition of the subject was given by K.Prachar (1957).

Since then several texts appeared which are either exclusively devoted to prime numbers or present their theory in a broader aspect. Here is our selection of them in chronological order: Estermann (1952), Ayoub (1963), Davenport (1967), Chandrasekharan (1968,1970), Schwarz (1969), Huxley (1972a), Ellison (1976), Tenenbaum (1990a), Tenenbaum, Mendés-France (1997).

Some aspects of the prime number theory and analytical or sieve-theoretical tools related to it are presented in the following treatises: Hua (1947), Trost (1953), Turán (1954), Walfisz (1963), Halberstam, Roth (1966), Blanchard (1969), Montgomery (1971), Bombieri (1974), Halberstam,Richert (1974), Apostol (1976), Motohashi (1983), Ivić (1983,1985), Patterson (1988), Ribenboim (1988,1991,1994,1996), Karatsuba, Voronin (1992).

Broad surveys of the theory and accompanying analytical tools can be found in the encyclopedia articles of H.Bohr, H.Cramér (1923) and Loo Keng Hua (1959).

Various methods of factorizations into primes and primality testing were described, together with their history, in the book by H.Riesel (1985) so we felt free not to touch this subject.

3. This book starts with various proofs of the infinitude of primes, commencing with the classical argument of Euclid. Passing through Euler's discovery of primitive roots and the divergence of the series of reciprocals of primes we conclude the first chapter with a survey of various formulas for prime numbers and related functions. In the second chapter we present Dirichlet's theorem on the infinitude of primes in arithmetical progressions, while the third chapter is devoted to the work of Čebyšev. We recall there the first nontrivial upper and lower bounds for the number $\pi(x)$ of primes not exceeding x as well as some of the applications of these bounds.

Although the application of the zeta-function $\zeta(s) = \sum_{n=1}^{\infty} n^{-s}$ to the study of prime numbers appears already in the work of Dirichlet and Čebyšev, it was Riemann who first considered this function in the complex plane, extended it to a meromorphic function and found its intimate connection with the behaviour of the function $\pi(x)$. His work is surveyed in the fourth chapter, where we also consider the analytical properties of Dirichlet's L-functions, describe the pioneering work of Cahen on the theory of Dirichlet series and consider Stieltjes' study of the expansion of $\zeta(s)$ into power series.

The fifth chapter is devoted to the first proofs of the Prime Number Theorem due to Hadamard and de la Vallée-Poussin. Before presenting them we give an exposition of two theorems proved by von Mangoldt confirming assertions made by Riemann, the first concerning the number of non-real roots of $\zeta(s)$ in a strip and the second giving an explicit formula for $\pi(x)$. Both Hadamard's and de la Vallée-Poussin's arguments use the non-vanishing of the zeta-function on the line Re $s = 1$. For this fact we also present several later proofs. We close this chapter with de la Vallée-Poussins bounds for the error term in the Prime Number Theorem.

In the last chapter we treat the developments in the first twenty years of our century. Here the main role is played by Landau, who essentially simplified the theory of prime numbers and published, in 1909, the first book devoted exclusively to that theory. We present his proofs of theorems of von Mangoldt as well as of the Prime Number Theorem, survey the study of the roots of the zeta-function and introduce Littlewood's result about the sign of the difference $\pi(x) - \mathrm{li}(x)$. We conclude with a list of conjectures posed by Hardy and Littlewood in 1923 and give a short survey of the research concerning them. We stop our story at that point. The following years brought a great leap forward in our knowledge of prime numbers determined by the birth of new powerful methods, but this should be, possibly, a subject of another book.

4. I would like to thank to my friends who helped me to obtain sources that were difficult to access. In particular my sincere thanks go to Franz Halter-Koch, Christian U. Jensen, Tauno Metsänkylä, Štefan Porubský, Ladislav Skula and Michel Waldschmidt.

I am also very grateful to Aleksandar Ivić, Jerzy Kaczorowski and Andrzej Schinzel who read early versions of my manuscript and proposed several improvements. Special thanks go to Mark Coleman who very carefully examined the text, corrected the language, pointed out several errors and inconsistencies and many printing errors and suggested many changes.

November 1999

Władysław Narkiewicz
Institute of Mathematics
Wrocław University
E-mail: narkiew@math.uni.wroc.pl

Table of Contents

Notation

1. The letter p with or without indices will be reserved for prime numbers, except when explicitly stated. The letter $\mathbf{C}$ will denote the complex field, $\mathbf{R}$ the field of real numbers, $\mathbf{Q}$ the field of rationals, $\mathbf{Z}$ the ring of all rational integers, $\mathbf{N}$ the set of all positive integers and $\mathbf{P}$ the set of all rational primes. $\mathbf{F}_q$ will denote the finite field of q elements. By $[x]$ and $\{x\}$ we shall denote the integral resp. fractional part of x. For prime p the notation $p^a \parallel n$ means that p^a is the largest power of p dividing n. By p_n we shall denote the nth consecutive prime number. The cardinality of a set A will be denoted by $\#A$.

We shall adhere to the old illogical tradition, which goes back to Riemann, and according to which one denotes the complex variable by $s = \sigma + it$, $\sigma = \mathrm{Re}\ s$ and $t = \mathrm{Im}\ s$ being the real and the complex part of s, respectively.

2. If f, g are two functions defined on a set S, then we shall write $f(x) = O(g(x))$ if the ratio $|f(x)/g(x)|$ is bounded for all $x \in S$. If the ratio $f(x)/g(x)$ tends to zero for x tending to a value x_0 (which may be infinite) then we shall write $f(x) = o(g(x))$.

The symbols $O(\cdot)$ and $o(\cdot)$ are called usually the *Landau symbols*. This name is only partially correct[1], since it seems that the first of them appeared first in the second volume of P.Bachmann's treatise on number theory (Bachmann 1894). In any case Landau (1909a,p.883) states that he had seen it for the first time in Bachmann's book. The symbol $o(\cdot)$ appears first in Landau (1909a). Earlier this relation has been usually denoted by $\{\cdot\}$.

The symbol $f(x) = \Omega(g(x))$ for x tending to some x_0 (which may be infinite) means the falsity of the relation $f(x) = o(g(x))$, i.e. it implies the existence of a positive constant C and a sequence x_n tending to x_0 such that

$$|f(x_n)| \geq C|g(x_n)|$$

for all n.

Its variants $f(x) = \Omega_+(g(x))$ and $f(x) = \Omega_-(g(x))$ mean that

---

[1] This kind of misnaming has an old tradition in mathematics: it is well-known that among the names *Pellian equation* and *non-Pellian equation* only the second is correct, since in fact Pell has nothing to do with either of them. Also the *Möbius inversion formula* was found by Dedekind and Liouville, not by Möbius. It seems that this list may be continued indefinitely.

$$f(x_n) \geq C|g(x_n)|$$

or

$$f(x_n) \leq -C|g(x_n)|,$$

hold with a positive C for some sequence x_n tending to x_0.

1. Early Times

1.1. The Infinitude of Prime Numbers

1. The distinction between prime and composite integers as well as the notions of the greatest common divisor and least common multiple already appear in the seventh book of Euclid's[1] *Elementa* (Euclid 300 B.C.). We find there the definition of prime and composite numbers (Definitions 11 and 13) and the statement (Proposition 30) that if a prime number divides a product then it must divide one of the factors, for which, contrary to the often expressed belief (see Hardy[2],Wright (1960), pp.181–182, Hasse[3] (1950, Sect. 1.2.10)) Euclid gave only a fallacious argument, as already pointed out by H.Zeuthen[4] [5] (1896) (see Hendy 1975, Knorr 1976). The first correct proof of this assertion seems to be that given by Gauss[6] (1801, Sect.13–14).

2. The first result dealing with the distribution of prime numbers appears as Proposition 20 in the ninth book of Euclid's *Elementa*:

Theorem 1.1. *There are infinitely many prime numbers.*

Actually Euclid (300 B.C.) stated his assertion in the following way "*Prime*[7] *numbers are more than any assigned quantity of prime numbers*", and gave the proof only in the case when this quantity was equal 3. In fact he never considered arbitrary finite sets of numbers, restricting himself usually to the case of three numbers.

---

[1] Euclid, around 300 B.C., lived in Alexandria.

[2] Hardy, Godfrey Harold (1877–1947), Professor at Oxford (1919–1931) and Cambridge (1931–1947).

[3] Hasse, Helmut (1898–1979), Professor in Halle, Marburg, Göttingen, Berlin and Hamburg.

[4] Zeuthen, Hieronimus Georg (1839–1920), Professor in Copenhagen.

[5] See Bašmakova (1948) for a polemics with Zeuthen's assertion. Her claim that the missing steps can be completed without great effort using the Euclidean algorithm and some properties of the greatest common divisor is correct but irrelevant, since Zeuthen put in question not the truth of Euclid's proposition but only his proof of it.

[6] Gauss, Carl Friedrich (1777–1855), from 1807 Professor in Göttingen.

[7] We use Heath's translation.

We present now several proofs of this theorem and start with the well-known Euclid's proof.

First proof of Theorem 1.1. If there are only finitely many primes, and D denotes their product, then the number $D+1>1$ cannot be divisible by any prime divisor of D, hence it has no prime divisors at all, which is a clear contradiction. □

This proof may be used to generate a sequence $\{a_n\}$ of primes as follows: put $a_1 = 2$ and if $a_1, a_2, \dots, a_{n-1}$ are already defined then let a_n be the largest prime divisor of $P_n = 1 + a_1 a_2 \cdots a_{n-1}$. The first few terms of this sequence are thus

$$2, 3, 7, 43, 139, 50\,207, 340\,999, 2\,365\,347\,734\,339, \dots .$$

This sequence was considered by A.A.Mullin (1963) who asked whether it contains all primes and is monotonic. The answer to both parts turned out to be negative. After a few terms of this sequence were computed (Korfhage 1964, Guy, Nowakowski 1975 and Naur 1984) it turned out that $a_{10} < a_9$ (Naur 1984), and it is easy to see that for no n one may have $a_n = 5$. In fact, if this would happen for some $n = n_0$ then, in view of $(P_n, 6) = 1$ for $n \geq 3$, the equality $P_{n_0} = 5^k$ with a certain positive k would follow, implying the divisibility of $P_{n_0} - 1 = 5^k - 1$ by 4. This contradicts the fact that $P_{n_0} - 1$ is a product of distinct primes.

It is still unknown whether $\{a_n\}$ contains all sufficiently large primes. C.D.Cox and A.J.van der Poorten (1968) proved that 2, 3, 7 and 43 are the only primes not exceeding 47 belonging to that sequence. It was stated (Naur 1984) that their proof is incorrect but this assertion seems to be based on some misunderstanding. We present now the arguments given by Cox and van der Poorten. They use certain simple properties of quadratic residues, which we now recall:

If a is not divisible by the prime p and the congruence $X^2 \equiv a \pmod p$ is solvable, then a is called a *quadratic residue* mod p, otherwise it is a *quadratic non-residue*. Since quadratic residues form the kernel of the homomorphism $a \mapsto a^2$ of $\mathbf{F}_p^*$ into $\mathbf{F}_p^*$, it follows that the *Legendre symbol* [8]

$$\left(\frac{a}{p}\right) = \begin{cases} 1 & \text{if } a \text{ is a quadratic residue mod } p \\ -1 & \text{if } a \text{ is a quadratic non-residue mod } p \end{cases}$$

satisfies

$$\left(\frac{ab}{p}\right) = \left(\frac{a}{p}\right)\left(\frac{b}{p}\right).$$

Let now $k \geq 2$ be a fixed integer and let $K > k$ be arbitrary. Write

[8] This symbol was introduced by Adrien Marie Legendre (1752–1833), Professor at l'École Militaire and l'École Normale in Paris (Legendre 1798).

$$P_K = 1 + a_1 a_2 \cdots a_{K-1} = \prod_{j=1}^{r} p_j^{x_j}$$

with $x_j \geq 0$. As P_K is odd we have $x_1 = 0$. Putting

$$a_{1j} = \begin{cases} 0 & \text{if } p_j \equiv 1 \pmod 4 \\ 1 & \text{if } p_j \equiv 3 \pmod 4 \end{cases}$$

for $2 \leq j \leq r$ and

$$a_{ij} = \begin{cases} 0 & \text{if } \left(\frac{p_j}{a_i}\right) = 1 \\ 1 & \text{if } \left(\frac{p_j}{a_i}\right) = -1 \end{cases}$$

for $2 \leq i \leq k$, $2 \leq j \leq r$ we get

$$\sum_{j=2}^{r} a_{1j} x_j \equiv 1 \pmod 2$$

$$\sum_{j=2}^{r} a_{ij} x_j \equiv 0 \pmod 2 \qquad (2 \leq i \leq k).$$

Indeed, since the a_i's are all distinct, we have

$$P_K = 1 + 2a_2 \cdots a_{K-1} \not\equiv 1 \pmod 4$$

hence the number of prime factors of P_K which are congruent to 3 mod 4 is odd. This implies the first of the above congruences and the second follows from the multiplicative property of the Legendre symbol. Taking $k = 6$ and $r = 16$ we get $p_r = 53$ and one sees after a short calculation that any solution of this system must satisfy $x_{16} \neq 0$. This implies the assertion made by Cox and van der Poorten. Indeed, if for some prime $p \leq 47$ distinct from 2,3,7 and 43 we would have $a_K = p$ with a suitable K then the above system should have a solution with $x_{16} = 0$, contradiction.

If in the construction of the sequence $\{a_n\}$ one chooses at each step instead of the largest prime factor the smallest one then another sequence results, starting with

$$2, 3, 7, 43, 13, 53, 5, 6\,221\,671, \ldots .$$

The first 43 terms of this sequence were calculated by S.S.Wagstaff Jr. (1993) and his results give some support to the conjecture of D.Shanks[9] (1991) stating that every prime occurs in that sequence.

Elements of the sequence $p_1 p_2 \cdots p_n + 1$ were tested for primality in Borning (1972), Templer (1980), Buhler, Crandall, Penk (1982) and Caldwell (1995). There are 18 primes of this form with $p_n < 35000$ and it is not known whether there are infinitely many of them.

R.Guy and R.Nowakowski (1975) considered a related set A of primes, consisting of all prime divisors of the sequence b_n defined by $b_0 = 2$, $b_{n+1} = 1 + \prod_{j=0}^{n} b_j$.

---

[9] Shanks, Daniel (1917–1996), Professor at the University of Maryland.

They established that no prime congruent to 5 mod 6 can appear in A. Using algebraic number theory R.W.K.Odoni (1985) showed that if $A(x)$ denotes the number of elements of A not exceeding x then the ratio $A(x)/\pi(x)$, where $\pi(x)$ is the number of all primes $\leq x$, is $O(1/\log\log\log x)$, hence tends to zero. He pointed out that in view of $b_{n+1} = b_n^2 + b_n$ the sequence $\{b_n \bmod p\}$ is, for any prime p, ultimately periodic and hence it is a finite task to determine whether a given prime belongs to A. There are 23 elements of A below 10000 and those below 1500 are[10] 2,3,7,13,43,73,139,181,547,607,1033 and 1459.

3. There are several variants of Euclid's proof of Theorem 1.1. The simplest of them, which according to H.Brocard (1915) is due to C.Hermite[11], runs as follows:

Second proof of Theorem 1.1. For $n = 1, 2, \ldots$ denote by q_n the smallest prime divisor of $n! + 1$. The evident inequality $q_n > n$ implies that this sequence contains infinitely many distinct elements. □

The first few terms of the sequence $\{q_n\}$ are 2, 3, 7, 5, 11, 7, 71, 61, 19, 11, 39916801, 13, 83, 23, 59, 17, 661, 19, 71, 20639383, 43, ... Wilson's theorem, which states that for every prime p one has

$$(p-1)! \equiv -1 \pmod{p},$$

implies that this sequence contains every prime number. This fact gives this proof a certain advantage over Euclid's proof.

It is not known whether the sequence $n! + 1$ contains infinitely many primes. There are 17 primes known of this form with $n \leq 4580$, the largest being $1477! + 1$ (Caldwell 1995). For previous searches see Borning (1972) and Buhler, Crandall, Penk (1982).

Another variant of Euclid's argument was presented by T.J.Stieltjes[12] (1890):

Third proof of Theorem 1.1. Assume the set of all primes to be finite, denote by D their product and let $D = mn$ be any factorization of D with positive integral m, n. For any prime p either m or n is divisible by p, but not both. This shows that $m+n$ cannot have any prime divisor, but obviously we have $m + n > 1$ and this gives a contradiction. □

Essentially the same idea occurs in Thompson (1953).

A similar idea has been used by J.Braun (1899):

---

[10] In Odoni (1985) the number 97 is also listed but this seems to be a computational error. This number seems to be rather unlucky; for a long time the quadratic field $Q(\sqrt{97})$ was believed to be Euclidean, until E.S.Barnes and H.P.F.Swinnerton-Dyer (1952) showed that this is not the case.

[11] Hermite, Charles (1822–1901), Professor in Paris.

[12] Stieltjes, Thomas Jan (1856–1894), Professor in Toulouse.

Fourth proof of Theorem 1.1. If the set of all primes $\{p_1, p_2, \dots, p_n\}$ is finite, then let D denote their product and write

$$\sum_{j=1}^{n} \frac{1}{p_i} = \frac{a}{D},$$

with $a = \sum_{j=1}^{n} D/p_j$. Since $a/D > 1/2 + 1/3 + 1/5 = 31/30 > 1$, the number a must have a prime divisor, say p_k, but then p_k must divide D/p_k, which is not possible. □

4. Leonhard Euler[13] gave two proofs of Theorem 1.1, differing in an essential way from that of Euclid. Euler was, for a long time, interested in prime numbers. He discovered a primality criterion for numbers congruent to unity mod 4 (Euler 1762/63) and later published a table of the least prime divisor of every number below 10^6 (Euler 1775). His first proof of the infinitude of primes (Euler 1737, Theorem 7) does not meet our criteria for rigour but hides a good idea. Trying to show that the product

$$\prod_p \frac{p}{p-1}$$

is infinite he put

$$x = 1 + \frac{1}{2} + \frac{1}{3} + \frac{1}{4} + \cdots$$

and successively eliminated all nonunit terms on the right-hand side. He divided this formula by 2, giving

$$\frac{1}{2}x = \frac{1}{2} + \frac{1}{4} + \cdots$$

and, subtracting, obtained

$$\frac{x}{2} = 1 + \frac{1}{3} + \frac{1}{5} + \frac{1}{7} + \frac{1}{9} \cdots.$$

Division by 3 and subtraction led to

$$\frac{1 \cdot 2}{2 \cdot 3} x = 1 + \frac{1}{5} + \frac{1}{7} + \cdots = \sum_{(n, 2\cdot 3)=1} \frac{1}{n}.$$

Repeating this procedure, Euler finally reached the equality

$$\frac{1 \cdot 2 \cdot 4 \cdot 6 \cdot 10 \cdots}{2 \cdot 3 \cdot 5 \cdot 7 \cdot 11 \cdots} x = 1.$$

He concluded his argument by noting that since the divergence of harmonic series implies $x = \infty$, one must have

[13]Euler, Leonhard (1707–1783), Professor at the Academies of Sciences in Sankt Petersburg (1730–1741 and from 1766 on) and Berlin (1741–1766).

$$\prod_p \frac{p}{p-1} = \frac{2\cdot 3\cdot 5\cdot 7\cdot 11\cdots}{1\cdot 2\cdot 4\cdot 6\cdot 10\cdots} = \infty. \tag{1.1}$$

A correct realization of Euler's idea was presented by L.Kronecker[14] in his lectures in 1875/76 (Kronecker 1901):

Fifth proof of Theorem 1.1. The argument is based on a formula proved by L.Euler (1737, Theorem 8) (see also Euler (1748, Chap.15)) in which the connection of prime numbers with the zeta function defined for $s > 1$ by[15]

$$\zeta(s) = \sum_{n=1}^{\infty} \frac{1}{n^s}$$

made its first appearance. We present it with Euler's original proof:

Lemma 1.2. *(Euler's product formula) For $s > 1$ one has*

$$\zeta(s) = \prod_p \frac{1}{1-p^{-s}}. \tag{1.2}$$

Proof. In his proof Euler used the same idea as in his argument leading to (1.1) but this time his reasoning was correct:

Put

$$A = 1 + \frac{1}{2^s} + \frac{1}{3^s} + \cdots,$$

observe that

$$\frac{A}{2^s} = \frac{1}{2^s} + \frac{1}{4^s} + \frac{1}{6^s}\cdots$$

and deduce

$$\frac{2^s-1}{2^s}A = 1 + \frac{1}{3^s} + \frac{1}{5^s}\cdots.$$

Divide the obtained equality by 3^s and subtract, getting

$$\frac{2^s-1}{2^s}\frac{3^s-1}{3^s}A = 1 + \frac{1}{5^s} + \frac{1}{7^s} + \frac{1}{11^s}\cdots.$$

Continuing in this way one arrives (using an obvious passage to the limit, which Euler omitted) at

$$A\prod_p \frac{p^s-1}{p^s} = 1,$$

which immediately implies (1.2). □

[14]Kronecker, Leopold (1823–1891), Professor in Berlin (1883–1891).
[15]By an old tradition the letter s is used to denote the variable here.

If there were only finitely many primes, then the right-hand side of (1.2) would, as x tends to 1, tend to a finite limit, say c. But then for every natural number N one would have

$$c = \lim_{x\to 1+0} \zeta(s) \geq \lim_{s\to 1+0} \sum_{n=1}^{N} \frac{1}{n^s} = \sum_{n=1}^{N} \frac{1}{n},$$

and the last sum may be made as large as one wishes by a suitable choice of N, due to the divergence of the harmonic series $\sum_{n=1}^{\infty} 1/n$. □

A variant of this proof, based on the evaluation $\sum_{n\leq x} 1/n \geq \log x$, was given by J.J.Sylvester[16] (1888a):

Sixth proof of Theorem 1.1. Obviously one has

$$\prod_{p\leq x} \left(1 - \frac{1}{p}\right)^{-1} = \prod_{p\leq x} \left(1 + \frac{1}{p} + \frac{1}{p^2} + \cdots\right) \geq \sum_{n\leq x} \frac{1}{n} \geq \log x \tag{1.3}$$

and since x may be arbitrarily large the set of primes must be infinite. □

5. Euler's second proof of Theorem 1.1 (Euler[17] 1849, Sect.135) uses the *totient function* $\varphi(n)$, defined as the number of positive integers not exceeding n and coprime to n, and uses the following two simple properties of this function: its multiplicativity (i.e. $\varphi(mn) = \varphi(m)\varphi(n)$ for coprime m,n) and the equality $\varphi(p) = p - 1$ for prime p. The last property is evident, and the first was established[18] by Euler (1760/61) in the following way: noting that if $a_1, a_2, \ldots, a_k$ (with $k = \varphi(m)$) are all the positive integers prime to m and not exceeding m then the integers $< mn$ and prime to mn lie in $\varphi(m)$ arithmetic progressions $a_i + km$ $(k = 0, 1, \ldots, n-1)$ and in view of $(m, n) = 1$ every such progression contains exactly $\varphi(n)$ integers prime to n.

Seventh proof of Theorem 1.1. Assume the set of all primes to be finite and let D denote their product. Then

$$\varphi(D) = \prod_{p} (p-1) \geq 2 \cdot 4 \cdots > 2$$

and thus there must be an integer a in the interval $[2, D]$ which is coprime to D. This integer cannot therefore have any prime divisor and so must equal 1. This contradicts $a \geq 2$. □

A variant of this argument appears in Kummer[19] (1878) and Perott (1888).

---

[16]Sylvester, James Joseph (1814–1897), Professor at the Military School at Woolwich (1855–1876), John-Hopkins University in Baltimore (1876–1883 and after 1892) and Savilian Professor in Oxford (1883–1892).

[17]Published posthumously.

[18]See Euler 1780.

[19]Kummer, Ernst Eduard (1810–1893), Professor in Breslau and Berlin.

The next proof uses a simple lower bound for the number of *square-free integers* (integers not divisible by the square of any integer exceeding 1) less than a given number:

Eighth proof of Theorem 1.1. (Perott 1881,1888) Observe that if $N > 2$ then the number of integers $\leq N$ which are divisible by a square exceeding 1 is bounded by

$$\sum_{2\leq n\leq N} \left[\frac{N}{n^2}\right] < N\left(\frac{1}{4}+\sum_{n=3}^{\infty}\frac{1}{n^2}\right) \leq N\left(\frac{1}{4}+\int_2^{\infty}\frac{dt}{t^2}\right) = \frac{3}{4}N,$$

hence there are at least $N/4$ square-free integers in the interval $[1, N]$.

If there are only k primes, say $p_1, p_2, \dots, p_k$, then every square-free number > 1 has the form $p_{i_1} p_{i_2} \cdots p_{i_s}$ with $1 \leq i_1 < \dots < i_s \leq k$ and thus there are at most 2^k square-free numbers exceeding 1. This implies the inequality $2^k > N/4$ which is evidently false for sufficiently large N. □

See also Gegenbauer[20] (1888).

The following argument uses the same idea in a slightly simpler form. It appears in different variants in Thue[21] (1897), Auric (1915) and Rubinstein (1993a).

Ninth proof of Theorem 1.1. If $p_1, p_2, \dots, p_k$ are all primes then every integer $N > 1$ can be written in the form $N = \prod_{i=1}^{k} p_i^{\alpha_i}$ with $\alpha_i \geq 0$. If $N \leq x$, then evidently none of the exponents α_i may exceed $\log x/\log 2$ and we have thus at most $f(x) = (\log x/\log 2)^{1+k}$ integers in the interval $[2, x]$, hence $f(x) \geq [x] - 1$ which is clearly impossible for large x.

6. Any infinite sequence of pairwise coprime positive integers leads to a proof of Theorem 1.1. Such a proof first appears in a letter of C.Goldbach[22] to Euler dated[23] July 20, 1730 (see Fuss[24] 1843, I, 32–34; Euler,Goldbach 1965) and is sometimes attributed to G.Pólya[25] (e.g. in Hardy, Wright (1960), Chandrasekharan (1968). P.Ribenboim (1994) wrote that this attribution appears in an unpublished list of exercises of A.Hurwitz[26] preserved in ETH

---

[20]Gegenbauer, Leopold (1849–1903), Professor in Czernowitz, Innsbruck and Vienna.

[21]Thue, Axel (1863–1922).

[22]Goldbach, Christian (1690–1764).

[23]The original date is July 20/31, the double dating being a consequence of the use of the Julianic calendar in Russia before 1918. It seems that this was the first proof of the infinitude of primes which essentially differed from that of Euclid.

[24]Fuss, Pavel Nikolaevič, (1798–1853), Euler's grandson. Secretary of the Academy of Sciences in Sankt Petersburg.

[25]Pólya, George (1887–1985), Professor in Zürich, at Brown University and in Stanford.

[26]Hurwitz, Adolf (1859–1919), Professor in Königsberg and Zürich.

in Zürich.) This proof was published in the well-known collection of exercises of G.Pólya and G.Szegö[27] (1925).

Tenth proof of Theorem 1.1. Let $F_n = 2^{2^n} + 1$ be the n-th *Fermat number*. If $m < n$, then $F_n - 2 = 2^{2^n} - 1$ is divisible by $2^{2^{m+1}} - 1 = \left(2^{2^m} - 1\right) F_m$, hence F_m divides $F_n - 2$. Thus if $d > 1$ is an integer dividing both F_n and F_m then $d = 2$. Yet F_n is odd and so this shows that every two distinct Fermat numbers are coprime. □

Several other sequences of this type leading to proofs of Theorem 1.1 appear in Bellman[28] (1947), Edwards (1964), Lambek, Moser[29] (1957), Mohanty (1978) and Subbarao (1966).

Still another way of proving Theorem 1.1 was shown by M.Wunderlich (1965).

Eleventh proof of Theorem 1.1. Let $\{a_n\}$ be a sequence of distinct positive integers having the property that $(m,n) = 1$ implies $(a_m, a_n) = 1$. Assume that there are only finitely many primes, say $\{p_1, p_2, \dots, p_k\}$ and consider the numbers $a_{p_1}, a_{p_2}, \dots, a_{p_k}$. They are pairwise coprime and at most one of them can be equal to 1. If one could show the existence of two prime indices p such that a_p has at least two distinct prime factors then the product $a_{p_1} a_{p_2} \cdots a_{p_k}$ would have more than k distinct prime divisors, contrary to the assumption.

The following lemma, which goes back to H.Siebeck (1846), shows that one can take for a_n the nth *Fibonacci number*[30] T_n defined for $n \geq 1$ by $T_1 = T_2 = 1$, $T_{n+1} = T_n + T_{n-1}$.

Lemma 1.3. *If* $(M,N) = 1$ *then* $(T_M, T_N) = 1$.

Proof. Observe first that for all m we have $(T_m, T_{m+1}) = 1$. Indeed, every common divisor d of T_m and T_{m+1} divides T_{m-1}, and by recurrence we arrive at $d \mid T_{m-2}, T_{m-3}, \dots, T_1 = 1$.

Let now $n > m$ be given and let d be a common divisor of T_n and T_m. Easy induction leads to

$$T_n = T_{j+1} T_{n-j} + T_j T_{n-j-1} \quad (j = 1, 2, \dots, n-2)$$

and, putting $j = m$, we obtain the divisibility of $T_{m+1} T_{n-m}$ by d. Since T_m and T_{m+1} are relatively prime we get $(d, T_{m+1}) = 1$, hence d divides T_{n-m} and by recurrence we obtain the divisibility of $T_{n-2m}, T_{n-3m}, \dots, T_{n \bmod m}$ by d.

Applying now the division algorithm to the pair $a_0 = N$, $a_1 = M$ we get for suitable s and q_i

[27] Szegö, Gabor (1895–1985), Professor in Königsberg, St.Louis and Stanford.
[28] Bellman, Richard (1920–1984).
[29] Moser, Leo (1921–1970).
[30] They are usually denoted by F_n but we do not want to clash with the notation for Fermat numbers.

$$\begin{aligned} N &= q_0 M + a_2 \qquad (0 < a_2 < a_1) \\ M &= q_1 a_2 + a_3 \qquad (0 < a_3 < a_2) \\ &\cdots\cdots \\ a_{j-2} &= q_{j-2} a_{j-1} + a_j \qquad (0 < a_j < a_{j-1} \\ &\cdots\cdots \\ a_{s-2} &= q_{s-2} a_{s-1} + a_s \quad (0 < a_s < a_{s-1}) \end{aligned}$$

with $a_s = 1$.

Since for $j = 2, 3, \ldots, s$ we have $a_{j-2} \bmod a_{j-1} = a_j$ hence the preceding argument shows that every divisor of T_M and T_N must divide $T_{a_1}, T_{a_2}, \ldots, T_{a_s} = T_1 = 1$ and $(T_M, T_N) = 1$ follows. □

It suffices now to observe that $T_{19} = 4181 = 113 \cdot 37$ and $T_{31} = 1\,346\,269 = 557 \cdot 2417$. □

Lemma 1.3 implies that the set of all prime divisors of the Fibonacci sequence is infinite. This is a particular case of a much general situation: it was shown by G.Pólya (1921) that the same happens for a large class of linear recurrences. See Hasse (1966), Lagarias (1985), Stephens (1976) and Ward (1954,1961).

For a simple proof of a generalization of Lemma 1.3 see Morton (1995).

7. We conclude our selection of proofs of infinitude of primes with a neat topological proof found by H.Furstenberg (1955):

Twelfth proof of Theorem 1.1. Topologize the set of all rational integers **Z** by taking for the basis of open sets the family of all arithmetic progressions. One easily sees that the resulting topology is normal[31] and every arithmetic progression is an open and closed set. For every prime p denote by A_p the set of all integers divisible by p and put $A = \bigcup_p A_p$. Since the set $\{-1, 1\} = \mathbf{Z} \setminus A$ is not open, A is not closed, but if we had only finitely many primes then A would be closed, being a finite union of closed sets. □

See Brown (1953), Golomb (1959) and Kirch (1969).

For other proofs of Theorem 1.1 see Barnes (1976) (where a witty argument is given, which uses the theory of non-Pellian equations $X^2 - dY^2 = -1$ and continued fractions), Cohen (1969), Harris (1956) and Sándor (1988).

Several other are listed in Dickson[32] (1919) and Ribenboim (1988,1994).

---

[31] A topological space is said to be normal if any two distinct points have disjoint neighbourhoods.

[32] Dickson, Leonard Eugene (1874–1954), Professor in Chicago.

1.2. Sum of Reciprocals of Primes

1. A slight modification of Euler's argument in his first proof of Theorem 1.1 leads to a stronger assertion (Euler 1737, Theorem 19):

Theorem 1.4. *The series*

$$\sum_p \frac{1}{p},$$

where p runs over all primes, diverges.

Euler used freely divergent series and proceeded as follows:

He put $A = \sum 1/p$, $B = \sum 1/p^2$, $C = \sum 1/p^3$, etc. and then used the 'equalities'

$$e^{A+B/2+C/3+\cdots} = \frac{2\cdot 3\cdot 5\cdots}{1\cdot 2\cdot 4\cdots} = 1 + \frac{1}{2} + \frac{1}{3} + \cdots = \infty ,$$

the first being a consequence of

$$\sum_{k=1}^{\infty} \frac{1}{kp^k} = \log\left(\frac{p}{p-1}\right),$$

and the second being taken from the 'proof' of the infinity of primes, which we presented in the preceding section. Since the series

$$\frac{B}{2} + \frac{C}{3} + \cdots = \sum_p \sum_{k=2}^{\infty} \frac{1}{kp^k}$$

converges, one must have $A = \infty$.

It is not difficult to rectify Euler's argument:

Proof of Theorem 1.4. Assume that the series of reciprocals of primes converges. Then the product in (1.2) converges for $s = 1$ to a finite value, say c. For $s > 1$ we have

$$\zeta(s) = \prod_p \frac{1}{1-p^{-s}} \le \prod_p \frac{1}{1-p^{-1}} = c,$$

but due to the divergence of the harmonic series the function $\zeta(s)$ is unbounded when s approaches unity (see the fifth proof of Theorem 1.1). □

A modification of this argument which uses the convergence of the series $\sum_{n=1}^{\infty} 1/n^2$ appears in Sylvester (1888a):

Second proof of Theorem 1.4. Using (1.3) one obtains

$$\prod_{p\le x}\left(1+\frac{1}{p}\right)=\frac{\prod_{p\le x}(1-p^{-2})}{\prod_{p\le x}(1-p^{-1})}\ge \log x\prod_{m=2}^{\infty}\left(1-\frac{1}{m^2}\right),$$

and it suffices to observe that the the right-hand side of this inequality tends to infinity. □

Another proof was found by C.Vandeneynden (1980):

Third proof of Theorem 1.4. For fixed $x \ge 2$ let $\mathcal{P}$ be the set of all primes not exceeding x and denote by $\mathcal{M}$ the set of all integers whose prime factors belong to $\mathcal{P}$. In view of the identity

$$\left(1+\frac{1}{p}\right)\left(1+\frac{1}{p^2}+\frac{1}{p^4}+\cdots+\frac{1}{p^{2k}}+\cdots\right)=\sum_{j=0}^{\infty}\frac{1}{p^j},$$

holding for every prime p, we obtain

$$\prod_{p\in\mathcal{P}}\left(1+\frac{1}{p}\right)\sum_{n\in\mathcal{M}}\frac{1}{n^2}=\sum_{n\in\mathcal{M}}\frac{1}{n}.$$

The divergence of the harmonic series and convergence of $\sum_{n=1}^{\infty}1/n^2$ imply that the product

$$\prod_{p\in\mathcal{P}}\left(1+\frac{1}{p}\right)$$

tends to infinity with x. As for every positive c the inequality $e^c > 1+c$ holds and the set of all primes is infinite, we get

$$\prod_{p\in\mathcal{P}}\left(1+\frac{1}{p}\right)<\prod_{p\in\mathcal{P}}e^{1/p}=\exp\left(\sum_{p\in\mathcal{P}}\frac{1}{p}\right),$$

and this shows that the sum

$$\sum_{p\in\mathcal{P}}\frac{1}{p}$$

tends to infinity for $x\to\infty$. The assertion follows immediately. □

For other proofs of Theorem 1.4 see Bellman (1943), Clarkson (1966), Dux (1956), Erdős[33] (1938) (reproduced in Hardy,Wright 1960), Moser (1958) and Treiber (1995).

We point out an immediate consequence of Theorem 1.4:

---

[33]Erdős, Paul (1913–1996), Professor of the Hungarian Academy of Sciences. Authored more than 1300 papers in number theory, combinatorics and analysis, which seems to be a world record.

Corollary 1. *One has*

$$\lim_{x\to\infty}\prod_{p\le x}\left(1-\frac{1}{p}\right)=0. \qquad \square$$

The next corollary was first stated by L.Euler (1737, Corollary 3 to Theorem 7) without a proof. We present a simple proof due to G.Fousserau (1892).

Corollary 2. *For x tending to infinity one has $\pi(x)=o(x)$.*

Proof. For any $k>1$ we have obviously

$$\pi(x)\le k+\sum_{\substack{n\le x\\(n,k)=1}}1,$$

hence writing $x=qk+r$ with $0\le r<k$ we get

$$\pi(x)\le k+q\varphi(k)+r\le 2k+\left[\frac{x}{k}\right]\varphi(k).$$

This implies

$$\limsup_{x\to\infty}\frac{\pi(x)}{x}\le\frac{\varphi(k)}{k}.$$

If we now choose $k=k_r=p_1p_2\cdots p_r$, the product of the first r primes, then $\varphi(k_r)=\prod_{j=1}^{r}(p_j-1)$ and thus

$$\limsup_{x\to\infty}\frac{\pi(x)}{x}\le\prod_{p\le p_r}\left(1-\frac{1}{p}\right)$$

for all $r\ge 1$. Corollary 1 shows that the right-hand side of this inequality tends to zero as r tends to infinity. $\square$

(See Mamangakis 1962.)

2. Both Euler (1737, 1744, Satz 19) and Gauss (1791) stated that the sum $s(x)$ of reciprocals of primes not exceeding x grows like $\log\log x$. An attempt to prove this assertion was made by P.L.Čebyšev[34] (1848) but the first correct

[34]Čebyšev, Pafnutij Lvovič (1821–1894), Professor in Sankt Petersburg. His name has been spelt in a record number of ways. I am aware of at least fifteen different spellings: Chebychef (Dressler, van der Lune 1975), Chebychev (Lou, Yao 1980), Chebyshev (this seems to be the usual spelling, see e.g. Niven et al. 1991), Tchebicheff (Poincaré 1892), Tchebychef (Dickson 1919, Ellison 1975, Hardy, Wright 1960), Tchebycheff (Sierpiński 1987), Tchébycheff (Markov 1895), Tchébychew (Césaro 1885b), Tchebyschef (Ricci 1934), Tschebyschef (Landau 1902), Tschebysheff (Ribenboim 1994), Tschebyschew (Der Grosse Brockhaus, 15th ed., 1934). Čebyšev himself used the spellings Tchebichef (Čebyšev 1848, 1850), Tchébychev (Čebyšev 1853) and Tchebichev (Čebyšev 1857).

proof was found by F.Mertens[35] (1874b) and we shall present it in Chap.3 (see Theorem 3.11). V.Brun[36] (1917) pointed out that the upper bound

$$s(x) = \sum_{p \le x} \frac{1}{p} = O(\log\log x)$$

can be obtained by the following argument which is completely elementary:

Let $x > 2$ be given and let $p_1 < p_2 < \ldots < p_m$ be all the primes contained in the interval $(\sqrt{x}, x]$. Consider the set $\mathcal{M}$ of all integers not exceeding x which are divisible by at least one of the p_i's. They can be uniquely represented in the form rp_j with $1 \le r \le x/p_j$ and $j = 1, 2, \ldots, m$, since from $rp_j = sp_k$ with $j < k$, $1 \le r \le x/p_j$ and $1 \le s \le x/p_k$ one gets $p_j | s$ and

$$p_j \le s \le \frac{x}{p_k} < \frac{x}{p_j}$$

thus $p_j^2 < x$, contradiction. It follows that $\mathcal{M}$ has at least

$$\sum_{j=1}^{m} \left[\frac{x}{p_j}\right] \ge x \sum_{j=1}^{m} \frac{1}{p_j} - m \ge x \sum_{j=1}^{m} \frac{1}{p_j} - \pi(x)$$

elements but, clearly, $\#\mathcal{M} \le x$ and so we obtain

$$\sum_{j=1}^{m} \frac{1}{p_j} \le 1 + \frac{\pi(x)}{x} \le 2.$$

In view of

$$\sum_{j=1}^{m} \frac{1}{p_j} = s(x) - s(\sqrt{x})$$

we get

$$s(x) - s(\sqrt{x}) \le 2.$$

If M is the minimal integer satisfying $x < 2^{2^M}$ then

$$s(x) = (s(x) - s(x^{1/2})) + (s(x^{1/2}) - s(x^{1/2^2})) + \cdots + (s(x^{1/2^j}) - s(x^{1/2^{j+1}})) + \cdots + (s(x^{1/2^{M-1}}) - s(x^{1/2^M})) < 2M$$

and in view of $M = \log\log x/\log 2 + O(1)$ the assertion results.

[35]Mertens, Franz (1840–1927), Professor in Kraków, Graz and Vienna.
[36]Brun, Viggo (1885-1978), Professor in Trondheim and Oslo.

1.3. Primitive Roots

1. Let N be an integer exceeding 1. If there exists an integer a such that the set of residues mod N of its powers is equal to the set of all residues mod N, coprime to N then a is called a *primitive root* mod N. J.H.Lambert[37] (1769) made the following assertion:

Theorem 1.5. *For every prime p there exist primitive roots* mod p.

Four years later L.Euler (1773) published a proof of this theorem. Gauss (1801, Sect.56) considered it to be incomplete and wrote:

"*... the author assumed tacitly that the congruence $x^n \equiv 1$ has n distinct solutions, although he established only that it cannot have more than n solutions. Moreover the formula in section 34 has only a proof by induction.*"

Gauss's criticism seems to be exaggerated. Although one finds the false assertion quoted by Gauss in Sections 31 and 32 of Euler's paper, it is rather obvious that this is only a *lapsus mani* because just before, in the *Scholion* in Sect.28 Euler gives examples to show that the said congruence may have less than n 'real' solutions, the others being called 'imaginary'. In any case this assertion is used only in the case when n equals $p-1$ or one of its divisors, in which case Fermat's theorem (which states that for all integers a not divisible by p the congruence $a^{p-1} \equiv 1 \pmod p$ holds) implies that the number of solutions of the considered congruence equals its degree[38].

After defining primitive roots Euler shows in Sect.28 that $x^n - 1$ can be divisible by p for at most n values of $x < p$. He starts with the evident case $n = 1$, while for $n = 2$ notes that if p divides $x^2 - 1$ then it must divide either $x - 1$ or $x + 1$, thus x equals either 1 or $p - 1$. In the case $n = 3$ he argues as follows:

If p divides $x^3 - 1 = (x-1)(x^2+x+1)$, then either $x = 1$ or p divides $x^2 + x + 1$. If $x = a, b$ where a, b are integers satisfying $1 < a < b < p$ for which p divides $x^2 + x + 1$ then, on subtraction, p divides $(a-b)(a+b+1)$, thus $p|a+b+1$ and we get $p|(a^2+a+1) - (a+b+1) = a^2 - b$ which gives only one possibility for b.

In the general case Euler writes:

"*If $x = a$ is a solution* [of $p|x^n - 1$], *then $x - a$ divides $x^n - 1 - mp$ with suitable m and dividing we get a form*[39] *of a lower degree. If $x = b$ is another solution then again we are led to a form of lower degree. Thus we cannot get more than n solutions.*"

[37]Lambert, Johann Heinrich (1728–1777), physicist and philosopher. He was the first to prove the irrationality of the number π.

[38]Cf. the comments of A.Rudio in Euler's *Opera omnia*, I.3, pp.XXVIII–XXX and Weil (1983)

[39]Euler used the word 'form' to mean 'polynomial'.

This argument is not quite exact, since in his induction step Euler assumes the truth of the assertion for all polynomials of degree n, not only for $x^n - 1$. This however can be easily repaired by replacing $x^n - 1$ by an arbitrary polynomial of degree n and noting the truth of this more general assertion in the trivial case $n = 1$. In this way one is lead to a result proved few years earlier by J.L.Lagrange[40] (1768). We state it using modern terminology:

Lemma 1.6. *If p is a prime, $f(X) = \sum_{i=0}^{n} a_j X^j$ a polynomial with integral coefficients and $p \nmid a_n$ then the congruence*

$$f(X) \equiv 0 \pmod{p} \tag{1.4}$$

has at most n solutions distinct mod p.

The proof of this assertion given by Gauss (1801, Sect.43) uses a similar idea:

Proof. The assertion being evident in case $n = 0$, assume its truth for all polynomials of degree $< n$ and suppose that for a polynomial $f(X)$ of degree n the congruence (1.4) has at least $1+n$ distinct solutions mod p. If p divides $f(0)$ then we may assume $a_0 = 0$, divide $f(X)$ by X and use the induction assumption. If however $f(0) \not\equiv 0 \pmod{p}$ then let α be any solution of (1.4). The polynomial $g(X) = f(X + \alpha)$ is of degree n, and $g(X) \equiv 0 \pmod{p}$ has $n + 1$ solutions distinct mod p, but $g(0)$ is divisible by p and this brings us to the previous case. □

Corollary. *If p is a prime then the congruence*

$$X^n \equiv 1 \pmod{p} \tag{1.5}$$

has at most n distinct solutions. □

In the last step of his proof Euler observed that if n divides $p - 1$ then the congruence (1.5) has exactly n solutions. He did not give a proof of that, but this gap is easily filled:

For $n = p - 1$ this assertion is a consequence of Fermat's theorem and if n is a divisor of $p - 1$ and we write $p - 1 = mn$, then for $j = 1, 2, \dots, p - 1$ the residue $a_j = j^m \bmod p$ is a solution of (1.5). If now $j^m \equiv k^m \pmod{p}$ and l is the reciprocal of k mod p then $(jl)^m \equiv 1 \pmod{p}$, and thus by Lemma 1.6 there are at most m possibilities for l. This shows that there are at least $(p - 1)m = n$ distinct a_i's and, invoking again Lemma 1.6 we are done.

In Sect.30 of his paper Euler calls a solution x of (1.5) *proper*, if for every $0 < k < n$ the number $x^k - 1$ is not divisible by p and then, in Sect.34 poses

[40]Lagrange, Joseph Louis de (1736–1813), Professor at the Academy of Sciences in Berlin (1766–1787) and later at l'École Normale and l'École Polytechnique in Paris.

the question of enumerating proper solutions. He computes their number S_n for $n = 1, 2, \dots, 9$ and then gives the following formula, implying $S_n = \varphi(n)$:

If $n = \prod_{j=1}^{t} q_j^{c_j}$ is the canonical factorization of n, then

$$S_n = \prod_{j=1}^{t} q_j^{c_j - 1}(q_j - 1).$$

This is the formula 'proved only by induction' according to Gauss. However the argument given in the case of small values of n seems to indicate that Euler knew how to proceed further, but for some reason did not do that explicitly. From this formula one obtains immediately $S_{p-1} \neq 0$ and this implies the existence of primitive roots.

2. Gauss gave two proofs of Theorem 1.5 which we shall now present. His first proof (Gauss 1801, Sect.52–54) is based on two preliminary results dealing with Euler's totient function $\varphi(n)$:

Lemma 1.7. *For every N one has*

$$\sum_{d|N} \varphi(d) = N.$$

Proof. For every divisor d of N let

$$A_d = \{mN/d : 1 \le m \le d : \quad (d, m) = 1\}.$$

Note that A_d has $\varphi(d)$ elements and observe that the sets A_d are pairwise disjoint. Indeed, if $a \in A_{d_1} \cap A_{d_2}$ then for $i = 1, 2$ write $a = m_i N/d_i \in A_{d_i}$ with $1 \le m_i \le d_i$ and $(d_i, m_i) = 1$. This implies that m_1 divides $m_2 d_1$ and hence $m_1 | m_2$. By symmetry we also obtain $m_2 | m_1$ and thus $m_1 = m_2$ and $d_1 = d_2$ results.

If we put $A = \bigcup_{d|N} A_d$ then $\sum_{d|N} \varphi(d) = \#A \le N$ and it remains to show that A contains all integers from $[1, N]$, but if $1 \le k \le N$ and $d = (k, N)$ then $(k/d, N/d) = 1$ and thus $k \in A_d$. □

Another proof[41] of Lemma 1.7 was given by E.Catalan[42] (1839):

Catalan's proof. Since $\varphi(mn) = \varphi(m)\varphi(n)$ holds for all relatively prime integers m, n and every divisor d of $n = \prod_{i=1}^{t} p_i^{\alpha_i}$ has the form $d = \prod_{i=1}^{t} p_i^{\beta_i}$ with $0 \le \beta_i \le \alpha_i$, we get

$$\sum_{d|n} \varphi(d) = \sum_{\substack{0 \le \beta_i \le \alpha_i \\ i=1,2,\dots,t}} \varphi(p_1^{\beta_1}) \cdots \varphi(p_t^{\beta_t}) = \prod_{i=1}^{t} (1 + \varphi(p_i) + \cdots + \varphi(p_i^{\alpha_i})).$$

[41]A variant of it can be found in Sylvester (1860).
[42]Catalan, Eugéne Charles (1814–1894), Professor in Liége.

It suffices now to use the equality

$$1+\varphi(p_i)+\cdots+\varphi(p_i^{\alpha_i})=1+(p_i-1)(1+p_i+\cdots+p_i^{\alpha_i-1})=p_i^{\alpha_i}$$

to obtain the assertion. □

Still another proof is contained in a survey paper of A.L.Crelle[43] (1845). If one replaces in it the words 'and so on' by a formal induction then this proof can be presented in the following way:

Crelle's proof. The assertion being evident for prime n, assume its truth for all integers with less than $t \geq 2$ distinct prime divisors. If $n=\prod_{i=1}^{t} p_i^{\alpha_i}$ has t prime divisors then choose a divisor D of n not divisible by p_1. Then for any $\beta \geq 1$ we have $\varphi(p_1^{\beta}D)=(p_1-1)p_1^{\beta-1}\varphi(D)$, and thus

$$\sum_{\beta=0}^{\alpha_1}\varphi(p_1^{\beta}D)=p_1^{\alpha_1}\varphi(D).$$

Since every divisor of n can be written uniquely in the form $p_1^{\beta}D$ with $D|n$, $(p_1,D)=1$ and $0\leq\beta\leq\alpha_1$ it follows that

$$\sum_{d|n}\varphi(d)=\sum_{\substack{D|n\\(p_1,D)=1}}\sum_{\beta=0}^{\alpha_1}\varphi(p_1^{\beta}D)=p_1^{\alpha_1}\sum_{\substack{D|n\\(p_1,D)=1}}\varphi(D)=p_1^{\alpha_1}\sum_{D|n/p_1^{\alpha_1}}\varphi(D)$$

and using the induction assumption we get the assertion. □

As noted by G.A.Miller[44] (1905) this lemma can be easily obtained from elementary group theory (Miller's paper also contains other applications of group theory to proofs of number theoretical results) in the following way:

If d divides n then the cyclic group $G=\{e,a,a^2,\dots,a^{n-1}\}$ of order n contains exactly one subgroup of order d, namely $H=\{e,a^m,a^{2m},\dots,a^{(d-1)m}\}$ where $m=n/d$ and as every generator of H is of the form a^{jm} with $(j,d)=1$ there are exactly $\varphi(d)$ of them. Since every element of G generates a cyclic subgroup, the lemma follows. □

For other proofs of Lemma 1.7 see Laguerre[45] (1872/73), Baker[46] (1889/90), Zsigmondy (1893), Vahlen (1895), Kronecker (1901, pp.241–244), Bauer (1971), Chao[47] (1982).

The second lemma in Gauss's proof of Theorem 1.5 can be used to give an alternative definition of $\varphi(n)$:

---

43 Crelle, August Leopold (1780–1855), railway engineer. He was the founder and first editor of *Journal für reine und angewandte Mathematik*, the first volume of which appeared in 1826.

44 Miller, George Abram (1863–1951), Professor at the University of Illinois.

45 Laguerre, Edmond Nicolas (1834–1886), Professor in Paris.

46 Baker, Henry Frederick (1866–1956), Fellow of St.John's College in Cambridge.

47 Chao's proof is based on the enumeration theorems of G.Pólya (1937) and N.G.de Bruijn (1971).

Lemma 1.8. *For any prime p and positive n denote by B_n the set of all integers $x \in [1,p]$ which satisfy*

$$x^n \equiv 1 \pmod p), \quad x^k \not\equiv 1 \pmod p) \quad \textit{for} \quad 1 \le k < n.$$

Let $\psi(n)$ be the number of elements of B_n. Then $\psi(n) = \varphi(n)$.

(Elements $m \in B_n$ are said to *belong to n* with respect to p. In modern language one says rather that n is the *order of m* mod p).

Proof. It follows easily from Fermat's theorem that $\psi(n) = 0$ holds for all n not dividing $p-1$ and moreover

$$\sum_{d|p-1} \psi(d) = p - 1. \tag{1.6}$$

Now one establishes the inequality

$$\psi(n) \le \varphi(n) \tag{1.7}$$

for every divisor n of $p-1$. If $\psi(n) = 0$ then (1.7) is evident. Otherwise choose $a \in B_n$. The numbers $a, a^2, \dots, a^n$ all satisfy the congruence (1.5) and as they all are distinct mod p, Corollary to Lemma 1.6 implies that there are no other solutions of this congruence. Observe now that if $(k,n) = \delta > 1$ then $a^k \notin B_n$. In fact, in this case we have

$$(a^k)^{n/\delta} \equiv (a^n)^{k/\delta} \equiv 1 \pmod p,$$

and $0 < n/\delta < n$. Thus every solution of (1.5) is congruent to a^k mod p with a suitable k prime to n, proving (1.7).

From (1.6), (1.7) and Lemma 1.7 we get now $\psi(n) = \varphi(n)$ for every n dividing $p-1$. □

Theorem 1.5 now follows immediately by applying the Lemma 1.8 to the case $n = p - 1$. □

Actually Gauss's argument leads to a stronger result: there are $\varphi(p-1)$ primitive roots mod p.

3. The second Gauss's proof also uses Lemma 1.6:

Second Gauss's proof of Theorem 1.5. (Gauss 1801, Sect.55–56). Write $p-1 = q_1 q_2 \cdots q_s$, where the q_i's are powers of distinct primes. We show first that for every $i = 1, 2, \dots, s$ there exists a number a_i which belongs to q_i with respect to p. Let q_i be a power of a prime p_i and observe that Lemma 1.6 implies the existence of an integer $1 \le b_i \le p-1$ for which

$$b_i^{(p-1)/p_i} \not\equiv 1 \pmod p.$$

Putting

$$a_i = b_i^{(p-1)/q_i}$$

we get $a_i^{q_i} \equiv 1 \pmod{p}$, and one sees easily that a_i cannot belong to a proper divisor of q_i. Let $g = a_1 a_2 \cdots a_s$. We shall show that if for some positive M one has $g^M \equiv 1 \pmod{p}$ then M must be equal to $p-1$. Indeed, Fermat's theorem implies that M divides $p-1$. If $M < p-1$ then for some i the number q_i does not divide M and so M is a divisor of $M_1 = M/p_i$. This leads to

$$1 \equiv g^{M_1} \equiv (a_1 \cdots a_s)^{M_1} \equiv a_i^{M_1} \pmod{p},$$

which is however not possible, since M_1 is not divisible by q_i, to which a_i belongs. Hence $M = p-1$ and we see that g is a primitive root mod p. □

There is also a simple modern algebraic proof of the existence of primitive roots for any prime:

Observe that if G is a finite subgroup of the multiplicative group of a field k then G must be cyclic. Indeed, every non-cyclic finite Abelian group contains a subgroup of the form $C_p \oplus C_p$ with a prime p and therefore if G would be non-cyclic then the equation $X^p = 1$ would have in k at least $p^2 > p$ solutions, which is not possible, as k is a field. Since the group of non-zero residue classes mod p is isomorphic to the multiplicative group of the finite field of p elements, the assertion follows.

4. Primitive roots exist also for certain composite integers, as pointed out by Gauss (1801, Sect.82–92). He proved in fact that this happens for all powers of odd primes, their doubles and also for the numbers 2 and 4. After proving in a rather cumbersome way the analogue of Lemma 1.6, which states that if p is an odd prime and $k \geq 1$, then the congruence

$$X^n \equiv 1 \pmod{p^k}$$

has exactly $(n, p^{k-1}(p-1))$ distinct solutions mod p^k, he noted in Sect.89 that his second proof of Theorem 1.5 works also for an arbitrary power p^k of an odd prime. The case of numbers of the form $2p^k$ is then shortly mentioned in Sect.92, where it is also shown that no other integers have primitive roots.

We give now a proof of these results:

Theorem 1.9. *Primitive roots* mod N *exist if and only if* $N = 2, 4, p^k, 2p^k$ *where* p *denotes an odd prime and* $k = 1, 2, \ldots$.

Proof. The existence of primitive roots for all powers of an odd prime follows from the next lemma. The first part of it is due to C.G.J.Jacobi[48] (1839) and the second to V.A.Lebesgue (1867).

Lemma 1.10. *(i) If p is an odd prime and g is a primitive root for p^2 then g is a primitive root for every power of p.*

[48]Jacobi, Carl Gustav Jacob (1804–1851), Professor in Königsberg and Berlin.

(ii) If $g < p$ is a primitive root mod p *and* $1 < h < p$ *satisfies*

$$gh \equiv 1 \pmod{p}$$

then either g *or* h *is a primitive root* mod p^2.

Proof. (i) Write $g^{p-1} = 1 + pa_1$ with a_1 not divisible by p. An easy recurrence argument, based on the congruence $(x+y)^p \equiv x^p + y^p \pmod{p}$, leads now to

$$g^{p^{j-1}(p-1)} = 1 + a_j p^j$$

with $p \nmid a_j$ for $j = 1, 2, \dots$ and this shows that the order of g with respect to p^k equals $(p-1)p^{k-1} = \varphi(p^k)$ for every k, as asserted.

(ii) Assume that g is a primitive root mod p but neither g nor h is a primitive root mod p^2. Then we must have

$$g^{p-1} \equiv 1 \pmod{p^2}, \quad h^{p-1} \equiv 1 \pmod{p^2}$$

Fermat's theorem allows us to write

$$h \equiv g^{p-2} \pmod{p}$$

hence with a suitable integer A we have

$$h = g^{p-2} + pA$$

and the congruences

$$1 \equiv h^{p-1} \equiv g^{(p-2)(p-1)} + Ap(p-1)g^{(p-2)(p-2)} \pmod{p^2}$$

follow. This implies the divisibility of $Ap(p-1)g^{(p-2)(p-2)}$ by p^2 and thus p divides A and this leads to

$$h \equiv g^{p-2} \pmod{p^2}.$$

Finally we obtain

$$h \equiv h^p \equiv g^{(p-2)p} \equiv g^{(p-1)^2-1} \equiv g^{-1} \pmod{p^2},$$

that is

$$gh \equiv 1 \pmod{p^2}.$$

However this is not possible because from $1 < g, h < p$ we get

$$1 < gh < p^2. \qquad \square$$

For a generalization of part (ii) of the lemma see Maxfield, Maxfield (1959/60).

The assertion concerning the numbers of the form $2p^k$ results now from the observation that if g is a primitive root for p^k and h is a solution of the system

$$X \equiv 1 \pmod{2}, \qquad X \equiv g \pmod{p},$$

whose solubility follows from the Chinese Remainder Theorem, then h is a primitive root mod $2p$. Indeed, it suffices to note that $\varphi(2p) = \varphi(p) = p - 1$.

The cases $N = 2, 4$ being trivial, it remains to show that only the numbers listed in the statement of the theorem possess primitive roots. To do this observe that if M is a divisor of N and M does not have primitive roots then N also cannot have them. In fact, were g a primitive root for N, then $h = g \bmod M$ would be a primitive root for M. If now p, q are distinct odd primes, then $\varphi(pq) = (p-1)(q-1)$ but if we put $A = (p-1)(q-1)/2$ then A will be a multiple of both $p - 1$ and $q - 1$, hence for every n prime to pq we would have $n^A \equiv 1 \pmod{p}$ and $n^A \equiv 1 \pmod{q}$, thus

$$n^A \equiv 1 \pmod{pq},$$

showing that there are no primitive roots for pq.

Similarly if p is an odd prime then there cannot be any primitive roots for $4p$ as we have $n^{p-1} \equiv 1 \pmod{4p}$ for all n prime to $4p$. Since obviously there are no primitive roots for 8 and every integer which is not a prime power nor a double of a prime power must be either divisible by a product of two distinct odd primes or by $4p$ with some odd prime p, we are finished. □

The case of powers of 2 was considered in Sect.90 and Sect.91 of Gauss (1801). The results obtained in Sect.90 imply that for all $k \geq 3$ the number 5 is of order 2^{k-2} with respect to 2^k. This can be seen directly by noting that $5^2 = 1 + 2^3 + 2^4$ and an easy recurrence argument leads to

$$5^{2^j} = 1 + 2^{j+2} + a_j 2^{j+3}$$

with integral a_j for all $j \geq 1$. Hence the congruence

$$5^{2^j} \equiv 1 \pmod{2^k}$$

cannot hold for $j \leq k - 3$.

5. The results of Gauss described above were used by Dirichlet[49] (1837c) to construct a system of indices for all[50] moduli, which he used to prove the infinitude of primes in arithmetic progressions (see Chap.2). In the case of prime powers such construction had been done earlier by Gauss (1801,Sect.57–59).

Let N be a given positive integer and let

[49]Dirichlet, Peter Gustav Lejeune (1805–1859), Professor in Berlin (1831–1855) and later in Göttingen.

[50]Actually Dirichlet considered only moduli divisible by 8, since this restriction allowed him to make a unified treatment and in any case, for the application he had in mind, this was sufficient.

$$N = \prod_{i=0}^{t} p_i^{\alpha_i}$$

be its canonical factorization into prime powers with $p_0 = 2$, $0 \le \alpha_0$, $0 < \alpha_i$ ($i = 1, 2, \dots, t$). For $i = 1, 2, \dots$ choose a primitive root g_i for $p_i^{\alpha_i}$ and put

$$g_0 = \begin{cases} 5 & \text{if } \alpha_0 \ge 3 \\ 1 & \text{otherwise,} \end{cases}$$

and

$$g_{-1} = \begin{cases} -1 & \text{if } \alpha_0 \ge 2 \\ 1 & \text{otherwise.} \end{cases}$$

If n is an integer prime to N then define $\gamma_i(n)$ by

$$n \equiv g_i^{\gamma_i(n)} \qquad (0 \le \gamma_i(n) < \varphi(p^{\alpha_i}) \quad \text{for } i = 1, 2, \dots, t. \tag{1.8}$$

In the case $\alpha_0 \ge 3$ define $\gamma_{-1}(n)$ and $\gamma_0(n)$ by

$$n \equiv (-1)^{\gamma_{-1}(n)} \pmod 4, \qquad (\gamma_{-1}(n) = 0 \textit{ or } 1) \tag{1.9}$$

and

$$(-1)^{\gamma_{-1}(n)} n \equiv 5^{\gamma_0(n)} \pmod{2^{\alpha_0}}, \qquad (0 \le \gamma_0(n) < 2^{\alpha_0} - 2). \tag{1.10}$$

In the case $\alpha_0 = 2$ put $\gamma_0(n) = 0$, define $\gamma_{-1}(n)$ by (1.9). If $\alpha_0 = 0$ or 1 put $\gamma_{-1}(n) = \gamma_0(n) = 0$.

The sequence $\{\gamma_i(n)\}$ is called the *system of indices* of the number n with respect to the modulus N. From the Chinese Remainder Theorem it follows immediately[51] that to every such sequence there corresponds a uniquely determined residue class mod N, prime to N.

1.4. Prime Number Formulas

1. The search for simple functions leading to formulas producing primes has been pursued since ancient times. There are three kinds of such formulas to be considered:

a) Functions taking prime values for all arguments lying in a prescribed set which in most cases is the set of all positive integers.

b) Functions f for which $f(n)$ gives the nth prime p_n.

Formulas of this kind are closely related to formulas giving the value of $\pi(x)$, the number of primes not exceeding x. Indeed, if $f(n) = p_n$ and g denotes the inverse of f, then clearly one has $\pi(x) = g(x)$.

[51]This assertion is implicitly contained in the last sentence of Sect.7 of Dirichlet (1837c).

c) Functions producing infinitely many primes.

A not very enlightening example of such function is given by $f(n) = n$. It is much less trivial and was shown first by Dirichlet (see Theorem 2.1) that for relatively prime integers a, b the function $f(n) = an + b$ produces infinitely many primes.

2. The history of functions of the first type begins with P.Fermat's[52] statement asserting that for all positive integers n the nth *Fermat number*

$$F_n = 2^{2^n} + 1$$

is prime. This assertion appears for the first time in a letter to Frenicle[53] in August 1640 (Fermat 1640a), where we read that all numbers

$$3, 5, 17, 257, 65537, 4\,294\,967\,297, \quad 18\,446\,744\,073\,709\,551\,617$$

are primes. Fermat did not claim to have any proof of that but stated that he excluded many possible divisors of these numbers.

He repeated his assertion in the next letter to Frenicle in October 1640 (Fermat 1640b) and to Digby[54] in June 1658 (Fermat 1658) stating again that he has no proof. Several years later (Fermat 1679) he again stated his assertion, writing[55]:

"*Quum autem numeros a binario quadratice in se ductos et unitate auctos esse semper numeros primos apud ne constet et jamdudum Analystis illius theorematis veritas fuerit significata, nempe esse primos 3,5,17,257,65537 etc. in infinitum* ... ,"

and in a letter to Carcavi[56] (Fermat 1659) he claimed to have a proof based on the infinite descent. We do not know how he tried to construct such a proof but in any case there must be an error in his reasoning since it was noted by L.Euler[57] (1732/33) that although the numbers $F_1 = 5$, $F_2 = 17$, $F_3 = 257$, $F_4 = 65537$ are prime, the next term, F_5, is composite, being equal to $4\,294\,967\,297 = 641 \cdot 6\,700\,417$.

In fact no other prime Fermat number has been found and it has been checked that, for $5 \leq n \leq 30$ Fermat's formula gives exclusively composite numbers. (For

[52]Fermat, Pierre (1601–1665), counsellor in Toulouse and Bordeaux.
[53]Frenicle de Bessy, Bernard (1605–1675).
[54]Digby, Sir Kenelm (1603-65).
[55]"*As before I regard as certain that the numbers obtained by adding a unit to consecutive squares which are formed starting with 2 are always prime, a theorem which I since long announced to analysts; I want to say that the numbers 3,5,17,257,65537 ... up to infinity are primes* ... " .
[56]Carcavi, Pierre de (1600–1684), counsellor in Paris and custodian of the Royal Library.
[57]Euler first tried unsuccesfully to prove Fermat's assertion. See the exchange of letters between him and Goldbach between December 1729 and June 1730 in Fuss (1843).

$n = 22$ this was done by R.E.Crandall, J.Doenias, C.Norrie, J.Young (1995) and for $n = 24$ by E.Mayer, J.Papadopuolos, R,E.Crandall in September 1999.) The same is known for several other values of n, the largest being $n = 382\,447$ (J.B.Cosgrave, August 1999). Complete prime factorization of F_n are known only for $n = 5$ (Euler 1732/33b), $n = 6$ (Landry 1880a,b, Williams 1993), $n = 7$ (Morrison, Brillhart 1971,1975), $n = 8$ (Brent, Pollard 1981), $n = 9$ (Lenstra, Lenstra, Manasse, Pollard 1993), $n = 10$ (Brent 1999) and $n = 11$ (Brent 1989,1990). For several further values of n at least one proper factor is known. It is also known (Golomb 1955) that Fermat numbers tend to have a rather large maximal prime divisor.

The next results concerning our problem were rather negative. C.Goldbach wrote to Euler on 28th September 1743 (see Fuss 1843, I, letter 65, pp.258–265; Euler,Goldbach 1965) that it is easy to show that no algebraic formula $a + bx + cx^2 + dx^3 + \cdots$ can led to exclusively prime values. One finds a short indication of the proof in Euler's letter to Goldbach dated 28th October 1752 (Euler writes there "*Jedoch*[58] *gibt es gewis keine series algebraica, deren omnes termini numeri primi seyn können. Denn es sey X der terminus indici x repondens, und daher A der terminus indici dato a respondens, so wird, wenn man nimmt $x = nA+a$, der terminus X divisibilis per A und also nicht primus.*") and the letter of Goldbach to Euler written 18th November 1752 contains a detailed proof in the case of a cubic polynomial. Goldbach pointed out that the same argument applies in the general case as well. The first published proof appears in Euler (1762/63) (see Fuss 1843, I, 586–595; Euler, Goldbach 1965).

Theorem 1.11. *There is no non-constant polynomial with integral coefficients which assumes prime values at all positive integers.*

Proof. We present first the original argument of Goldbach. It omits two cases ($e = 0, -1$) and forgets to mention, that the number $p^3 + bpp + cp + 1$ should be $\neq 0, 1, -1$ but this is easy to repair.

"*Dass keine formula algebraica lauter numeros primos geben könne, hatte ich schon in einem meiner vorigen Briefe angemerket; denn es sey z.Ex. die formula $x^3 + bxx + cx + e$, so ist offenbar, das so oft x ein multiplus des termini absoluti e ist, so oft auch (und folglich infinitis modis) die formula einen numerum non primum geben wird, sollte aber $e = 1$ seyn, so setze ich nur $x + p$ anstatt x, alsdann wird die formula transmutirt in*

$$x^3 + 3pxx + 3ppx + p^3 + bxx + 2bpx + bpp + cx + cp + 1$$

und so oft x ein multiplus numeri $p^3 + bpp + cp + 1$ ist, so oft gibt die formula einen numerum non primum. Da nun dieser casus, wo die potestas maximus ipsius $x = 3$ ist, auf alle andere lusus naturae, quaecunque fuerit potestas

[58] "*But certainly there is no algebraic series all terms of which could be prime. Because, if X is the term corresponding to index x and A the term corresponding to a given index a then, if one takes $x = nA + a$, the term X would be divisible by A and hence not prime.*"

ipsius x, appliciret werden kann, so ist es unmöglich eine seriem algebraicam anzugeben, in welcher nicht infiniti termini aus numeris non primis bestehen sollten[59] ".

The modern proof uses Goldbach's idea and runs as follows:

Let $f \in \mathbf{Z}[X]$ be a non-constant polynomial and choose x_0 with $f(x_0) \neq 0, 1, -1$. Then $f(x_0)$ has a prime divisor, say p and thus

$$f(x_0 + np) \equiv f(x_0) \equiv 0 \pmod{p}$$

holds for all $n \in \mathbf{Z}$. If f would assume only prime values, then we would have $f(x_0 + np) = \pm p$ for all n, which is clearly impossible. □

This theorem of Goldbach was later extended by R.C.Buck (1946) to the following form:

Theorem 1.12. *There is no non-constant rational function $R(X)$ with integral coefficients which assumes prime values at all sufficiently large integers.*

Proof. Let $R(X) = f(X)/g(X)$, where f, g are relatively prime polynomials with integral coefficients and assume that for $n = N, N+1, \dots$ one has $R(n) = q_n$, where $\{q_n\}$ is a sequence of primes. Assume also that $R(X)$ is non-constant. By increasing N, if necessary, we may assume that the primes q_n $(n \geq N)$ are all distinct.

There exist polynomials $A(X), B(X) \in \mathbf{Z}[X]$ and a non-zero integer c such that

$$A(X)f(X) + B(X)g(X) = c$$

holds. For $n \geq N$ we get

$$A(n)q_n g(n) + B(n)g(n) = c$$

and hence $g(n)$ divides c for all large values of n which is possible only when $g(X)$ is constant, say $g(X) = c_1 \in \mathbf{Z}$. Thus for large n we have $f(n) = c_1 q_n$. If we now choose a sufficiently large integer m with $q_m \nmid c_1$ then for $k = 1 + q_m$ we have

$$km \equiv m \pmod{q_m}$$

---

[59] "*I observed already in one of my previous letters that no algebraic formula can give only prime values; thus if e.g. the formula $x^3 + bxx + cx + e$ is given, then it is clear that if x is a multiple of the free term e, then (and hence infinitely often) that formula gives a nonprime number. If however e $= 1$, then I put $x+p$ in place of x and the formula will transform into*

$$x^3 + 3pxx + 3ppx + p^3 + bxx + 2bpx + bpp + cx + cp + 1$$

and if x is a multiple of the number $p^3 + bpp + cp + 1$ then the formula gives a nonprime number. As that case, in which the highest power x equals 3 can be applied to all others, with arbitrary powers of x, therefore it is impossible to produce an algebraic formula which would not give infinitely many nonprime numbers." .

and thus

$$c_1 q_{km} = f(km) \equiv f(m) \equiv 0 \pmod{q_m},$$

which is not possible since the prime q_m divides neither c_1 nor q_{km}. □

Corollary. *The assertion of Theorem 1.11 holds also for polynomials with rational coefficients.* □

A stronger result was obtained by D.Sato and E.G.Straus (1970). They proved that the only algebraic functions which are prime-valued at integers are constants (another proof may be found in Jones, Sato, Wada, Wiens (1976) where the case of algebraic functions in several variables was also treated). However there exist infinitely many (even nondenumerably many[60]) entire functions which attain prime values at positive integers. In fact for any given sequence a_n of complex numbers there exist infinitely many entire functions f with $f(n) = a_n$ $(n = 1, 2, \ldots)$ (see e.g. Osgood[61] (1928, pp.575–577), Sheffer (1927)).

3. The use of known formulas of type *a*) is rather limited, since they usually involve an encoding of an infinite set of prime numbers, which nullifies their applicability. The first such formula was found by W.H.Mills (1947) who established the existence of a real number θ having the property that for $n = 1, 2, 3, \ldots$ the number

$$\left[\theta^{3^n}\right]$$

is prime. Later L.Kuipers (1950) showed that a small modification of the argument of Mills permits to replace in his result the number 3 by any integer ≥ 3. I.Niven (1951) and A.R.Ansari (1951) noted that integrality is not essential in this result and Ansari showed that any real number exceeding 77/29 will do. These results are connected with an upper bound for the difference of consecutive primes. We prove now Mills' result in a more general version (see Dudley 1969, Tee 1974):

Theorem 1.13. *Assume there exists constants $\beta < 1$ A and c such that $p_{n-1} < p_n$ are two consecutive primes exceeding A then one has*

$$p_n - p_{n-1} < c p_{n-1}^{\beta}.$$

Then for all $\alpha > \frac{1}{1-\beta}$ there exists a real number θ such that

$$\left[\theta^{\alpha^n}\right] \tag{1.11}$$

is prime for $n = 1, 2, 3, \cdots$.

Proof. The assumptions imply that for every sufficiently large integer N there is a prime number in the interval $[N^\alpha, (N+1)^\alpha - 1)$. Indeed, let

---

[60] Sato,Straus (1966a,b).

[61] Osgood, William Fogg (1864–1943), Professor at Harvard.

$$\epsilon = (\alpha(1-\beta)-1)/2,$$

let N be a number satisfying $N^\alpha > A$, $N^\epsilon > c$, denote by p the largest prime not exceeding N^α and let q be the next prime. Then

$$p \le N^\alpha < q < p + cp^\beta < N^\alpha + N^{\alpha\beta+\epsilon}$$

but in view of $\alpha\beta < \alpha - 1$ we get for sufficiently large N, say $N \ge N_0(\alpha,\beta)$,

$$(N+1)^\alpha - N^\alpha \ge \alpha N^{\alpha-1} > 1 + N^{\alpha\beta+\epsilon}$$

and the inequality $N^\alpha \le q < (N+1)^\alpha - 1$ results.

Choose now a prime q_1 exceeding both $N_0(\alpha,\beta)$ and $c^{1/\epsilon}$. If the primes $q_1, q_2, \ldots, q_n$ are already chosen, then choose a prime q_{n+1} in the interval $[q_n^\alpha, (q_n+1)^\alpha - 1)$. For $n = 1, 2, \ldots$ put $u_n = q_n^{\alpha^{-n}}$ and $v_n = (1+q_n)^{\alpha^{-n}}$. Then $u_n < v_n$ and since u_n increases and v_n decreases both limits

$$\theta = \lim_{n\to\infty} u_n, \quad \Theta = \lim_{n\to\infty} v_n$$

exist. In view of $u_n < \theta \le \Theta < v_n$ we get

$$u_n^{\alpha^n} < \theta^{\alpha^n} \le \Theta^{\alpha^n} < v_n^{\alpha^n},$$

and finally

$$q_n < \theta^{\alpha^n} < q_n + 1$$

giving

$$q_n = \left[\theta^{\alpha^n}\right]. \qquad \square$$

Since it is now known (see Baker, Harman 1996b) that the assumptions of the last theorem are satisfied for any $\beta \ge 107/200 = 0.535$, a suitable θ may be found for every $a > 200/93 = 2.15053\ldots$. Mills used a result of Ingham[62] (1937) showing that one can take $\beta = 5/8$ and this allowed him to take $a = 3$. I.Niven (1951) obtained a variant of this result stating that for any $\theta > 1$ there exists a real α such that for all n the formula (1.11) produces prime numbers. See Ore[63] (1952) and Wright (1954).

4. A similar result which however does not use heavy analytic tools in its proof was given by E.M.Wright (1951). It is based on *Bertrand's postulate* asserting that there is always a prime between x and $2x$, provided $x \ge 2$. We shall prove this in Chap.3 (see Theorem 3.7), and as a corollary we will obtain Wright's result which we now state in a conditional form:

---

[62]Ingham, Albert Edward (1900–1967), Professor in Leeds and Cambridge.
[63]Ore, Oystein (1899–1968), Professor at Yale.

Theorem 1.14. *Put $F_1(t) = 2^t$ and $F_{i+1}(t) = 2^{F_i(t)}$ for $i = 1, 2, \dots$. If for every $x \geq 2$ there is a prime between x and $2x$ then there exists a real constant ϑ such that for all n the number $[F_n(\vartheta)]$ is prime.*

Proof. We follow Dudley (1969). Define a sequence $q_1 = 2, q_2, \dots$ of primes by choosing for q_{i+1} any prime contained in the interval $(2^{q_i}, 2^{1+q_i})$ and denote by $L_n(t)$ the n-th iterate of the function $L_1(t) = \frac{\log t}{\log 2}$. Since $L_n(t)$ is increasing, we get

$$q_i < L_1(q_{i+1}) < L_1(1 + q_{i+1}) \leq 1 + q_i$$

and

$$L_n(q_i) < L_{n+1}(q_{i+1}) < L_{n+1}(1 + q_{i+1}) \leq L_n(1 + q_i)$$

for all i and n. In particular, for $i = n$ we get

$$L_n(q_n) < L_{n+1}(q_{n+1}) < L_{n+1}(1 + q_{n+1}) \leq L_n(1 + q_n)$$

from which it follows that the sequence $\{L_n(q_n)\}$ increases and its nth term is majorized by the nth term of the non-increasing sequence $\{L_n(1 + q_n)\}$. This shows that the limit

$$\vartheta = \lim_{n\to\infty} L_n(q_n)$$

exists and for all n one has

$$L_n(q_n) < \vartheta < L_n(1 + q_n).$$

In view of $F_n(L_n(t)) = t$ the last inequality leads to

$$q_n < F_n(\vartheta) < 1 + q_n$$

and thus $q_n = [F_n(\vartheta)]$, as asserted. □

5. There are several formulas of type *b*), giving the value of the nth prime but most of them either suffer from the same inadequacy as the formula of Mills given above (see e.g. Bang 1952, Moser 1950, Sierpiński[64] 1952a and Härtter 1961) or they involve complicated and long computations. More interesting are formulas and procedures which permit us to obtain the value of the nth prime using the knowledge of smaller primes.

The oldest procedure of this sort goes back to Eratosthenes[65] and is commonly known[66] as the *sieve of Eratosthenes*. H.J.Smith[67] (1857) described it in the following way:

---

[64] Sierpiński, Wacław (1882–1969), Professor in Lwów (1910–1918) and Warsaw (1918–1960).
[65] Eratosthenes (276 B.C.–197 B.C.), director of the Alexandrian Library.
[66] The assertion made by E.Hoppe (1911) that this method was used already by Socrates seems unsubstantiated.
[67] Smith, Henry John Stephen (1826–1883), Professor in Oxford.

"*It is . . . essentially a method of exclusion, by which all composite numbers are succesively erased from the series of natural numbers, and the primes alone are left remaining. It requires only one kind of arithmetic operation; that is to say, the formation of consecutive multiples of given numbers, or, in other words, addition only. Indeed it may be said to require no arithmetic operation whatever; for if the natural series of numbers be represented by points set off at equal distances along a line, by using a geometrical compass we can determine without calculation the multiples of any given number.*"

A quantitative form of the sieve of Eratosthenes was given by A.M.Legendre (1830). In a rather unprecise version it already occurs in Waring[68] (1770; p.333 of English translation). Some alleged applications of it were given without proof by A.de Polignac[69] (1849b), e.g. the existence of primes between x and its double.

The original formulation of Legendre was rather cumbersome as he did not use the *Möbius function* $\mu(n)$ defined by the formula

$$\mu(n) = \begin{cases} 1 & \text{if } n = 1 \\ (-1)^k & \text{if } n \text{ is a product of } k \text{ distinct primes} \\ 0 & \text{otherwise.} \end{cases}$$

The function μ occurs explicitly for the first time in a paper of A.F.Möbius[70] (1832), who denoted it by b_n. The notation μ appeared for the first time in Mertens (1874a).

Möbius used this function to show that under certain assumptions the conditions

$$f(x) = \sum_{n=1}^{\infty} g(x^n)$$

and

$$g(x) = \sum_{n=1}^{\infty} \mu(n) f(x^n)$$

are equivalent. If we put $F(x) = f(e^x)$ and $G(x) = g(e^x)$ then these conditions take the simpler form:

$$F(x) = \sum_{n=1}^{\infty} G(nx),$$

and

$$G(x) = \sum_{n=1}^{\infty} \mu(n) F(nx).$$

Möbius' arguments disregarded convergence questions and a correct form for his results was given several years later in papers of E.Hille[71], O.Szász[72] (1936) and

---

[68] Waring, Edward (1734–1778), Professor in Cambridge.
[69] de Polignac, Alphonse (1826–1863).
[70] Möbius, August Ferdinand (1790–1868), since 1816 Professor of astronomy and from 1844 also of mathematics in Leipzig.
[71] Hille, Einar (1894–1908), Professor at Yale.
[72] Szász, Otto (1884–1952), Professor in Frankfurt and Cincinnati.

E.Hille (1937). (See Loxton, Sanders (1980) for the history of this kind of inversion and its applications.) An interesting nontypical application of Möbius' formulas can be found in Lüroth[73] (1905), where a series of complex functions is constructed which inside the unit circle sums up to 1 and outside to -1.

Before formulating Legendre's formula we recall the main properties of $\mu(n)$, including inversion formulas for arithmetic functions. These formulas are called usually after Möbius, although they do not appear explicitly in his work in the form used today.

For a general theory of Möbius-type inversion formulas see Delsarte[74] (1948), Rota (1964), Weisner (1935) and Wiegandt (1959). F.Mertens (1897b) formulated a conjecture about the behaviour of the sum $M(x) = \sum_{n\leq x}\mu(n)$. We shall say more about it in Chap.4. Here we would like only to point out that this sum had already been considered by E.Césaro[75] (1883, p.157) who asserted that the limit of the ratio $M(x)/x$, for x tending to infinity, equals $36/\pi^4$ using a nice looking but completely fallacious argument. In fact this limit exists and equals zero.

Theorem 1.15. *(i) One has*

$$\sum_{d|n}\mu(d) = \begin{cases} 1 & \text{if } n = 1, \\ 0 & \text{if } n > 1. \end{cases}$$

(ii) If f, g are arbitrary arithmetic functions, then the following three statements are equivalent:

(α) *For $n \geq 1$ one has $\sum_{d|n} f(d) = g(n)$,*
(β) *For $n \geq 1$ one has $\sum_{d|n}\mu(d)g(n/d) = f(n)$,*
(γ) *For $x > 0$ one has $\sum_{n\leq x} g(n) = \sum_{d\leq x} f(d)[x/d]$.*

(iii) If f, g are arbitrary functions defined for $x > 0$ then the following two statements are equivalent:

(δ) *For all $x > 0$ one has $\sum_{n\leq x} g(x/n) = f(x)$,*
(ϵ) *For all $x > 0$ one has $\sum_{n\leq x}\mu(n)f(x/n) = g(x)$.*

The implication (α) $\Rightarrow$ (β) appears for the first time in Dedekind[76] (1857) and[77] Liouville[78] (1857) and the equivalence of (α) and (γ) in von Sterneck (1893).

---

[73]Lüroth, Jacob (1844–1910), Professor in Karlsruhe, München and Freiburg.
[74]Delsarte, Jean (1903–1968), Professor in Nancy.
[75]Césaro, Ernesto (1859–1906), Professor in Naples.
[76]Dedekind, Richard (1831–1916), Professor in Braunschweig.
[77]Liouville regarded this result as a part of folklore, since he wrote: "*...je me garde bien ... de* [le] *présenter comme nouvelle, car quiconque l'aura cherchée a dû la trouver.*" ("*I refrain ... from presenting* [this result] *as new, since everybody who would seek it, would be forced to find it*" .) Dedekind also did not give a proof, writing only "*Es bedarf nur eines Blickes ...*" . ("*It needs only a look ...*").
[78]Liouville, Joseph (1809–1882), Professor in Paris. In 1836 he founded the journal *Journal de mathématiques pures et appliquées* and edited it until 1872.

See also Čebyšev (1851,1879), Césaro (1885a), Dirichlet (1879) and the literature quoted in Dickson (1919, vol.I, Chap.19). Note also that the properties stated in Theorem 1.15 characterize the function μ (Apostol 1965, Satyanarayana 1963).

Proof. An arithmetic function f is called *multiplicative* if for relatively prime m, n one has $f(mn) = f(m)f(n)$. Such functions are determined by their values at prime powers, due to the formula

$$f(\prod p^{\alpha_p}) = \prod f(p^{\alpha_p}).$$

We need a simple lemma dealing with *Dirichlet convolution* of arithmetic functions defined by the formula

$$(f * g)(n) = \sum_{d|n} f(d)g(n/d).$$

Lemma 1.16. *If f, g are multiplicative functions then their convolution $h = f * g$ is also multiplicative.*

Proof. If $(m, n) = 1$ then every divisor d of mn can be uniquely written in the form $d = d_1 d_2$, where $d_1|m$ and $d_2|n$. Thus

$$h(mn) = \sum_{d|mn} f(d)g(mn/d) = \sum_{\substack{d_1|m \\ d_2|n}} f(d_1 d_2)g(m/d_1 \cdot n/d_2)$$

$$= \sum_{d_1|m} \sum_{d_2|n} f(d_1)g(m/d_1)f(d_2)g(n/d_2) = h(m)h(n). \qquad \square$$

To prove (i) observe that μ is a multiplicative function and thus $\sum_{d|n} \mu(d)$ is also multiplicative by Lemma 1.16. Hence it suffices to check the equality in (i) for prime powers in which case it is a simple exercise.

The equivalence of (α) and (β) follows immediately from (i). If (α) holds then

$$\sum_{n \le x} g(n) = \sum_{n \le x} \sum_{d|n} f(d) = \sum_{d \le x} f(n) \sum_{\substack{n \le x \\ d|n}} 1 = \sum_{n \le x} f(d) \left[\frac{x}{d}\right],$$

hence (γ). Conversely if (γ) holds, then with $S(n) = \sum_{m \le x} f(m) \left[\frac{x}{m}\right]$ we get

$$g(n) = S(n) - S(n-1) = f(n) + \sum_{m \le n-1} f(m) \left(\left[\frac{n}{m}\right] - \left[\frac{n-1}{m}\right] \right),$$

and it remains to observe that

$$\left[\frac{n}{m}\right] - \left[\frac{n-1}{m}\right] = \begin{cases} 1 & \text{if } m \text{ divides } n, \\ 0 & \text{otherwise.} \end{cases}$$

Finally to prove (iii) note that if (δ) holds, then using (i) we get

$$\sum_{n\le x}\mu(n)f\left(\frac{x}{n}\right)=\sum_{n\le x}\mu(n)\sum_{x\le m/n}g\left(\frac{x}{mn}\right)=\sum_{mn\le x}\mu(n)g\left(\frac{x}{mn}\right)$$
$$=\sum_{r\le x}g\left(\frac{x}{r}\right)\sum_{n|r}\mu(n)=g(x),$$

and similarly, if (ϵ) holds then

$$\sum_{n\le x}g\left(\frac{x}{n}\right)=\sum_{n\le x}\sum_{m\le x/n}\mu(m)f\left(\frac{x}{mn}\right)$$
$$=\sum_{n\le x}\mu(n)\sum_{x\le m/n}g\left(\frac{x}{mn}\right)=\sum_{mn\le x}\mu(n)f\left(\frac{x}{mn}\right)=f(x).\qquad\square$$

For future use we note an easy corollary:

Corollary. *The characteristic function of the set of all integers n prime to a given integer D equals*

$$\sum_{d|(n,D)}\mu(d).$$

Proof. Follows immediately from part (i) of the theorem. $\square$

6. Now we consider the formula of Legendre which we state in the modern form using the function μ:

Theorem 1.17. *If Δ denotes the product of all primes $\le\sqrt{n}$, then*

$$\pi(n)-\pi(\sqrt{n})=\sum_{d|\Delta}\mu(d)\left[\frac{x}{d}\right]-1.$$

Proof. The main step of the proof is embodied in the following auxiliary result:

Lemma 1.18. *Let D be a positive integer and denote by $\Phi(x,D)$ the number of positive integers not exceeding x which are prime to D. Then*

$$\Phi(x,D)=\sum_{d|D}\mu(d)\left[\frac{x}{d}\right].$$

In particular for the number $\Phi_m(x)$ of integers $\le x$ which are not divisible by any of the first m primes $p_1=2,p_2,\dots,p_m$ one has

$$\Phi_m(x) = \sum_{d|p_1\cdots p_m} \mu(d) \left[\frac{x}{d}\right].$$

Proof. The classical proof of this formula uses the inclusion-exclusion principle:

Let $\{q_1, q_2, \dots, q_m\}$ be the set of all prime divisors of D. From the $n = [x]$ positive integers not exceeding x we delete all which are divisible by a prime q_j. This makes $\sum_{j=1}^{m}[n/q_j]$ deletions but since integers divisible by two distinct primes are deleted twice we have to remove $\sum_{i\neq j}[n/q_iq_j]$ deletions. One sees however that now numbers divisible by three distinct primes remain undeleted. Hence we have to make $\sum_{i,j,k}[n/q_iq_jq_k]$ deletions etc. Finally one obtains the asserted formula with $n = [x]$ in place of x. It remains to observe that $[x/d] = [[x]/d]$. □

This argument of Legendre is not an example of perfect precision, as it does not really show that at the end every element which should be deleted will be deleted exactly once. This gap was filled out by R.Lipschitz[79] (1882) (see also Lipschitz 1883a,b) and E.de Jonquières[80] (1882a,b,1883).

The first based his proof on the formula

$$\sum_{n=1}^{\infty} \mu(n) \left[\frac{x}{n}\right] = 1,$$

established earlier by E.Meissel[81] (1854) (see Exercise 12 (i)) and the second counted the number of deletions using Newton's binomial formula. Also J.J.Sylvester (1883) gave a proof.

Nowadays one usually applies the Corollary to Theorem 1.15 which leads to

$$\Phi(x,D) = \sum_{m\leq x} \sum_{d|(m,D)} \mu(d) = \sum_{\substack{d\leq x\\ d|D}} \mu(d) \sum_{\substack{m\leq x\\ d|m}} 1 = \sum_{\substack{d\leq x\\ d|D}} \mu(d) \left[\frac{x}{d}\right]$$

as asserted.

To prove the theorem it suffices now to observe that a positive integer $m \leq x$ is prime to Δ if and only if either m is a prime satisfying $\sqrt{x} < m \leq x$ or $m = 1$ and apply the lemma with $D = \Delta$. □

Legendre's formula is not very convenient for computation. Therefore E.Meissel (1870) presented a modification of it which permitted him (Meissel 1871,1885) to calculate $\pi(20\,000)$, $\pi(500\,000)$, $\pi(10^6)$, $\pi(10^7)$, $\pi(10^8)$ and $\pi(10^9) = 50\,847\,478$. Unfortunately, the last result is in error, the correct

[79] Lipschitz, Rudolf (1832–1903), Professor in Breslau and Bonn.

[80] Jonquières, Ernest Jean Philippe Fauque de (1820–1901), vice-admiral in the French Navy.

[81] Meissel, Daniel Friedrich Ernst (1826–1895), Professor in Kiel.

value being 50 847 534 (Lehmer[82] 1959). (See Gram[83] (1893), where results of computation of $\pi(2 \cdot 10^7)$ and $\pi(9 \cdot 10^7)$ made by Bertelsen are stated.)

We present now Meissel's formula with a proof given by A.Brauer (1946):

Theorem 1.19. (Meissel's formula). *For $x \geq 4$ put $n = \pi(\sqrt{x})$ and let Δ_n be the product of the first n primes. Then we have*

$$\pi(x) = \Phi(x, \Delta_m) + m(1+s) + \frac{1}{2}s(s-1) - 1 - \sum_{j=1}^{s} \pi\left(\frac{x}{p_{m+j}}\right),$$

where $m = \pi(\sqrt[3]{x})$, p_j denotes the j-th prime and $s = n - m$.

Proof. Since exactly $\Phi(x/p_k, \Delta_{k-1})$ integers $\leq x$ and prime to the first $k-1$ primes are divisible by p_k we obtain

$$\Phi(x, \Delta_k) = \Phi(x, \Delta_{k-1}) - \Phi(\frac{x}{p_k}, \Delta_{k-1}).$$

Addition of these equalities for $k = m+1, m+2, \ldots, n$ leads to

$$\Phi(x, \Delta_n) = \Phi(x, \Delta_m) - \sum_{j=1}^{s} \Phi(\frac{x}{p_{m+j}}, \Delta_{m+j-1}).$$

The inequalities

$$x^{1/3} < p_{m+j} \leq x^{1/2} < \frac{x}{p_{m+j}} < x^{2/3},$$

holding for $j = 1, 2, \ldots, s$ imply

$$\Phi(x, \Delta_n) = 1 + \pi(x) - \pi(\sqrt{x}) = \pi(x) - n + 1$$

and

$$\Phi\left(\frac{x}{p_{m+j}}, \Delta_{m+j-1}\right) = 1 + \pi\left(\frac{x}{p_{m+j}}\right) - \pi(p_{m+j-1}) = \pi\left(\frac{x}{p_{m+j}}\right) - (m+j-2),$$

hence

$$\begin{aligned}\pi(x) = \; & \Phi(x, \Delta_n) + n - 1 = \Phi(x, \Delta_m) - \sum_{j=1}^{s}\left(\pi\left(\frac{x}{p_{m+j}}\right) - m - j + 2\right) + n - 1 \\ = \; & \Phi(x, \Delta_m) - \sum_{j=1}^{s} \pi\left(\frac{x}{p_{m+j}}\right) + m(1+s) + \frac{s(s-1)}{2} - 1,\end{aligned}$$

as asserted. □

---

[82]Lehmer, Derrick Henry (1905–1991), Professor in Berkeley.

[83]Gram, Jorgen Pedersen (1850–1916), Professor in Copenhagen and director of an insurance company.

For other proofs of Meissel's formula and of similar formulas see Baranowski[84] (1895), Cipolla[85] (1905), Comessatti[86] (1906), Hoffmann (1879), de Jonquiéres (1882a,b, 1883), Laurent[87] (1898) (rediscovered in Langmann (1974)), Lebon (1904), D.H.Lehmer (1959), Lipschitz (1882, 1883a,b), Lugli (1888), Rogel (1889, 1890, 1900), Sylvester (1883) and Torelli[88] (1901). See also Brocard (1880) and Gegenbauer (1899).

Analogues of Meissel's formula for the number of primes in arithmetic progressions $kx + l$ were obtained for $k = 4, 6$, $l = \pm 1$ in Hudson, Brauer (1977) and in the general case in Hudson (1977).

A modern approach to Meissel's method was proposed by D.H.Lehmer (1959) and a variant of it can be found in Mapes (1963). J.Bohman (1972) used this approach to calculate $\pi(10^{13})$. Further development of Meissel's idea is contained in Lagarias,Miller,Odlyzko (1985) (where it is shown that the value of $\pi(10^{13})$ obtained by Bohman is too small by 941), Lagarias, Odlyzko (1984,1987) and Deléglise,Rivat (1996). An analytic method for computing $\pi(x)$ was presented in Lagarias,Odlyzko (1987). The largest known values of $\pi(x)$ seem to be

$$\pi(10^{18}) = 24\,739\,954\,287\,740\,860$$

(Deléglise,Rivat 1996) and

$$\pi(10^{19}) = 234\,057\,667\,276\,344\,607, \quad \pi(10^{20}) = 2\,220\,819\,602\,560\,918\,840$$

computed by M.Deléglise.

A new interpretation of the sieve of Eratosthenes which led to several important results in number theory was sketched by J.Merlin (1911) (see Hadamard[89] (1915) for an exposition of Merlin's idea) but his arguments were only of heuristical nature, to be dressed in a proper mathematical form in a series of papers of Viggo Brun (1915,1916,1919a,b,1920,1925). For an exposition of the development of Brun's sieve method and its applications the reader should consult the book of H.Halberstam and H.-E.Richert (1974).

7. For various formulas giving the value of the nth prime number, the number $\pi(x)$ of primes not exceeding x or similar functions consult Almansa, Prieto (1994), Brun (1935), Gegenbauer (1892b,1896), Hardy (1906), Hayashi[90] (1901a,b), von Koch[91] (1894), Koessler (1917), Kronecker (1891), Laurent (1898,1899), Levi-Civita (1895), Lindgren (1963), Ortega-Costa (1950), Papadimitriou (1975), Regimbal (1975) (rediscovered in Quesada Chaverri

[84]Baranowski, Antoni (1835–1902), Roman Catholic bishop in Sejny.
[85]Cipolla, Michele (1880–1947), Professor in Catania.
[86]Comessati, Annibale (1886–1945), Professor in Padua.
[87]Laurent, Matthieu Paul Hermann (1841–1908).
[88]Torelli, Gabriele (1849–1931), Professor in Naples.
[89]Hadamard, Jacques Salomon (1865–1963), Professor in Bordeaux (1893–1897) and in Paris. He created the modern theory of entire functions.
[90]Hayashi, Tsuruichi (1873–1935).
[91]Koch, Niels Fabian Helge von (1870–1924), Professor in Stockholm.

(1980)), Rogel (1895), Tee (1972), Tsangaris, Jones (1992), Wigert (1895), Willans (1964) and Wormell (1961).

These formulas do not have any practical importance. So, for example, the formula of Papadimitriou uses the fact that the characteristic function of the set of all odd primes equals 1 if and only if the difference

$$2\frac{(x-1)!}{x} - \left[2\frac{(x-1)!}{x}\right]$$

is positive, and the formula of Willans which has the form

$$p_n = 1 + \sqrt[n]{n}\sum_{m=1}^{2^n}\left(\sum_{x=1}^{m}\left(\cos^2\left(\pi\frac{(x-1)!+1}{x}\right)\right)^{-1/n}\right),$$

is based on a similar argument. (For a discussion of the last formula see Neill, Singer (1965) and Goodstein (1967)).

To the same category belongs the following formula for the largest prime divisor $a(N)$ of a given integer $N > 0$, given by G.H.Hardy (1906):

$$a(N) = \lim_{r\to\infty}\lim_{m\to\infty}\lim_{n\to\infty}\sum_{j=0}^{m}(1-\cos((j!)^r\pi/N))^{2n}.$$

The proof of it is based on the easily shown formula

$$\lim_{m\to\infty}\lim_{n\to\infty}(\cos((m!)\pi x))^{2n} = \begin{cases}1 & \text{if } x\in\mathbf{Q}\\ 0 & \text{otherwise.}\end{cases}$$

An iterative procedure for calculating the next prime was proposed by C.Isenkrahe (1900):

For any positive integer n denote by $\nu_p(n)$ the exponent of the maximal power of the prime p dividing n and put $\Psi(m,n) = \prod_{p\le m} p^{\nu_p(n!)}$.

Note that $\Psi(m,n)$ can be easily computed, since there is a simple formula for $\nu_p(n!)$, given for the first time in the second edition of Legendre[92] (1798) which appeared in 1808:

Lemma 1.20. *For all $n \ge 2$ and prime p one has*

$$\nu_p(n!) = \sum_{k=1}^{\infty}\left[\frac{n}{p^k}\right].$$

Proof. The assertion follows from the observation that $\nu_p(n!) = \sum_{m=1}^{n}\nu_p(m)$ and in the interval $[1,n]$ there are $[n/p]$ integers divisible by p, $[n/p^2]$ integers divisible by p^2, and in general $[n/p^k]$ integers divisible by p^k. Every such integer contributes one to $\nu_p(n!)$. □

[92]The attribution of this formula to Čebyšev and de Polignac in Hardy, Wright (1960, p.373) is unjustified.

A more formal argument runs as follows (see e.g. Tenenbaum (1990a)):

$$\nu_p(n!) = \sum_{m=1}^{n} \nu_p(m) = \sum_{m=1}^{n} \sum_{k=1}^{\nu_p(m)} 1 = \sum_{k=1}^{\infty} \sum_{\substack{m\leq n \\ \nu_p(m)\geq k}} 1$$

$$= \sum_{k=1}^{\infty} \sum_{\substack{m\leq n \\ p^k \mid m}} 1 = \sum_{k=1}^{\infty} \left[\frac{n}{p^k}\right].$$

Another proof is given in E.Landau[93] (1927a). It is based on the equality

$$\log d = \sum_{p^k \mid d} \log p$$

expressing the fact that the exponent of a prime p appearing in the factorization of d equals the number of powers of p dividing d. Using this equality one gets

$$\log n! = \sum_{d\leq n} \log d = \sum_{d\leq n} \sum_{p^k \mid d} \log p = \sum_{p^k\leq n} \log p \sum_{\substack{d\leq n \\ p^k \mid d}} 1$$

$$= \sum_{p\leq n} \left(\sum_{k=1}^{\infty} \left[\frac{n}{p^k}\right] \right) \log p.$$

This implies

$$n! = \prod_{p\leq n} p^{\alpha_p}$$

with

$$a_p = \sum_{k=1}^{\infty} \left[\frac{n}{p^k}\right]$$

and the assertion follows.

The last argument can be slightly generalized to yield the identity

$$\sum_{d\leq n} a_d \log d = \sum_{p^k\leq n} A(n,p^k) \log p$$

valid for any sequence a_d with

$$A(n,r) = \sum_{\substack{d\leq n \\ r\mid d}} a_d.$$

This identity can be used for many purposes in number theory. Several applications of it were discussed in Friedlander (1985).

[93]Landau, Edmund (1877–1938), Professor in Göttingen (1909–1933).

For other proofs of Lemma 1.20 see Cohen (1969) and Niven,Zuckerman,Montgomery (1991).

Theorem 1.21. (Isenkrahe's procedure.) *Let q be a prime and for integral x put*

$$D(x) = \frac{x!}{\Psi(q,x)} + \frac{\Psi(q,x)}{(x-1)!} - \left[\frac{(x-1)!}{\Psi(q,x)}\right].$$

If one starts with $x = q$ and iterates the function D then this procedure will stop after finitely many steps at the smallest prime exceeding q.

Proof. It suffices to observe that if q' is the smallest prime larger than q, then for $q \le x < q'$ one has $D(x) = 1 + x$ while $D(q') = q'$. □

Clearly this method is of no practical value.

8. The simplest known recurrence formula for the nth prime seems to be the following one proposed by J.M.Gandhi (1971):

Theorem 1.22. *If we denote by Q_k the product of the first k primes $p_1, p_2, \ldots, p_k$ then p_{k+1} is equal to the unique integer p satisfying the inequality*

$$1 < 2^p \left(\sum_{d|Q_k} \frac{\mu(d)}{2^{d-1}} - \frac{1}{2} \right) < 2.$$

The original proof was rather complicated but C.Vandeneynden (1972) gave a simpler argument and then S.W.Golomb (1974) produced a really transparent proof which we now present:

Proof. If, for positive integers n, we put $p(n) = 2^{-n}$ then the set $\mathbf{N}$ of all positive integers becomes a measure space with a countably additive measure p^* defined by

$$p^*(A) = \sum_{n \in A} p(n).$$

If for any given integer d we denote by A_d the set of all multiples of it, then evidently

$$p^*(A_d) = \sum_{n=1}^{\infty} p(nd) = \frac{1}{2^d - 1}$$

and using the inclusion-exclusion principle we obtain that the set of all integers prime to n has its measure equal to

$$1 - \sum_{p_1|n, p_1 \in \mathcal{P}} \frac{1}{2^{p_1} - 1} + \sum_{p_1p_2|n, p_i \in \mathcal{P}} \frac{1}{2^{p_1p_2} - 1} - \cdots = \sum_{d|n} \frac{\mu(d)}{2^d - 1}.$$

In particular the measure of the set of all integers prime to Q_k, the product of the first k primes, equals

$$\sum_{d|Q_k} \frac{\mu(d)}{2^d - 1}.$$

On the other hand this measure equals

$$\sum_{(n,Q_k)=1} p(n) = \frac{1}{2} + \frac{1}{2^{p_{k+1}}} + \cdots,$$

the sum of omitted terms being less than

$$\sum_{j=p_{k+1}+1}^{\infty} \frac{1}{2^j} \leq \frac{1}{2^{p_{k+1}}},$$

and thus we obtain finally

$$1 < 2^{p_{k+1}} \left(\sum_{d|Q_k} \frac{\mu(d)}{2^d - 1} - \frac{1}{2} \right) < 2.$$

This immediately implies the assertion, since the integer p_{k+1} is uniquely determined by the last formula. □

Another proof was given by C.Baxa (1992). For a generalization see Namboodiripad (1971). See also Ellis (1981). Other recurrent formulas may be found in Ernvall (1975), Golomb (1976), Knopfmacher (1977/78), Regimbal (1975), Rogel (1897), Srinivasan (1961,1962), Teuffel (1954) and Venugopalan (1983a,b). For a survey see Dudley (1983).

The following explicit formula for p_n which however is too complicated to be useful, was proved by J.P.Jones (1975):

For integers a, b put $a \diamond b = \max\{a - b, 0\}$. Then

$$p_n = \sum_{i=0}^{n^2} (1 \diamond (\sum_{j=1}^{i} (((j \diamond 1)!)^2 \bmod j) \diamond n)).$$

We do not mention here various approximative formulas for the nth prime number, the sum of the first n primes and related expressions obtained in an empirical way in earlier times. A discussion of many of them can be found in the survey prepared by G.Torelli[94] (1901).

---

[94]This survey of prime number theory contains plenty of information. However it should be read with extreme prudence since many results quoted there are false, like the formula

$$p_{n+1} - p_n = \log n + \log\log n + o(1)$$

for the difference of two consecutive primes appearing on p.59. Also several proofs presented there are incorrect, e.g. that on p.156 where a fallacious argument leading to the absence of positive real zeros of Dirichlet's L-functions with real characters is presented.

9. Although no formula for primes can be given using polynomials and rational functions in one variable, one can nevertheless describe primes with the use of polynomials in several variables. This follows from the result of M.Davis, H.Putnam, J.Robinson (1961) (see also Yu.V.Matijasevič 1970,1971a) who showed that the set of all prime numbers is diophantine. (A set A of nonnegative integers is called a *diophantine set* if there exists a polynomial $f(X_1, X_2, \ldots, X_n)$ with integral coefficients such that the equation $f(X_1, X_2, \ldots, X_{n-1}, a) = 0$ has integral solutions if and only if $a \in A$.) Since H.Putnam (1960) proved that a set A is diophantine if and only if it coincides with the set of positive values of a suitable polynomial $g(X_1, \ldots, X_n)$ over $\mathbf{Z}$ taken at nonnegative integers thus it follows that there exists a polynomial whose positive values at integers $X_i \geq 0$ are primes and every prime can be obtained in that way. The first example of such polynomial (in 24 variables and of degree equal to 37) was constructed by Yu.V.Matijasevič (1971b). From a general result of Yu.Matijasevič and J.Robinson (1975) it follows that the number of variables may be reduced to 14 at the cost of an increase of the degree. In Jones, Sato, Wada, Wiens (1976) one finds an explicit example of degree 25 in 26 variables as well as the proof of the existence of such a polynomial in 12 variables (with degree being equal to 13 697) and of another polynomial, in 42 variables, having degree 5. Later Yu.V.Matijasevič (1977) proved that one can find such a polynomial in 10 variables and R.J.McIntosh (1995) pointed out that if a conjecture connected with Wolstenholme's theorem is true then one can construct such a polynomial in seven variables. For a similar assertion about the set of Fermat primes see Jones (1979).

10. Now we turn to functions of type (c) and will restrict our attention to polynomials with integral coefficients. In the next chapter we shall consider a theorem of Dirichlet stating that every linear polynomial $ax + b$ with relatively prime coefficients represents infinitely many primes. V.Bouniakowsky (1857)[95] conjectured that if $f \in \mathbf{Z}[X]$ is irreducible, has no fixed divisor (which means that there is no integer $d > 1$ dividing all values of f at integers) and its leading coefficient is positive then f represents infinitely many primes. In the case $f(X) = X^2 + 1$ this assertion is called also *Landau's conjecture* but actually its origin goes back to a letter of L.Euler to Goldbach dated October 28th 1752 (see Fuss (1843),I, letter 149, pp.586–591; Euler, Goldbach 1965). Euler also computed all primes $p < 1500$ of that form.*Landau's conjecture*

The assertion stating that $x^2 + y^2 + 1$ represents infinitely many primes was deduced from Extended Riemann Hypothesis by C.Hooley (1957) and then proved unconditionally by B.M.Bredihin (1962,1963a) who later (Bredihin 1963b,1964) established the infinity of primes of the form $\varphi(x, y) + c$, where $\varphi(x, y)$ is a quadratic form whose discriminant is not a square and c is a non-zero fixed integer (see also Motohashi 1970a,1971). More generally,

[95]Bouniakowsky, Victor (1804-1889), Professor in Sankt Petersbourg.

every quadratic polynomial in two variables with integral coefficients, which satisfies certain necessary conditions represents infinitely many primes, as shown in Iwaniec (1972a,b,1973/74).

Since every prime congruent to unity mod 4 is a sum of two squares, Dirichlet's result quoted above shows that there are infinitely many primes of the form $p = x^2 + y^2$. There is a vast literature concerning possible restrictions on x and y here and we quote only two recent results: M.Coleman (1993) proved that infinitely often one can have here $x < p^{0.1631}$ and quite recently E.Fouvry and H.Iwaniec (1997) showed that one can restrict here x to be a prime. At the Number Theory Conference in Zakopane in July 1997 J.B.Friedlander and H.Iwaniec announced the existence of infinitely many primes of the form $x^2 + y^4$ (Friedlander,Iwaniec 1997).

Bouniakowsky's conjecture was extended to the case of several polynomials simultaneously representing prime numbers in Schinzel, Sierpiński (1958) (see also Schinzel (1961)). G.H.Hardy and J.E.Littlewood[96] (1923a) proposed a quantitative version of several special cases of Bouniakowsky's conjecture and in the general form such a version was formulated by P.T.Bateman and R.A.Horn (1962,1965). See Chap.6 for more information on this topic.

Although no non-linear polynomial representing infinitely many primes is known there do exist methods which produce polynomials representing many prime numbers. The first such example was given by C.Goldbach who noted in a letter to Euler of September 28th 1743 (Fuss 1843, I,p.255), which we have already quoted, that among the first 47 terms of the sequence $X^2 + 19X - 19$ there are only four composite integers (Goldbach counted the number 1 as a prime) and conjectured that at powers of 2 this polynomial attains exclusively prime values. This does happen for very small arguments but for $X = 2^6$ one already gets the composite number $5293 = 67 \cdot 79$ (letter of Euler to Goldbach of 15th October 1743 (Fuss 1843, I, letter 66, pp.258–265)). Later Euler (1772) observed that the polynomial $X^2 - X + 41$ represents primes for $X = 0, 1, \ldots, 40$. This fact was later related to imaginary quadratic fields whose rings of integers have unique factorization (Frobenius[97] (1912), Rabinowitsch 1913): the polynomial $X^2 - X + m$ assumes prime values for $X = 0, 1, 2, \ldots, m - 1$ if and only if the ring of integers of the quadratic field $Q(\sqrt{-4m - 1})$ is a unique factorization domain (see Ayoub,Chowla[98] 1981, Lehmer 1936, Mitchell 1925 and Szekeres 1974). It is now known (Heegner (1952), Baker (1966), Stark (1967b)) that this implies that m is one of the numbers $2, 3, 5, 11, 17, 41$. Note that Heegner's proof had been regarded incomplete till Deuring[99] (1968) and Stark (1969) showed its essential correctness. There is a large literature dealing with generalizations and analogues

[96]Littlewood, John Edensor (1885–1977), Professor in Cambridge.
[97]Frobenius, Ferdinand Georg (1849–1917), Professor in Berlin.
[98]Chowla, Saravadaman (1907–1995).
[99]Deuring, Max (1907–1984), Professor in Marburg, Hamburg and Göttingen.

of the Frobenius-Rabinowitsch theorem. See e.g. Halter-Koch (1991), Hendy (1974), Möller (1976), Mollin (1995,1996) and Mollin,Williams (1989).

It seems that the first examples of non-linear polynomials of fixed degree representing arbitrary many primes were given by W.Sierpiński (1964). He showed that with a suitable $c = c(n)$ the polynomial $X^2 + c$ represents at least n primes (see Ageev 1994, Garrison 1990). A more general theorem was later proved by U.Abel and H.Siebert (1993). It implies that if $f \in \mathbf{Z}[X]$ is nonlinear and $N > 0$ is given, then for a proper choice of $c = c(N)$ the sequence $f(n) + c$ contains at least N primes. We shall present their proof in Chap.3 (see the Corollary to Theorem 3.12). See also Forman (1992) and Brown,Shiue,Yu (1998). Examples of polynomials, with small degree and coefficients, that represent many primes appear in Fung,Williams (1990), Goetgheluck (1989) and Karst (1973). For example the polynomial

$$2x^3 - 489x^2 + 39487x - 1\,084\,553$$

represents 267 primes for $x \in [1, 499]$ (Goetgheluck 1989) and among the values of $x^2 + x + 27941$ below 10^6 there are 27941 primes (Beeger, see Lukes,Patterson,Williams (1995)).

A search for quadratic polynomials assuming many prime values in intervals of length 100 was performed by N.Boston and M.L.Greenwood (1995). They discovered the polynomial $41x^2+3x-43321$ representing, in the interval $[-57, 42]$, 90 distinct primes.

There are polynomials with integral coefficients which assume prime values at many consecutive integers. The first result of such type is due to G.J.Chang and K.W.Lih (1977). They showed that for every prime p there exists a polynomial $f(X) \in \mathbf{Z}[X]$ whose values in the interval $[0, p-1]$ are primes. This was extended by A.Balog (1990) who constructed for every k infinitely many polynomials of degree k which assume prime values in the interval $[0, 2k]$.

M.Filaseta (1988) showed that a positive proportion of irreducible polynomials of a fixed degree N ordered according to their height (defined as the maximal absolute value of the coefficients) represent at least $N + 5$ primes. Among quadratic polynomials the current record holder is the polynomial $36X^2 - 810X + 2753$ which assumes distinct prime values for 45 consecutive integers ($X = 0, 1, \dots, 44$) (see Mollin (1995,1996), where it is called the *Ruby polynomial*).

Going in the other direction B.Garrison (1981) and K.S.McCurley (1984b, 1986a,b) constructed irreducible polynomials which do not assume prime values in large intervals. For example the polynomial $X^{12} + 488\,669$ assumes composite values for all integers x satisfying $|x| < 616\,980$.

Exercises

1. (Hacks 1893) Prove that $n \in \mathbf{Z}$ is a prime if and only if

$$\sum_{a=1}^{n-1}\sum_{b=1}^{n-1}\left[\frac{ab}{n}\right] = \frac{(n-1)^2(n-2)}{4}.$$

2. (Carmichael[100] 1905) (a) Let n be odd and put $k = (n-1)/2$. Prove that n is a prime if and only if it divides $(k!)^2 + (-1)^k$.

(b) Prove that integers $1+n$ and $1+2n$ are both primes if and only if their product divides $(n!)^4 - 1$.

3. (a) (Rados; see Dickson 1919, vol.I p.428) Prove that $n \in \mathbf{Z}$ is a prime if and only if

$$(2!\cdot 3!\cdots(n-2)!(n-1)!)^4 \equiv 1 \pmod{n}.$$

(b) (Patrizio 1976)[101] Prove that an integer $n \geq 11$ is prime if and only if

$$4\left[\frac{(n-1)!}{n}\right] \equiv -3 \pmod{n-4}.$$

(c) (Popa 1982) Show that an integer $n > 1$ is prime if and only if, for each $k = 0, 1, \ldots, n-1$, it divides $k!(n-k-1)! + (-1)^k$.

(d) (Subbarao 1974) Let $n \neq 4, 6, 22$ and denote by $\sigma(n)$ the sum of all positive divisors of n. Prove that n is prime if and only if

$$n\sigma(n) \equiv 2 \pmod{\varphi(n)}.$$

(e) (Vecchi 1913) Prove that an integer $n > 2$ is prime if and only if it can be written in the form

$$n = 2^a N - P_m$$

where P_m denotes the product of all odd primes not exceeding $\sqrt{n}$, all prime divisors of N exceed $\sqrt{n}$ and $a \geq 0$.

4.(a) (Clement 1949) Prove that integers n and $n+2$ are both primes if and only if $4((n-1)!+1)+n$ is divisible by $n(n+2)$.

(b) (Pellegrino 1963) Prove that integers n and $n+2$ are both primes if and only if

$$4\left[\frac{(n-3)!}{n-2}\right] \equiv -5 \pmod{n}.$$

5. (Mohanty 1978) Let $(a,b)=1$ and define the sequence A_n recursively by putting $A_0 = a+b$ and $A_{n+1} = A_n^2 + bA_n + b$. Prove that for $i \neq j$ we have $(A_i, A_j) = 1$.

---

[100] Carmichael, Robert Daniel (1879–1967), Professor in Urbana.

[101] For a generalization see Smarandache (1991).

6. (Mohanty 1978) Let $p > 3$ be a prime. Prove that every prime divisor of $(2^p + 1)/3$ exceeds p.

7. (Gauss 1801, sect. 78) Show that the product

$$\prod_{\substack{1 \le a \le n \\ (a,n)=1}} a$$

is congruent to -1 mod n if n has a primitive root and to 1 in all other cases.

8. (Arndt 1846) Let p be an odd prime, $k \ge 2$ and $n \ge 1$. Determine the residue mod p^n of the product of all residue classes a mod p^n prime to p for which the congruence

$$X^k \equiv a \pmod{p^n}$$

is solvable.

9. (Wolstenholme's theorem) (i) Prove that if $p \ge 5$ is a prime then the numerator of

$$\sum_{j=1}^{p-1} \frac{1}{j}$$

is divisible by p^2 and one has

$$\binom{2p-1}{p-1} \equiv 1 \pmod{p^3}.$$

(ii) Show that the assertion in (i) is false if $p \ge 5$ is not a prime.

10. (Laguerre 1872/73) Prove the following generalization of Lemma 1.7: If $\Phi(n/k, n)$ denotes the number of integers $1 \le m \le n/k$ which are prime to n, then for $k = 1, 2, \ldots$ one has

$$\sum_{d|n} \Phi(d/k, d) = \left[\frac{n}{k}\right].$$

11. (Sancery 1875) Let p be an odd prime, let n be an integer not divisible by p and denote by a the order of n mod p. Let $k \ge 2$ and finally let $p^t \parallel n^a - 1$. Prove that the order of n mod p^k equals a if $t \ge a$ and equals ap^{k-t} otherwise.

12. (i) (Meissel 1854, Bougaïeff[102] 1876, Lipschitz 1879) Prove that for all $x \ge 1$ one has

$$\sum_{n=1}^{\infty} \mu(n) \left[\frac{x}{n}\right] = 1.$$

(ii) (Gram 1884) Prove that for $x \ge 2$ one has

$$\sum_{n \le x} \frac{1}{n} \sum_{k \le x/n} \frac{\mu(k)}{k} = 1.$$

---

[102] Bougaïeff, Nikolai Vasilevič (1837–1903), Professor in Moscow.

(iii) (Sylvester 1860) Prove that

$$\sum_{k\leq n}\left[\frac{n}{k}\right]\varphi(k)=\frac{1}{2}n(n+1).$$

13. (Gauss 1846 (for prime n), Schönemann 1846 (for prime powers), Serret 1855) Show that if a and n are positive integers then

$$\sum_{d|n}a^d\mu(n/d)\equiv 0 \pmod{n}.$$

14. (Golomb 1962; see also Shiu 1991) Prove that for every integer $k>1$ there exist an integer n such that $k\pi(n)=n$.

15. (Hardy 1929, Ward 1930) (i) Let $f(x_1,x_2,\dots,x_k)$ be a polynomial with rational integral coefficients. Prove that if for every sufficiently large integer n the value $f(n,2^n,3^n,\dots,k^n)$ is prime, then the set of values of f consists of one element.

(ii) Give an example showing that in (i) the polynomial f may be non-constant.

(iii) (G.Pólya) Show that the assertion (i) remains true if f is a polynomial with rational coefficients.

16. (Braun 1899) Let $r(X)$ be a rational function with rational coefficients. Prove that if for all sufficiently large integers n the value $[r(n)]$ is prime, then r is a constant.

17. (Langmann 1974) Let $a>b\geq 5$ be integers and denote by C the circle in the complex plane with center at $(a+b)/2$ and radius $(a-b+1)/2$. Prove that the number of primes in the interval $[a,b]$ equals

$$-\frac{1}{2\pi i}\oint_C\frac{z\sin(\pi\Gamma(z)/z)}{\sin(\pi z)}dz.$$

18. (i) Show that the sum of reciprocals of all primes not exceeding x is greater than $\log\log x-C$ for a suitable constant C.

(ii) Prove that the number $\pi(n)$ of primes not exceeding n satisfies

$$\pi(n)<3^{1+n}.$$

(iii) (Sierpiński 1952a) Prove that there exists a real number ϑ such that the nth prime p_n is given by the formula

$$p_n=\left[10^{2^n}\theta\right]-10^{2^{n-1}}\left[10^{2^{n-1}}\theta\right].$$

19. (Elliot 1983) Prove that the formula

$$f(n)=(n-1)\cdot\left[\left[\frac{n!+1}{n+1}\right]-\frac{n!-n}{n+1}\right]+2$$

attains for $n = 1, 2, \dots$ only prime values and every prime can be obtained in this way.

20. (Regimbal 1975) Prove that for every positive integer n the sum

$$\sum_{i=1}^{n-1} \left[\frac{[n/i]}{n/i} \right]$$

equals 1 if n is a prime and that this sum exceeds 1 in all other cases. Use this fact to find an explicit formula for the n-th prime.

21. (i) (Brun 1931) For a given integer n define $n_1 = \pi(n)$ and for $j = 2, 3, \dots$ put

$$n_j = n - \pi(n + n_1 + n_2 + \cdots + n_{j-1}).$$

Prove that if k is the first index for which $n_k = 0$ then the nth prime equals $n_1 + \cdots + n_{k-1}$.

(ii) (van der Corput[103], Schaake 1932) Show that in (i) one can replace the sequence of primes by any increasing sequence of integers and $\pi(n)$ by its counting function.

[103] van der Corput, Johannes Gvaltherus (1890–1975), Professor in Fribourg, Groningen, Amsterdam, Berkeley, Stanford and Madison.

2. Dirichlet's Theorem on Primes in Arithmetic Progressions

2.1. Progressions with a Prime Difference

1. The first use of analytic methods in number theory was made by P.G.Dirichlet who, in a series of papers (1837a,b,c, 1838a,b, 1839), obtained several new important results. He proved in particular the infinitude of primes in arithmetic progressions with coprime first element and difference.

Theorem 2.1. *If k, l are relatively prime integers then there are infinitely many primes congruent to l mod k.*

It seems that this assertion was first stated explicitly by L.Euler (1785) for the case $l = 1$ and by A.M.Legendre (1798) in the general case[1]. The first attempt to prove it was made by A.M.Legendre who in the second edition of his book (Legendre 1798), published in 1808, gave a fallacious proof. He repeated the same error in the third edition (Legendre 1830). This error is concealed in the words "*It is easy to see...*" at the end of Sect[2].409 where he sketched a proof of the general case of the following lemma stated in Sect.410 and proved (correctly), in a few special cases, in the preceding sections:

If $q_1, \dots, q_r$ are given distinct odd primes and q denotes the r-th prime then among any q consecutive terms of an arithmetic progression $kx + l$ with $(k, l) = 1$ there is at least one not divisible by any of the q_i's.

The inadequacy of the proof given by Legendre was pointed out by Dirichlet (1837c,1838b) who wrote about it:

"*Cette*[3] *démonstration*[4] *dont le principe est très ingénieux, ne semble pas complète; en la considérant avec beaucoup d'attention, on reconnait que l'auteur y fait usage d'un théorème qu'il ne fonde que sur l'induction, et qui*

[1] Cf. Gauss's comments on it in Gauss (1801, Sect.296).
[2] The section numbers refer to Legendre (1830).
[3] Dirichlet 1838b.
[4] "*This proof whose principle is very ingenious seems to be incomplete; considering it with great attention one sees that the author uses a theorem which he bases only on induction and which is possibly as difficult to prove as the statement which the author deduces from it. In any case my efforts to complete the study of Legendre did not succeed and I had to find completely different means*".

n'est peut-être pas moins difficile à prouver que la proposition que l'auteur en déduit. Du moins les tentatives que j'ai faites pour compléter les recherches de Legendre, ne m'ont pas réussi et j'ai été obliger de recourir à des moyens tout à fait différents.".

Later A.Desboves (1855) deduced several consequences of Legendre's assertion showing in particular that its truth would imply the existence of a prime between any two positive squares. Even today it is unknown whether this assertion is true.

The falsity of Legendre's assertion was finally established by A.Dupré[5] who showed[6] that it already fails in the case when $k = 1$ and the primes $q_1, q_2, \ldots, q_r$ are the first r odd primes with $q_r = 23, 37$ or $43 \leq q_r \leq 113$ (see also Moreau (1873), Piltz (1884), Bachmann[7] (1894), Backlund[8] (1929)). It seems that Dupré's result was not widely known since as late as in 1893 a fallacious proof of Legendre's lemma appeared (Scheffler 1893) receiving a favourable review[9] in *Zeitschrift für Mathematik und Physik* signed by Jahnke. Only several months later did the editor of that journal, O.Schlömilch[10], point out[11] its falsity referring to Piltz's habilitation thesis (Piltz 1884). Cf. Schlömilch (1895).

The conjecture, put forward by A.Piltz (1884), that Legendre's assertion fails in the case considered by Dupré (i.e. $k = 1$ and $q_1, \ldots, q_r$ are the first odd primes) for every sufficiently large r was established by A.Brauer and H.Zeitz[12] (1930) (cf. Zeitz 1930) for all $r \geq 13$ (i.e. $q_r > 41$). They provided a proof for $q_r \geq 103$ and for primes from the interval $[43, 101]$ referred to the paper of Dupré. We give now their argument in the special case $p_r = 103$:

Proof of the falsity of Legendre's lemma: Denote by D the product of all odd primes not exceeding 47 and let x be an odd positive integer divisible by D. Then the only members of the sequence of the 101 odd numbers

$$x - 100, x - 98, \ldots, x - 2, x, x + 2, \ldots, x + 98, x + 100$$

which might be prime to D are those of the form $x \pm 2^\alpha$. Indeed, if $|j| \leq 50$ is not a power of two then j has an odd prime divisor ≤ 47 which divides $(D, x \pm 2j)$. Now note that we have 12 numbers of the form $x \pm 2^a$ in our

[5] Dupré, Athanase Louis (1808–1869), Professor in Rennes.

[6] In a paper submitted to the Paris Academy; see "*Rapport sur le concours pour le Grand Prix de Sciences Mathématiques, proposé pour 1858*", Comptes Rendus Acad. Sci. Paris, **48**, 1859, 487–488. Cf. Dupré (1859).

[7] Bachmann, Paul Gustav Heinrich (1837–1920), Professor in Münster and Weimar.

[8] Backlund, Ralf Josef (1888–1949), worked in the insurance company Kaleva. One of the founders of the Actuarial Society of Finland.

[9] Zeitschr. Math. Phys., **39**, 1894, Hist.-Litt.Abt., 221–222.

[10] Schlömilch, Oskar (1823–1901), Professor in Jena and Dresden. He founded in 1856 the journal *Zeitschrift für Mathematik und Physik* and remained its editor until 1898.

[11] Zeitschr. Math. Phys., **40**, 1895, p.125.

[12] Zeitz, Hermann (1870–1939).

sequence, say $x - a_1, x - a_2, \ldots, x - a_{12}$ and also exactly 12 prime numbers in $(47, 103]$, namely $s_1 = 53, \ldots, s_{12} = 103$. Requiring x to satisfy additionally the congruences

$$x \equiv a_i \pmod{s_i} \qquad (i = 1, 2, \ldots, 12)$$

we are led to a sequence of 101 consecutive odd integers, each having a prime factor not exceeding 103. □

As we mentioned already, the first correct proof of the infinitude of primes in any progression $kn + l$ with $(k, l) = 1$ was given by P.G.Lejeune Dirichlet (1837b,c) who presented it on July 27th 1837 at the meeting of the Royal Prussian Academy of Sciences in Berlin. He gave a complete proof for progressions with prime difference and also developed the general case up to a crucial point. A detailed proof was presented later (Dirichlet 1839).

Dirichlet's proof shows that if l_1, l_2 are integers prime to a given integer k then the ratio

$$\frac{\sum_{p \equiv l_1 \pmod{k}} p^{-x}}{\sum_{p \equiv l_2 \pmod{k}} p^{-x}}, \qquad x > 1,$$

tends to 1 when x tends to 1. This shows that primes are in a certain sense uniformly distributed in reduced residue classes with respect to a fixed modulus. His methods were however inadequate to obtain uniform distribution in the proper sense, i.e. he could not prove that

$$\lim_{x \to \infty} \frac{\pi(x; k, l_1)}{\pi(x; k, l_2)} = 1$$

with $\pi(x; k, l)$ denoting the number of primes $p \leq x$ lying in the progression $l \bmod k$. This limit appeared in Legendre (1830) accompanied by a fallacious proof. A correct proof was finally obtained in 1896 independently by J.Hadamard (1896c) and C.de la Vallée-Poussin[13] (1896a) (see Chap.5).

In the following sections we shall give details of Dirichlet's original proof in the case of a prime difference and then sketch his argument in the general case.

2. At the beginning of his paper Dirichlet (1837c) introduces a class of functions which are now called *Dirichlet's L-functions*. We shall follow him rather closely, although for the sake of clarity we change slightly the order of his argument.

Fix an odd prime number p and choose a primitive root $g \bmod p$. For any positive integer n which is not divisible by p denote by γ_n the *index* of n with respect to g, i.e. the unique integer $m \in [0, p-1]$ satisfying

[13] Vallée-Poussin, Charles Jean Gustave Nicolas de la (1866–1962), Professor in Louvain.

$$g^m \equiv n \pmod{p}.$$

Denote by ω any, not necessarily primitive, root of unity of order $p-1$ and put, for real s,

$$L_\omega(s) = \sum \frac{\omega^{\gamma_n}}{n^s},$$

the sum being taken over all positive integers n not divisible by p. This series clearly converges for $s > 1$. It is usually written in the form

$$L_\omega(s) = \sum_{m=1}^{\infty} \chi_\omega(m) m^{-s},$$

where

$$\chi_\omega(m) = \begin{cases} \omega^{\gamma_m} & \text{if } p \nmid m, \\ 0 & \text{otherwise.} \end{cases}$$

The functions $\chi_\omega(m)$ are now called *Dirichlet character* corresponding to the modulus p.

One sees directly that for all $m_1, m_2 \in \mathbf{N}$ one has

$$\chi_\omega(m_1 m_2) = \chi_\omega(m_1)\chi_\omega(m_2)$$

and

$$|\chi_\omega(m)| \leq 1.$$

This is used by Dirichlet to obtain a product representation for $L_\omega(s)$ analogous to Euler's product formula for $\zeta(s)$ (Lemma 1.2) :

$$L_\omega(s) = \prod_{\substack{q \in \mathbf{P} \\ q \neq p}} \left(1 - \frac{\chi_\omega(q)}{q^s}\right)^{-1}, \tag{2.1}$$

valid for $s > 1$. Dirichlet's argument can be easily adapted to furnish a proof of the following more general result:

Lemma 2.2. *Let s_0 be real and let f be a complex-valued function which is multiplicative, i.e. for all relatively prime integers m, n the equality $f(mn) = f(m)f(n)$ holds. Assume further that the series*

$$F(s) = \sum_{n=1}^{\infty} \frac{f(n)}{n^s}$$

is absolutely convergent for $s = s_0$ and put

$$P(s) = \prod_p \sum_{j=0}^{\infty} \frac{f(p^j)}{p^{js}}. \tag{2.2}$$

Then in the half-plane $\mathrm{Re}\, s \geq s_0$ both the series $F(s)$ and the product $P(s)$ converge and one has there

$$F(s) = P(s).$$

Furthermore the series converges there absolutely.

Proof. Let Re $s \geq s_0$. The absolute convergence of the series results from $|n^s| = n^{\mathrm{Re}s} \geq n^{\mathrm{Re}s_0}$. For a given $x > 2$ denote by A_x the set of all integers $n \geq 1$ whose prime divisors do not exceed x. Since for every prime p the series

$$\sum_{j=0}^{\infty} \frac{f(p^j)}{p^{js}}$$

converges absolutely we get

$$F_x(s) = \sum_{n \in A_x} \frac{f(n)}{n^s} = \prod_{p \leq x} \sum_{j=0}^{\infty} \frac{f(p^j)}{p^{js}}.$$

For x tending to infinity the right-hand side of the last equality tends to $P(s)$ and it suffices to note that

$$|F(s) - F_x(s)| \leq \sum_{n \notin A_x} \left|\frac{f(n)}{n^s}\right| = \sum_{n \notin A_x} \frac{|f(n)|}{n^{\mathrm{Re}s}} \leq \sum_{n \geq x} \frac{|f(n)|}{n^{\mathrm{Re}s}} \longrightarrow 0. \qquad \square$$

Corollary 1. *If the function f is completely multiplicative, i.e. for all m, n one has $f(mn) = f(m)f(n)$, and the series $\sum_{n=1}^{\infty} |f(n)|n^{-s}$ converges at some real $s = s_0$ then for* Re $s \geq s_0$ *one has*

$$F(s) = \prod_p \frac{1}{1 - f(p)p^{-s}}.$$

Proof. In this case we have $f(p^n) = f(p)^n$ for all primes p and $n = 1, 2, \ldots$ hence the terms on the right-hand side of (2.2) can be summed up. $\square$

Corollary 2. *The equality (2.1) holds for all s in the open half-plane* Re $s >$ 1.

Proof. Immediate from Corollary 1, $\square$

If Ω denotes a primitive root of unity of order $p - 1$ then ω is one of the numbers $1, \Omega, \Omega^2, \ldots, \Omega^{p-2}$. In this way one gets $p - 1$ functions $L_\omega(s)$. For brevity we shall follow Dirichlet and write

$$L_j(s) = L_{\Omega^j}(s) \qquad (j = 0, 1, \ldots, p-2).$$

3. The next lemma is crucial in determining the behaviour of the function $L_0(s)$ to the right of $s = 1$:

Lemma 2.3. *If for positive c and ρ we put*

$$S(c,\rho) = \sum_{j=0}^{\infty} \frac{1}{(c+j)^{1+\rho}}$$

then the limit

$$\lim_{\rho\to 0}\left(S(c,\rho) - \frac{1}{\rho}\right)$$

exists.

Proof. The argument is based on the formula

$$\frac{\Gamma(s)}{c^s} = \int_0^1 x^{c-1}\log^{s-1}\left(\frac{1}{x}\right)dx \tag{2.3}$$

(valid for all positive c and $s > 1$), which follows from the equality (essentially due to L.Euler (1730))

$$\Gamma(s) = \int_0^\infty e^{-t}t^{s-1}dt$$

by the change of variables $x = e^{-t/c}$.

Equality (2.3) (with $s = 1+\rho$ and $c+j$ in place of c) leads to

$$\sum_{j=0}^{N} \frac{1}{(c+j)^{1+\rho}} = \frac{1}{\Gamma(1+\rho)}\int_0^1 x^{c-1}\log^\rho\left(\frac{1}{x}\right)\frac{1-x^{N+1}}{1-x}dx$$

and

$$S(c,\rho) = \frac{1}{\Gamma(1+\rho)}\int_0^1 \log^\rho\left(\frac{1}{x}\right)\frac{x^{c-1}}{1-x}dx.$$

In view of

$$\frac{1}{\rho} = \frac{\Gamma(\rho)}{\Gamma(1+\rho)} = \frac{1}{\Gamma(1+\rho)}\int_0^1 \log^{\rho-1}\left(\frac{1}{x}\right)dx$$

we obtain

$$S(c,\rho) = S(c,\rho) - \frac{1}{\rho} + \frac{1}{\rho} = \frac{1}{\rho} + \frac{1}{\Gamma(1+\rho)}\int_0^1\left(\frac{x^{c-1}}{1-x} + \frac{1}{\log x}\right)\log^\rho\left(\frac{1}{x}\right)dx.$$

Now observe that the integral in the last formula has a limit for ρ tending to 0. Indeed, the substitution $t = -\log x$ transforms our integral into

$$\int_0^\infty\left(\frac{e^{-ct}}{1-e^{-t}} - \frac{e^{-t}}{t}\right)t^\rho dt$$

and a standard computation shows that for ρ tending to zero it converges to

$$\int_0^\infty \left(\frac{e^{-ct}}{1-e^{-t}} - \frac{e^{-t}}{t} \right) dt = -\frac{\Gamma'}{\Gamma}(c).$$

(For the last equality see e.g. Whittaker[14], Watson (1902), 3rd ed.,p.247). □

Corollary 1. *If for positive a, b, ρ we put*

$$S(a,b,\rho) = \sum_{j=0}^{\infty} \frac{1}{(b+ja)^{1+\rho}}$$

then the difference

$$S(a,b,\rho) - \frac{1}{a\rho}$$

tends to a finite limit as ρ tends to 0 through positive values and

$$\lim_{\rho \to 0} \rho S(a,b,\rho) = \frac{1}{a}.$$

Proof. The first assertion follows from the equality

$$S(a,b,\rho) = \frac{1}{a^{1+\rho}} S(b/a,\rho)$$

and Lemma 2.3. The second is an immediate consequence of the first. □

Proofs of the last part of this corollary not making use of the Γ-function can be found in Heine[15] (1846) and Dirichlet (1856). See also Hölder[16] (1882), Pringsheim[17] (1890) for generalizations.

This corollary has an easy consequence for the zeta-function although this function does not occur explicitly in Dirichlet's paper:

Corollary 2. *The difference*

$$\zeta(s) - \frac{1}{s-1}$$

tends to a finite limit, when $s > 1$ tends to 1.

---

[14]Whittaker, Edmund Taylor (1873–1956), Fellow of Trinity College in Cambridge, Professor in Dublin and Edinburgh.

[15]Heine, Heinrich Eduard (1821–1881), Professor in Bonn and Halle.

[16]Hölder, Ludwig Otto (1859–1937), Professor in Göttingen, Tübingen, Königsberg and Leipzig.

[17]Pringsheim, Alfred (1850–1941), Professor in München. Father-in-law of the writer Thomas Mann.

Proof. This is a special case ($a = b = 1$) of the preceding corollary. □

Corollary 3. *The limit*

$$\lim_{\rho\to 0}\left(L_0(1+\rho) - \frac{p-1}{p}\frac{1}{\rho}\right)$$

exists.

Proof. Observe that

$$L_0(1+\rho) = \sum_{b=1}^{p-1} S(p,b,\rho)$$

and apply Corollary 1. □

Corollary 4. *One has*

$$\lim_{\rho\to 0}(\log L_0(1+\rho) + \log\rho) = \log\left(1-\frac{1}{p}\right).$$

Proof. This follows immediately from Corollary 3. □

4. After determining the behaviour of $L_0(s)$ Dirichlet turned to the series defining the remaining L-functions and showed that each of them converges for all positive s and its sum represents a function which is continuous and has a continuous derivative.

Actually for the proof of Theorem 2.1 only the behaviour of L-functions in a neighbourhood of $s = 1$ is of importance.

Dirichlet's original proof was also in this case based on the formula (2.3). Two years later he proved (Dirichlet 1839, Sect.1) essentially in the same way a more general result. The first part of it is a particular case of the well-known convergence criterion proved first, presumably, by P.du Bois-Reymond[18] (1870) but often named after Dirichlet, possibly because it appears in the third edition of his lectures on number theory, edited by Dedekind (Dirichlet 1879, p.376.)

We present now the latter proof.

Lemma 2.4. *Let $c_1, c_2, \ldots$ be a periodic sequence of complex numbers, denote by k its period, and assume that the sum $c_1 + c_2 + \cdots + c_k$ vanishes. If $s > 0$ then the series*

$$\sum_{n=1}^{\infty} \frac{c_n}{n^s} \tag{2.4}$$

[18]Bois-Reymond, Paul du (1831–1889), Professor in Freiburg, Tübingen and Charlottenburg.

converges and its sum defines a function $\Phi(s)$ which has a continuous derivative. Furthermore the equality

$$\Phi(s) = \frac{1}{\Gamma(s)} \int_0^1 \frac{\sum_{i=1}^k c_i x^i}{1-x^k} \log^{s-1}\left(\frac{1}{x}\right) dx$$

holds for positive s.

Proof. Denote by $S_N(s)$ the N-th partial sum of (2.4) and let $N > k$. For a suitable $h \geq 0$ we have $N = kh + i$ with $0 \leq i < k$. Put $C = \max\{|c_1|, |c_2|, \dots, |c_k|\}$ and observe that

$$|S_N(s) - S_{kh}(s)| \leq C \sum_{n=kh}^{kh+i} \frac{1}{n^s} \leq C\frac{k}{(kh)^s} < C\frac{k}{(N-k)^s} \longrightarrow 0$$

when N tends to infinity. Thus it suffices to establish the relation

$$\lim_{h\to\infty} S_{kh}(s) = \Phi(s).$$

Put $N = hk$. Using (2.3) we may write

$$\begin{aligned} S_N(s) =& \frac{1}{\Gamma(s)} \int_0^1 \left(\sum_{j=0}^{h-1} \sum_{i=1}^{k} c_i x^{jk+i-1} \right) \log^{s-1}\left(\frac{1}{x}\right) dx \\ =& \frac{1}{\Gamma(s)} \int_0^1 \left(\sum_{i=1}^{k} c_i x^{i-1} \sum_{j=0}^{h-1} x^{jk} \right) \log^{s-1}\left(\frac{1}{x}\right) dx \\ =& \frac{1}{\Gamma(s)} \int_0^1 \frac{\sum_{i=1}^k c_i x^i}{x(1-x^k)} \log^{s-1}\left(\frac{1}{x}\right) dx \\ &- \frac{1}{\Gamma(s)} \int_0^1 x^{kh} \frac{\sum_{i=1}^k c_i x^i}{x(1-x^k)} \log^{s-1}\left(\frac{1}{x}\right) dx. \end{aligned}$$

Observe that our assumption implies the divisibility of the polynomial $\sum_{i=1}^k c_i x^i$ by $x - 1$, hence with a suitable constant K we have

$$|\sum_{i=1}^k c_i x^i/(1-x^k)| \leq K$$

for $0 \leq x \leq 1$. This shows that the second summand of the last term is bounded in absolute value by

$$\frac{K}{\Gamma(s)} \int_0^1 x^{kh-1} \log^{s-1}\left(\frac{1}{x}\right) dx = \frac{K}{(hk)^s},$$

thus tends to zero for h tending to infinity.

In this way we arrive at the equality

$$\lim_{N\to\infty} S_N(s) = \frac{1}{\Gamma(s)} \int_0^1 \frac{\sum_{i=1}^k c_i x^i}{x(1-x^k)} \log^{s-1}\left(\frac{1}{x}\right) dx,$$

and the assertion follows, if we observe that the right-hand side of this equality is continuous and has a continuous derivative for $s > 0$. □

Corollary. *For $j \neq 0$ the series defining $L_j(s)$ converges, for $s > 0$, to a continuous function having a continuous derivative and one has*

$$L_j(1) = \int_0^1 \frac{f_j(x)}{x(1-x^p)} dx,$$

where

$$f_j(x) = \sum_{n=1}^{p-1} \Omega^{j\gamma_n} x^n.$$

Proof. One applies the lemma with

$$c_n = \begin{cases} \Omega^{j\gamma_n} & \text{if } p \nmid n \\ 0 & \text{otherwise,} \end{cases}$$

and uses $\Gamma(1) = 1$. □

5. The use of integrals in the proof of convergence of the series for Dirichlet L-functions $L_j(s)$ $(i \neq 0)$, for positive s, is not necessary. The convergence is now usually proved using the following elementary lemma due to N.H.Abel[19] (1826, Lehrsatz III):

Lemma 2.5. (Partial summation) *If $a_1, a_2, \ldots, a_m$ and $b_1, b_2, \ldots, b_m$ are arbitrary complex numbers and for $j = 1, 2, \ldots, m$ we put*

$$S(j) = a_1 + a_2 + \cdots + a_j,$$

then

$$\sum_{j=1}^{m} a_j b_j = S(m) b_m + \sum_{j=1}^{m-1} S(j)(b_j - b_{j+1}).$$

Proof. The assertion follows from the following chain of equalities:

[19] Abel, Niels Henrik (1802–1829).

$$\begin{aligned}\sum_{j=1}^{m} a_j b_j &= a_1 b_1 + \sum_{j=2}^{m} (S(j) - S(j-1)) b_j \\ &= S(m) b_m + a_1 b_1 + \sum_{j=2}^{m-1} S(j) b_j - \sum_{j=2}^{m} S(j-1) b_j \\ &= S(m) b_m + a_1 b_1 + \sum_{j=2}^{m-1} S(j) b_j - \sum_{j=1}^{m-1} S(j) b_{j+1} \\ &= S(m) b_m + a_1 b_1 - a_1 b_2 + \sum_{j=2}^{m-1} S(j)(b_j - b_{j+1}) \\ &= S(m) b_m + \sum_{j=1}^{m-1} S(j)(b_j - b_{j+1}).\end{aligned}$$ □

To establish convergence of the series defining $L_j(s)$ $(j \neq 0)$ for all positive values of s observe that for all positive N one has

$$|S(N)| = |\sum_{m=1}^{N} \chi_\omega(m)| \leq p.$$

In fact, if $N = Ap + r$ with $0 \leq r < p$, then

$$S(N) = \sum_{m=1}^{Ap} \chi_\omega(m) + \sum_{m=Ap+1}^{Ap+r} \chi_\omega(m),$$

and since $\chi_\omega(m)$ is periodic with period p one gets

$$\sum_{m=1}^{Ap} \chi_\omega(m) = A \sum_{m=1}^{p} \chi_\omega(m) = A \sum_{k=0}^{p-1} \omega^k = 0,$$

because ω is a primitive root of unity of order $p-1$. Thus in view of $|\chi_\omega(m)| \leq 1$ we arrive at

$$|S(N)| \leq |\sum_{m=Ap+1}^{Ap+r} \chi_\omega(m)| \leq r \leq p.$$

Then it suffices to apply Lemma 2.5. to get, for M tending to infinity and $N > M$,

$$|\sum_{m=M}^{N} \frac{\chi_\omega(m)}{m^s}| = |\frac{S(N)}{N^s} + \sum_{k=M}^{N-1} S(k) \left(\frac{1}{n^s} - \frac{1}{(n+1)^s}\right)| \leq \frac{p}{M^s} \longrightarrow 0.$$

6. After establishing the convergence of the L-series with $\omega \neq 1$ to the right of 0, Dirichlet proved that they do not vanish at $s = 1$. We shall present his way of doing it in the next section but now turn to the final part of his paper, where the assertion of Theorem 2.1 is shown to be a consequence of the non-vanishing of L-functions at 1.

The next lemma permits us to separate the prime powers lying in a prescribed residue class and forms the essence of Dirichlet's proof. Its idea may have its roots in the theory of Fourier series with which Dirichlet was occupied at that time. It expresses essentially the fact that the functions $\chi_\omega(m)$ form an orthogonal basis of the space of all functions on $\{1, 2, \ldots, p-1\}$ with the standard inner product.

Lemma 2.6. *Let l be an integer not divisible by p. Then for $s > 1$ the following identity holds:*

$$\sum_{k=1}^{\infty} \frac{1}{k} \sum_{\substack{q \text{ prime} \\ q^k \equiv l \pmod p}} \frac{1}{q^{ks}} = \frac{1}{p-1} \sum_{j=0}^{p-2} \Omega^{-j\gamma_l} \log L_j(s),$$

provided we use that branch of the complex logarithm which is real for positive reals.

Proof. Corollary 2 to Lemma 2.2 implies, that for $j > 0$ and $s > 1$ we have the equality

$$L_j(s) = \prod_{\substack{q \text{ prime} \\ q \neq p}} \left(1 - \frac{\Omega^{j\gamma_q}}{q^s}\right)^{-1},$$

from which it follows that $L_j(s)$ does not vanish for $s > 1$. We may thus write

$$\log L_j(s) = \sum_{k=1}^{\infty} \frac{1}{k} \sum_{\substack{q \text{ prime} \\ q \neq p}} \frac{\Omega^{jk\gamma_q}}{q^{ks}},$$

which implies for $l = 1, 2, \ldots, p-1$ the equality

$$\sum_{j=0}^{p-2} \Omega^{-j\gamma_l} \log L_j(s) = \sum_{k=1}^{\infty} \frac{1}{k} \sum_{\substack{q \text{ prime} \\ q \neq p}} \frac{1}{q^{ks}} \sum_{j=0}^{p-2} \Omega^{j(k\gamma_q - \gamma_l)}. \tag{2.5}$$

Now observe that the sum $\sum_{j=0}^{p-2} \Omega^{j(k\gamma_q - \gamma_l)}$ vanishes except when $k\gamma_q - \gamma_l$ is divisible by $p-1$, in which case it equals $p-1$. Noting that the condition

$$k\gamma_q \equiv \gamma_l \pmod{p-1}$$

is equivalent to

$$q^k \equiv l \pmod p$$

we see that (2.5) implies

$$\sum_{j=0}^{p-2} \Omega^{-j\gamma_l} \log L_j(s) = (p-1) \sum_{k=1}^{\infty} \frac{1}{k} \sum_{\substack{q \text{ prime} \\ q^k \equiv l \pmod p}} \frac{1}{q^{ks}}$$

as asserted. □

Corollary 1. *If $p \nmid l$ then for $s > 1$ one has*

$$\sum_{\substack{q \text{ prime} \\ q \equiv l \pmod p}} \frac{1}{q^s} = \frac{1}{p-1} \sum_{j=0}^{p-2} \Omega^{-j\gamma_l} \log L_j(s) + H(s),$$

where $H(s)$ is a function bounded for $s > 1$.

Proof. It suffices to observe that for $s \geq 1$ one has

$$\sum_{k=2}^{\infty} \frac{1}{k} \sum_{q \text{ prime}} \frac{1}{q^{ks}} \leq \sum_{k=2}^{\infty} \frac{1}{k} \sum_{m=1}^{\infty} \frac{1}{m^k} < \infty.$$ □

Corollary 2. *To prove Theorem 2.1 in the case of a prime difference $k = p$ it suffices to establish that for $j = 1, 2, \ldots, p-2$ none of the functions L_j vanishes at $s = 1$.*

Proof. Corollary 3 to Lemma 2.3 implies that for s tending to 1 the function $L_0(s)$ is unbounded. If $L_j(1) \neq 0$ for $j \neq 0$, then, for $s \to 1$, $\log L_j(s)$ tends to a finite limit. Hence the preceding corollary implies

$$\lim_{s \to 1+0} \sum_{\substack{q \text{ prime} \\ q \equiv l \pmod p}} \frac{1}{q^s} = \infty,$$

and thus there must be infinitely many primes congruent to l mod p. □

2.2. The Non-vanishing of $L_j(1)$ in Case of a Prime Difference

1. Corollary 2 to Lemma 2.6 shows that to complete the proof of Theorem 2.1 in the case $k = p$, a prime, it suffices to establish that for $j \neq 0$ the functions L_j do not vanish at $s = 1$. This is the most difficult part of the proof. We show now how Dirichlet overcame this difficulty:

Theorem 2.7. *If $j \neq 0$ then $L_j(1) \neq 0$.*

Proof: Applying the Corollary to Lemma 2.4 one gets

$$L_j(1) = -\int_0^1 \frac{f_j(x)}{x(x^p-1)}\,dx,$$

where

$$f_j(x) = \sum_{n=1}^{p-1} \Omega^{j\gamma_n} x^n.$$

Here the choice of the primitive root Ω of unity of order $p-1$ is irrelevant but we shall assume henceforth that Ω is chosen to be equal to $\zeta_{p-1} = \exp(2\pi i/(p-1))$.

The next lemma gives an explicit expression for $L_j(1)$:

Lemma 2.8. *For $j \neq 0$ one has*

$$L_j(1) = -\frac{1}{p} f_j(\zeta_p) \sum_{m=1}^{p-1} \Omega^{-j\gamma_m} \left[\log\left(2\sin\left(\frac{m\pi}{p}\right)\right) + \frac{\pi i}{2}\left(1 - \frac{2m}{p}\right)\right],$$

ζ_p denoting the p-th primitive root of unity.

Proof. Since $f_j(x)/x(x^p-1)$ is a rational function, hence decomposing it into partial fractions we obtain

$$L_j(1) = -\frac{1}{p} \sum_{m=1}^{p-1} f_j(\zeta_p^m) \int_0^1 \frac{dx}{x - \zeta_p^m}.$$

Observe now that

$$f_j(\zeta_p^m) = \Omega^{-\gamma_m} f_j(\zeta_p).$$

Indeed it suffices to put $h \equiv km \pmod p$ in the formula

$$f_j(\zeta_p^m) = \sum_{k=1}^{p-1} \Omega^{j\gamma_k} \zeta_p^{km}$$

and use $\gamma_k \equiv \gamma_h - \gamma_m \pmod{p-1}$.

Since for positive a we have

$$\int_0^1 \frac{dx}{x - e^{2\pi i a}} = \log(2\sin a\pi) + \frac{\pi(1-2a)}{2} i$$

the assertion results. □

2. The last lemma does not seem to be of particular help in establishing the non-vanishing of $L_j(1)$ in the general case. However Dirichlet succeeded in doing this in the important special case $j = (p-1)/2$, i.e. when $\omega = \Omega^j = -1$. Put $t = (p-1)/2$ and observe that since γ_m is even if and only if m is a square mod p, i.e. the Legendre symbol $\left(\frac{m}{p}\right)$ equals one then

$$\omega^{-\gamma_m} = \left(\frac{m}{p}\right)$$

and Lemma 2.8 shows that the value of the corresponding L-function at $s = 1$ equals

$$L_t(1) = \sum_{\substack{n \\ (n,p)=1}} \left(\frac{n}{p}\right) \frac{1}{n}$$

$$= -\frac{1}{p} f_t(\zeta_p) \sum_{m=1}^{p-1} \left(\frac{m}{p}\right) \left[\log\left(2\sin\left(\frac{m\pi}{p}\right)\right) + \frac{\pi i}{2}\left(1 - \frac{2m}{p}\right)\right]. \tag{2.6}$$

If $p \equiv 3 \pmod 4$, then $\left(\frac{-1}{p}\right) = -1$ and hence for all j one has

$$\left(\frac{j}{p}\right) = -\left(\frac{p-j}{p}\right).$$

But we have $\sin(j\pi/p) = \sin(\pi(1 - j/p))$ thus (2.6) leads to

$$L_t(1) = \frac{\pi i}{p^2} f_t(\zeta_p)\left(\sum r - \sum n\right),$$

where r and n runs over all quadratic residues resp. non-residues mod p in the interval $[1,p]$. Since

$$\sum r + \sum n = 1 + 2 + \cdots + (p-1) = p(p-1)/2$$

is odd one gets $\sum r - \sum n \neq 0$ and it remains to establish the non-vanishing of $f_t(\zeta_p)$.

To deal with this problem Dirichlet observes that $f_t(\zeta_p)$ is equal to a sum studied by Gauss (1811) and called now the *Gaussian sum* corresponding to the Legendre symbol. Gauss proved the equalities

$$f_t(\zeta_p) = \begin{cases} \sqrt{p} & \text{if } p \equiv 1 \pmod 4 \\ i\sqrt{p} & \text{if } p \equiv 3 \pmod 4 \end{cases} \tag{2.7}$$

thus

$$f_t(\zeta_p) \neq 0. \tag{2.8}$$

There are several proofs of (2.7), the first being due to Gauss (1811) and the second to Dirichlet (1835,1837a). A simplified version of it appears in Dirichlet (1839). In the first version Dirichlet uses the formula

$$\frac{f(0)+f(N)}{2}+\sum_{j=1}^{N-1} f(j\pi)=\frac{1}{\pi}\lim_{k\to\infty}\int_0^N f(t)\frac{\sin((2k+1)t)}{\sin t}dt$$

valid for integral N and every function f which is continuous in $[0,N]$ and has there a finite number of maxima and minima. He obtained this formula as a consequence of his earlier research on the convergence of Fourier series (Dirichlet 1829). The second version is based on Poisson's formula and is reproduced in Davenport[20] (1967) (see Exercise 4).

A survey of various proofs of (2.7) was given by B.C.Berndt and R.J.Evans (1981).

Actually the exact value of $f_t(\zeta_p)$ is irrelevant for our purpose and it is enough to know that it is non-zero, which in our case can be seen directly as follows: if ζ_p were a zero of f_t then it would be also a root of the polynomial $f_t(X)/X$ which has rational coefficients and is of degree $p-2$. However ζ_p is a root of the irreducible polynomial $(X^p-1)/(X-1)$ which divides every polynomial with rational coefficients having ζ_p as a root. This leads to the impossible inequality $p-1\le p-2$. Observe that we did not use here the assumption $p\equiv 3 \pmod 4$ and thus the inequality (2.8) holds for every odd prime p.

In the case $p\equiv 1 \pmod 4$ Dirichlet takes another way. This time one has

$$\left(\frac{j}{p}\right)=\left(\frac{p-j}{p}\right)$$

and this implies

$$L_t(1)=\frac{1}{p}f_t(\zeta_p)\log\left(\frac{\prod_n \sin(n\pi/p)}{\prod_r \sin(r\pi/p)}\right), \tag{2.9}$$

where, as in the previous case, r runs through quadratic residues and n through quadratic non-residues mod p. We know already that $f_t(\zeta_p)$ does not vanish hence it remains to show that the two products in (2.9) are distinct.

To obtain this Dirichlet utilizes two formulas discovered by C.F.Gauss in his study of cyclotomy (Gauss 1801, Sect.357):

$$A(X)=\prod_r (X-\zeta_p^r)=\frac{Y(X)-Z(X)\sqrt{p}}{2}$$

and

$$B(X)=\prod_n (X-\zeta_p^n)=\frac{Y(X)+Z(X)\sqrt{p}}{2},$$

where $Y(X)$ and $Z(X)$ are polynomials with rational integral coefficients.

The simplest way to prove them is the following: Note that the polynomials $A(X)$ and $B(X)$ are invariant under the automorphism of the field $Q(\zeta_p)$ defined

[20] Davenport, Harold (1907–1969), Professor in Bangor, London and Trinity College, Cambridge.

by $\zeta_p \mapsto \zeta_p^j$ when j is a quadratic residue mod p and are permuted when j is a quadratic non-residue. This shows that the coefficients of $A(X)$ and $B(X)$ lie in the unique quadratic subfield of $Q(\zeta_p)$ which equals $Q(\sqrt{p})$ because of (2.7). Since these coefficients are algebraic integers, the asserted formulas follow from the observation that algebraic integers contained in $Q(\sqrt{p})$ with $p \equiv 1 \pmod 4$ are of the form $(a + b\sqrt{p})/2$ where a, b are rational integers of the same parity.

Multiplying $A(X)$ and $B(X)$ one obtains

$$4(1 + X + X^2 + \cdots + X^{p-1}) = Y^2(X) - pZ^2(X)$$

and putting here $X = 1$ and $Y(1) = g$, $Z(1) = h$ one is led to $g^2 - ph^2 = 4p$.

Now

$$\frac{\prod_n \sin(n\pi/p)}{\prod_r \sin(r\pi/p)} = \frac{\prod_n \left(e^{in\pi/p} - e^{-in\pi/p}\right)}{\prod_r \left(e^{ir\pi/p} - e^{-ir\pi/p}\right)}$$
$$= \zeta_p^{(R-N)/2} \frac{\prod_n (\zeta_p^n - 1)}{\prod_r (\zeta_p^r - 1)} = \frac{B(1)}{A(1)} \zeta_p^{(R-N)/2}$$

where $R = \sum_r r$ and $N = \sum_n n$.

If

$$\frac{B(1)}{A(1)} \zeta_p^{(R-N)/2} = 1,$$

then $\zeta_p^{\frac{R-N}{2}}$ must be real, hence equal to 1 or -1. In the first case we get

$$1 = \frac{B(1)}{A(1)} = \frac{g + h\sqrt{p}}{g - h\sqrt{p}},$$

hence $h = 0$ and $4p = g^2$, a contradiction. In the second case we get

$$-1 = \frac{B(1)}{A(1)} = \frac{g + h\sqrt{p}}{g - h\sqrt{p}},$$

hence $g = 0$ and we arrive at $0 < 4p = -ph^2 < 0$, again a contradiction.

Finally

$$\frac{\prod_n \sin(n\pi/p)}{\prod_r \sin(r\pi/p} \neq 1$$

and in view of (2.9) this shows that the L-function in question does not vanish at 1.

3. It remains to deal with the case when $\omega = \Omega^j \neq 1, -1$. Taking logarithms one gets for $x > 1$

$$\log L_j(x) = \sum_{k=1}^{\infty} \frac{1}{k} \sum_{q \text{ prime}} \omega^{k\gamma_q} \frac{1}{q^{kx}}.$$

Summing over $\omega = 1, \Omega, \ldots, \Omega^{p-2}$ one obtains

$$\log(L_0 L_1 \cdots L_{p-2}) = (p-1) \sum_{k=1}^{\infty} \frac{1}{k} \sum_{q_k} \frac{1}{q_k^{kx}} \geq 0, \tag{2.10}$$

where q_k runs for $k = 1, 2, \ldots$ over all primes satisfying $q_k^k \equiv 1 \pmod{p}$.

Since $\overline{L_j(x)} = L_{p-j-1}(x)$ the product $L_j(x)L_{p-j-1}(x)$ is real and we may write

$$\log(L_0 L_1 \cdots L_{p-2}) = \log L_0 + \log L_{(p-1)/2} + \sum_{k=1}^{(p-3)/2} \log(L_k L_{p-k-1}). \tag{2.11}$$

From Corollary 4 to Lemma 2.3 it follows that the first term equals $-\log(x-1) + g_0(x)$, where $g_0(x)$ is a function tending to a definite limit when x tends to 1. Previous arguments lead now to the existence of the limit $\lim_{x \to 1+0} \log L_{(p-1)/2}(x)$.

If $j \neq 0, (p-1)/2$ and $L_j(1) = 0$ then

$$\lim_{x \to 1} \frac{L_j(x)}{x-1} = L_j'(1),$$

because L_j has a continuous derivative. Moreover $L_{p-j-1}(1) = 0$ and we see that in a neighbourhood of $x = 1$ the inequality $\log(L_j(x)L_{p-j-1}(x)) \leq 2\log(x-1) + c_j$ holds with a certain constant c_j. If however $L_j(1) \neq 0$ then in a neighbourhood of $x = 1$ one has $\log(L_j(x)L_{p-j-1}(x)) \leq c_j$ with a suitable constant c_j.

Putting everything together we obtain that the left-hand side of (2.11) does not exceed $2M\log(x-1) + c$, where $2M$ is the number of functions L_j vanishing at i and c is a constant. This shows that if M is positive then $\log(L_0 L_1 \cdots L_{p-2})$ tends to $-\infty$ when x approaches 1, contradicting (2.10). This concludes Dirichlet's proof of Theorem 2.1 in the case of a prime modulus. □

2.3. The Case of an Arbitrary Difference

1. Let now $k > 0$ be a composite integer divisible by 8. Clearly it is sufficient to prove Theorem 2.1 in this case. To do this Dirichlet first defined the system of indices for such k by a construction which we presented at the end of Sect.3 of the preceding chapter. Let

$$k = \prod_{j=0}^{t} p_i^{\alpha_i}$$

(with $p_0 = 2$, $\alpha_0 \geq 3$, $\alpha_i \geq 1$ for $i \geq 1$) be the canonical factorization of k into prime-powers, and define $\gamma_i(n)$ by the formulas (1.8), (1.9) and (1.10).

For $i = 1, \ldots, t$ choose a primitive root of unity Ω_i of order $\varphi(p_i^{\alpha_i})$, put $\Omega_{-1} = -1$ and let Ω_0 be a primitive root of unity of order $2^{\alpha_0 - 2}$. Then define for $s > 1$ and any system $B = \{b_{-1}, b_0, \ldots, b_t\}$ satisfying $b_{-1} = 0, 1$, $0 \le b_0 < 2^{\alpha_0 - 2}$, $0 \le b_i < \varphi(p_i^{\alpha_i})$ $(i = 1, 2, \ldots, t)$

$$L_B(s) = L_{b_{-1}, b_0, \ldots, b_t}(s) = \sum_{m=1}^{\infty} \chi_B(m) m^{-s}$$

where $\chi_B(m) = \prod_{i=-1}^{t} \Omega_i^{b_i \gamma_i(n)}$.

The argument proceeds now similarly to the case of prime difference. The same reasoning which led to Corollary 3 to Lemma 2.3 shows that in the case $B = B_0 = \{0, 0, \ldots, 0\}$ the limit

$$\lim_{\rho \to 0} \left(L_{B_0}(1 + \rho) - \frac{\varphi(k)}{k} \cdot \frac{1}{\rho} \right)$$

exists and using Lemma 2.3 we obtain that for all other choices of B the series defining $L_B(s)$ converges for $s > 0$ to a continuous function. Finally Dirichlet sketched the proof (following closely the argument used in the prime case) that the following analogue of Lemma 2.6 with its Corollary 2 holds also for composite values of k:

Lemma 2.9. *Let l be an integer prime to k. Then for $s > 1$ the following identity holds:*

$$\sum_{m=1}^{\infty} \frac{1}{m} \sum_{\substack{q \text{ prime} \\ q^m \equiv l \pmod{k}}} \frac{1}{q^{ms}} = \frac{1}{\varphi(k)} \sum_{B} \overline{\chi_B(l)} \log L_B(s). \qquad \square$$

Corollary. *To prove Theorem 2.1 in the case of an arbitrary prime k it suffices to establish that for $B \ne B_0$ none of the functions L_B vanishes at $s = 1$.* $\square$

Thus everything has been reduced to the non-vanishing of $L_B(1)$ for $B \ne B_0$. One does not find the proof of this assertion in Dirichlet (1837c) who only sketched the proof of the following implication:

If for all systems B for which the function $\chi_B(m)$ is real-valued one has $L_B(1) \ne 0$ then the same holds for all $B \ne B_0$.

He did this using essentially the same argument as used in subsection 3 of the preceding section, which leads to an analogue of (2.10).

With respect to L-functions with real coefficients Dirichlet wrote:

"*In*[21] *der Abhandlung, so wie sie der Akademie ursprünglich vorgelegt wurde, hatte ich diese Eigenschaft durch indirecte und ziemlich complicierte Betrachtungen bewiesen. Ich habe mich aber später überzeugt, dass man denselben Zweck auf einem andern Wege weit kürzer erreicht. Die Principien, von welchen wir hier ausgegangen sind, lassen sich auf mehrere andere Probleme anwenden, zwischen denen und dem hier behandelten Gegenstande man zunächst keinen Zusammenhang vermuthen sollte.*

Namentlich kann man mit Hülfe dieser Principien die sehr interessante Aufgabe lösen, die Anzahl der verschiedenen quadratischen Formen zu bestimmen, welche einer beliebigen positiven oder negativen Determinante entsprechen, und man findet, dass diese Anzahl ... als Product von zwei Factoren dargestellt werden kann, wovon der erste eine sehr einfache Function der Determinante ist, welche für jede Determinante einen endlichen Wert hat, während der andere Factor durch eine Reihe ausgedrückt ist, die mit der obigen ... zusammenfällt. Auss diesem Resultat folgt dann unmittelbar, dass die Summe ... nie Null sein kann, da sonst für die entsprechende Determinante die Anzahl der quadratischen Formen sich auf Null reduciren würde, während diese Anzahl wirklich $\overset{=}{>} 1$ *ist.*"

2. The promised formula for the number of classes of quadratic forms was introduced in Dirichlet (1838b) and proved with all details in Dirichlet (1839). We shall state now this formula but refrain from giving a proof which would lead us far away from our main topic. The reader may find it in e.g. Davenport (1967) or, in a version dealing with class-numbers of quadratic fields, in books on algebraic number theory[22].

Dirichlet follows Gauss (1801) and considers the set $\mathbf{f}_{\mathfrak{G}}(D)$ of all binary quadratic forms $f(x,y) = ax^2 + 2bxy + cy^2$ with integral a, b, c of a given *determinant* $D(f) = b^2 - ac$, satisfying $(a, 2b, c) = 1$. Such forms are called *properly primitive*. Assume also that in case of negative D the form f is *positive definite*, i.e. one has $a > 0$ and $c > 0$.

Earlier J.L.Lagrange (1773) considered forms $F(x,y) = Ax^2 + Bxy + Cy^2$ whose middle coefficient could have any parity and used the *discriminant* $d(F) = B^2 -$

[21] "*In the paper presented originally to the Academy I proved this property* [namely $L_B(1) \neq 0$] *by indirect and very complicated arguments. However later I became convinced that one can obtain this aim much quicker in another way. The principles from which we started here can be applied to several other problems, having seemingly no connection with these treated here. Namely one can solve, with the help of these principles, a very interesting problem consisting of the determination of the number of various quadratic forms* [Dirichlet means here the number of classes of equivalent forms], *corresponding to an arbitrary, positive or negative determinant. One finds that that number ... can be represented as a product of two factors, the first of which is a very simple function of the determinant and the second given by a series, coinciding with the considered above* [the series $L_B(1)$ in case of real-valued $\chi_B(m)$]. *From this result it follows directly that this sum can never be zero, because otherwise for the corresponding determinant the number of forms would be reduced to zero, whereas this number is in reality always* $\overset{=}{>} 1$."

[22] See e.g. Borevič, Šafarevič 1964 or Narkiewicz 1990.

$4AC$. This approach is more general since one sees easily that Gaussian forms coincide with Lagrangian forms with discriminants divisible by 4.

On the set $\mathbf{f}_{\mathfrak{G}}(D)$ the group $SL_2(\mathbf{Z})$ of 2×2 matrices with integral coefficients and unit determinant acts by

$$\mathfrak{A} \cdot f(x,y) = f(a_{11}x + a_{12}y, a_{21}x + a_{22}y),$$

for $\mathfrak{A} = [a_{ij}] \in SL_2(\mathbf{Z})$.

Two forms $f, g \in \mathbf{f}_{\mathfrak{G}}(D)$ are called *equivalent*, if for some $\mathfrak{A} \in SL_2(\mathbf{Z})$ one has $g = \mathfrak{A} \cdot f$. The number of so defined equivalence classes is denoted by $H(D)$ and called the *class-number*[23] *of determinant* D.

Dirichlet found a formula expressing the class-number $H(D)$ in terms of the value at $s = 1$ of the L-function corresponding to a particular real character mod $|D|$. His original statement was rather complicated as he did not have at his disposal Kronecker's generalization of the symbol of Legendre and worked instead with the *Jacobi symbol* (defined in Jacobi (1837)), which forced him to consider separately several cases.

Jacobi's symbol $\left(\frac{k}{D}\right)$ is defined in the following way:

If D is an odd integer and $D = p_1 p_2 \cdots p_r$ is its factorization into prime factors (not necessarily distinct) then for every integer k prime to D one puts

$$\left(\frac{k}{D}\right) = \prod_{j=1}^{r} \left(\frac{k}{p_j}\right),$$

with Legendre symbols on the right-hand side. In the case $D = \pm 1$ one puts

$$\left(\frac{k}{D}\right) = 1.$$

The Jacobi symbol has all formal properties of the Legendre symbol, viz.

$$\left(\frac{mn}{D}\right) = \left(\frac{m}{D}\right)\left(\frac{n}{D}\right),$$

$$\left(\frac{n}{D_1 D_2}\right) = \left(\frac{n}{D_1}\right)\left(\frac{n}{D_2}\right),$$

$$\left(\frac{-1}{D}\right) = (-1)^{(D-1)/2},$$

$$\left(\frac{2}{D}\right) = (-1)^{(D^2-1)/8}$$

and

[23]If one defines in the same way equivalence in the set $\mathbf{F}_{\mathfrak{L}}(d)$ of Lagrangian forms of discriminant d and denotes by $h(d)$ the number of resulting equivalence classes then $H(D) = h(4D)$. Dirichlet's class-number formula remains true also for Lagrangian forms of discriminant not divisible by 4, as shown by L.Kronecker (1885).

$$\left(\frac{D_1}{D_2}\right)\left(\frac{D_2}{D_1}\right) = (-1)^{[(D_1-1)/2][(D_2-1)/2]},$$

the last equality holding for odd and relatively prime D_1,D_2.

The proofs of these properties follow easily from the corresponding properties of the Legendre symbol and Dirichlet omits their proofs, referring instead to §133 of Gauss (1801), where one finds the last of the above equalities in disguise. If one puts $\chi_D(n) = \left(\frac{n}{D}\right)$, then the properties listed above imply that χ_D is a real Dirichlet character mod $|D|$.

Kronecker's symbol extends the symbol of Jacobi by allowing the denominators to be even. He does this by defining

$$\left(\frac{k}{2}\right) = \begin{cases} (-1)^{(k^2-1)/8} & \text{if } k \text{ is odd} \\ 0 & \text{if } k \text{ is even,} \end{cases}$$

taking into account the sign of the denominator by redefining

$$\left(\frac{k}{-1}\right) = \begin{cases} 1 & \text{if } k \geq 0 \\ -1 & \text{if } k < 0 \end{cases}$$

and extending it by multiplicativity in the denominator.

Using Kronecker's symbol one may state Dirichlet's class-number formulas in the most important case of a fundamental discriminant d (i.e. when d is a product of relatively prime factors of the form $-4, \pm 8, (-1)^{(p-1)/2}p$ for odd primes p) in the following way:

In case of positive discriminant d one has

$$H(d) = \frac{\sqrt{d}}{2\log\epsilon} L(1,\chi),$$

and for negative D one has

$$H(d) = \frac{1}{\pi}\sqrt{|d|}L(1,\chi) = \frac{1}{d}\sum_{j=1}^{|d|} j\chi(j) \quad (d \neq -3, -4).$$

Here one has

$$\chi(n) = \left(\frac{d}{n}\right),$$

the Kronecker's symbol and (in case $d > 0$) $\epsilon = (T + U\sqrt{D})/2$, where

$$D = \begin{cases} d & \textit{if } d \textit{ is odd} \\ d/4 & \textit{if } d \textit{ is even,} \end{cases}$$

and T, U are the smallest positive integers satisfying the condition

$$|T^2 - DU^2| = 4.$$

These formulas immediately imply the non-vanishing of $L(1,\chi)$ in the considered case. In the general case one needs the following observation which is implicit in Dirichlet's paper:

If χ is a real character mod k *then there exists a fundamental discriminant D dividing k such that the equality*

$$\chi(n) = \begin{cases} \left(\frac{D}{n}\right) & \textit{if } (n,D)=1 \\ 0 & \textit{otherwise} \end{cases}$$

holds.

The quickest proof of this assertion is based on the notion of a *primitive* or *proper character* which appeared in an explicit form only much later. The distinction between primitive and imprimitive characters is also important in the study of analytic properties of $L(s,\chi)$ made by R.Lipschitz (1889). However, contrary to the quotation in Landau (1909a, p.898), no explicit classification of characters appears in Lipschitz's paper.

Landau (1909a, §125) defines a character χ mod k to be *improper* mod k, if there exists a proper divisor d of k with the following property:

If $m \equiv n$ (mod d) *and both m and n are prime to k, then $\chi(m) = \chi(n)$.*

Otherwise χ is called *proper* or *primitive*. (Note that in Hasse (1950) a slightly different definition of a proper character is given.)

The proof of the assertion above consists of two steps: first one has to show that primitive real characters mod k exist if and only if either k or $-k$ is a fundamental discriminant and then one needs to establish that if k is a fundamental discriminant then the only primitive real characters corresponding to k are $\left(\frac{k}{n}\right)$ and possibly $\left(\frac{-k}{n}\right)$, the second only if $-k$ is also a fundamental discriminant (see Exercise 7).

The above assertion immediately implies the equality

$$L(1,\chi) = \prod \left(1 - \frac{1}{p}\right) L(1,X),$$

the product being taken over all primes p which divide k but do not divide D and $X(n) = \left(\frac{D}{n}\right)$. Thus $L(1,\chi)$ does not vanish. This observation achieves the proof of Theorem 2.1. □

3. A similar method has been utilized by Dirichlet in two later papers (Dirichlet 1840a, 1841) to deal with related questions. In the first paper he sketched a proof of a rather special case of the following theorem, a complete proof of which was later given by H.Weber[24] (Weber 1882):

---

[24]Weber, Heinrich (1842–1913), Professor in Heidelberg, Zürich, Königsberg, Marburg, Göttingen and Strassburg.

Every quadratic form $f(x, y) = ax^2 + 2bxy + cy^2$ $(a, b, c \in \mathbf{Z}, (a, 2b, c) = 1)$ *represents infinitely many primes.*

Dirichlet himself considered only the case when the determinant $d = b^2 - 4ac$ is negative, $-d$ is a prime congruent to 3 mod 4 and moreover, in modern terminology, has a cyclic class-group. A proof of this theorem, which is usually called the *theorem of Dirichlet–Weber* was also found[25] by E.Schering[26] in 1863 who however did not publish it[27] and it appeared only in his Collected Works in 1909 (see Schering 1909a). H.Weber (1882, pp.301–302) pointed out that neither Schering's proof nor the proof given by Mertens (1874a) contains any details of the proof of non-vanishing at $s = 1$ of the analogues of Dirichlet's L-functions which appear in the argument. His assertion concerning Mertens' proof seems to be not quite accurate. A proof of the quantitative form of this result was given by C.de la Vallée-Poussin (1896a III,1897). See also Landau (1915b).

Elementary proofs of the Dirichlet-Weber theorem were found, much later, by W.E.Briggs (1954) and H.Ehlich (1959).

The following extension of this result is stated in Dirichlet (1840b) without proof:

Every positive definite quadratic form

$$f(x, y) = ax^2 + 2bxy + cy^2 \qquad (a, b, c \in \mathbf{Z}, (a, 2b, c) = 1)$$

represents infinitely many primes lying in a prescribed residue class $l \bmod k$, *provided certain necessary compatibility conditions are fulfilled.*

A proof was later given by A.Meyer (1888). See also Mertens (1895a,1900) for elementary proofs. H.Weber (1897) deduced Dirichlet's assertion from a more general result in algebraic number theory. A quantitative version of this result was proved by C.de la Vallée-Poussin (1897).

In Dirichlet (1841) one finds the following analogue of Theorem 2.1 in the ring of complex integers (i.e. complex numbers with integral real and imaginary parts):

If α, β *are relatively prime complex integers then there are infinitely many primes of the form* $z\alpha + \beta$, *where* z *is a complex integer.*

The proof follows largely the pattern of Dirichlet's proof in the rational case. The proof of non-vanishing at $s = 1$ of the corresponding series, forming the central point

---

[25]In a letter to Kronecker (see Schering 1887) dated 14th May 1863 Schering stated that he was able to reduce the problem to Dirichlet's L-functions and in Schering (1869) he wrote "*Habe ich ... allgemein nachwiesen, dass jede eigentlich primitive Form unendlich viele Primzahlen darstellt*". ("*I showed ... in full generality that every properly primitive form represents infinitely many prime numbers.*")

[26]Schering, Ernst Christian Julius (1833–1897), Professor in Göttingen.

[27]The reasons for that are explained in his letter to P.Bachmann of 10th May 1893. See Schering 1909b.

of the argument, was established in a later paper (Dirichlet 1842a) which contains a formula for the number of classes of binary quadratic forms with complex integer coefficients of a given determinant. See also Mertens (1899b).

Complex integers form the ring of integers of the quadratic field $Q(i)$ hence Dirichlet's result is a special case of the following assertion:

Let K be a finite extension of the rationals and Z_K its ring of integers. If $\alpha, \beta \in R$ generate coprime ideals, i.e. $\alpha R + \beta R = R$ then for infinitely many $z \in R$ the number $z\alpha + \beta$ is prime, i.e. generates a prime ideal.

This assertion was proved in the case $K = Q(\sqrt{-3})$ by E.Fanta (1901a) and in the general case by P.Furtwängler[28] (1907). This was put into a more general setting by E.Hecke[29] (1917b). (See e.g. Narkiewicz (1990, Corollary 6 to Proposition 7.9), where also further literature concerning that topic is listed).

4. Most modern analytic proofs of Theorem 2.1 follow the pattern of Dirichlet's original argument and differ only in the way of establishing the non-vanishing of $L(1, \chi)$. Moreover, instead of considering systems of indices and roots of unity one uses now characters of the multiplicative group of invertible residue classes mod k, usually called *Dirichlet characters*. It should be pointed out that Dirichlet himself used the word character in a sense different from that used now (see e.g. Dirichlet 1839, sect. 2). He followed Gauss (1801) in speaking of a *character of a quadratic form* f with a given determinant d, meaning by that the set of conditions expressed by the values of quadratic characters corresponding to prime power divisors of d, which are satisfied by any n represented by the form f. It seems that general Dirichlet characters in a modern sense[30] appear for the first time in the third edition of Dirichlet's lectures, prepared by R.Dedekind (Dirichlet 1879).

Nowadays one defines Dirichlet characters in the following way:

A complex-valued function χ defined on the set $\mathbf{Z}$ of all rational integers is called a *Dirichlet character* mod k (or *character* mod k for short) if it has the following three properties:

(i) $\chi(n) = 0$ holds if and only if $(k, n) > 1$,
(ii) $\chi(n + k) = \chi(n)$,
(iii) $\chi(n_1 n_2) = \chi(n_1)\chi(n_2)$.

Here n, n_1, n_2 are arbitrary rational integers.

Alternatively, one can define a character mod k using elementary theory of abelian groups: if $G(k)$ denotes the multiplicative group of residue classes mod k prime to k, T denotes the circle group $\{z \in \mathbf{C} : |z| = 1\}$ and

$$\psi : G(k) \longrightarrow T$$

---

[28] Furtwängler, Philipp (1869–1940), Professor in Vienna.
[29] Hecke, Erich (1887–1947), Professor in Basel, Göttingen and Hamburg.
[30] The development of the notion of the group characters, whose first examples were the Dirichlet characters, is described in Hawkins (1970/71).

is a homomorphism then the Dirichlet character χ associated with ψ is defined for all $n \in \mathbf{Z}$ by

$$\chi(n) = \begin{cases} \psi(n \bmod k) & \text{if} \quad (n,k) = 1 \\ 0 & \text{otherwise.} \end{cases}$$

It is not difficult to show that these two definitions are in perfect agreement.

The function $\chi_0(n)$ which assumes for all n prime to k the value 1 and vanishes for other n's is obviously a Dirichlet character. It is called the *principal character* (mod k).

5. The next lemma lists the main properties of characters:

Lemma 2.10. *(i) Let χ be a character* mod k. *Then $\chi(1) = 1$ and the non-zero values of χ are roots of unity of order dividing $\varphi(k)$.*

(ii) The set $\Gamma(k)$ of all characters mod k *forms a group under multiplication. If $k = MN$ with $(M,N) = 1$, then $\Gamma(k)$ is isomorphic with the direct sum $\Gamma(M) \oplus \Gamma(N)$.*

(iii) There are exactly $\varphi(k)$ characters mod k.

(iv) The functions χ_B occuring in the previous section are characters mod k *and every character is of this form.*

(v) If χ runs over all characters mod k, *then*

$$\sum_{\chi} \chi(a) = \begin{cases} \varphi(k) & \textit{if } a \equiv 1 \pmod{k}, \\ 0 & \textit{otherwise.} \end{cases}$$

(vi) One has

$$\sum_{a=1}^{k} \chi(a) = \begin{cases} \varphi(k) & \textit{if } \chi \textit{ is principal} \\ 0 & \textit{otherwise.} \end{cases}$$

Proof. (i) The first assertion results from $\chi(1) = \chi(1)\chi(1) \neq 0$. If $(k,n) = 1$ then Euler's theorem leads to

$$1 = \chi(1) = \chi(n^{\varphi(k)}) = \chi(n)^{\varphi(k)},$$

and the second assertion follows.

(ii) The first assertion is obvious. To prove the second observe first that every pair $[\chi_1, \chi_2]$ with $\chi_1 \in \Gamma(M)$, $\chi_2 \in \Gamma(N)$ induces a character $\chi \in \Gamma(MN)$ by the formula

$$\chi(x) = \chi_1(x)\chi_2(x) \qquad \text{for } x \text{ prime to } MN.$$

The map $\Psi : \Gamma(M) \oplus \Gamma(N) \longrightarrow \Gamma(MN)$ given by $[\chi_1, \chi_2] \mapsto \chi$ is clearly a group homomorphism. To show that its kernel is trivial use for given m, n with $(m, M) = (n, N) = 1$ the Chinese Remainder Theorem to get an integer x satisfying $x \equiv m \pmod{M}$ and $x \equiv n \pmod{N}$. If now $\chi = \Psi([\chi_1, \chi_2])$

equals the principal character χ_0, then $1 = \chi(x) = \chi_1(m)\chi_2(n)$, and putting first $m = 1$ and then $n = 1$ we obtain that both χ_1 and χ_2 must be principal.

The surjectivity of Ψ results now from the observation that due to the equality $\varphi(MN) = \varphi(M)\varphi(N)$ the cardinalities of $\Gamma(M) \oplus \Gamma(N)$ and $\Gamma(MN)$ coincide.

(iii) Let first $k = p^n$ be a power of an odd prime p. Since every character χ mod k is determined by its value at a fixed primitive root g mod k and that value may be equal to an arbitrary $\varphi(k)$th root of unity we get $\varphi(k)$ distinct characters, as asserted. In the case $k = 2^n$ (with $n \geq 3$) Dirichlet's construction of indices shows that every odd integer is congruent mod 2^k to a number of the form $(-1)^\alpha 5^\beta$ with $\alpha = 0, 1$ and $0 \leq \beta < 2^{n-2}$. Thus every character is determined by its values attained at -1 (either 1 or -1) and at 5 (here we can have any root of unity of order 2^{n-2}). Hence we get $2^{n-1} = \varphi(2^n)$ distinct characters in this case. The cases $k = 2, 4$ being trivial it suffices now to apply (ii).

(iv) The first assertion is immediate and the second follows from (iii) and the observation that there are exactly $\varphi(k)$ functions χ_B.

(v) Put $S = \sum_\chi \chi(a)$. If $a \equiv 1 \pmod{k}$ then we get $S = \varphi(k)$ from (i) and (iii). In the remaining case we shall first show that there exists a character X for which $X(a) \neq 1$. Indeed, there must exist a prime power $p^n \parallel k$ such that $a \not\equiv 1 \pmod{p^n}$. If $p \neq 2$ and g is a primitive root mod p^n then write

$$a \equiv g^r \pmod{p^n}$$

with $0 < r < \varphi(p^n)$. If we now construct the function χ_B for a system B in which all b_i's except that corresponding to the factor p^n are equal to 0 and the remaining one equals 1 then the resulting character will attain at a a value $\neq 1$. In case $p = 2$ this construction has to be modified in an evident way.

Now observe that

$$X(a)S = \sum_\chi X(a)\chi(a) = \sum_\chi (X\chi)(a) = S$$

because when χ runs over all characters mod k so does $X\chi$. Since $X(a) \neq 1$, S must be zero.

(vi) The assertion is obvious for the principal character. If χ is non-principal, then choose b with $(b, k) = 1$ and $\chi(b) \neq 1$ and put $S = \sum_{n=1}^k \chi(n)$. Then $\chi(b)S = \sum_{n=1}^k \chi(bn) = S$ and $S = 0$ follows. $\square$

For later use we point out an easy corollary:

Corollary. *If χ is a non-principal character* mod k *then for all $x \geq y \geq 1$ one has*

$$\left|\sum_{y \leq n \leq x} \chi(n)\right| \leq k$$

and

$$|\sum_{n\le x} n\chi(n)| \le 2kx.$$

Proof. The first assertion follows from the periodicity of χ and part (vi) of the lemma. To obtain the second note that it suffices to deal with integral x, put $S(m) = \sum_{n=1}^{m} \chi(n)$, write

$$\sum_{n\le x} n\chi(n) = \sum_{n\le x} n(S(n) - S(n-1)) = -\sum_{n\le x-1} S(n) + xS(x)$$

and apply the bound for $S(m)$ just obtained. □

From the lemma one immediately obtains for the sum

$$F_r(x) = \sum_p p^{-x},$$

(where p runs over all primes congruent to $r \pmod{k}$, $(r,k) = 1$ and $x > 1$), the equality

$$F_r(x) = \frac{1}{\varphi(k)} \sum_\chi \overline{\chi(r)} \sum_p \frac{\chi(p)}{p^x},$$

where the first sum runs over all characters χ mod k and the second runs over all prime numbers p.

If now one puts for any character χ and $s > 1$

$$L(s,\chi) = \sum_{n=1}^{\infty} \frac{\chi(n)}{n^s}$$

then Lemma 2.10 (iv) implies that these functions coincide with L-functions, considered in the preceding section. In particular, for non-principal χ they are well-defined for $s > 0$ and do not vanish for $s > 1$.

It is now an easy task to transcribe Dirichlet's proof into modern language.

Another analytic proof of Dirichlet's theorem which uses power series instead of Dirichlet series was given in Koshiba,Uchiyama (1967).

2.4. Elementary Proofs of $L(1,\chi) \neq 0$

1. In this section we shall present a selection of proofs of the non-vanishing of L-functions at $s = 1$ which can be regarded as elementary in that they do not involve complex analysis or the theory of quadratic forms[31]. In the case

[31] For a selection of analytic proofs see Chap.5.

of a prime modulus such was Dirichlet's original proof (see the proof of Theorem 2.7). For real characters elementary proofs were provided by F.Mertens (1895a, 1895b, 1897c, 1899a). (The first paper contains also an elementary proof for non-real characters which avoids the use of integrals occuring in Dirichlet's proof.) We shall now give two of his proofs in slightly simplified versions which, however, preserve the original ideas. The first proof of Mertens (1895b,1897c) we present in a version simplified by H.N.Shapiro (1950b, II) who used it for an elementary proof of Theorem 2.1, not utilizing Dirichlet series. In the second proof we follow its exposition in Landau (1906a).

Both proofs depend on certain simple facts from elementary number theory which we now prove.

Lemma 2.11. *If χ is a character, then the function $f(n,\chi) = \sum_{d|n} \chi(d)$ is multiplicative. If $n = \prod p^{\alpha_p}$ is the canonical factorization of n, then*

$$f(n,\chi) = \prod_p (1 + \chi(p) + \cdots + \chi(p)^{\alpha_p}).$$

If the character χ is real then $f(n,\chi)$ is non-negative and for all $n \geq 1$ one has $f(n^2,\chi) \geq 1$.

Proof. The first part follows from Lemma 1.16, the second is a consequence of the equality $\chi(p^r) = \chi(p)^r$ for $r = 1, 2, \ldots$ and for the last it suffices to observe that for real χ we have $\chi(p) = 0, 1$ or -1. □

The next lemma gives a simple formula describing the asymptotic behaviour of the sum of reciprocals of quadratic roots of positive integers. It is obtained by using a formula which is the simplest case of the *Euler–Maclaurin summation formula* stated by Euler (1732/33a) and proved few years later by himself (Euler 1736a) and independently by C.Maclaurin[32] (1742) (for proofs see also Euler 1755, Jacobi 1834, Kronecker 1889, 1901 (pp.317–319), Wirtinger[33] 1902, Barnes[34] 1903, Jordan[35] 1922 and Hardy 1949). This formula permits us to replace finite sums by integrals . We shall have the opportunity later to use this result in other cases.

Lemma 2.12. *Let A be an integer and $F(t)$ a function having a continuous derivative in the interval $[A,x]$. Then*

$$\sum_{A \leq n \leq x} F(n) = \int_A^x F(t)dt + \int_A^x \{t\}F'(t)dt + F(A) - \{x\}F(x),$$

[32] Maclaurin, Colin (1698–1746), Professor in Aberdeen and Edinburgh.
[33] Wirtinger, Wilhelm (1865–1945), Professor in Innsbruck and Vienna.
[34] Barnes, Ernest William (1874–1953), 1898–1916 Fellow of the Trinity College, Cambridge, 1924–52 Bishop of Birmingham.
[35] Jordan, Károlyi (Charles) (1871–1959), Professor in Budapest.

where $\{x\}$ denotes the fractional part of x.

Proof. Integration by parts gives

$$\int_A^x tF'(t)dt = xF(x) - AF(A) - \int_A^x F(t)dt$$

and on the other hand we have

$$\int_A^x [t]F'(t)dt = \sum_{n=A}^{[x]-1} n \int_n^{n+1} F'(t)dt + [x] \int_{[x]}^x F'(t)dt$$

$$= \sum_{n=A}^{[x]-1} n(F(n+1) - F(n)) + [x](F(x) - F([x]))$$

$$= - \sum_{n=A+1}^{[x]} F(n) - AF(A) + [x]F([x]) + [x](F(x) - F([x]).$$

The assertion follows now by subtracting the obtained equalities. □

Corollary. *For $x \geq 2$ one has*

$$\sum_{n \leq x} \frac{1}{\sqrt{n}} = 2\sqrt{x} + c + \frac{\alpha(x)}{\sqrt{x}},$$

where c is a constant and the function $\alpha(x)$ is uniformly bounded.

Proof. Take $A = 1$ and $F(t) = 1/\sqrt{x}$ and notice that the integral

$$\int_1^\infty \frac{\{t\}}{t^{3/2}} dt$$

converges and

$$\sqrt{x} \int_x^\infty \frac{\{t\}}{t^{3/2}} dt$$

is bounded. □

Finally one needs an evaluation of the remainder arising by approximating the value of the series $L(s, \chi)$ by one of its partial sums:

Lemma 2.13. *If χ is a non-principal Dirichlet character* mod k *and $s > 0$ then for $x \geq 2$ one has*

$$\left| L(s, \chi) - \sum_{n \leq x} \frac{\chi(n)}{n^s} \right| \leq \frac{2k}{x^s}.$$

Proof. The Corollary to Lemma 2.10 along with Lemma 2.5 show that for $2 \leq x < y$ one has

$$|\sum_{x \leq n \leq y} \frac{\chi(n)}{n^s}| \leq \frac{k}{y^s} + k \sum_{x \leq n \leq y} \left(\frac{1}{n^s} - \frac{1}{(n+1)^s} \right) \leq \frac{2k}{x^s},$$

and the assertion follows immediately. □

Now let χ be a real non-principal character mod k and assume that $L(s,\chi)$ vanishes at $s = 1$. For $n = 1, 2, \ldots$ let $f(n,\chi)$ be the function occuring in Lemma 2.11 and put

$$S(x,\chi) = \sum_{n \leq x} \frac{f(n,\chi)}{\sqrt{n}}.$$

The said lemma implies that the series

$$\sum_{n=1}^{\infty} \frac{f(n,\chi)}{\sqrt{n}} \tag{2.12}$$

diverges.

Now Mertens utilizes a method of Dirichlet (1849) (who used it to obtain the mean value of the number of divisors of integers in an interval) to show that $S(x,\chi)$ is bounded in which case the series (2.12) is certainly convergent as its terms are nonnegative. This will provide the needed contradiction.

One has

$$S(x,\chi) = \sum_{d \leq x} \sum_{m \leq x/d} \frac{\chi(d)}{\sqrt{dm}} = S_1(x) + S_2(x),$$

where

$$S_1(x) = \sum_{d \leq \sqrt{x}} \sum_{m \leq x/d} \frac{\chi(d)}{\sqrt{dm}}$$

and

$$S_2(x) = \sum_{\sqrt{x} < d \leq x} \sum_{m \leq x/d} \frac{\chi(d)}{\sqrt{dm}}.$$

Corollary to Lemma 2.12 leads now to

$$\begin{aligned} S_1(x) &= \sum_{d \leq \sqrt{x}} \frac{\chi(d)}{\sqrt{d}} \sum_{m \leq x/d} \frac{1}{\sqrt{m}} \\ &= 2\sqrt{x} \sum_{d \leq \sqrt{x}} \frac{\chi(d)}{d} + c \sum_{d \leq \sqrt{x}} \frac{\chi(d)}{\sqrt{d}} + \frac{1}{\sqrt{x}} \sum_{d \leq \sqrt{x}} \chi(d)\alpha(x/d) \end{aligned}$$

and using Lemma 2.13 with $s = 1$ we get

$$S_1(x) = 2\sqrt{x}L(1,\chi) + \beta(x),$$

with bounded $\beta(x)$. Hence the vanishing of $L(1,\chi)$ implies the boundedness of $S_1(x)$.

Applying once again Lemma 2.13, this time with $s = 1/2$, we get

$$S_2(x) = \sum_{m\le\sqrt{x}} \frac{1}{\sqrt{m}} \sum_{\sqrt{x}\le d\le x/m} \frac{\chi(d)}{\sqrt{d}}$$

$$= \sum_{m\le x} \left(L(1/2,\chi) - \sum_{d\le\sqrt{x}} \frac{\chi(d)}{\sqrt{d}} - L(1/2,\chi) + \sum_{d\le x/m} \frac{\chi(d)}{\sqrt{d}} \right)$$

$$= \sum_{m\le\sqrt{x}} \frac{1}{\sqrt{m}} \left(O(x^{-1/4}) + O((m/x)^{1/2}) \right)$$

and thus $S_2(x)$ is bounded and so is $S(x,\chi) = S_1(x) + S_2(x)$. □

Shapiro's version differs from the original proof in three aspects:

a) Mertens used the formula given in the Corollary to Lemma 2.12 in a slightly different form:

$$\sum_{j=k+1}^{n} \frac{1}{\sqrt{j}} = 2(\sqrt{n} - \sqrt{k}) - \frac{4\rho}{\sqrt{k+1}},$$

with a suitable $\rho = \rho(k,n)$ satisfying $0 < \rho < 1$. He proved this without using Euler's formula but by adding the inequalities

$$0 < 2(\sqrt{m} - \sqrt{m-1}) - \frac{1}{\sqrt{m}} = \frac{1}{\sqrt{m}(\sqrt{m} + \sqrt{m-1})^2} \le 4\left(\sqrt{m} - \sqrt{m+1}\right)$$

for $m = k+1, k+2, \ldots n$. The assertion of the corollary follows immediately by applying the formula to the inner sum in the identity

$$\sum_{n\le x} \frac{1}{\sqrt{n}} = 1 + \sum_{k=0}^{r-1} \sum_{n=4^r+1}^{4^{r+1}} \frac{1}{\sqrt{n}} + \sum_{4^{r+1}<n\le x} \frac{1}{\sqrt{n}},$$

where r is determined by $4^r < x \le 4^{r+1}$.

b) Instead of splitting the sum $S(x,\chi)$ into two parts Mertens used the identity

$$\sum_{i=1}^{n} u_i \sum_{j=1}^{n} v_j$$

$$= \sum_{ij\le n} u_i v_j + \sum_{\substack{ij>n \\ i\le\sqrt{n}}} u_i v_j + \sum_{\substack{ij>n \\ j\le\sqrt{n}}} u_i v_j + \sum_{i,j>\sqrt{n}} u_i v_j$$

$$= A_1 + A_2 + A_3 + A_4,$$

in the case $u_i = 1/\sqrt{i}$, $v_i = \chi(i)/\sqrt{i}$. This led to a splitting of $S(x,\chi)$ into four summands instead of two, which made the argument more complicated.

c) The original argument of Mertens did not utilize Lemma 2.13 explicitly. He worked with partial sums instead and proved that from some point on they exceed a certain positive limit.

2. Dirichlet's proof does not give any information about the smallest prime $p(k,l)$ occuring in the progression $kn+l$. The argument presented in the preceding subsection was utilized by F.Mertens (1897a) to overcome this inadequacy and to give the first bound for such a prime. He proved in fact a stronger result, dealing with primes in intervals:

If $(k,l)=1$ *and*

$$M(k) = \sum_{\chi\neq\chi_0} \frac{1}{|L(1,\chi)|},$$

then there exists an absolute constant c such that for any given $N \geq 1$ *the interval* $[N, N\exp(ckM(k))]$ *contains at least one prime* $p \equiv l \bmod k$.

In particular

$$p(k,l) \leq \exp(ckM(k)).$$

Essentially the same result appears in Kronecker's lectures (Kronecker 1901).

Note that in view of the easily proved evaluation $L(1,\chi) = O(\log k)$ one cannot obtain from this result a bound better than $p(k,l) = O(\exp(ck^2/\log k))$ with a suitable constant $c > 0$. Later S.Chowla (1934a) proved that if k is a prime congruent to 3 mod 4 then one has

$$\log p(k,l) = O(k^{3/2}\log^6 k)$$

and in the case $l = 1$ he obtained the bound

$$\log p(k,1) = O(k).$$

Later a very simple proof (based on the consideration of values of L-functions at real arguments $x \geq 1$) of the bound

$$p(k,l) = O(\exp(ck\log^{11} k))$$

was given by Sz.Gy.Révesz (1980).

All this was superseded by *Linnik's theorem* which gives the bound $p(k,l) = O(k^L)$ with a certain constant L and was proved[36] by Yu.V.Linnik[37] (1944). Linnik's proof was very complicated and subsequently has been simplified due to the work of W.Haneke (1970), K.A.Rodosskij (1954) (this proof is reproduced in Prachar (1957)), P.Turán (1961), S.Knapowski[38] (1962a) and M.Jutila (1969). It is now known (Heath-Brown 1992) that one can take for L any number larger than 5.5. Previous results showed that L can be any number exceeding 5448 (Pan[39] 1957), 777 (Chen 1965), 550 (Jutila 1970), 168 (Chen 1977), 80 (Jutila 1977a), 17 (Chen 1979b), 13.5 and 11.5 (Chen,Liu 1989,1991), and 8 (Wang Wei 1991). The *Extended Riemann Hypothesis*, stating that there are no zeros of the functions $L(s,\chi)$ in the

---

[36] Earlier Paul Turán ((1910–1976), Professor in Budapest) (1938/1940) deduced this assertion from the still unproved assumption that there is a rectangle containing $s = 1$ in which no L-function has a zero. Another proof of this implication appears in Prachar (1957).

[37] Linnik, Yurij Vladimirovič (1915–1972), Professor in Leningrad.

[38] Knapowski, Stanisław (1931–1967).

[39] Pan Cheng Dong (1934–1997), Professor in Shangdong.

open half-plane Re $s > 1/2$ implies that one can take for L any number exceeding 2 (Titchmarsh[40] 1930a, Chowla 1934a). Recently E.Bach and J.Sorenson (1994,1996) deduced from that hypothesis the explicit bound

$$p(k,l) < 2k^2 \log^2 k.$$

They showed also that for sufficiently large k the coefficient 2 may be replaced by any number exceeding 1.

D.R.Heath-Brown (1990) showed that the assumption of a hypothesis going in the opposite direction, namely the existence of a real zero close to 1 of a L-function corresponding to a character mod k can also lead to strong upper bounds for $p(k,l)$.

If $k = p^r$ is a prime-power then one can reduce considerably the bound for $p(k,l)$. This forms the main result of Barban[41], Linnik, Čudakov[42] (1964) where the bound $p(p^r,l) = O(p^{rM})$ (with the implied constant depending on p) was obtained with any $M > 8/3 = 2.666\dots$. Later M.Jutila (1972) showed that this bound holds with any $M > M_0 = (27+3\sqrt{17})/16 = 2.46058\dots$ (cf. Motohashi 1970b) and a more general result is due to H.Iwaniec (1974) who proved that if D denotes the product of all distinct primes dividing k then $p(k,l) = O(k^M)$ holds for every $M > M_0$, with the implied constant being dependent on D.

Note that for a positive fraction of progressions this bound may be considerably improved. In fact P.Turán (1936/37) deduced from the Extended Riemann Hypothesis that for every $\epsilon > 0$ and almost all $l \bmod k$ one has

$$p(k,l) \le \varphi(k) \log^{2+\epsilon} k.$$

(The words 'almost all' in this context mean that if $A(k)$ denotes the number of integers $1 \le l < k$ with $(k,l) = 1$ for which the last inequality fails, then the ratio $A(k)/\varphi(k)$ tends to zero.)

Without assuming any unproved hypotheses P.Erdős (1949c) showed that for a positive fraction of residue classes $l \bmod k$ and an absolute constant B one has

$$p(k,l) \le B\varphi(k)\log k$$

and one cannot replace the right-hand side of this inequality by an essentially smaller function. A similar result was obtained by P.D.T.A.Elliott and H.Halberstam (1971) who proved that for every positive ϵ the inequality

$$p(k,l) \le B\varphi(k)\log^{1+\epsilon} k$$

holds for almost all k and $(1+o(1))\varphi(k)$ residue classes $l \bmod k$.

Recently R.C.Baker and G.Harman (1996a) showed for any fixed l and most values of k one has

$$p(k,l) < k^{1.9231}.$$

Regarding lower bounds for $p(k,l)$ it has been shown that with an absolute constant $c > 0$ for every fixed l and infinitely many k one has (Prachar 1961, Schinzel 1962)

$$p(k,l) \ge c\frac{k \log k \log\log k \log\log\log\log k}{(\log\log\log k)^2}.$$

---

[40] Titchmarsh, Edmund Charles (1899–1963), Professor in Liverpool and Oxford.

[41] Barban, Mark Borisovič (1935–1968), worked in Tashkent.

[42] Čudakov, Nikolai Grigorievič (1905–1986), Professor in Saratov, Moscow and Leningrad.

For further studies of lower bounds for $p(k,l)$ see Wagstaff (1978,1979), Pomerance (1980), Granville (1989) and Granville,Pomerance (1990).

3. Now we turn to the second proof of $L(1,\chi)\neq 0$ for real characters given by Mertens (1899a).

Proof. Let again $f(n,\chi)$ be defined by $f(n,\chi)=\sum_{d|n}\chi(d)$ and put, for $x\geq 1$,

$$g(x)=g(x,\chi)=2\sum_{n\leq x}(x-n)f(n,\chi).$$

The last part of Lemma 2.11 implies the following lower bound for $g(x)$:

$$g(x)\geq 2\sum_{m\leq\sqrt{x/2}}(x-m^2)\geq 2\sum_{m\leq\sqrt{x/2}}\frac{x}{2}\geq\frac{1}{\sqrt{2}}x^{3/2}+O(x).$$

To obtain an upper bound observe that

$$g(x)=2\sum_{\substack{m,n\\ mn\leq x}}(x-mn)\chi(n)$$

and split this sum into three parts

$$g(x)=S_1+S_2-S_3,$$

where the summation in S_1 is carried out over all pairs m,n with $m\leq x^{1/3}$, that in S_2 is carried out over $\{(m,n)) : n\leq x^{2/3}\}$ and finally

$$S_3=2\sum_{\substack{m\leq x^{1/3}\\ n\leq x^{2/3}}}(x-mn)\chi(n).$$

From the Corollary to Lemma 2.10 one easily gets $S_1=O(x^{4/3})$ and $S_3=O(x^{4/3})$ and so it remains to bound S_2. To do this write

$$S_2=2\sum_{n\leq x^{2/3}}\chi(n)\sum_{m\leq x/n}(x-mn)$$

$$=2x\sum_{n\leq x^{2/3}}\chi(n)\left[\frac{x}{n}\right]-\sum_{n\leq x^{2/3}}n\chi(n)\left[\frac{x}{n}\right](1+\left[\frac{x}{n}\right])$$

$$=2x\sum_{n\leq x^{2/3}}\chi(n)(\frac{x}{n}-\left\{\frac{x}{n}\right\})$$

$$-\sum_{n\leq x^{2/3}}n\chi(n)(\frac{x}{n}-\left\{\frac{x}{n}\right\})(1+\frac{x}{n}-\left\{\frac{x}{n}\right\})$$

$$=x^2\sum_{n\leq x^{2/3}}\frac{\chi(n)}{n}-x\sum_{n\leq x^{2/3}}\chi(n)+\sum_{n\leq x^{2/3}}n\left\{\frac{x}{n}\right\}\chi(n)(1-\left\{\frac{x}{n}\right\}).$$

Corollary to Lemma 2.10 shows that the coefficient of x in the last expression is bounded and since the last term is $O(x^{4/3})$ we get

$$S_2 = x^2 \sum_{n\le x^{2/3}} \frac{\chi(n)}{n} + O(x^{4/3}).$$

As Lemma 2.13 gives

$$\sum_{n\le x^{2/3}} \frac{\chi(n)}{n} = L(1,\chi) + O(x^{-2/3})$$

we arrive finally at

$$S_2 = x^2 L(1,\chi) + O(x^{4/3})$$

thus

$$\frac{1}{\sqrt{2}} x^{3/2} \le g(x) = x^2 L(1,\chi) + O(x^{4/3}).$$

This finishes the proof since if we would have $L(1,\chi) = 0$ then $x^{3/2} = O(x^{4/3})$ would result which is not possible. □

This proof is reproduced in Landau (1927a, vol.1).

4. In the case of a real character modulo a prime a very simple proof was given by S.Chowla, L.J.Mordell[43] (1961):

Proof. The argument is based on the identity

$$\exp\left(\sum_{m=1}^{\infty} \frac{z^m}{m}\right) = \frac{1}{1-z} \tag{2.13}$$

whose validity for $|z| \le 1$, $z \ne 1$ follows from Abel's theorem[44].

Let p be an odd prime and let χ be the unique non-principal real character mod p. Thus $\chi(n) = \left(\frac{n}{p}\right)$. Choose $c \in [1, p-1]$ with $\chi(c) = -1$ and observe that by Euler's criterion we have

$$c^{(p-1)/2} \equiv -1 \pmod{p}. \tag{2.14}$$

Put

$$f(X) = \prod_r \frac{1 - X^{cr}}{1 - X^r},$$

where r runs over all quadratic residues mod p, i.e. $\chi(r) = 1$. Observe that f is a polynomial with rational integral coefficients and note that (2.14) implies

[43]Mordell, Louis Joel (1888–1972), Professor in Manchester and Cambridge.
[44]Abel 1826.

$$f(1) = \prod_r c = c^{(p-1)/2} \equiv -1 \pmod p.$$

Using (2.13) for $z = \zeta_p$ we obtain

$$f(\zeta_p) = \exp\left(\sum_r \log(1-\zeta_p^{cr}) - \sum_r \log(1-\zeta_p^r)\right)$$
$$= \exp\left(\sum_{m=1}^{\infty} \frac{1}{m}\left(\sum_r \zeta_p^{rm} - \sum_r \zeta_p^{cmr}\right)\right).$$

Since $\chi(r) = 1$ and $\chi(cr) = -1$ we have

$$\sum_r \zeta_p^{rm} - \sum_r \zeta_p^{cmr} = \sum_{j=1}^{p-1} \chi(j)\zeta_p^{jm},$$

and thus

$$f(\zeta_p) = \exp\left(\sum_{m=1}^{\infty} \frac{1}{m}\tau_m(\chi)\right),$$

where we denoted by $\tau_m(\chi)$ the Gaussian sum

$$\sum_{n=1}^{p-1} \chi(n)\zeta_p^{nm},$$

which has already occured in Dirichlet's proof.

Observe now that for m divisible by p Lemma 2.10 (vi) implies $\tau_m(\chi) = 0$ and for remaining m's we have

$$\tau_m(\chi) = \chi(m)\tau_1(\chi). \tag{2.15}$$

Indeed, one has

$$\chi(m)\tau_m(\chi) = \sum_{n=1}^{p-1} \chi(m)\chi(n)\zeta_p^{nm} = \sum_{n=1}^{p-1} \chi(mn)\zeta_p^{nm} = \tau_1(\chi),$$

since when n runs over all residues mod p, prime to p, mn does the same.

Finally we arrive at

$$f(\zeta_p) = \exp(\tau_1(\chi)L(1,\chi)),$$

and this implies that if $L(1,\chi) = 0$ then ζ_p is a root of $f(X) - 1$. Since ζ_p is the root of the irreducible polynomial $g(X) = 1 + X + \cdots + X^{p-1}$ the equality

$$f(X) - 1 = g(X)h(X)$$

results with a certain $h \in \mathbf{Z}[X]$. Putting $X = 1$ we get

$$ph(1) = f(1) - 1 \equiv -2 \pmod p,$$

i.e. $p|2$, contradiction. □

5. In the general case a simple proof appears in the book of A.O.Gelfond[45] and Yu.V.Linnik (1962) (cf. Gelfond 1953,1956). For real characters their argument was simplified by P.Monsky (1993) and in this form we present it here:

Proof. Let χ be a real non-principal character mod k and consider, as before, $f(n,\chi) = \sum_{d|n} \chi(d)$ for $n = 1, 2, \ldots$.

Put

$$F(t) = \sum_{n=1}^{\infty} \chi(n) \frac{t^n}{1-t^n},$$

the series being convergent in the interval $(0,1)$, and observe that

$$\frac{t^n}{1-t^n} = \sum_{k=1}^{\infty} t^{kn},$$

thus

$$F(t) = \sum_{n=1}^{\infty} \chi(n) \sum_{k=1}^{\infty} t^{kn} = \sum_{n=1}^{\infty} f(n,\chi) t^n.$$

The last part of Lemma 2.11 implies

$$\lim_{t \to 1-0} F(t) = \infty.$$

Now assume $L(1,\chi) = 0$, put

$$b_n(t) = \frac{1}{n(1-t)} - \frac{t^n}{1-t^n}$$

and observe that

$$F(t) = -\sum_{n=1}^{\infty} b_n(t)\chi(n) \quad (0 < t < 1).$$

Lemma 2.14. *For* $0 < t < 1$ *we have* $b_1(t) \geq b_2(t) \geq \ldots \geq b_n(t) \geq \ldots$.

Proof. One has

$$(1-t)(b_n(t) - b_{n+1}(t)) = \frac{1}{n(n+1)} - \frac{t^n}{(1+t+\cdots+t^{n-1})(1+t+\cdots+t^n)}$$

and since the inequality between arithmetical and geometrical means implies

$$1 + t + \cdots + t^{n-1} \geq nt^{(n-1)/2} \geq nt^{n/2}$$

and

$$1 + t + \cdots + t^n \geq (n+1)t^{n/2}$$

[45]Gelfond, Aleksandr Osipovič (1906–1968), Professor in Moscow.

we arrive at

$$(1-t)(b_n(t)-b_{n+1}(t)) \geq 0,$$

implying our assertion. □

Since by Corollary to Lemma 2.10 the sums $\sum_{n=1}^{N} \chi(n)$ are uniformly bounded and the sequence $b_n(t)$ monotonically decreases to zero, partial summation (Lemma 2.5) leads to a uniform bound for $F(t)$ in the open unit interval, but we know already that this cannot happen. This contradiction implies our assertion. □

The same idea can be used to obtain the explicit lower bound

$$|L(1,\chi)| \geq \frac{1}{33\sqrt{k}\log^2 k}$$

for real characters χ mod k (Gelfond 1953). (Later J.Pintz (1971) showed that with the same method one can replace the right-hand side of this inequality by $c/\sqrt{k}\log k$ with a suitable $c > 0$.) In the same paper Gelfond proves that if $\theta(n)$ is a completely multiplicative function which assumes only the values $0, 1$ and -1 and satisfies

$$\limsup_{n\to\infty} \frac{1}{\sqrt{n}} |\sum_{k=1}^{n} \theta(k)| < \frac{1}{12}$$

then

$$\sum_{n=1}^{\infty} \frac{\theta(n)}{n} \neq 0$$

unless θ vanishes identically. In the case when $\theta(n)$ is a completely multiplicative complex-valued function with $|\theta(n)| \in \{0,1\}$ and $\sum_{k=1}^{n} \theta(n) = O(1)$ the same assertion was earlier proved by N.G.Čudakov,K.A.Rodosskij (1950). Functions satisfying these assumptions were studied also in Čudakov,Linnik (1950), Čudakov (1953) Bombieri (1960) and Goldsmith (1965).

See also Bateman (1959,1966), Teege (1901) for other elementary proofs of the non-vanishing of $L(1,\chi)$ in the case of a real character χ.

2.5. Elementary Approaches for Particular Moduli

1. For many particular values of k and l one can obtain simple elementary (i.e. not using analytic means) proofs of Theorem 2.1. Several of them are listed in Dickson (1919). We give here a selection and start with the case of the progression 1 mod n.

Theorem 2.15. *For every $n > 1$ there are infinitely many primes congruent to* 1 mod n.

Algebraic proof. The shortest proof uses some algebraic number theory:

According to a theorem of E.E.Kummer (1856) (see e.g. Narkiewicz 1990, th.4.16) a rational prime p splits in the n-th cyclotomic field $\mathbf{Q}(\zeta_n)$ if and only if it is congruent to 1 mod n. Since in any given finite extension of $\mathbf{Q}$ there are infinitely many splitting primes the assertion follows. □

This argument, although very simple, certainly cannot be regarded as completely elementary. Now we present a selection of more elementary approaches.

In all of them the first step consists in the following observation:

Lemma 2.16. *If for every $n > 1$ there exists at least one prime congruent to* 1 mod n *then there exist also infinitely many of them.*

Proof. If for $m = 2, 3, \ldots$ there exists a prime $p(m)$ congruent to 1 mod m, then in view of the evident inequality $p(kn) > kn$ the sequence $p(kn)$ ($k = 1, 2, \ldots$) contains infinitely many distinct primes which are all congruent to 1 mod n. □

It remains thus to show the truth of the following assertion:

Theorem 2.17. *For every $n > 1$ there exists a prime* $\equiv 1 \bmod n$.

To prove this theorem it suffices to show that for a given n there exists a prime p and an integer a such that the order of $a \bmod p$ equals n, as then n divides $p-1$, i.e. $p \equiv 1 \pmod{n}$. This is very easy (Lebesgue (1843), footnote on p.51., see also Cahen (1911)) when n is a prime since in that case if p is a prime divisor of $2^n - 1$ then the order of 2 mod p equals n. In the general case this assertion follows from a result of A.S.Bang (1886) who showed that for any integer $a > 1$ and $n \geq 2$, $n \neq 6$ there exists at least one *primitive prime divisor*, i.e. a prime p dividing $a^n - 1$ but not dividing $a^d - 1$ for $d \leq n - 1$. Assuming this result it remains to observe that if p is a primitive prime divisor of $a^n - 1$, then the order of a mod p equals n.

The result of Bang was generalized by K.Zsigmondy (1892) to sequences $a^n - b^n$, where a, b are coprime integers, with explicitly listed exceptional cases:

(i) $n = 1$, $a = b + 1$,

(ii) $n = 2$, $a + b$ is a power of 2,

(iii) $n = 6$, $a = 2$, $b = 1$.

This assertion is now usually called the *Birkhoff–Vandiver Theorem*, because it was rediscovered by G.D.Birkhoff[46] and H.S.Vandiver[47] (1904). Although many proofs of this result as well as of Bang's theorem occur in the literature (see Artin[48] 1955, Carmichael 1909, Dickson 1905, Hering 1974, Kanold 1950, Leopoldt 1966, Lüneburg 1981, Rédei 1958, and Rotkiewicz 1960 (this proof is reproduced in

[46]Birkhoff, George David (1884–1944), Professor at Harvard.

[47]Vandiver, Harry Schultz (1882–1973), Professor at the University of Texas.

[48]Artin, Emil (1898–1962), Professor in Hamburg, Princeton and at Notre Dame and Indiana.

Narkiewicz (1986) Chap.1)) none of them can be regarded as really simple. The idea used by Birkhoff and Vandiver was used by R.A.Smith (1981) to give a proof of Theorem 2.17.

The first simple proof of Theorem 2.17 was given by E.Lucas[49] (1878). See also Landry (1853), Lefébure (1884,1885), Kraus (1886). Later E.Wendt (1895) gave an even simpler argument and we present it now:

Proof of Wendt. Fix n, denote by $g(X)$ the least common multiple of polynomials $X^d - 1$ for all proper divisors d of n and put

$$f(X) = \frac{X^n - 1}{g(X)}.$$

Clearly both polynomials f and g have integral coefficients and $f(X)$ is non-constant since the primitive root of unity ζ_n is a root of $f(X)g(X) = X^n - 1$, without being a root of $g(X)$.

The following lemma was proved by Wendt with the use of the resultant of f and g but the argument given below is simpler:

Lemma 2.18. *For infinitely many integral x one has $(f(x), g(x)) = 1$.*

Proof. As $X^n - 1$ does not have multiple roots, the greatest common divisor of f and g equals unity. Therefore one can find polynomials $A(X)$ and $B(X)$ with rational coefficients, satisfying

$$A(X)f(X) + B(X)g(X) = 1.$$

After multiplying by the common denominator of the coefficients of A and B we arrive at the equality

$$a(X)f(X) + b(X)g(X) = c, \tag{2.16}$$

where a, b are polynomials with integral coeffficients, and c is a non-zero integer. If for some $x \in \mathbf{Z}$ we would have $(f(x), g(x)) = d > 1$, then putting $X = x$ in (2.16) we would get $d|c$. This implies our assertion since for every integral multiple mc of c one has $f(mc)|(mc)^n - 1$, thus $f(mc)$ is prime to c and $(f(mc), g(mc)) = 1$ follows. □

Now choose $x \in \mathbf{Z}$ with $(f(x), g(x)) = 1$ and $f(x) \neq 0, \pm 1$ and let q be a prime dividing $f(x)$, thus $x^n \equiv 1 \pmod{q}$. As $q \nmid g(x)$ it follows that for every proper divisor d of n we have $q \nmid x^d - 1$. This shows that the order of $x \bmod q$ equals n. □

An elegant proof of Theorem 2.17 was given by I.Niven and B.Powell (1976). It uses only elementary divisibility properties and the fact that the number of roots of

[49]Lucas, François Edouard Anatole (1842–1891), Professor at Lycée St.Louis and Lycée Charlemagne in Paris.

a non-zero polynomial cannot exceed its degree. Another elementary proof of Theorem 2.15 was found by A.Rotkiewicz (1961) and it was simplified by T.Estermann[50] (1962/63). It is more complicated than the proof given by Wendt but avoids the use of polynomials. For other proofs see Sedrakian, Steinig (1998) and Srinivasan (1984).

A similar argument leading to the existence of infinitely many primes congruent to $-1 \bmod n$ is given in Hasse (1950). A short but not quite elementary proof in this case can be found in M.Bauer[51] (1906). Cf. Genocchi[52] (1868/69), Bauer (1903), Carmichael (1913) and Powell (1977).

R.D.von Sterneck (1896) considered integer-valued functions $f(n)$ satisfying the following three conditions:

(i) $f(1) = 1$,

(ii) $(f(m), f(n)) = f((m,n))$ for all $m, n = 1, 2, \ldots$,

(iii) For almost all (i.e. with exception of at most finitely many) primes p one has $p | f(p-1)f(p+1)$,

and showed that if

$$F(n) = \prod_{d|n} f(n/d)^{\mu(d)},$$

then every prime dividing $F(n)$ divides $f(n)$ but does not divide $f(i)$ for $i = 1, 2, \ldots, n-1$. He observed that (ii)–(iii) are fulfilled if $f(1) = 1$ and f satisfies the recurrence $f(n) = f(n-1) + cf(n-2)$ with a positive integer c and used this remark to obtain an elementary proof of the existence of infinitely many primes $\equiv -1 \pmod{q}$ for prime-powers q.

An elementary proof of the infinitude of primes in the progression $1 \bmod p^m$ for prime p and $m = 1, 2, \ldots$ was given by F.Hartmann (1931).

2 For progressions with small differences one can prove Dirichlet's theorem with arguments modelled upon Euclids proof of the infinitude of primes. The infinitude of primes in progression $4k + 3$ can be established in this way by observing that an integer congruent to 3 mod 4 must have a prime factor in the same residue class mod 4. For progressions with differences 8 and 12 this was essentially done by J.A.Serret[53] (1852) who used the Gaussian law of quadratic reciprocity. We present now his argument for progressions with difference 8 omitting the progression 1 mod 8 as it is covered by Theorem 2.15 and Serret's argument in this case is incomplete.

Theorem 2.19. *Each of the progressions* $8n + 3, 8n + 5, 8n + 7$ *contains infinitely many primes.*

Proof. Let $x > 4$ and denote by A_x the product of all odd primes not exceeding x.

If a prime p divides $a = A_x^2 + 2$ then -2 is a quadratic residue mod p and thus, as shown by Gauss (1801, Sect.112–114), p is congruent to either 1

[50] Estermann, Theodor (1902–1991), Professor in London.

[51] Bauer, Michael (1874–1945), Professor in Budapest.

[52] Genocchi, Angelo (1817–1889), Professor of law in Piacenza and later Professor of algebra and analysis in Torino.

[53] Serret, Joseph Alfred (1819–1885), Professor in Paris.

or 3 mod 8. However A_x is odd and thus $a \equiv 3 \pmod 8$ showing that not all prime factors of a can be congruent to unity mod 8, so at least one of them is of the form $8k+3$ and clearly it exceeds x. As x is arbitrary we get infinitely many primes in the progression $8k+3$.

In the remaining cases one proceeds similarly: Put $b = A_x^2 + 4$ and let p be a prime dividing b. As $2 \nmid A_x$, we see that -1 is a quadratic residue mod p and thus $p \equiv 1 \pmod 4$, i.e. $p \equiv 1 \pmod 8$ or $p \equiv 5 \pmod 8$. Since b is congruent to $5 \pmod 8$ it must have a prime divisor q in the progression $8k+5$. Since $q > x$ this settles the case 5 mod 8. In the last case consider $c = A_x^2 - 2$. Here the previous argument shows that prime divisors of c are congruent to ± 1 mod 8 and as $c \equiv 7 \pmod 8$ this leads to the infinitude of primes in the progression $8k+7$. □

Other elementary proofs of this theorem were given by V.A.Lebesgue (1856) and J.J.Sylvester (1888b) who also treated progressions of difference 12. Cf. Sylvester (1888a). In Sylvester (1871b) progressions with differences 4 and 6 were treated with the use of certain identities between rational functions. A simple proof for the progression $4k+1$ based on some properties of Fibonacci numbers was presented by N.Robbins (1994).

I.Schur[54] (1912/13) extended Serret's approach and showed in particular that if $l^2 \equiv 1 \pmod k$ and there is a prime exceeding $\varphi(k)/2$ in the progression $kn + l$, then this progression contains infinitely many primes. In this way he obtained elementary proofs of Theorem 2.1 for infinitely many arithmetic progressions, including the cases $k = 2^m$, $l = 2^{m-1} \pm 1$ $(m \geq 1)$ and $k = 8m$ (with m being an odd positive square-free integer) and $l = 2m+1, 4m+1, 6m+1$.

A similar method was later used by A.S.Bang (1937) who proved Dirichlet's theorem in the following cases:

$k = 4p^m$, $l = 2p^m + 1$, with prime $p \equiv 3 \pmod 4$,

$k = 6p^{2n+1}$, $l = 2p^{2n+1} + 1$, with prime $p \equiv 2 \pmod 3$,

and

$k = 6p^{2n}$, $l = 4p^{2n} + 1$, with prime $p \equiv 2 \pmod 3$.

Later P.T.Bateman and M.E.Low (1965) used this approach to prove that every progression $24n + k$ with $(k, 24) = 1$ represents infinitely many primes.

We describe now the idea of their proof which stems from algebraic number theory:

The extension $\mathbf{Q}(\zeta_k)/\mathbf{Q}$ (where ζ_k denotes a primitive k-th root of unity) is normal and its Galois group is isomorphic to the group $G(k)$ of residue classes mod k, prime to k, the isomorphism being given by assigning to the residue class l mod k that automorphism φ_l of $\mathbf{Q}(\zeta_k)$ which maps ζ_k to ζ_k^l. Denote by K_l the fixed field of φ_l, ie.

$$K_l = \{\alpha \in \mathbf{Q}(\zeta_k) : \varphi_l(\alpha) = \alpha\},$$

[54]Schur, Issai (1875–1941), Professor in Bonn and Berlin.

and let $F_l \in \mathbf{Z}[X]$ be the minimal polynomial (assumed to be monic) of any integral generator of K_l.

According to a theorem of Kummer (see e.g. Narkiewicz 1990, Theorem 4.12) apart of finitely many possible exceptional primes a prime p splits in $K_l/\mathbf{Q}$ if and only if for some $a \in \mathbf{Z}$ the value $F_l(a)$ is divisible by p. On the other hand p splits in $K_l/\mathbf{Q}$ if and only if for a suitable integer $r \geq 0$ we have $p \equiv l^r \pmod{k}$. If now $l^2 \equiv 1 \pmod{k}$, then we obtain an infinite set of primes which are congruent either to 1 or l mod k, because the set of primes dividing some value of F_l at integers is infinite.

Actually for any non-constant polynomial $f \in \mathbf{Z}[X]$ the set of prime divisors of values attained by f at integers is infinite as the following extension of Euclid's proof of Theorem 1.1 shows:

Write $f(X) = \sum_{j=0}^{n} a_j X^j$ and assume that for $j = 0, 1, 2, \ldots$ the number $f(j)$ has no prime divisors outside a finite set U of primes. Without restricting the generality assume a_0 to be non-zero and denote by D the product of all primes from U. Now choose $m \geq 1$ to be sufficiently large, so that $f(0) = a_0$ is not divisible by any m-th power of an integer. Then a_0 divides D^m hence we may write $D^m = a_0 \Delta$. For $n \geq 1$ write

$$f(nD^{2m}) = \sum_{j=1}^{n} a_j n^j D^{2mj} + a_0 = a_0(\sum_{j=1}^{n} a_j n^j \Delta^{2mj} a_0^{2mj-1} + 1)$$

and observe that for sufficiently large n the term in brackets must have a prime divisor p but clearly $p \nmid a_0 \Delta$ and thus p does not belong to U, a contradiction.

For a variant of this argument see Pólya (1921).

It remains to show that the set so constructed contains infinitely many primes congruent to l mod k and this can be achieved by elementary congruence arguments.

In this way one may treat progressions $kn + l$ for which $l^2 \equiv 1 \pmod{k}$ holds. Since every integer l prime to $k = 24$ satisfies this condition, this idea is succesfully applicable to $k = 24$ but one cannot proceed similarly in the case of a bigger modulus, as 24 is the largest integer with that property.

M.Ram Murty (1988) proved that this approach cannot be adapted to progressions l mod k which do not satisfy $l^2 \equiv 1 \pmod{k}$. In fact he showed that for given k and l there exists a polynomial $f \in \mathbf{Z}[X]$ such that apart of finitely many primes the congruence

$$f(x) \equiv 0 \pmod{p}$$

is insolvable for primes p with $p \not\equiv 1 \pmod{k}$ or $p \not\equiv l \pmod{k}$ if and only if $l^2 \equiv 1 \pmod{k}$.

J.Wójcik[55] (1966,1966/67,1968/69) extended Schur's approach providing in a series of papers purely algebraic proofs of various special cases of Theorem 2.1 including all cases considered in Schur (1912/13) and Bang (1937).

[55]Wójcik, Jan (1936–1994), worked in Warsaw.

He established in this way the following result which implies in particular Dirichlet's theorem for all progressions l mod k with $l^2 \equiv 1 \pmod{k}$:

If G is a subgroup of the multiplicative group of residue classes mod k *and H is a proper subgroup of G, then there exist infinitely many primes in $G\setminus H$.*

Wójcik used a rather advanced machinery from algebraic number theory and his arguments cannot be regarded as elementary in the usual sense (cf. Lenstra, Stevenhagen 1991).

For a list of older elementary proofs of many other special cases see Chap.18 in the first volume of Dickson (1919). In the case when

$$\sum_{\substack{p<k \\ k\not\equiv 0 \pmod{p}}} \frac{1}{p} < 1$$

such a proof was given by P.Erdős (1935b) (cf. Moree 1993).

For progressions with difference 4,5,6,9,10,12,15,18 and 30 a short proof based on the Prime Number Theorem was given by G.Ricci[56] (1933a) (cf. Ricci 1934).

Note that there exist elementary proofs of the full extent of Theorem 2.1, the first being due to H.Zassenhaus (1949), A.Selberg (1949a) and A.O.Gelfond (1956) but these lie already outside the scope of this book.

Exercises

1. (Apostol 1975) Let $f(n)$ be a periodic function with minimal positive period k and assume that it is completely multiplicative, i.e. satisfies $f(mn) = f(m)f(n)$ for all integers m, n. Prove that f is a Dirichlet character mod k.

2. (Chowla 1936a) Let χ be a real non-principal character mod k and put

$$S_1(x) = \sum_{n\le x} \chi(n)$$

and

$$S_m(x) = \sum_{n\le x} S_{m-1}(x).$$

Prove that for $m = 1, 2, \dots$ and positive s one has

$$\begin{aligned} L(s,\chi) &= \sum_{n=1}^{\infty} S_m(n) \sum_{r=0}^{m} (-1)^r \binom{m}{r} (n+r)^{-s} \\ &= s(s+1)\cdots(s+m-1) \sum_{n=1}^{\infty} S_m(n) I(m,n), \end{aligned}$$

[56]Ricci, Giovanni (1904–1973), Professor in Pisa and Milan.

where

$$I(m,n)=\int_0^1\int_0^1\cdots\int_0^1(n+u_1+\cdots+u_m)^{-s-m}du_1\cdots du_m$$

and deduce that if there is an index m with $S_m(x)\geq 0$ for all $x\geq 1$ then the function $L(s,\chi)$ does not have any positive real zeros.

3. Let N be an integer, χ a non-principal character mod N and put

$$\tau(\chi)=\sum_{n=1}^{N-1}\chi(n)\zeta_N^n,$$

where $\zeta_N=\exp(2\pi i/N)$. Prove that if $N=p$ is a prime, then

$$|\tau(\chi)|=\sqrt{p}.$$

4. (Evaluation of the Gaussian sum) Let p be an odd prime and let $\chi(n)=\left(\frac{n}{p}\right)$.

(i) Show that

$$\tau^2(\chi)=p.$$

(ii) Prove

$$\tau(\chi)=\sum_{n=0}^{p-1}\exp(2\pi in^2/p).$$

(iii) (Dirichlet 1835) Applying Poisson's formula

$$\frac{f(0)}{2}+\frac{f(p)}{2}+\sum_{j=1}^{p-1}f(j)=\sum_{n=-\infty}^{\infty}\int_0^p f(t)\exp(2\pi int)dt$$

to the functions

$$f(t)=\cos\left(\frac{2\pi t^2}{p}\right)$$

and

$$f(t)=\sin\left(\frac{2\pi t^2}{p}\right)$$

prove the formula (2.7)

$$\tau(\chi)=\begin{cases}\sqrt{p} & \text{if } p\equiv 1 \pmod 4\\ i\sqrt{p} & \text{if } p\equiv 3 \pmod 4\end{cases}$$

of Gauss.

(iv) Let $N>2$ be an integer and put

$$S(N)=\sum_{n=0}^{N-1}\exp(2\pi in^2/N).$$

Use the method of (iii) to prove

$$S(N) = \begin{cases} (1+i)\sqrt{N} & \text{if } 4|N \\ \sqrt{N} & \text{if } N \equiv 1 \pmod 4 \\ 0 & \text{if } N \equiv 2 \pmod 4 \\ i\sqrt{N} & \text{if } N \equiv 3 \pmod 4. \end{cases}$$

5. Let $N = p_1 p_2 \cdots p_r$ be the factorization into prime factors of a square-free odd integer N and let $\chi, \chi_1, \ldots, \chi_r$ be the Jacobi symbols corresponding to $N, p_1, \ldots, p_r$.

(i) Prove

$$\tau(\chi) = \prod_{i=1}^{r} \chi_i \left(\frac{N}{p_i}\right) \tau(\chi_i).$$

(ii) Show that

$$\tau(\chi) = \begin{cases} \sqrt{N} & \text{if } \chi(-1) = 1 \\ i\sqrt{N} & \text{if } \chi(-1) = -1. \end{cases}$$

6. (i) Let k_1 be a multiple of k, let χ be a character mod k and put

$$\psi(n) = \begin{cases} \chi(n \bmod k) & \text{if } (n, k_1) = 1 \\ 0 & \text{otherwise.} \end{cases}$$

(ψ is called the character mod k_1 *induced* by χ). Show that ψ is a character mod k_1.

(ii) For $i = 1, 2$ let χ_i be a character mod k_i. The characters χ_1 and χ_2 are called *equivalent* if there exists a character induced by both χ_1 and χ_2. Show that equivalence of characters is an equivalence relation.

(iii) Let χ be a character mod k and let d be a proper divisor of k. Prove that χ is equivalent to a character mod d if and only if for every n prime to k and congruent to 1 mod d one has $\chi(n) = 1$.

(iv) Prove that every character is equivalent to a primitive character.

7. Let $F(k)$ denote the number of primitive characters mod k.

(i) (Spira 1969) Prove the equality

$$F(k) = \sum_{d|k} \mu(d) \varphi\left(\frac{k}{d}\right).$$

(ii) (Jager 1973) Evaluate the sum $\sum_{k<x} F(k)$.

8. (i) Observe that the set $X(k)$ of all characters mod k is an Abelian group under multiplication and prove that if $N = n_1 n_2 \cdots n_r$ with $(n_i, n_j) = 1$ for $i \neq j$ then the groups $X(N)$ and $\bigoplus_{j=1}^{r} X(n_j)$ are isomorphic. Prove also that if $\chi \in X(N)$ corresponds under this isomorphism to

$$[\chi_1, \ldots, \chi_r] \in \bigoplus_{j=1}^{r} X(n_j)$$

then the character χ is primitive if and only if each character χ_j is primitive.

(ii) Show that if p^n is a prime power then there exists a real primitive character mod p^n if and only if either $p > 2$ and $n = 1$ or $p = 2$ and $n \leq 3$.

(iii) Prove that if p is an odd prime then $\chi(p) = \left(\frac{n}{p}\right)$ is the unique real character mod p.

(iv) Prove that $(-1)^{\frac{p-1}{2}}$ is the unique real primitive character mod 4.

(v) Show that there are two real primitive characters mod 8 and determine their form.

(vi) Deduce that real primitive characters mod k exist if and only if either k or $-k$ is a fundamental discriminant and determine their form.

9. Show that 24 is the largest integer k with the property that $(m, k) = 1$ implies $m^2 \equiv 1 \pmod{24}$.

10. (Murty 1988) Let $k > 2$ be given and let H be a subgroup of the multiplicative group of residue classes mod k prime to k.

(i) Prove the existence of an irreducible polynomial $f \in \mathbf{Z}[X]$ such that if p is a prime for which the congruence $f(x) \equiv 0 \pmod{k}$ is solvable then apart of finitely many cases $p \bmod k$ lies in H.

(ii) Show that the polynomial in (i) can be chosen in such a way that p is a prime satisfying $p \bmod k \in H$ then the congruence $f(x) \equiv 0 \pmod{k}$ is solvable.

11. (Durand 1856) Show without the use of the Chinese Remainder Theorem that if $(k, l) = 1$ and an integer N is given then the arithmetic progression $l \bmod k$ contains infinitely many integers prime to N.

3. Čebyšev's Theorem

3.1. The Conjecture of Legendre

1. Euler (1762/63) stated that the number $\pi(x)$ of primes not exceeding x equals approximately $x/\log x$ and several years later the young Gauss wrote the same assertion on the margin of the collection of mathematical tables of J.C.Schulze (1778) (Gauss 1791). A few years later A.M.Legendre in his book (Legendre 1798) conjectured that a good approximation to $\pi(x)$ is given by

$$\frac{x}{A\log x + B} \tag{3.1}$$

where A and B are certain constants. In the second edition in 1808 he proposed for these constants the values $A = 1$ and $B = -1.08366$ and repeated this in the third edition (Legendre 1830). A slightly different value for A was later proposed in Drach (1844) (see Torelli (1901,p.16) for a discussion).

The fact that the number $\pi(x)$ is well approximated by the *integral logarithm* defined by

$$\mathrm{li}(x) = \lim_{\epsilon\to 0}\left(\int_0^{1-\epsilon} + \int_{1+\epsilon}^{x}\right)\frac{dt}{\log t} = \int_2^x \frac{dt}{\log t} - 1.04\ldots$$

was apparently mentioned for the first time in the letter of F.W.Bessel[1] to W.Olbers[2] of September 1st 1810 (Erman 1852, I, 235–238). Bessel, who at that time was interested in properties of the function $\mathrm{li}(x)$, quotes there a remark of Gauss that for $x = 400\,000$ one has $\pi(x) = 33859$ and $\mathrm{li}(x) = 33922.621995$ which makes the relative difference smaller than 0.2%. (See Bessel (1811). For other early studies of the function $\mathrm{li}(x)$ see Plana[3] (1821/22) and Lacroix[4] (1819)).

Gauss himself did not publish anything on that subject and only in a letter to Encke[5] (Gauss 1849) of December 24th 1849 recalled his belief based on

---

[1] Bessel, Friedrich Wilhelm (1784–1846), astronomer, Professor in Königsberg.

[2] Olbers, Wilhelm (1758–1840), non-professional astronomer, known for his method of determining comet orbits.

[3] Plana, Giovanni Antonio Amadeo (1781–1864), Professor in Turin.

[4] Lacroix, Sylvestre François, (1765–1843), Professor in Rochefort and Paris.

[5] Encke, Johann Franz (1791–1865), student of Gauss, astronomer, director of the Berlin Observatory (1825–1863).

the examination of the table of all primes not exceeding 3 000 000, made several years earlier, that $\pi(x)$ is well approximated by $x/\log x$ and that even a better approximation to it is given by $\mathrm{li}(x)$. For possibly the first unsuccessfull attempt to prove this assertion see Hargreave (1849).

Dirichlet believed that his analytic methods would allow him to prove Legendre's conjecture. He wrote (Dirichlet 1838a):

"So[6] *hat namentlich Legendre durch Induction eine sehr merkwürdige Formel gefunden, welche auf eine sehr genaue Weise die Anzahl der Primzahlen ausdrückt, die eine gegebene Grenze nicht übersteigen. ... Die der Akademie vorgelegte Abhandlung hat den Zweck mehrere Methoden zu entwickeln, welche bei Untersuchungen dieser Art in vielen Fällen mit Erfolg benutzt werden können, und deren Anwendung ausser verschiedenen andern Resultaten auch die Legendresche Formel ... ergiebt"*.

However he never returned to this question.

The conjectural approximations to $\pi(x)$ proposed by Gauss and Legendre differ from each other by a function of order $x/\log^2 x$ and it turned out later that they are both asymptotically true, i.e. the ratio of $\pi(x)$ to the approximating function tends to unity when x grows indefinitely. This forms the content of the *Prime Number Theorem* the first proofs of which were found as late as 1896. We shall present them in Chap.5.

The first definite result concerning the asymptotic behaviour of $\pi(x)$ was obtained by P.L.Čebyšev (1848), who proved that if the limit

$$\lim_{x\to\infty}\left(\frac{x}{\pi(x)}-\log x\right) \tag{3.2}$$

exists, then it must be equal to -1. It follows from Čebyšev's result that neither the first Gaussian approximation $x/\log x$ nor Legendre's (3.1) can approximate $\pi(x)$ with an error of order lower than $x/\log^2 x$ (see Corollary 7 to Theorem 3.5).

2. We shall now present the main result of P.L.Čebyšev's first paper introducing only a slight change in the notation since, in place of $\pi(x)$, Čebyšev dealt with the number of primes less than x, denoting it by $\phi(x)$.

Theorem 3.1. *If for every positive integer k and $\rho > 0$ we put*

$$H_k(\rho)=\sum_{n=2}^{\infty}\left(\pi(n)-\pi(n-1)-\frac{1}{\log n}\right)\frac{\log^k n}{n^{1+\rho}}$$

[6] *"Legendre found by induction a very remarkable formula, expressing in a very precise manner the number of primes not exceeding a given limit. ... The purpose of the memoir presented to the Academy is the development of several methods, which could succesfully be used in research of this kind, and whose application beside various other results ... gives also Lagrange's formula".*

then the limit $\lim_{\rho\to 0} H_k(\rho)$ *exists and is finite.*

Proof. Čebyšev's main tool was the following lemma which extends Dirichlet's Corollary 2 to Lemma 2.3.

Lemma 3.2. *For positive* ρ *put*

$$F(\rho) = \zeta(1+\rho) - \frac{1}{\rho}$$

and

$$G(\rho) = \sum_p \frac{1}{p^{1+\rho}} - \log\frac{1}{\rho}.$$

The derivatives of every order of these functions exist and have finite limits for ρ *tending to* 0. *Moreover*

$$\lim_{\rho\to 1+0} G(\rho) = -\sum_p \sum_{k=2}^{\infty} \frac{1}{kp^k}.$$

Proof of Čebyšev. The existence of the derivatives of F and G is clear and to obtain the assertions about their limits observe first that since for positive ρ one has[7]

$$\frac{1}{n^{1+\rho}}\Gamma(1+\rho) = \frac{1}{n}\int_0^\infty e^{-x}\left(\frac{x}{n}\right)^\rho dx = \int_0^\infty e^{-tn}t^\rho dt$$

the equality

$$\Gamma(1+\rho)\sum_{n=2}^{\infty}\frac{1}{n^{1+\rho}} = \int_0^\infty t^\rho \sum_{n=2}^{\infty} e^{-tn}dt = \int_0^\infty \frac{e^{-t}}{e^t-1}t^\rho dt$$

follows and using $\Gamma(1+\rho) = \rho\Gamma(\rho)$ we get

$$F(\rho) = 1 + \frac{1}{\Gamma(1+\rho)}\int_0^\infty \frac{t-e^t+1}{t(e^t-1)}e^{-t}t^\rho dt.$$

An easy induction argument now shows that the k-th derivative of the left hand-side equals a fraction with denominator equal to $\Gamma(1+\rho)^k$ and numerator being a polynomial in

$$\int_0^\infty e^{-t}t^\rho \log^\alpha t\, dt, \quad (\alpha = 0, 1, \ldots, k),$$

and

[7] Note that Čebyšev did not utilize the notation $\Gamma(x)$ for the Gamma-function occuring in the proof and used its integral representation instead.

$$\int_0^\infty \frac{t-e^t+1}{t(e^t-1)} e^{-t} t^\rho \log^\alpha t\, dt \quad (\alpha = 0, 1, \dots, k).$$

Since both these integrals have finite limits for ρ tending to 0 the first assertion results. To obtain the second note that Euler's formula (1.2) implies

$$\log\zeta(1+\rho) = -\sum_p \log(1 - \frac{1}{p^{1+\rho}}) = \sum_p \frac{1}{p^{1+\rho}} + A(\rho),$$

with

$$A(\rho) = \sum_p \sum_{k=2}^\infty \frac{1}{kp^{(1+\rho)k}} .$$

As

$$\zeta(1+\rho) = \frac{1+\rho F(\rho)}{\rho}$$

we can write $\log\zeta(1+\rho) = g(\rho) - \log\rho$ with $g(\rho) = \log(1+\rho F(\rho))$. Now note that $A(\rho)$ and $g(\rho)$ have all derivatives for $\rho > 0$ tending to finite limits when ρ tends to zero. It remains to use the equality

$$\sum_p \frac{1}{p^{1+\rho}} + \log\rho = g(\rho) - A(\rho).$$ □

To prove the theorem observe that for positive ρ one has

$$H_k(\rho) = \sum_p \frac{\log^k p}{p^{1+\rho}} - \sum_{n=2}^\infty \frac{\log^{k-1} n}{n^{1+\rho}}.$$

The first sum equals the k-th derivative of

$$(-1)^k \sum_p \frac{1}{p^{1+\rho}},$$

and the second equals the $(k-1)$-st derivative of $(-1)^{k-1}\zeta(1+\rho)$. The assertion follows now from Lemma 3.2 and the observation that $(-1)^{k-1}H_k(\rho)$ equals the $(k-1)$st derivative of $\frac{dG(\rho)}{d\rho} - F(\rho)$. □

3. Theorem 3.1 has immediate applications to the distribution of primes:

Corollary 1. *For every $k = 2, 3, \dots$ and any positive constant C one has*

$$\pi(m) < \mathrm{li}(m) + C\frac{m}{\log^k m} \tag{3.3}$$

and

$$\pi(n) > \mathrm{li}(n) - C\frac{n}{\log^k n} \tag{3.4}$$

for infinitely many integers m and n.

Proof. Denote by $S_N(\rho)$ the Nth partial sum of the series defining $H_k(\rho)$. Applying Lemma 2.5 we get

$$S_N(\rho) = (\pi(N) - \alpha(N))\frac{\log^k N}{N^{1+\rho}} + \sum_{m=2}^{N-1}(\pi(m) - \alpha(m))c_m, \tag{3.5}$$

where

$$\alpha(m) = \sum_{j=2}^{m} \frac{1}{\log j}$$

and

$$c_m = \frac{\log^k m}{m^{\rho+1}} - \frac{\log^k(m+1)}{(m+1)^{\rho+1}} .$$

One easily sees that for $m > e^k$ one has $c_m > 0$ and that with suitable constants $0 < C_1 \le C_2$ the inequalities

$$C_1\frac{\log^k m}{m^{2+\rho}} \le c_m \le C_2\frac{\log^k m}{m^{2+\rho}}$$

hold.

Now assume that for all sufficiently large m one has

$$\pi(m) \ge \mathrm{li}(m) + C\frac{m}{\log^k m} .$$

Observe that the difference $|\mathrm{li}(x) - \alpha(x)|$ is bounded. Indeed, with suitable $j \le \theta_j \le j+1$ $(j = 2, 3, \dots, [x]-1)$ one has

$$|\mathrm{li}(x) - \alpha(x)| \le \sum_{j=2}^{[x]-1}\left|\int_j^{j+1}\frac{dt}{\log t} - \frac{1}{\log j}\right| + O(1)$$

$$= \sum_{j=2}^{[x]-1}\left|\frac{1}{\log\theta_j} - \frac{1}{\log j}\right| + O(1) \le \sum_{j=2}^{[x]-1}\left(\frac{\log(j+1) - \log j}{\log^2 j}\right) + O(1)$$

$$= O(\sum_{j=2}^{[x]-1}\frac{1}{j\log^2 j}) + O(1) = O(1).$$

Our assumption implies that for large m, say for $m \ge M_0$, we have

$$\pi(m) \ge \alpha(m) + C_3\frac{m}{\log^k m}$$

with a certain positive constant C_3.

If now N exceeds $M_1 = \max\{M_0, e^k\}$ then we get from (3.5)

$$S_N(\rho) \geq \sum_{m=1}^{M_1} (\pi(m) - \alpha(m))c_m + C_3 \sum_{m=1+M_1}^{N-1} \frac{mc_m}{\log^k m}$$

thus

$$H_k(\rho) \geq \sum_{m=1}^{M_1} (\pi(m) - \alpha(m))c_m + C_1 C_3 \sum_{m=1+M_1}^{\infty} \frac{1}{m^{1+\rho}}.$$

The right-hand side of the last inequality tends to infinity for ρ tending to zero and thus $H_k(\rho)$ cannot be bounded, contradiction.

Reversing the inequalities in the preceding argument we obtain the second part of the assertion. □

Corollary 2. *If the limit*

$$\lim_{x\to\infty} \frac{\pi(x)\log x}{x}$$

exists then it equals 1.

Proof. Since the ratio of $\mathrm{li}(x)$ and $x/\log x$ tends to 1 for $x \to \infty$ the assertion follows from Corollary 1. □

For an elementary proof of the last corollary, utilizing properties of the binomial coefficient $\binom{2n}{n}$, see Kanold (1984). An analytic proof appears in Césaro (1893). An analogue of this corollary for the number $\pi(x;4,1)$ of primes $p \leq x$ congruent to 1 mod 4 was obtained by H.Poincaré[8] (1892) who showed that if the limit

$$\lim_{x\to\infty} \frac{\pi(x;4,1)\log x}{x}$$

exists then it must be equal 1/2. Another proof of Poincaré's result was given by V.Stanievitch (1892) who also noted that his method applies in principle to primes in any arithmetic progression. E.Phragmén[9] (1892) pointed out that Poincaré's result can be deduced from his theorem concerning the behaviour of integrals $\int_a^\infty f(x)x^{-s-1}dx$ (see Theorem 4.13) and another analytic proof is contained in Landau (1902) as a special case of a theorem concerning the distribution of prime ideals in algebraic number fields.

4. From the preceding corollary Čebyšev deduced the following result:

Corollary 3. *If the limit (3.2) exists, then it equals* -1.

Proof. Assume that the limit in question exists and denote it by L. Moreover denote by M the set of all integers m satisfying (3.3) with $k = 3$ and $C = 1$. Corollary 1 shows that M is infinite. Now observe that for $m \in M$ one has

[8] Poincaré, Henri (1854–1912), Professor in Paris.
[9] Phragmén, Lars Edvard (1863–1937), Professor in Stockholm.

$$\frac{m}{\pi(m)} - \log m \geq \frac{m}{\mathrm{li}(m) + m/\log^3 m} - \log m$$

and since this implies

$$L \geq \frac{m}{\mathrm{li}(m) + m/\log^3 m} - \log m$$

for sufficiently large $m \in M$ we obtain

$$L\ \mathrm{li}(m) + O(\frac{m}{\log^3 m}) \geq m - \mathrm{li}(m)\log m + O(\frac{m}{\log^2 m}).$$

Since by partial integration we get

$$\mathrm{li}(x) = \frac{x}{\log x} + \frac{x}{\log^2 x} + O(\frac{x}{\log^3 x}),$$

the last inequality leads to

$$L\frac{m}{\log m} \geq -\frac{m}{\log m} + O(\frac{m}{\log^2 m}),$$

and thus $L \geq -1$.

The same argument utilizing (3.4) leads to $L \leq -1$ and thus $L = -1$ results. □

Further assertions of this type appear in Glaisher[10] (1879–1883, part III), Césaro (1893), Ajello (1896), and Torelli (1901, pp.45–51) though the arguments used in their proofs are not always convincing.

3.2. True Order of $\pi(x)$

1. In his second paper devoted to prime numbers Čebyšev (1850) obtained rather close upper and lower bounds for the number of primes below x and established that its order is comparable with $x/\log x$. This allowed him to prove a conjecture put forward by J.Bertrand[11] (1845) which states that for $x > 6$ there is always a prime in the interval $(x/2, x-2]$ (*Bertrand's Postulate*). Bertrand noted that this assertion holds for all $x \leq 6 \cdot 10^6$ and he used it to show that if f is a polynomial in n variables then the set

$$\{f(x_{s(1)}, \ldots, x_{s(n)}) : s \in S_n\},$$

S_n denoting the group of all permutations of the set $\{1, 2, \ldots, n\}$, contains either one or two or at least n different polynomials. This was later established

[10] Glaisher, James Whitbread Lee (1848–1928), lecturer in Cambridge.
[11] Bertrand, Joseph (1822–1900), Professor in Paris, and from 1856 perpetual secretary of the Académie des Sciences there.

unconditionally by J.A.Serret (1850). Stated in modern language: the index of a proper subgroup of the symmetric group on n letters either equals 2 or is $\geq n$.

Note however that G.N.Matvievskaja (1961) pointed out (see Matvievskaja, Ožigova 1983) that Bertrand's conjecture already appears in one of the unpublished manuscripts of Euler written possibly around 1750. He checked the truth of this assertion in some simple cases by applying the Eratosthenian sieve.

In contrast to his proof of Theorem 3.1 Čebyšev obtained his bounds for $\pi(x)$ with completely elementary tools. He introduced the functions

$$\theta(x) = \sum_{p\leq x} \log p$$

and

$$\psi(x) = \sum_{n=1}^{\infty} \theta(x^{1/n})$$

which still play an important role in prime number theory. He also used the notation

$$T(x) = \sum_{n\leq x} \log n = \log([x!]).$$

2. The next lemma establishes fundamental relations between the functions $\psi(x)$, $\theta(x)$, $T(x)$ and $\pi(x)$ and gives an important property of $\psi(x)$.

Lemma 3.3. *(i) One has*

$$\psi(x) = \sum_{\substack{p^n \leq x \\ n\geq 1}} \log p = \sum_{p\leq x} \left[\frac{\log x}{\log p}\right] \log p,$$

thus $\exp\psi(x)$ *is the least common multiple of all integers not exceeding* x.

(ii) For $x \geq 2$ *one has*

$$T(x) = \sum_{n\leq x} \psi\left(\frac{x}{n}\right) = \sum_{n=1}^{\infty} \psi\left(\frac{x}{n}\right) = \sum_{m,n\geq 1} \theta\left(\left(\frac{x}{n}\right)^{1/m}\right).$$

(iii) $\qquad \psi(x) - \sqrt{x}\log x \leq \theta(x) \leq \psi(x).$

(iv) For every $\epsilon > 0$ *one has*

$$\frac{\theta(x)}{\log x} \leq \pi(x) \leq \frac{1}{1-\epsilon}\frac{\theta(x)}{\log x} + x^{1-\epsilon}.$$

Proof. (i) One has

$$\psi(x) = \sum_{n=1}^{\infty} \theta(x^{1/n}) = \sum_{n=1}^{\infty} \sum_{p \le x^{1/n}} \log p = \sum_{\substack{p^n \le x \\ n \ge 1}} \log p.$$

(ii) It suffices to consider the case of integral x. If we write

$$x! = \prod_{p \le x} p^{\alpha_p}$$

and use Lemma 1.20 then we get

$$T(x) = \log x! = \sum_{p \le x} \alpha_p \log p = \sum_{p \le x} \sum_{k=1}^{\infty} \left[\frac{x}{p^k}\right] \log p,$$

and on the other hand (i) implies

$$\sum_{n \le x} \psi\left(\frac{x}{n}\right) = \sum_{n \le x} \sum_{\substack{p^k \le x/n \\ k \ge 1}} \log p = \sum_{k,n \ge 1} \sum_{p \le (x/n)^{1/k}} \log p$$

but the last sum equals both

$$\sum_{p \le x} \sum_{k=1}^{\infty} \left[\frac{x}{p^k}\right] \log p$$

and

$$\sum_{k,n \ge 1} \theta\left(\left(\frac{x}{n}\right)^{1/k}\right)$$

which establishes (ii).

(iii) Using (i) one gets

$$|\psi(x) - \theta(x)| = \sum_{k \ge 2} \sum_{p^k \le x} \log p \le \sum_{p \le \sqrt{x}} \sum_{2 \le k \le \log x / \log p} 1$$

$$\le \sum_{p \le \sqrt{x}} \frac{\log x}{\log p} \log p \le \sqrt{x} \log x.$$

(iv) The inequality $\theta(x) \le \pi(x) \log x$ is obvious and to obtain the second assertion write

$$\theta(x) \ge \sum_{x^{1-\epsilon} \le p \le x} \log p \ge (1-\epsilon)(\pi(x) - \pi(x^{1-\epsilon}) \log x)$$

and use the trivial bound $\pi(x^{1-\epsilon}) \le x^{1-\epsilon}$. □

Part (ii) of the last lemma was also proved by de Polignac (1851, 1854). For generalizations see Césaro (1885b) and Gegenbauer (1892a).

Corollary 1. *If the limit* $\lim_{x\to\infty} \theta(x)/x$ *exists, then it equals 1.*

Proof. Follows from (iv) and Corollary 2 to Theorem 3.1. □

Corollary 2. *If the limit* $\lim_{x\to\infty} \psi(x)/x$ *exists, then it equals 1.*

Proof. Follows from (iii) and the preceding corollary. □

A short elementary proof of the last corollary was presented by N.Costa Pereira (1985b):

Proof of Costa Pereira. Let $2 \le m < n$ be integers and put $r = \left[\frac{n}{m}\right]$. Then Lemma 3.3 (ii) implies

$$\sum_{k=1}^{r} \frac{1}{k}\frac{\psi\left(\frac{n}{k}\right)}{n/k} = \frac{1}{n}\sum_{k=1}^{r}\psi\left(\frac{n}{k}\right) \le \frac{T(n)}{n} \le \sum_{k=1}^{r} \frac{1}{k}\frac{\psi\left(\frac{n}{k}\right)}{n/k} + \sum_{k=r+1}^{n} \frac{1}{k}\frac{\psi\left(\frac{n}{k}\right)}{n/k}. \tag{3.6}$$

Since for $k = 1, 2, \dots, r$ one has $n/k \ge m$ the first part of that inequality implies

$$\inf_{x\ge m} \frac{\psi(x)}{x}\sum_{k=1}^{r}\frac{1}{k} \le \frac{\log n!}{n},$$

hence using Stirling's formula in its simplest form, viz. $\log n! = n\log n + O(n)$, and utilizing the equality $\sum_{k\le x} \frac{1}{k} = \log x + O(1)$ we get

$$\inf_{x\ge m}\frac{\psi(x)}{x}(\log n - \log m + O(1)) \le \log n + O(1)$$

hence

$$\inf_{x\ge m}\frac{\psi(x)}{x} \le \frac{\log n + O(1)}{\log n - \log m + O(1)},$$

from which we get

$$\liminf_{x\to\infty}\frac{\psi(x)}{x} \le 1.$$

To deal with the upper limit use the second part of (3.6) which leads to

$$\frac{\log n!}{n} \le \sup_{x\ge m}\frac{\psi(x)}{x}\sum_{k=1}^{r}\frac{1}{k} + \sup_{1\le x<m}\frac{\psi(x)}{x}\sum_{k=r+1}^{n}\frac{1}{k},$$

and the same argument as in previous case shows

$$\sup_{x\ge m}\frac{\psi(x)}{x} \ge \frac{\log n - \sup_{1\le x<m}\left(\frac{\psi(x)}{x}\right)\log m + O(1)}{\log n - \log m + O(1)}$$

implying

$$\limsup_{x\to\infty}\frac{\psi(x)}{x} \ge 1.$$ □

3. Another proof of the assertion (ii) of Lemma 3.3 was given by R.E.Dressler and J.van der Lune (1975) who obtained it as a very special case of the following general principle:

Lemma 3.4. *Let f be a function which is positive, continuous and decreasing in $[0,\infty]$ and assume that for sufficiently large x we have $f(x) < 1$. Let g be the function inverse to f. Then*

$$\sum_{n=1}^{\infty}[f(n)] = \sum_{m<f(0)}[g(m)].$$

Proof. The assertion results from the equalities

$$\sum_{n=1}^{\infty}[f(n)] = \sum_{n=1}^{\infty}\sum_{1\le m\le f(n)} 1 = \sum_{n=1}^{\infty}\sum_{\substack{m \\ g(m)\ge n}} 1$$
$$= \sum_{m<f(0)}\sum_{n\le g(m)} 1 = \sum_{m<f(0)}[g(m)]. \qquad \square$$

If we denote by $L(x)$ the least common multiple of all positive integers $\le x$, then

$$L(x) = \prod_{p\le x} p^{[\log x/\log p]},$$

and Lemma 3.3 (i) gives

$$L(x) = \exp\psi(x).$$

Putting in Lemma 3.4 $f(t) = x/p^t$ for fixed $x > 0$ and $p > 1$ we obtain

$$\sum_{n=1}^{\infty}\left[\frac{x}{p^n}\right] = \sum_{r<x}\left[\frac{\log(x/r)}{\log p}\right] = \theta_p,$$

say, and this leads to

$$\sum_{n\le x}\psi\left(\frac{x}{n}\right) = \sum_{n\le x}\log L\left(\frac{x}{n}\right) = \log\prod_{n\le x} L\left(\frac{x}{n}\right)$$
$$= \log\prod_{n\le x}\prod_{p\le x/n} p^{[\log(x/n)/\log p]} = \log\prod_{n\le x}\prod_{p\le x} p^{[\log(x/n)/\log p]} = \log\prod_{p\le x} p^{\theta_p}.$$

In view of Lemma 1.20 the assertion (ii) follows. $\square$

4. H.Poincaré (1892) used Lemma 3.3 (ii) to give a simple direct proof of Corollary 2 to Lemma 3.3:

Poincaré's proof of Corollary 2 to Lemma 3.3. Put $S_n = \sum_{m=1}^{n} 1/m$ and $V(x,n) = \sum_{m=1}^{n} \left[\frac{x}{m}\right]$. In view of

$$\left[\frac{x}{m}\right] \le \frac{[x]}{m} \le \left[\frac{x}{m}\right] + 1$$

we get

$$[x]S_n - n \le V(x,n) < [x]S_n,$$

which in view of

$$\log(1+n) < S_n < 1 + \log n$$

leads to

$$[x]\log(n+1) - n \le V(x,n) \le [x](1+\log n).$$

If we now let

$$V(x) = \sum_{m=1}^{\infty} \left[\frac{x}{m}\right] = V(x,[x]),$$

then the preceding inequalities imply

$$\lim_{x\to\infty} \frac{V(x)}{x\log x} = 1.$$

Now assume that for a certain $c > 1$ and all sufficiently large x one has $\psi(x) \ge cx$. Then with a suitable b and all $x > 1$ one has

$$\psi(x) > c(1+x) - b \ge c[x] - b,$$

thus

$$T(x) = \sum_{n=1}^{\infty} \psi\left(\frac{x}{n}\right) > cV(x) - b[x].$$

Dividing by $x\log x$ and letting x go to infinity we get

$$1 = \lim_{x\to\infty} \frac{T(x)}{x\log x} \ge c,$$

contradicting the asumption $c > 1$. This establishes

$$\liminf_{x\to\infty} \frac{\psi(x)}{x} \le 1,$$

and to get the inequality

$$\limsup_{x\to\infty} \frac{\psi(x)}{x} \ge 1$$

one proceeds in the same way. □

The quoted paper of Poincaré contains also another proof of this corollary.

5. The main result of Čebyšev's second paper is the following theorem:

Theorem 3.5. *(i) There exists a constant A_1 such that for all $x \geq 2$ one has*

$$\psi(x) \leq A_1 x.$$

(ii) There exists a positive constant A_2 such that for all $x \geq 2$ one has

$$\psi(x) \geq A_2 x.$$

Čebyšev's proof of both theorems is based on the identity given in Lemma 3.3 (ii) and the inequality given in Lemma 3.6 below. We shall give essentially Čebyšev's argument though, to avoid messy calculations, we shall use a simpler form of Stirling's formula which leads to worse values of the constants A_1 and A_2, namely $A_1 = 1.12224$ and $A_2 = 0.9072$ (under the assumption $x \geq 350$).

Lemma 3.6. *If*

$$\alpha(x) = T(x) - T\left(\frac{x}{2}\right) - T\left(\frac{x}{3}\right) - T\left(\frac{x}{5}\right) + T\left(\frac{x}{30}\right)$$

then for $x \geq 2$ one has

$$\alpha(x) < \psi(x) < \psi\left(\frac{x}{6}\right) + \alpha(x).$$

Proof. Note first that Lemma 3.3 (ii) implies that one can write

$$\alpha(x) = \sum_{n=1}^{\infty} A_n \psi\left(\frac{x}{n}\right) \tag{3.7}$$

with suitable coefficients A_n. We shall establish that the A_n's are given by the formula

$$A_n = \begin{cases} 1 & \text{if } (n,30) = 1, \\ -1 & \text{if } (n,30) \text{ has exactly two distinct prime divisors,} \\ 0 & \text{otherwise.} \end{cases} \tag{3.8}$$

Indeed, Lemma 3.3 (ii) gives

$$\alpha(x) = \sum_{m=1}^{\infty} \psi\left(\frac{x}{m}\right) - \sum_{m=1}^{\infty} \psi\left(\frac{x}{2m}\right)$$
$$- \sum_{m=1}^{\infty} \psi\left(\frac{x}{3m}\right) - \sum_{m=1}^{\infty} \psi\left(\frac{x}{5m}\right) + \sum_{m=1}^{\infty} \psi\left(\frac{x}{30m}\right),$$

and thus in the case when $(n,30) = 1$ we get $A_n = 1$, as asserted. If $(n,30)$ equals 2,3 or 5 then $\psi(x/n)$ appears in the first summand of the above equality with coefficient 1, in one of the last three summands with coefficient -1, and in all remaining summands with coefficient 0, so $A_n = 0$. Similar counts lead to the assertion in all remaining cases.

Now let $c_0 = 1 < c_1 < c_2 < \dots$ be the sequence of all integers n for which A_n does not vanish and observe that by a happy coincidence we have $A_{c_n} = (-1)^n$, which follows easily from (3.8) by checking all cases. Thus (3.7) takes the form

$$\alpha(x) = \sum_{n=1}^{\infty} (-1)^n \psi\left(\frac{x}{c_n}\right)$$

and since $\psi(x)$ is non-decreasing we arrive at

$$\psi(x) = \psi\left(\frac{x}{c_0}\right) > \alpha(x) > \psi\left(\frac{x}{c_0}\right) - \psi\left(\frac{x}{c_1}\right) = \psi(x) - \psi\left(\frac{x}{6}\right)$$

as asserted. □

Proof of Theorem 3.5. Applying[12] Lemma 2.12 to the function $F(t) = \log t$ we get

$$T(x) = x \log x - x + S(x)$$

with

$$S(x) = \int_2^x \frac{\{t\}}{t} dt - \{x\} \log x - \log 2$$

and since obviously

$$|S(x)| \le \log x + 1$$

we obtain the following evaluation:

$$|\alpha(x) - cx| \le 5 \log x$$

where

$$c = \log\left(2^{14} \cdot 3^9 \cdot 5^5\right)/30 = 0.921292\cdots. \tag{3.9}$$

Observe now that for $x > 3000$ we have $5 \log x / x \le 0.014$ and finally we obtain

$$0.9072x \le \alpha(x) \le 0.9353x.$$

---

[12] In his proof Čebyšev used Stirling's formula (with $n = [x]$) in the following form:

$$0 < \log n! - (\log \sqrt{2\pi} + n \log n - n + \frac{1}{2} \log n) < \frac{1}{12n}$$

and this allowed him to obtain the evaluations

$$cx - 2.5 \log x - 1 < \psi(x) < \frac{6}{5} cx + \frac{5}{4 \log 6} \log^2 x + \frac{5}{4} \log x + 1$$

(with c given by (3.9)) leading to the values $A_1 = 0.92129$ and $A_2 = 1.10555$, at least for sufficiently large x. For the Corollary 1 below this approach does not bring any improvement.

From Lemma 3.6 the assertion of Theorem 3.5 (ii) follows for $x > 3000$ with $A_2 = 0.9072$ and since $\psi(x)/x$ has a positive lower bound in the interval $[2, 3000]$ we obtain this part of the theorem with an unspecified value of A_2. A trivial computation shows that the value $A_2 = 0.9072$ can in fact be taken for $x \geq 350$.

To obtain (i) we use the same lemma, which gives

$$\psi(x) \leq \sum_{j\geq 0} \left(\psi(\frac{x}{6^j}) - \psi(\frac{x}{6^{j+1}})\right) \leq \sum_{j\geq 0} \alpha(\frac{x}{6^j})$$
$$\leq 0.9353(1 + \frac{1}{6} + \frac{1}{6^2} \cdots) \leq 1.1224$$

for $x > 350$ and again one can see by computation that the same inequality holds for all $x \geq 2$, thus (i) holds with $A_1 = 1.1224$. □

6. It is easy to observe that if one is interested only in behaviour of $\psi(x)$ for large x then the proof presented above leads to the following assertion:

Corollary 1. *For every $\epsilon > 0$ and sufficiently large x one has*

$$(c - \epsilon)x \leq \psi(x) \leq (\frac{6}{5}c + \epsilon)x,$$

c being given by (3.9). □

The next corollary deals with $\theta(x)$:

Corollary 2. *(i) For all $x \geq 2$ one has*

$$\theta(x) \leq A_1 x,$$

where A_1 is as in Theorem 3.5. In particular one can take $A_1 = 1.1224$.

(ii) For all $x \geq 2$ one has

$$\theta(x) \geq A_3 x$$

with a certain positive A_3. For $x \geq 37$ one can take $A_3 = 0.73$.

Proof. The assertion (i) results immediately from Theorem 3.5 (i) and Lemma 3.3 (iii). To get assertion (ii) assume $x \geq 350$ and use part (ii) of that theorem and the same lemma to get

$$\theta(x) \geq 0.9x - \sqrt{x}\log x \geq 0.9x - 0.15x = 0.75x.$$

A direct check of the range $[37, 350]$ does the rest. □

Corollary 3. *If $P(x)$ denotes the product of all primes not exceeding x then with suitable constants $B_2 > B_1 > 1$ one has*

$$B_1^x \le P(x) \le B_2^x$$

for all $x \ge 2$.

Proof. Immediate from the preceding corollary. □

Corollary 4. *There exist constants* $C_1 > C_2 > 0$ *such that for all* $x \ge 2$ *one has*

$$C_2 \frac{x}{\log x} \le \pi(x) \le C_1 \frac{x}{\log x}.$$

Proof. Follows from the theorem and Lemma 3.3 (iv). □

In an effort to use only elementary tools T.S.Broderick (1939) presented a proof of the last corollary in which the use of logarithms has been eliminated, $\log n$ being replaced by $\lambda(n) = \sum_{k=1}^{n} 1/k$. His arguments were simplified by F.A.Behrend[13] (1940).

Corollary 5. *There exist constants* $D_1 > D_2 > 0$ *such that if* p_n *denotes the* n*-th consecutive prime then*

$$D_2 n \log n \le p_n \le D_1 n \log n.$$

Proof. In view of $\pi(p_n) = n$ the preceding corollary implies

$$\frac{1}{C_1} n \log p_n \le p_n \le \frac{1}{C_2} n \log p_n.$$

This implies $\log p_n - \log\log p_n = O(\log n)$ and we get $\log p_n = O(\log n)$ which, jointly with the trivial inequality $\log p_n > \log n$ gives the assertion. □

Theorem 3.5 was used by J.Pintz (1980a) for a short proof of Corollary 3 to Theorem 3.1. He established first a corresponding result for the function $\psi(x)$, the said corollary following immediately by partial summation:

Corollary 6. *If with some constants* C, D *one has*

$$\psi(x) = Cx + \frac{(D + o(1))x}{\log x}$$

then $C = 1$ *and* $D = 0$.

Proof. By partial summation one gets

$$\sum_{p^k \le x} \frac{\log p}{p^k} = \sum_{n \le x} \frac{\psi(n)}{n^2} + O(1) = \int_2^x \frac{\psi(t)}{t^2} dt + O(1),$$

---

[13] Behrend, Felix Adalbert (1911–1962), Professor in Melbourne.

and using the assumed approximation to $\psi(x)$ one deduces that the last integral equals $C \log x + D \log\log x + o(\log\log x)$.

Stirling's formula gives

$$\log x + O(1) = \frac{1}{x}\log([x]!) = \frac{1}{x}\sum_{p^k \le x}\left[\frac{x}{p^k}\right]\log p = \sum_{p^k \le x}\frac{\log p}{p^k} + O\left(\frac{\psi(x)}{x}\right)$$

and by Theorem 3.5 (i) the error term here is $O(1)$, thus

$$\sum_{p^k \le x}\frac{\log p}{p^k} = \log x + O(1)$$

and one is led to $C = 1$, $D = 0$, as asserted. □

The next corollary shows that neither Legendre's nor the first Gaussian approximation to $\pi(x)$ can have a particularly good error term:

Corollary 7. *If there exists a constant C and a number $\delta > 2$ such that*

$$|\pi(x) - \frac{x}{\log x - C}| \le B\frac{x}{\log^\delta x}$$

holds for all large x, then $C = 1$.

Proof. It suffices to observe that the assumption along with Corollary 4 implies

$$\lim_{x\to\infty}\left(\frac{x}{\pi(x)} - \log x\right) = -C$$

and apply Corollary 3 of Theorem 3.1. □

Čebyšev's proof of Theorem 3.5 gives in its Corollary 4 the values 0.921292 and 1.1055 for the constants. Better values were obtained by J.J.Sylvester (1881) who followed Čebyšev's approach but used a more complicated function in place of Čebyšev's $\alpha(x)$. He obtained for sufficiently large x the bounds

$$0.956 < \frac{\pi(x)}{x/\log x} < 1.045$$

and

$$0.9926 < \frac{\psi(x)}{x} < 1.0667.$$

(See Gram 1897/98). E.Landau (1906c) observed that using Čebyšev's method one can deduce in an elementary way the asymptotic relation

$$\lim_{x\to\infty}\frac{\psi(x)}{x} = 1$$

from the equality

$$\sum_{n=1}^{\infty} \frac{\mu(n) \log n}{n} = -1$$

which however at that time did not have an elementary proof. (The first deduction of the last equality from the Prime Number Theorem (i.e. $\pi(x) = (1+o(1))x/\log x$) appears in Landau (1899b).)

A simpler proof of the weaker evaluation

$$0.92\frac{x}{\log x} \leq \pi(x) \leq 1.26\frac{x}{\log x}$$

valid for $x \geq 11$ can be found in H.Harborth,H.J.Kanold,A.Kemnitz (1981). See also Deshouillers (1977).

The question, as to whether Čebyšev's original method may lead to bounds

$$(1-\epsilon)\frac{x}{\log x} \leq \pi(x) \leq (1+\epsilon)\frac{x}{\log x}$$

for every fixed $\epsilon > 0$ and sufficiently large x was answered positively by H.G.Diamond and P.Erdős (1980). They established that to obtain such bounds one has to know the values of the Möbius function μ in an interval $[1, T]$, where T depends on ϵ. Their proof uses the Prime Number Theorem and so this approach does not lead to a new proof of that theorem but, in any case, their result does show that the bounds of Čebyšev and Sylvester could be improved using their methods but at the cost of cumbersome calculations.

Analogues of Theorem 3.5 and its corollaries for primes in progressions were first considered by A.de Polignac (1859b) who showed that for sufficiently large x the interval $[2, x]$ contains at least $cx/\log x$ primes in any of the progressions $4k+1$ and $4k+3$. For certain other progressions with small differences this was done in Roux (1956).

R.S.Hall (1973) noted that Corollary 4 to Theorem 3.5 may be not true for systems of *Beurling generalized primes*. A sequence $1 < \pi_1 < \pi_2 < \ldots$ of real numbers tending to infinity is said to form such a system if the number $N(x)$ of elements $\leq x$ of the multiplicative semigroup generated by the π_i's satisfies

$$N(x) = Ax + O\left(\frac{x}{\log^\gamma x}\right)$$

with certain positive A and γ. A.Beurling (1937) established that if $\gamma > 3/2$ then one has

$$P(x) = \sum_{\pi_i \leq x} 1 = (1+o(1))\frac{x}{\log x}.$$

(This may fail if $\gamma \leq 3/2$ (Diamond 1970). See the book of J.Knopfmacher (1975) for more on this topic.)

The analogue of Corollary 4 to Theorem 3.5 for such systems in the case $\gamma > 1$ was established by Diamond (1973) and in Hall (1973) a system was constructed with $\gamma < 1$ for which one has

$$\liminf_{x\to\infty} \frac{P(x)\log x}{x} = 0$$

and

$$\limsup_{x\to\infty} \frac{P(x)\log x}{x} = \infty$$

(see Zhang Wen-Bin 1987).

7. After obtaining his main result Čebyšev deduced Bertrand's postulate, which we present in its modern version:

Theorem 3.7. *For $x \geq 2$ there is a prime in the interval $(x, 2x]$.*

Proof. We may safely assume that $x \geq 37$ since for smaller x one of the primes

$$2, 3, 5, 7, 13, 23, 43$$

will do. If there are no primes in $(x, 2x]$ then clearly

$$\theta(2x) = \theta(x),$$

however Corollary 2 to Theorem 3.5 implies

$$1.46x \leq \theta(2x) = \theta(x) \leq 1.13x$$

which gives a clear contradiction. □

Note that a slight change in the presented argument leads to the proof of Bertrand's postulate in its original form:

We may assume that $x \geq 10$. If the interval $(x/2, x-2]$ does not contain primes then

$$1.13\frac{x}{2} \geq \theta(x/2) = \theta(x-2) \geq 0.73(x-2)$$

thus $x < 10$, a contradiction.

S.Ramanujan[14] (1919) gave a direct proof of Bertrand's postulate:

Proof of Ramanujan. The argument is based on the formula established in Lemma 3.3 (ii). Using it we get immediately

$$\log([x]!) - 2\log([x/2]!) = \psi(x) - \psi(x/2) + \psi(x/3) - \cdots. \tag{3.10}$$

Since

$$\psi(x) - 2\psi(\sqrt{x}) = \theta(x) - \theta(\sqrt{x}) + \theta(\sqrt[3]{x}) - \cdots$$

then, as both ψ and θ are non-decreasing the inequalities

$$\psi(x) - 2\psi(\sqrt{x}) \leq \theta(x) \leq \psi(x) \tag{3.11}$$

and

$$\psi(x) - \psi(x/2) \leq \log([x]!) - 2\log([x/2]!) \leq \psi(x) - \psi(x/2) + \psi(x/3)$$

result.

Stirling's formula implies the bounds

$$\frac{2x}{3} < \log([x]!) - 2\log([x/2]!) < \frac{3x}{4},$$

[14]Ramanujan, Srinivasa (1887–1920), worked in Cambridge.

the lower bound holding for $x > 300$ and the upper bound holding for all positive x. Hence

$$\psi(x) - \psi(x/2) < \frac{3x}{4} \tag{3.12}$$

for all positive x and

$$\psi(x) - \psi(x/2) + \psi(x/3) > \frac{2x}{3} \tag{3.13}$$

for $x > 300$.

Substituting consecutively $x/2, x/4, \dots$ in place of x in (3.12) and adding one arrives at

$$\psi(x) < \frac{3}{2}x$$

for all positive x. Using (3.11) we now obtain

$$\psi(x)-\psi(\frac{x}{2})+\psi(\frac{x}{3}) \le \theta(x)+2\psi(\sqrt{x})-\theta(\frac{x}{2})+\psi(\frac{x}{3}) < \theta(x)-\theta(\frac{x}{2})+\frac{x}{2}+3\sqrt{x},$$

which with (3.13) gives

$$\theta(x) - \theta(\frac{x}{2}) > \frac{x}{6} - 3\sqrt{x}$$

for x exceeding 300.

Since for $x \ge 324$ one has $\frac{x}{6} \ge 3\sqrt{x}$ thus finally we obtain for $x \ge 162$ the inequality

$$\theta(2x) - \theta(x) > 0.$$

This shows that for $x \ge 162$ there exists a prime between x and $2x$ and since this is easily checked for $x < 162$, Bertrand's postulate results. □

For other proofs of Bertrand's postulate see Erdős[15](1932), Finsler[16](1945), Moser (1949), Pillai (1944), Singh (1977) and Stečkin[17](1968).

Bertrand's postulate forms one of several results of the form:

If $n > c(\epsilon)$ then there is a prime in the interval $[n, (1+\epsilon)n]$.

From the Prime Number Theorem (which we shall prove in Chap.5) it follows that this holds with a suitable $c(\epsilon)$ for every $\epsilon > 0$ hence the main interest here lies in finding explicit values for $c(\epsilon)$. Bertrand's postulate gives $c(1) = 1$ and Čebyšev's result gives the existence of $c(\epsilon)$ for every $\epsilon > \epsilon_0 = 1/5$. The value of ϵ_0 was consecutively diminished using elementary methods by J.Petersen[18](1882), J.P.Gram (1897/98), R.D.von Sterneck (1900), E.Fanta (1901b) and E.Waage (1913, 1914) who got $\epsilon_0 = 1/23$.

---

[15] This proof is reproduced in Hardy,Wright (1960,Th.418).
[16] Finsler, Paul (1894–1970), Professor in Zürich.
[17] Stečkin, Sergei Borisovič (1920–1995), Professor in Moscow.
[18] Petersen, Julius Peter Christian (1839–1910).

Using elementary methods several values of $c(\epsilon)$ were calculated in Schur (1929) ($c(1/4) = 24$), Gatteschi (1947) ($c(2/9) \leq 29$), Nagura (1952) ($c(1/5) = 25$) and Rohrbach,Weis (1964) ($c(1/13) = 118$). The last paper used the method of Waage (1913,1914) and revealed certain numerical inaccuracies therein. Using analytic tools R.Breusch (1932) obtained c(1/8)=48. Based on analytic estimates of L.Schoenfeld (1976), G.Giordano (1987,1989) obtained an explicit upper bound for $c(\epsilon)$ for every positive ϵ. A table of values of $c(\epsilon)$ for various small values of ϵ, e.g. $c(0.00006) = 2010761$ was given in Harborth,Kemnitz (1981).

The Prime Number Theorem implies that restricting the argument x to sufficiently large numbers one can in Theorem 3.5 and its corollaries put for A_1, B_1, C_1 arbitrary numbers exceeding 1 and for A_2, B_2, C_2 arbitrary numbers smaller than 1. Good values for these constants for various ranges of x were given by J.B.Rosser (1962,1975) and L.Schoenfeld (1976) using analytic methods and heavy computations[19]. In particular they showed that the function $\psi(x)/x$ attains it maximal value equal to $1.03883\ldots$ at $x = 113$ (Rosser,Schoenfeld 1962, Theorem 12; an elementary proof was given by N.Costa Pereira (1981,1989)). Moreover

$$\theta(x) < 1.002x$$

holds for all $x > 0$ and

$$\psi(x) \geq \theta(x) > 0.998x$$

holds for $x > 1\ 319\ 007$ (Rosser,Schoenfeld 1975, Theorem 6). One also has

$$\pi(x) > \frac{x}{\log x}$$

for $x \geq 17$,

$$\pi(x) < 1.25506\frac{x}{\log x}$$

for $x > 1$ and

$$\pi(x) < \frac{5x}{4\log x}$$

for all $x > 1$ with the exception of numbers x from the interval $[113, 113.6)$ (Rosser,Schoenfeld 1962, Corollaries 1 and 2 to Theorem 2). See also Massias,Robin (1996). The best known lower bound for p_k, the kth consecutive prime is due to P.Dusart (1999):

$$p_k > k(\log k + \log\log k - 1) \qquad (k \geq 2).$$

Refining the elementary method of Sylvester (1892) N.Costa Pereira (1989) obtained evaluations comparable with those achieved by analytic means. In particular he showed

$$\left|\frac{\psi(x)}{x - 1}\right| < \frac{1}{2976}$$

for $x \geq 10^8$,

$$\theta(x) < \frac{532}{531}x$$

for all positive x and

$$\theta(x) > \frac{499}{500}x$$

for $x \geq 487\ 381$.

---

[19]For previous bounds see Rosser (1941).

The best known elementary upper bound for $\theta(x)$ was obtained by W.E.L.Grimson and D.Hanson (1978) who proved $\theta(x) < 1.0508x$, i.e. $\prod_{p\le x} p < (2.86)^x$. Previous elementary approaches had led to $\theta(x) < 1.215x$ (Moser, 1959), $\theta(x) < 1.1152$ (J.L.Selfridge, unpublished) and $\theta(x) < x \log 3 < 1.0987x$ (Hanson 1972). The values of $\theta(x)$ at prime arguments were studied by G.Robin (1983) who proved that if p_n denotes the nth prime, then for $n \ge 2$ one has

$$\theta(p_n) \ge n(\log n + \log\log n - c),$$

where c is a constant equal approximately to 1.076868, with equality occuring for $n = 66$.

A related problem consists in the evaluation of $B(x)$, the least common multiple of all positive integers not exceeding x. Clearly one has $\theta(x) \le \log B(x)$. Elementary proofs of the bound $B(x) < 3^x$ were given by D.Hanson (1972) and U.Felgner (1991).

Similar questions for primes in arithmetic progressions were considered by R.Breusch (1932) who showed that for $x > 7$ the interval $[x, 2x]$ always contains primes of progressions $\pm 1 \bmod 3$ and $\pm 1 \bmod 4$. Using analytic tools he proved this for $x > 10^6$ and for the interval $[7, 10^6]$ he used tables of primes. An elementary proof of his result was given by P.Erdős (1935b) whose method is applicable to any progression $kn + l$ (with $(k, l) = 1$) for which the sum σ of the reciprocals of primes $< k$ and not dividing k is smaller than 1. In that case he obtained the existence of a prime of that progression lying in the interval $[x, \lambda x]$ with any λ exceeding $k/(k-1)(1-\sigma)$ for sufficiently large x (see Ricci 1933a,1934). Later K.Molsen (1941) showed that for $n \ge 199$ there are always primes of the form $3k+1$ and $3k+2$ in the interval $[n, 8n/7]$ and dealt also with progressions $12k + l$ with $l = \pm 1, \pm 5$. Recently P.Moree (1993) gave a completely effective version of the result of Erdős and listed all integers k's to which it is applicable, the largest being 840.

Using strong analytic tools T.Tatuzawa (1962) established the existence of a constant c such that for all sufficiently large k, all l prime to k and $x \ge \exp(c \log k \log\log k)$ there is a prime congruent to $l \bmod k$ lying in $(x, 2x]$.

An important generalization of Bertrand's postulate appears in Sylvester (1892) where it was shown that if $n > k > 0$ then there exists a prime exceeding k, a certain multiple of which lies in the interval $[n, n + k - 1]$. For $n = k$ this gives Bertrand's postulate. This result was later rediscovered by I.Schur (1929) and for that reason it is called the *Sylvester-Schur theorem* Proofs of it can be also found in Erdős (1934), Faulkner (1966), Ecklund,Eggleton (1972) and Hanson (1973). See Narkiewicz (1986, Sect.3.2) for more information on that theorem and its generalizations.

H.-E.Richert (1949a) used Bertrand's postulate to prove that every integer exceeding 6 is a sum of distinct primes. (See also Richert 1949b, Mąkowski 1960). A slightly stronger form stating that for $x \ge 17$ there is always a prime between x and $2x - 10$ (for a generalization of this assertion see Badea (1986)) permitted R.E.Dressler (1972) to obtain such a representation with odd primes for every positive integer except 1,2,4,6 and 9. See also Dressler (1973) for a stronger assertion.

8. There exist several proofs of Čebyšev's Theorem 3.5. We commence with a very short proof of the upper bound given by F.Mertens (1874a):

Second proof of Theorem 3.5 (i). In view of the equality (3.10) and the monotonicity of $\psi(x)$ one has

$$\psi(x) - \psi(x/2) < \log([x]!) - 2\log([x/2]!)$$

and an application of a strong form of Stirling's formula yields now

$$\psi(x) - \psi(x/2) < x \tag{3.14}$$

for $x > 4$. Clearly the inequality (3.14) holds also for $1 < x \leq 4$. Thus finally

$$\begin{aligned}\psi(x) &\leq \psi(x) - \psi(x/2) + \psi(x/2) - \psi(x/4) + \cdots \\ &< x(1 + \frac{1}{2} + \frac{1}{4} + \cdots) = 2x\end{aligned}$$

proving the upper bound. □

The use of Stirling's formula in Mertens' argument can be avoided in the following way: let $x > 14$ and define the integer N by $2(N-1) < x \leq 2N$. Then

$$\log([x]!) - 2\log([x/2]!) < \log((2N)!) - 2\log((N-1)!) = \log\left(\frac{(2N)!}{((N-1)!)^2}\right)$$

and since an inductive argument shows that for $N \geq 8$ the inequality

$$\frac{(2N)!}{((N-1)!)^2} < e^{2(N-1)}$$

holds one is lead to (3.14) for $x > 14$ and it its truth for smaller x can be checked by inspection.

9. Another proof, based on one of the inversion formulas of Theorem 1.15, was found by N.Spears (1970):

Third proof of Theorem 3.5 (i). Using the identity given in Lemma 3.3 (ii) and recalling Stirling's formula we can write

$$\sum_{n \leq x} \psi\left(\frac{x}{n}\right) = x \log x - x + O(\sqrt{x}).$$

Note that it would be unwise to use here Stirling's formula with the better error term $O(\log x)$ since this would give problems in the evaluation of the error term, arising in the last stage of the proof, due to the unboundness of of $1/\log(x/n)$ for n close to x.

Write

$$\psi(x) = a + bx + R(x),$$

where a, b are real numbers to be fixed later and $R(x)$ is a certain error term depending on a and b. Using the equality (which is a consequence of Lemma 2.12)

$$\sum_{n \leq x} \frac{1}{n} = \log x + \gamma + O\left(\frac{1}{x}\right)$$

with a certain constant[20] γ, we get

$$\begin{aligned} x\log x - x + O(\sqrt{x}) &= \sum_{n\le x}\left(a + \frac{bx}{n} + R\left(\frac{x}{n}\right)\right) \\ &= bx\log x + (a+b\gamma)x + \sum_{n\le x} R\left(\frac{x}{n}\right) + O(1). \end{aligned}$$

Put $g(x) = \sum_{n\le x} R\left(\frac{x}{n}\right)$. The choice $b=1$, $a=-1-\gamma$ now leads to $g(x) = O(\sqrt{x})$ and applying Theorem 1.15 (iii) we finally arrive at

$$|R(x)| = \Big|\sum_{n\le x}\mu(n)g\left(\frac{x}{n}\right)\Big| = O\left(\sum_{n\le x}\sqrt{\frac{x}{n}}\right) = O(x).$$

As $\psi(x) = -1-\gamma+x+R(x)$ the assertion follows. □

10. Now we present two direct proofs of the Corollary 2 to Theorem 3.5 without using the theorem itself. If one does not bother about the size of the constants occurring in the assertion of that corollary then the following argument seems to be the shortest possible:

Second proof of Corollary 2 to Theorem 3.5. Let $x \ge 2$ and let $k \ge 2$ be the integer satisfying $2^{k-1} \le x < 2^k$. Observe now that for every integer n the product of all primes from the interval $(n, 2n]$ divides the binomial coefficient $\binom{2n}{n}$ and thus

$$\theta(2n) - \theta(n) = \log \prod_{n<p\le 2n} p \le \log\binom{2n}{n} < n\log 4$$

because

$$\binom{2n}{n} < \sum_{j=0}^{2n}\binom{2n}{j} = (1+1)^{2n} = 4^n.$$

This gives

$$\theta(2^k) = \sum_{j=0}^{k-1}(\theta(2^{j+1}) - \theta(2^j)) < \log 4\sum_{j=0}^{k-1}2^j < \log 4\cdot 2^k$$

---

[20] In fact $\gamma = -\Gamma'(1) = 0.5772156649\ldots$ is *Euler's constant* which can be also defined by

$$\gamma = -\int_0^\infty \frac{\log t}{e^t}\,dt.$$

Euler himself computed it to 16 decimal places (Euler 1734/35b,1736,1769). See Glaisher (1870,1872), Adams (1878), Wrench (1952), Knuth (1962), Sweeney (1963) for improvements on his computation. It is not known yet, whether γ is irrational, although it is generally thought that it is.

and finally

$$\theta(x) \le \theta(2^k) < \log 4 \cdot 2^k < x \log 16,$$

establishing the upper estimate with $A_1 = \log 16 = 2.772588\ldots$.

To prove the lower estimate note first that for all n we have

$$\binom{2n}{n} \le \prod_{p \le 2n} p^{[\log(2n)/\log p]} \le (2n)^{\pi(2n)}.$$

Indeed, Lemma 1.20 implies that the exponent of the maximal power of a prime p dividing

$$\binom{2n}{n} = \frac{(2n)!}{(n!)^2}$$

equals

$$\sum_{k \le (\log 2n)/(\log p)} \left(\left[\frac{2n}{p^k}\right] - 2\left[\frac{n}{p^k}\right] \right),$$

and one sees easily that every summand here equals either 0 or 1.

Thus

$$2^n \le \frac{2n}{n} \cdot \frac{2n-1}{n-1} \cdots \frac{n}{1} = \binom{2n}{n} \le (2n)^{\pi(2n)}$$

and taking logarithms we get

$$n \log 2 \le \pi(2n) \log(2n)$$

from which the assertion follows using Lemma 3.3 (iv). One sees that for sufficiently large x one can take for A_3 any number smaller than $\frac{1}{2}\log 2 = 0.346\ldots$. □

The first part of this argument appears in Landau (1930, Satz 17). The part dealing with the lower bound can be found in Blanchard (1969). A more delicate treatment of the same idea permitted N.Costa Pereira (1985/86) to double the obtained lower bound.

11. The same idea realized more precisely by P.Erdős and L.Kálmár[21] (see Erdős (1932) for the first version of this proof) leads to a smaller constant[22], viz. $A_1 = \log 4 = 1.386294\ldots$. We present it with a small simplification due to M.Nair (1982).

Third proof of Corollary 2 (i) to Theorem 3.5. We shall prove by induction the inequality $\theta(n) < \log 4 \cdot n$ for all integers $n \ge 2$. For $n = 2$ the assertion is evident, so assume its truth for a certain $n \ge 2$. If $n+1$ is composite then the induction assertion follows from $\theta(n) = \theta(n+1)$, hence we may assume that $n+1$ is an odd prime. If we write $n+1 = 2m+1$ then

[21] Kálmár, Lászlo (1905–1976), Professor in Szeged.

[22] This proof is given in several textbooks.

$$\theta(n+1) = \theta(m+1) + \sum_{m+1<p\leq 2m+1} \log p$$

and since the product of all primes from the interval $[m+2, 2m+1]$ divides $\binom{2m+1}{m}$ we arrive at

$$\theta(n+1) \leq \theta(m+1) + \log \binom{2m+1}{m}.$$

Now note that $\binom{2m+1}{m} = \binom{2m+1}{m+1}$ and so

$$\binom{2m+1}{m} \leq \frac{1}{2}(1+1)^{2m+1} = 4^m.$$

Finally we get

$$\theta(n+1) < \log 4 \cdot (m+1+m) = \log 4 \cdot (n+1). \qquad \square$$

12. A few years after proving Theorem 3.5 Čebyšev made (in a letter to Fuss (Čebyšev 1853)) three statements about the distribution of prime numbers in residue classes mod 4:

(A) *The ratio*

$$\frac{\pi(x;4,3) - \pi(x;4,1)}{\sqrt{x}/\log x}$$

attains values arbitrarily close to 1.

(B) *One has*

$$\lim_{x\to 0} \sum_{p\geq 3} (-1)^{(p+1)/2} e^{-px} = \infty.$$

(C) *If $f(x)$ is a decreasing function defined for positive reals and the series*

$$\sum_{p\geq 3} (-1)^{(p+1)/2} f(p)$$

converges, then

$$\lim_{x\to\infty} \sqrt{x} f(x) = 0.$$

He did not give any indication how one might prove them. A few years after the appearance of Čebyšev's paper de Polignac (1859a) claimed to have a proof of the assertion (A) but gave only a heuristic argument to support his claim. Also Césaro (1896) tried unsuccesfully to establish this assertion, albeit in a less precise form[23]. The first correct proof of (A) was found by

[23]See Landau (1905b) for a discussion of Césaro's argument.

E.Phragmén (1891) and we shall sketch it later (Corollary 2 to Theorem 4.13). A simpler proof is contained in Landau (1905b).

The assertion (B) expresses the empirical fact that in long intervals primes congruent to 3 mod 4 are prevalent. R.H.Hudson (1980) presented a combinatorial explanation of this fact. Later S.Knapowski and P.Turán (1964,I) conjectured that if we denote by $N(x)$ the number of integers $n \leq x$ for which $\pi(n;4,3) < \pi(n;4,1)$, then

$$\lim_{x\to\infty} \frac{N(x)}{x} = 0.$$

Since the difference $\pi(n;4,3)-\pi(n;4,1)$ changes its sign infinitely often (Littlewood 1914, Hardy,Littlewood 1918) it follows that $N(x)$ is unbounded. J.Leech[24] (1957) and D.Shanks (1959) computed $N(x)$ for various values and obtained in particular $N(3 \cdot 10^6) = 3406$. The first integer n for which the difference $\pi(n;4,3) - \pi(n;4,1)$ becomes negative is 26861 (Leech 1957). See also Bays, Hudson (1978b,1979,1983), Grosswald[25] (1967), Hudson (1985) and Hudson,Bays (1977).

Similar investigations were carried out for the differences $\pi(x;5,4) - \pi(x;5,2)$ (Stark 1971), $\pi(x;24,13)-\pi(x;24,1)$, $\pi(x;3,2)-\pi(x;3,1)$ (note that this difference remains positive for more than 18 millions consecutive integers starting at $x = 2$ (Bentz 1982)) and $\pi(x;8,5)-\pi(x;8,1)$ (Bays,Hudson 1977,1978a,c,1979). It follows from a theorem of I.Kátai (1967) that if the product of all L-functions corresponding to characters mod k does not vanish in the interval $(0,1)$ then for distinct l_1, l_2 the difference $\pi(x;k,l_1) - \pi(x;k,l_2)$ changes its sign in every interval $[T^{1-\epsilon}, T]$ (with positive ϵ) for sufficiently large T.

Recently it was shown by J.Kaczorowski (1992) (see also Kaczorowski 1995b) that the Knapowski–Turán conjecture is incompatible with the Extended Riemann Hypothesis for Dirichlet's L-functions (stating that all zeros of $L(s,\chi)$ in the half-plane Re $s > 0$ lie on the line Re $s = 1/2$.) His result is in fact more general, as it shows that under this assumption one has

$$Ax \leq \#\{n \leq x : \ \pi(x;k,l) \geq \pi(x;k,1)\}$$

for $(k,l) = 1, l \not\equiv 1 \pmod k$ and sufficiently large x, with a suitable positive constant A depending on k and l.

There is still no proof of (B) but now it is known through the work of G.H.Hardy and J.E.Littlewood (1918) that (B) is a consequence of the conjectured non-vanishing of the function $L(s,\chi)$ in the half-plane Re $s > 1/2$ with χ being the only nonprincipal character mod 4 and it turned out that the converse implication is also true (Landau 1918a). An analogous result concerning primes in progressions mod 8 appears in Knapowski,Turán (1963 VIII,1965a).

---

[24]Leech, John (1926–1992), Professor in Stirling.
[25]Grosswald, Emil (1912–1989), Professor at Temple University.

Various assertions similar to (B) were considered in Bentz (1982), Bentz, Pintz (1980a,b,1982), Besenfelder (1979), Fujii (1988) and Knapowski, Turán (1964, 1965a,b).

Not much is known about the last of the three assertions of Čebyšev. It was proved by E.Landau (1918) that if (A) is true and for some positive δ one has $p^{1/2+\delta}f(p) = O(1)$ then the series in (C) converges.

3.3. Applications of Čebyšev's Theorems

1. Almost thirty years after the appearance of Čebyšev's papers F.Mertens realized that their results can be used to prove several new theorems concerning primes and obtained, in particular, asymptotic formulas for the sums $\sum_{p\le x}\log p/p$ and $\sum_{p\le x} 1/p$. The first of these had already been announced, without proof, by A.de Polignac (1857, part 3):

Theorem 3.8. (F.Mertens 1874a,b) *One has*

$$\sum_{p\le x}\frac{\log p}{p} = \log x + O(1).$$

Proof. The proof utilizes only the upper bound for $\theta(x)$ and to be able to use it later for a further proof of the lower bound we formulate it in a conditional form:

Lemma 3.9. *If $\theta(x) \le Ax$ holds for $x \ge x_0$ then in the same range one has*

$$\sum_{p\le x}\frac{\log p}{p} = \log x + R(x),$$

with $R(x)$ bounded by a constant depending on A and x_0.

Proof. Clearly it suffices to deal with the case when $x = n \ge x_0$ is an integer. Using Lemma 1.20 we may write

$$\log n! = \sum_p \left[\frac{n}{p}\right]\log p + \sum_p \left[\frac{n}{p^2}\right]\log p + \cdots$$

$$= n\sum_{p\le x}\frac{\log p}{p} + O(\theta(n)) + O(n\sum_p\sum_{k=2}^{\infty}\frac{\log p}{p^k}).$$

Since $\log n! = n\log n + O(n)$ and

$$\sum_p\sum_{k=2}^{\infty}\frac{\log p}{p^k} = O\left(\sum_p\frac{\log p}{p^2}\right) = O(1)$$

we get

$$\sum_{p\le n}\frac{\log p}{p}=\log n+O\left(\frac{\theta(n)}{n}\right)+O(1)$$

and the assertion follows. □

The theorem follows immediately from the lemma and Corollary 2 (i) to Theorem 3.5. □

The analogous result for primes in progressions stating that the sum

$$\sum_{\substack{p\le x\\ p\equiv l\,(\mathrm{mod}\ k)}}\frac{\log p}{p}$$

is asymptotic to $\log x/\varphi(k)$ can be obtained by analytic means from the quantitative form of Dirichlet's theorem (see Theorem 5.11). A proof using mostly elementary means and a tauberian theorem of Hardy-Littlewood may be found in Koshiba,Uchiyama (1966). For a study of the related sum

$$\sum_{\substack{p^m\le x\\ p^m\equiv l\,(\mathrm{mod}\ k)}}\frac{\log p}{p}$$

see Wirsing (1964) and Nevanlinna (1973).

2. As an application of Theorem 3.8 we now show that the proofs of Corollary 2 (i) to Theorem 3.5 presented in the previous section lead also to the proofs of part (ii) of that corollary. In fact it was observed by H.N.Shapiro (1950a) that a lower bound for $\theta(x)/x$ follows without difficulty from the existence of the upper bound:

Lemma 3.10. *If $\theta(x)\le Ax$ holds for $x\ge x_0$ then with a suitable positive $B=B(A,x_0)$ one has $\theta(x)\ge Bx$ for sufficiently large x.*

Proof. Lemma 3.9 shows that the assumption of our lemma implies

$$\sum_{p\le x}\frac{\log p}{p}=\log x+R(x)$$

with $|R(x)|\le c$ for $x\ge x_0$ and a suitable constant c. Put $a=\exp(-2c-1)$ and observe that if x exceeds $x_0\exp(2c+1)$ then on one hand we have

$$\sum_{ax<p\le x}\frac{\log p}{p}=\log\frac{1}{a}+R(x)-R(ax)\ge\log\frac{1}{a}-2c,$$

and on the other hand

$$\sum_{ax<p\le x}\frac{\log p}{p}\le\frac{\theta(x)}{ax}.$$

Thus the inequality

$$\theta(x) \geq a\left(\log\frac{1}{a} - 2c\right)x = ax$$

results for sufficiently large x and since $\theta(x)$ is positive for $x \geq 2$ the assertion follows. □

3. Theorem 3.8 was utilized by Mertens to evaluate the sum of reciprocals of primes not exceeding x. The obtained result confirmed a statement already made by Euler (1737, Satz 19) and Gauss (1791). Although Čebyšev (1848) attempted a proof of it, his deduction cannot be regarded as correct since he assumed that $\theta(x)$ is asymptotically equal to x (which was proved only fifty years later) and he did not bother about error terms.

Theorem 3.11. (Mertens 1874b) *For $x \geq 2$ one has*

$$\sum_{p\leq x}\frac{1}{p} = \log\log x + C + O(\frac{1}{\log x}),$$

where

$$C = \gamma - \sum_{k=2}^{\infty}\frac{1}{k}\sum_{p}\frac{1}{p^k}$$

and γ denotes Euler's constant.

The proof given by Mertens was based on a meticulous study of the sum

$$S(y,t) = \sum_{p>y}\frac{1}{p^{1+t}}$$

for $0 < t < 1$, which led to the evaluation

$$S(y,t) = -\log t - \log\log y - \gamma + O\left(\frac{1}{\log y}\right) + o(1),$$

uniformly in $t \to 0$. The comparison of this formula with that given in Lemma 3.2 concluded the proof.

Later E.Landau (1909a) using the same main idea eliminated the use of infinite series making the proof completely elementary.

Landau's proof: Clearly we may assume that x is an integer. Write

$$\sum_{p\leq x}\frac{1}{p} = \sum_{n\leq x}\frac{\epsilon_n\log n}{n}\frac{1}{\log n}$$

with

$$\epsilon_n = \begin{cases}1 & \text{if } n \text{ is prime}\\ 0 & \text{else}\end{cases}$$

and apply Lemma 2.5 to $a_n = (\epsilon_n \log n)/n$ and $b_n = 1/\log n$. Since Theorem 3.8 gives $\sum_{j=1}^{k} a_j = \log k + R(k)$ with $|R(k)| \leq c$ with a suitable constant c, we get

$$A(x) = \sum_{p \leq x} \frac{1}{p} = \frac{\log x + R(x)}{\log x} + \sum_{j=1}^{x-1} (\log j + R(j)) \frac{\log(1+1/j)}{\log j \log(j+1)}.$$

In view of $\log(1+1/j) = 1/j + R_1(j)$ (with $R_1(j) = O(1/j^2)$) this gives

$$A(x) = \sum_{j=1}^{x-1} \frac{1}{j \log(j+1)} + 1 + O(\frac{1}{\log x})$$

$$+ \sum_{j=1}^{x-1} \frac{R_1(j)}{\log(j+1)} + \sum_{j=1}^{x-1} \frac{R(j)(\frac{1}{j} + R_1(j))}{\log j \log(j+1)},$$

and it suffices to observe that the following equalities hold:

$$\sum_{j=1}^{x-1} \frac{R_1(j)}{\log(j+1)} = \sum_{j=1}^{\infty} \frac{R_1(j)}{\log(j+1)} + O(1/x)$$

$$\sum_{j=1}^{x-1} \frac{R(j)(\frac{1}{j} + R_1(j))}{\log j \log(j+1)} = \sum_{j=1}^{\infty} \frac{R(j)(\frac{1}{j} + R_1(j))}{\log j \log(j+1)} + O(\frac{1}{\log x})$$

and

$$\sum_{j=1}^{x-1} \frac{1}{j \log(j+1)} = \log \log x + B + O(\frac{1}{\log x})$$

with a suitable constant B, the last equality being a consequence of Lemma 2.12 applied to the function $1/t \log t$.

Landau's method does not lead to an explicit value for the constant C. To obtain it one can use the following short reasoning (given in Prachar (1957, Chap.3)) which however is not elementary:

Write $A(x) = \sum_{p \leq x} 1/p$ and $f(t) = \sum_p 1/p^t$ for $t > 1$. Then

$$f(t) = (t-1) \int_2^{\infty} \frac{A(x)}{x^t} dx$$

and using the statement of Theorem 3.11 with an undetermined value of C one splits this integral into three and on evaluating them one is led quickly to the formula

$$f(t) = -\log(t-1) - \gamma + C + o(1)$$

for $t \to 1$. On the other hand using Corollary 2 to Lemma 2.3 one obtains

$$f(t) = \log \zeta(t) - \sum_p \sum_{m=2}^{\infty} \frac{1}{mp^{mt}} = -\log(t-1) - \sum_p \sum_{m=2}^{\infty} \frac{1}{mp^{mt}} + o(1)$$

and the comparison of the last two equalities determines C. □

Corollary. *For x tending to infinity one has*[26]

$$\prod_{p\le x}\left(1-\frac{1}{p}\right)=\frac{e^{-\gamma}}{\log x}+O\left(\frac{1}{\log^2 x}\right)$$

and

$$\prod_{p\le x}\left(1-\frac{1}{p}\right)^{-1}=e^{\gamma}\log x+O(1).$$

Proof. If we put

$$A_x=\prod_{p\le x}\left(1-\frac{1}{p}\right)$$

then the theorem implies

$$\begin{aligned}\log A_x&=\sum_{p\le x}\log\left(1-\frac{1}{p}\right)=-\sum_{p\le x}\sum_{k=1}^{\infty}\frac{1}{kp^k}\\&=-\sum_{p\le x}\frac{1}{p}-\sum_{p}\sum_{k=2}^{\infty}\frac{1}{kp^k}+\sum_{p>x}\sum_{k=2}^{\infty}\frac{1}{kp^k}\\&=-\log\log x-\gamma+O\left(\frac{1}{\log x}\right).\end{aligned}$$

The first assertion results now by exponentiating both sides of the last equality and the second follows immediately. □

Another proof of the last corollary, based on a tauberian theorem was later given by G.H.Hardy (1927,1935). Mertens (1874b) also obtained the following analogue of Theorem 3.11 for primes in progressions (see Chap.5, Exercise 19):

$$\sum_{\substack{p\le x\\ p\equiv l \pmod k}}\frac{1}{p}=\frac{1}{\varphi(k)}\log\log x+c(k,l)+O(\frac{1}{\log x}).$$

This easily implies the analogue of the preceding corollary. See also Amitsur (1956/57), Grosswald (1987b), Uchiyama (1971), Vasilkovskaja (1977) and K.S.Williams (1974) on this topic. The error term in the first formula of the last corollary has been improved several times, the best known bound being of the order

$$O(\exp(-c\log^{0.6}x(\log\log x)^{-0.2}))$$

(A.I.Vinogradov 1963).

[26]The asymptotic equality of the product $\prod_{p\le x}p/(p-1)$ and $c\log x$ with $c=1.874\ldots$ had already been asserted by Gauss (1791). Actually $e^{\gamma}=1.781\ldots$.

4. Another application of Čebyšev's results, mentioned already in Chap.1, concerns primes in polynomial values. We shall prove here a result of U.Abel and H.Siebert (1993) whose corollary implies that a good choice of the constant term of an arbitrary non-constant polynomial with integral coefficients leads to a polynomial representing many primes:

Theorem 3.12. *Let $a_1, a_2, \ldots$ be an arbitrary sequence of distinct positive integers and assume that its counting function $A(x) = \sum_{a_n \leq x} 1$ satisfies*

$$\limsup_{x \to \infty} \frac{A(x)}{\log x} = \infty.$$

Then for every given positive integer N there exists $c = c(N)$ such that the number $a_n + c$ is prime for at least N values of n.

Proof. We shall use Corollary 4 to Theorem 3.5, which gives the existence of constants $0 < C_2 < C_1$ such that for $x \geq 2$ one has $C_2 < \pi(x) \log x / x < C_1$. If N is given then put $\lambda = 2/C_1$, $\mu = 2C_1/C_2$ and choose x sufficiently large, so that

$$A(x) > \lambda \mu N \log x$$

and

$$\frac{\log x}{\log(\mu x)} > \frac{3}{4}$$

holds. Moreover let $Z(x)$ be the set of all pairs (p, n) such that p is prime, $a_n \leq x$ and

$$0 < p - a_n \leq \mu x.$$

If a prime p lies in the interval $(x, \mu x]$ and $a_n \leq x$ then the pair (p, n) is counted by $Z(x)$ and hence one gets for the cardinality of $Z(x)$ the lower bound

$$A(x)(\pi(\mu x) - \pi(x)) \geq \lambda \mu N x \left(\frac{C_2 \mu \log x}{\log(\mu x)} - C_1 \right) \geq \frac{1}{2} C_1 \lambda \mu N x = \mu N x.$$

Since we have at most μx possibilities for the values of the difference $p - a_n$ it follows that some value, say c, must be attained at least N times. Thus $a_n + c$ is a prime for at least N values of n. □

Corollary. *If $f \in \mathbf{Z}[X]$ is non-constant with a positive leading coefficient and N is a given integer then with a suitable $c > 0$ the polynomial $f(X) + c$ represents at least N primes.*

Proof. If f increases for $x \geq N_0 \in \mathbf{Z}$ $(N_0 > 0)$ then it suffices to apply the theorem with $a_n = f(N_0 + n)$. □

As noted already in Chap.1 certain particular cases of this corollary were earlier established in Sierpiński (1964), Ageev (1994) and Garrison (1990).

Exercises

1. (Čebyšev 1850) Let a_n be a sequence of positive reals. Prove that the series

$$\sum_{n=2}^{\infty} \frac{a_n}{\log n}$$

converges if and only if the series

$$\sum_p a_p$$

converges.

2. Put

$$\theta_1(x) = \sum_{p^k \le x} \frac{1}{k}$$

and prove that for $N = 1, 2, \dots$ there exists a finite limit

$$\lim_{x \to 0+0} \sum_{n=2}^{\infty} (\theta_1(n+1) - \theta_1(n)) \frac{\log^N m}{n^x}.$$

3. (Gentile 1950) Let p_n denotes the n-th prime. Prove that there are at least $p_n - 1$ primes between p_n and $\prod_{k=1}^{n} p_k$.

4. (i) (Landau 1901a, Bonse 1907, Zeitz 1932) Use Bertrand's postulate to show that if $n \ge 4$ then

$$p_{n+1}^2 < p_1 p_2 \cdots p_n.$$

(ii) (Suzuki 1913) Show that for every m and sufficiently large n one has

$$p_{n+1}^m < p_1 p_2 \cdots p_n.$$

5. (i) (Šatunovskij[27] 1893, Laeng 1900, Wolfskehl[28] 1900,1901, Bonse 1907.) Deduce from the preceding exercise that 30 is the largest integer n with the property that every positive integer $< n$ and prime to n is a prime[29].

(ii) Prove that the largest integer n with the property that every positive integer $< n$ and prime to n is a prime power equals 42.

(iii) (Maillet 1900, Remak 1909, Hiroshi 1984) Let j be a fixed integer. Prove that there are only finitely many integers n with the property that every integer m satisfying $1 < m < n$ and prime to n has at most j distinct prime divisors.

---

[27] Šatunovskij, Samuil Osipovič (1859–1929), worked at Odessa University.
[28] Wolfskehl, Paul (1856–1906).
[29] A short proof of this assertion can be found in Landau (1901a).

6. (Santos 1976) Prove that there are only finitely many[30] integers n with the property that if $1 \le m \le n$ and $(m,n) = 1$, then $m+n$ is prime.

7. Put

$$H(n) = \sum_{k=1}^{n} \frac{1}{k}.$$

(i) Deduce from Bertrand's postulate that for $n \ge 2$ the numbers $H(n)$ and $\sum_{n=1}^{N} \mu(n)/n$ are not integers.

(ii) (Eswarathasan, Levine 1991) Write $H(n) = a(n)/b(n)$ with $a(n)$, $b(n)$ being relatively prime integers and for prime p denote by $J(p)$ the set of all integers n for which p divides $a(n)$. Prove that $J(2)$ is empty and the sets $J(p)$ are finite[31] for $p = 3, 5, 7$. Show also that for $p \ge 5$ the set $J(p)$ contains the numbers $p-1$, $p(p-1)$ and p^2-1.

8. (Costa Pereira 1985a, see also Cook 1991.) Show that for all $x > 0$ one has

$$\psi(x) - \theta(x) = \psi(x^{1/2}) + \psi(x^{1/3}) + R(x)$$

with

$$\psi(x^{1/7}) \le R(x) \le \psi(x^{1/5}),$$

hence $R(x) = O(x^{1/5})$.

9. (Landau 1908b) Prove that

$$\pi(x) - \frac{\theta(x)}{\log x} = O\left(\frac{x}{\log^2 x}\right)$$

and that this bound cannot be improved.

10. (Pintz 1980a) (i) Prove the equality

$$\int_2^x \frac{\psi(t)}{t^2} dt = \log x + O(\psi(x)/x) + O(1).$$

(ii) Deduce from (i) that if for some A, B one has

$$\pi(x) = \frac{Bx}{\log x - A + o(1)}$$

then $A = B = 1$.

11. (Shapiro 1950a) Let a_n be a sequence of nonnegative numbers and put

$$A(x) = \sum_{n \le x} \left[\frac{x}{n}\right] a_n.$$

Prove that if for x tending to infinity one has

$$A(x) = x \log x + O(x)$$

---

[30] H.G.Kopetzky and W.Schwarz (1979) proved that 12 is the maximal integer having this property.

[31] It is conjectured that $J(p)$ is finite for every prime p.

then there exist constants $0 < \alpha < \beta$ such that for all sufficiently large x one has

$$\alpha x \leq \sum_{n \leq x} a_n \leq \beta x.$$

12. Prove the equality

$$\int_0^\infty \left(\frac{1}{e^x - 1} - \frac{e^{-x}}{x} \right) dx = \gamma$$

used in the proof of Theorem 3.11.

13. Construct a quadratic polynomial which assumes more than 1000 distinct prime values.

14. (Abel,Siebert 1988) (i) Let a_n be a sequence of distinct positive integers and assume that

$$\limsup_{x \to \infty} \frac{\#\{n : x < a_n \leq 2x\}}{\log x} = \infty.$$

Prove that for every $N > 0$ there exists an integer $c = c(N)$ such that the sequence $a_n - c$ contains at least N primes.

(ii) Give an upper bound for the value of $c(N)$ in case $a_n = n^2$.

15. (Moree 1993) Show that 840 is the largest integer k with the property that the sum of reciprocals of primes $p < k$ not dividing k is smaller than 1.

16. (Kalecki[32] 1966/67) Prove that if N is an integer which is a product of r distinct primes then between N and $2N$ there are at least r integers which are products of r primes, not necessarily distinct.

17. (i) Let $a_1, a_2, \ldots$ be an arbitrary sequence and put

$$A(x,d) = \sum_{\substack{n \leq x \\ d \mid n}} a_n.$$

Prove the identity

$$\sum_{n \leq x} a_n \log n = \sum_{p^k \leq x} A(x, p^k) \log p.$$

(ii) Utilize (i) to obtain a direct proof of Theorem 3.8.

18. (Moser 1951) Prove that for $n \geq 91$ one has $\pi(n) > \varphi(n)$.

19. (Birch,Singmaster 1984) Deduce from Bertrand's postulate that if n has $r \geq 5$ distinct prime factors then $\pi(\sqrt{n}) \geq 2r$.

---

[32]Kalecki, Michał (1899–1970), Professor of economics in Warsaw.

4. Riemann's Zeta-function and Dirichlet Series

4.1. The Zeta-function; Riemann's Memoir

1. The function defined for $s > 1$ by

$$\zeta(s) = \sum_{n=1}^{\infty} \frac{1}{n^s} \tag{4.1}$$

and now called *Riemann's zeta-function* was first[1] considered seriously by L.Euler (1734/35,1740,1743,1748 Chap.15,1774) (see Stäckel[2] (1907/08)) who determined its value first[3] at $s = 2$ (for an analysis of Euler's arguments see McKinzie,Tuckey (1997)) and then at all even positive integers. He proved the following formula which expresses these values in terms of Bernoulli numbers:

$$\zeta(2n) = \frac{(-1)^{n-1} B_{2n}}{2(2n)!} (2\pi)^{2n}.$$

We recall that the *Bernoulli numbers* B_m are defined by the identity

$$\frac{z}{e^z - 1} = \sum_{n=0}^{\infty} \frac{B_n}{n!} z^n.$$

A proof in the cases $n = 1, 2, 3$ (see Fuss 1843, Eneström 1890, Weil[4] 1983, §§17–19) was also discovered by Johann Bernoulli[5]. For newer proofs of Euler's formula and related recurrence formulas see Apostol (1973a), Berndt (1975), Bhattacharjee, Bhattacharjee (1990), Bouwkamp (1933), Brika (1933), Carlitz (1961), Chen Ming Po (1975), Day (1952), Estermann (1947), Holme (1970), Kuo (1949), Papadimitriou (1973), E.L.Stark (1979), Stoica (1987), Titchmarsh (1926), Tsumura (1994), G.T.Williams (1953) and K.S.Williams (1971).

---

[1] For the early history of the theory of the zeta-function see Schuppener (1994).

[2] Stäckel, Paul (1862–1919), Professor in Berlin, Königsberg, Kiel, Hannover, Karlsruhe and Heidelberg.

[3] This result was highly appreciated at the time and in 1755 a monograph devoted to it was published (Meldercreutz 1755).

[4] Weil, André (1906–1998), Professor in Paris and Princeton.

[5] Bernoulli, Johann (1667–1748), brother of Jacob Bernoulli. Professor in Groningen (1695–1705) and then in Basel.

The simplest available proof is certainly that which appears in many text-books (e.g. in Serre (1970, Chap.7)) and consists in comparing the coefficients in the expansions

$$z\cot z = 1 + \sum_{k=1}^{\infty} (-1)^k \frac{4^k B_{2k}}{(2k)!} z^{2k}$$

and

$$z\cot z = 1 + 2\sum_{n=1}^{\infty} \frac{z^2}{z^2 - n^2\pi^2} = 1 - 2\sum_{n=1}^{\infty}\sum_{k=1}^{\infty} \frac{z^{2k}}{(\pi n)^{2k}} = 1 - 2\sum_{k=1}^{\infty} \zeta(2k)\frac{z^{2k}}{\pi^{2k}},$$

obtained by taking the logarithmic derivative of

$$\sin z = z\prod_{n=1}^{\infty}\left(1 - \frac{z^2}{\pi^2 n^2}\right).$$

The values of the zeta-function at odd positive integers still present a mystery. There are several formulas expressing them by various series and we quote only one nice looking formula of J.C.Kluyver (1896):

$$\zeta(3) = \frac{4\pi}{5}\left(\frac{1}{6} - \sum_{k=1}^{\infty} \frac{B_{2k}\pi^{2k}}{(2k+3)!}\right).$$

Only recently has it been established that $\zeta(3)$ is irrational (Apéry[6] 1979, Beukers 1979, Reyssat 1980) but the arithmetical properties of $\zeta(5), \zeta(7), \ldots$ are still unknown.

Euler (1737) also found the connection of $\zeta(s)$ with prime numbers given by the product formula (1.2) and used it to give a new proof for the existence of infinitely many primes. We presented it in Chap.1. In a later paper (Euler 1761) he considered the function

$$\Phi(s) = \sum_{n=1}^{\infty} \frac{(-1)^{n-1}}{n^s},$$

the series being convergent for all positive values of s. It is closely related to $\zeta(s)$ since for $s > 1$ one has

$$\zeta(s) + \Phi(s) = 2\sum_{n=1}^{\infty} \frac{1}{(2n-1)^s} = 2\left(\zeta(s) - \sum_{n=1}^{\infty} \frac{1}{(2n)^s}\right) = \left(2 - \frac{1}{2^{s-1}}\right)\zeta(s),$$

and thus

$$\Phi(s) = \left(1 - 2^{1-s}\right)\zeta(s).$$

Euler conjectured that $\Phi(s)$ satisfies the equality

$$\frac{\Phi(1-s)}{\Phi(s)} = -\Gamma(s)\frac{2^s - 1}{2^{s-1} - 1}\pi^{-s}\cos\left(\frac{\pi s}{2}\right), \tag{4.2}$$

for all real s which are not poles of $\Gamma(s)$ and he tried to prove it in the case when s is a positive integer. The left-hand side of (4.2) is well-defined only if $0 < s < 1$ but Euler freely utilized divergent series[7] extending the

[6] Apéry, Roger (1916–1994), Professor in Caen.
[7] See the discussion of Euler's usage of such series in the introduction to Hardy's book on divergent series (Hardy 1949) and Kline (1983).

usual definition of the sum of a series. He was well aware that such extension should be properly defined since he wrote (Euler 1761):

"... *il*[8] *faut donner au mot de somme une signification plus étendue et entendre par là une fraction ou autre expression analytique, laquelle étant développée selon les principes d'analyse produise la même série dont on cherche la somme*".

Also in a letter[9] of August 7th 1745 to Goldbach he affirmed his belief that this approach leads to a unique determination of the sum of any series. However, according to G.H.Hardy (1949,p.14), in the eighteenth century F.Callet had already produced the following example showing that this definition may lead to distinct values of so defined sum:

If $m < n$ are given positive integers, then

$$1 - x^m + x^n - x^{n+m} + \cdots + x^{kn} - x^{m+kn} + \cdots$$
$$= \frac{1 + x + \cdots + x^{m-1}}{1 + x + \cdots + x^{n-1}} = \frac{1 - x^m}{1 - x^n},$$

and for $x \to 1$ this tends to m/n hence using Euler's method one can obtain for the sum of the series $1 - 1 + 1 - 1 + \cdots$ any positive rational value.

Euler exposed his views on divergent series in a paper published in 1760 (Euler 1754). An analysis of its contents was given in Barbeau, Leah (1976) where one can also find a translation of the introduction to Euler's paper.

In his paper from the year 1761 Euler dealt with the series

$$\Phi(-t) = 1 - 2^t + 3^t - 4^t + \cdots$$

and

$$\Phi(u) = 1 - \frac{1}{2^u} + \frac{1}{3^u} - \frac{1}{4^u} + \cdots$$

defining their sum as the limit of functions given by the power series

$$x - 2^t x^2 + 3^t x^3 - 4^t x^4 + \cdots$$

resp.

$$x - \frac{x^2}{2^u} + \frac{x^3}{3^u} - \frac{x^4}{4^u} + \cdots$$

for x tending to 1, thus using in fact the Abelian method of summation. In this formulation Euler's assertion (4.2) was established by E. Landau (1906b).

Euler's research concerning the zeta-function is described in Ayoub (1974) and Weil (1983, Chap.3).

---

[8] "*... One has to give to the word sum a more extended meaning and understand by it a fraction or another analytic expression, which expanded according to the principles of analysis produces the same series, whose sum is sought.*"

[9] Fuss 1843, letter 83, 323–328.

For a deeper utilization of the formula (1.2) to the study of primes one had to wait until Dirichlet (1837c) who used the unboundeness of the zeta-function to the right of $s = 1$ in his proof of the infinitude of primes in arithmetic progressions (see Chap.2). In the course of that proof Dirichlet established the existence of a continuous derivative of $\zeta(s)$ to the right of $s = 1$ and showed that $(s-1)\zeta(s)$ tends to 1, when s tends to 1. Later Čebyšev (1848) calculated the derivatives of $\zeta(s)$ (for $s > 1$) of all orders in his proof of Theorem 3.1. They both considered $\zeta(s)$ only at real arguments.

The development of the theory of complex functions and in particular the invention of the analytic continuation of holomorphic functions by Karl Weierstrass[10] (1842)[11] and Bernhard Riemann[12] (1851) led to a very important breakthrough in the study of prime numbers. It was achieved by Riemann in his celebrated memoir[13] (Riemann 1860). He considered there for the first time the zeta-function at complex arguments and studied its properties. This paper[14] laid the way for future development of the prime number theory whose culmination in the XIXth century was the proof of the Prime Number Theorem by J.Hadamard (1896c) and C.de la Vallée-Poussin (1896a).

2. Now we turn to Riemann's paper of 1860. Wc do not enter into a detailed analysis of its contents since this has been done very carefully in the book of H.M.Edwards (1974) but restrict ourselves to certain highlights which are important for the theory of prime distribution. Riemann starts with an extension of the definition of the zeta-function to the whole plane (with exception of $s = 1$) which gives its continuation to a meromorphic function:

Theorem 4.1. *The function $\zeta(s)$ defined in the open half-plane* Re $(s) > 1$ *by the formula (4.1) can be continued to a meromorphic function in the complex plane and the difference $\zeta(s) - 1/(s-1)$ is an entire function.*

Proof. First of all observe that the series (4.1) converges absolutely and uniformly in every closed halfplane Re $s \geq 1 + \delta$ with $\delta > 0$. Hence it defines in the open half-plane Re $s > 1$ a regular function. To obtain its continuation Riemann starts with the identity

$$\int_0^\infty e^{-nx} x^{s-1} dx = \frac{\Gamma(s)}{n^s}, \tag{4.3}$$

[10] Weierstrass, Karl (1815–1897), Professor in Berlin.

[11] The paper of Weierstrass was published as late as in 1894 in the first volume of his collected papers but its results became known through Weierstrass's lectures at the Berlin University.

[12] Riemann, Bernhard (1826–1866), Professor in Göttingen.

[13] An English translation can be found in Edwards (1974).

[14] It is not clear how Riemann became interested in prime numbers. It was suggested by A.Weil (1989) that this may be due to his contacts with Gotthold Eisenstein (1823–1852), a student of Gauss.

valid for Re $s > 1$, which leads to

$$\Gamma(s)\zeta(s) = \int_0^\infty \frac{x^{s-1}\,dx}{e^x - 1}.$$

Then Riemann considers the integral

$$\int_C \frac{(-x)^{s-1}\,dx}{e^x - 1},$$

where C is described as a curve "*going from* $+\infty$ *to* $+\infty$ *around a region containing* 0 *but no other discontinuity of the integrand*" and states that it follows easily that this integral is for every complex $s \neq 1$ equal to

$$\left(e^{-\pi si} - e^{\pi si}\right) \int_0^\infty \frac{x^{s-1}\,dx}{e^x - 1}$$

and thus the formula

$$2\sin(\pi s)\Gamma(s)\zeta(s) = i \int_C \frac{(-x)^{s-1}\,dx}{e^x - 1} \tag{4.4}$$

results which can be utilized to define $\zeta(s)$ for every $s \neq 1$. □

Riemann did not bother about convergence questions but his idea is sound and its exact realization can be found e.g. in the fourth section of the second chapter of the treatise of Titchmarsh (1951). We present here a variant of his argument due[15] to Hermite (1885):

Proof of Hermite. Let $0 < \epsilon < 2\pi$ and write

$$\zeta(s) = \frac{1}{\Gamma(s)} \int_0^\infty \frac{x^{s-1}\,dx}{e^x - 1} = \frac{1}{\Gamma(s)} \left(F(s) + G(s)\right),$$

where $F(s)$ denotes the integral of $x^{s-1}/(e^x - 1)$ taken over the interval $[0, \epsilon]$ and $G(s)$ denotes the integral over $[\epsilon, \infty]$. The function $G(s)/\Gamma(s)$ is obviously entire and so it remains to deals with $F(s)$. In view of the equality

$$\frac{1}{e^x - 1} = \frac{1}{x} - \frac{1}{2} - \sum_{n=1}^{\infty} \frac{B_{2n}}{(2n)!} x^{2n-1},$$

valid for non-zero $x \in (-2\pi, 2\pi)$ one gets

$$F(s) = \int_0^\epsilon \frac{x^{s-1}\,dx}{e^x - 1} = \epsilon^{s-1} \left(\frac{1}{s-1} - \frac{\epsilon}{2s} - \sum_{n=1}^{\infty} \frac{B_{2n}}{n!} \frac{\epsilon^{2n}}{2n + s - 1} \right),$$

---

[15] See Baillaud,Bourget(1905),I,148–149, letter of Hermite to Stieltjes dated June 21st 1885.

and hence $F(s)$ is meromorphic with simple poles at 0 and negative odd integers. Since $\Gamma(s)$ has simple poles at all nonpositive integers it follows that $F(s)/\Gamma(s)$ is meromorphic with a unique pole at $s=1$ which is simple and has its residue equal to 1. Hence finally

$$\zeta(s) = \frac{1}{s-1} + H(s)$$

with $H(s)$ entire. □

Corollary. *For $n = 1,2,3\dots$ one has $\zeta(-2n) = 0$.*

Proof. Since $F(s) + G(s)$ is regular at negative even integers and $\Gamma(s)$ has poles there, the ratio $\zeta(s) = (F(s)+G(s))/\Gamma(s)$ must vanish at these points. □

Several other proofs of Theorem 4.1 are given in Titchmarsh (1951, Chap.2). See also Apostol (1954), Backlund (1918), Bohr[16](1909), Burlačenko (1968), Craig (1923), Jensen[17](1887), Prowse (1960), Ramaswami (1934), Sondow (1994), Srivastava (1987) and Wirtinger (1902).

E.Grosswald and F.J.Schnitzler (1978) constructed several functions defined for Re $s > 0$ by the formula

$$\zeta^*(s) = \prod \left(1 - \frac{1}{q_n^s}\right)$$

with q_n lying between the nth and $(n+1)$st prime. These functions have many properties in common with Riemann's zeta-function: they can be continued to the half-plane Re $s > 0$ with a simple pole at $s = 1$ and have in that half-plane the same zeros as $\zeta(s)$. However, all of them have the line Re $s = 0$ for their natural boundary thus the assertion of Theorem 4.1 fails for them.

3. One of the main results of Riemann's memoir is the functional equation of the zeta-function for which he gave two proofs. We state the functional equation in two forms, both of them appearing in Riemann's paper, although the second in a slightly disguised form.

Theorem 4.2. *(i) If we put for complex $s \neq 0,1$*

$$F(s) = \Gamma(\frac{s}{2})\pi^{-s/2}\zeta(s)$$

then

$$F(s) = F(1-s). \tag{4.5}$$

(ii) For $s \neq 0,1$ one has

[16] Bohr, Harald (1887–1951), Professor in Copenhagen.
[17] Jensen, Johann Ludwig William Valdemar (1859–1925), worked for a telephone company.

$$\zeta(1-s) = \frac{2}{(2\pi)^s} \cos\left(\frac{\pi s}{2}\right) \Gamma(s)\zeta(s). \tag{4.6}$$

Before presenting Riemann's proof we show first that the two forms of the functional equation are equivalent. We give the details only for the implication (i) $\Rightarrow$ (ii), since all steps in our argument will be reversible.

Applying the equality

$$\Gamma(w)\Gamma(w+\frac{1}{2}) = \frac{\sqrt{\pi}}{2^{2w-1}}\Gamma(2w)$$

for $w = 1/2 - s/2$ one gets from the equalities in (i)

$$\frac{\zeta(1-s)}{\zeta(s)} = \frac{1}{(2\pi)^s}\frac{\Gamma(s/2)\Gamma(1-s/2)}{\Gamma(1-s)},$$

and using the equality

$$\Gamma(w)\Gamma(1-w) = \frac{\pi}{\sin(\pi w)}$$

for $w = s/2$ and $w = s$ one obtains

$$\frac{\zeta(1-s)}{\zeta(s)} = \frac{\Gamma(s)}{(2\pi)^s}\frac{\sin(\pi s)}{\sin(\pi s/2)} = \frac{2\Gamma(s)\cos(\pi s/2)}{(2\pi)^s}.$$

First proof of Riemann. We repeat almost verbatim the original text:

If Re $s < 0$ then the integral on the right-hand side of (4.4) can be also taken in the reverse direction since the integrand vanishes at infinity. The integral is thus equal to the sum of integrals taken around the points $2k\pi i$ ($k = 0, \pm 1, \pm 2, \ldots$). These integrals are equal to $(-2k\pi i)^{s-1}(-2\pi i)$ and this leads to

$$\begin{aligned} 2\sin(\pi s)\Gamma(s)\zeta(s) &= (2\pi)^s \sum_{k=1}^{\infty} n^{s-1}\left((-i)^{s-1} + i^{s-1}\right) \\ &= (2\pi)^s\zeta(1-s)\left((-i)^{s-1} + i^{s-1}\right). \end{aligned}$$

This is clearly equivalent to the formula in (ii). □

Obviously something more should have been said here but the essential idea is correct. How the details could be filled out one can see in Edwards (1974) and Titchmarsh (1951).

Second proof of Riemann. It is based on a transformation formula[18] of the *theta-function* defined for positive x by

[18] A.F.Lavrik (1990a,b,1991) has shown that conversely this formula can be deduced from the functional equation (4.6).

$$\Omega(x) = \sum_{n=-\infty}^{\infty} e^{-n^2\pi x},$$

which appears already in Poisson[19] (1823).

Lemma 4.3. *For all positive x one has*

$$\Omega(x) = \frac{1}{\sqrt{x}}\Omega\left(\frac{1}{x}\right). \tag{4.7}$$

Proof. We sketch here a short proof due to G.Landsberg[20] (1893) which is also presented with all details in Landau (1909a).

Let l_+, l_- denote the horizontal lines in the complex plane passing through i resp. $-i$. A standard application of Cauchy's integral theorem gives

$$\Omega(x) = \int_{l_-} \frac{e^{-\pi x s^2} ds}{e^{2\pi i s} - 1} - \int_{l_+} \frac{e^{-\pi x s^2} ds}{e^{2\pi i s} - 1},$$

and in view of

$$\frac{e^{-\pi x s^2}}{e^{2\pi i s} - 1} = \sum_{n=1}^{\infty} e^{-\pi x s^2 - 2n\pi i s} = -\sum_{n=0}^{\infty} e^{-\pi x s^2 + 2n\pi i s}$$

one gets

$$\Omega(x) = \sum_{n=-\infty}^{-1} \int_{l_-} e^{-\pi x s^2 + 2\pi n i s} ds + \sum_{n=0}^{\infty} \int_{l_+} e^{-\pi x s^2 + 2\pi n i s} ds$$

$$= \sum_{n=-\infty}^{-1} e^{-\pi n^2/x} \int_{l_-} e^{-\pi x (s - ni/x)^2} ds + \sum_{n=0}^{\infty} e^{-\pi n^2/x} \int_{l_+} e^{-\pi x (s - ni/x)^2} ds.$$

For any fixed n the substitution $s = w + ni/x$ transforms the integrals occuring at the end of this formula into the integral

$$\int e^{-\pi x w^2} dw$$

taken along the lines $\operatorname{Im} w = 1 \mp \frac{n}{x}$ and applying again Cauchy's formula one sees that both integrals are equal to

$$\int_{-\infty}^{\infty} e^{-\pi x w^2} dw = \frac{1}{\sqrt{\pi x}} \int_{\infty}^{\infty} e^{-t^2} dt = \frac{1}{\sqrt{x}}.$$

---

[19] Poisson, Siméon Denis (1781–1840), Professor in Paris.
[20] Landsberg, Georg (1865–1912), Professor in Heidelberg, Breslau and Kiel.

This immediately implies the assertion. □

Another proof of this lemma was given in de la Vallée-Poussin (1896a,II):

Proof of Vallée-Poussin. Consider the function

$$\Omega(\alpha, x) = \sum_{n=-\infty}^{\infty} e^{-(n+\alpha)^2\pi x}$$

defined for positive x and real α. Its defining series is absolutely convergent and since obviously $\Omega(\alpha+1, x) = \Omega(\alpha, x)$ and $\Omega(\alpha, x)$ is even as a function of α, we may expand it for fixed x into a Fourier series

$$\Omega(\alpha, x) = \frac{a_0}{2} + \sum_{k=1}^{\infty} a_k(x)\cos(2k\pi\alpha)$$

with

$$a_k(x) = 2\int_0^1 \Omega(\alpha, x)\cos(2k\pi\alpha)d\alpha.$$

A short computation leads to

$$a_0(x) = 2\int_0^{\infty} \exp(-\alpha^2\pi x)d\alpha = \frac{2}{\sqrt{x}}$$

and

$$a_k(x) = 2\int_{-\infty}^{\infty} \exp(-\alpha^2\pi x)\cos(2k\pi\alpha)d\alpha = \frac{2}{\sqrt{x}}\exp(-k^2\pi/x).$$

This implies

$$\Omega(\alpha, x) = \frac{1}{\sqrt{x}} \sum_{k=-\infty}^{\infty} \exp(-k^2\pi/x)\cos(2k\pi\alpha)$$

and putting here $\alpha = 0$ we obtain the assertion. □

C.G.J.Jacobi (1828) furnished another proof of this lemma and E.Landau (1908b) ascribed this result to Cauchy[21]. For a modern exposition of the theory of theta-functions see the book of D.Mumford (1983), which contains in Sect.1.7 a proof of a generalization of Lemma 4.3.

Corollary. *If for $x > 0$ we put*

$$f(x) = \sum_{n=1}^{\infty} e^{-n^2\pi x}$$

[21] Cauchy, Augustin (1789–1857), Professor in Paris.

then

$$1+2f(x)=x^{-1/2}(1+2f(1/x)).$$

Proof. It suffices to note that $1+2f(x)=\Omega(x)$ and apply (4.7). □

Riemann uses now the equality

$$\frac{1}{n^s}\Gamma(\frac{s}{2})\pi^{-s/2}=\int_0^\infty e^{-n^2\pi x}x^{s/2-1}dx$$

(which results from (4.3) by the substitution $x\mapsto \pi n x$) to obtain

$$F(s)=\int_0^\infty f(x)x^{s/2-1}dx,$$

and applies the last corollary to obtain

$$\begin{aligned} F(s)&=\int_1^\infty f(x)x^{s/2-1}dx+\int_0^1 f(x^{-1})x^{(s-3)/2}dx\\ &\quad+\frac{1}{2}\int_0^1\left(x^{(s-3)/2}-x^{s/2-1}\right)dx \qquad (4.8)\\ &=\frac{1}{s(s-1)}+\int_1^\infty f(x)\left(x^{s/2-1}+x^{-(1+s)/2}\right)dx.\end{aligned}$$

Riemann's argument ends here. Note however that the correctness of this computation can be directly verified only for real $s>1$ and in the general case some further steps are needed. First of all note that for all positive T the integral

$$\int_1^T f(x)\left(x^{s/2-1}+x^{-(1+s)/2}\right)dx$$

represents an entire function and since for T tending to infinity it converges uniformly, its limit is also entire. Hence (4.8) gives a continuation of $\zeta(s)$ to a function meromorphic in the plane and the replacement of s by $1-s$ within it leads to the functional equation. □

If one puts[22]

$$\begin{aligned}\xi(s)&=\Gamma(1+\frac{s}{2})(s-1)\pi^{-s/2}\zeta(s)\\ &=\frac{1}{2}s(s-1)\Gamma(\frac{s}{2})\pi^{-s/2}\zeta(s)=\frac{1}{2}s(s-1)F(s) \qquad (4.9)\end{aligned}$$

and $\Xi(s)=\xi(1/2+is)$, then a short computation using (4.8) leads to the formula

[22] We adhere here to modern notation. Riemann himself denoted by $\xi(s)$ the function $\xi(1/2+is)$ which is now usually denoted by $\Xi(s)$.

$$\Xi(s) = \frac{1}{2} - (s^2 + \frac{1}{4}) \int_1^\infty f(t) t^{-3/4} \cos(\frac{s}{2} \log t) dt. \tag{4.10}$$

This formula was later used by J.Hadamard (1893) to obtain the order of growth of the sequence of non-real roots of the zeta-function. (See Theorem 5.1.)

As the Γ-function never vanishes it is clear that the zeros of $\zeta(s)$ and $\xi(s)$ coincide. Moreover there is a one-to-one correspondence between zeros α of $\Xi(s)$ and the non-real zeros ρ of $\zeta(s)$ given by $\alpha \leftrightarrow \rho = 1/2 + i\alpha$. Note that the functional equation implies that the only real zeros of the zeta-function on the negative half-line are exactly the poles $-2, -4, \ldots, -2n, \ldots$ of the Γ-function.

Observe that the second assertion of Theorem 4.2 (formula (4.6)) is easily shown to be equivalent to Euler's assertion (4.2). Note also that the assertion made by several authors (Bachmann 1894, Cahen 1894, Torelli 1901) that a proof of (4.2) was given already by O.Schlömilch (1849,1858) is incorrect, as pointed out by E.Landau (1906b). Both papers of Schlömilch actually deal with the series

$$\Psi(s) = \sum_{n=0}^{\infty} \frac{(-1)^n}{(2n+1)^s},$$

convergent for positive values of s. Its sum coincides with the L-function corresponding to the unique non-principal character mod 4. Euler (1761) stated that $\Psi(s)$ satisfies the identity

$$\frac{\Psi(1-s)}{\Psi(s)} = \frac{1}{\pi} \Gamma(s) 2^s \sin(\pi s/2)$$

and Schlömilch repeated this assertion in his first paper (in the section "Uebungsaufgaben[23] für Schüler" of *Archiv f. Mathematik und Physik*) and in the second presented a proof for $s \in (0,1)$ where both series $\Psi(s)$ and $\Psi(1-s)$ converge. Other proofs were sketched in Malmstén (1849), Lipschitz (1857), Clausen[24](1858) and Schlömilch (1878).

4. There exist several proofs of Theorem 4.2. The first to give a proof different from Riemann's was O.Schlömilch (1878). He considered for $t \in (0,1)$ the integral

$$U(t) = \int_0^\infty \left(\frac{1}{2x} - \frac{1}{e^x - e^{-x}} \right) x^{t-1} dx,$$

and computing it in two different ways obtained the formulas

$$U(t) = \frac{1}{2^{1-t} - 1} \left(1 - \frac{1}{2^t} \right) \Gamma(t) \Phi(t),$$

and

$$U(t) = \frac{\pi^t}{2 \cos(t\pi/2)} \Phi(1-t),$$

[23] "Exercises for students" .

[24] Clausen, Thomas (1801–1885), Professor in Dorpat.

where
$$\Phi(t) = \sum_{n=1}^{\infty} \frac{(-1)^{n-1}}{n^t}.$$

This leads to Euler's equation (4.2) from which the functional equation of $\zeta(s)$ results, as noted already at the end of the preceding subsection.

A variant of this argument appears as the fifth proof of (4.5) in Titchmarsh (1951). This book contains altogether nine proofs of the functional equation for $\zeta(s)$ including a proof due to G.H.Hardy (1922a,b) and an unpublished proof of Riemann completed by C.L.Siegel[25](1932). Further proofs may be found in Chakravarty (1974), Denjoy[26](1954), Hamburger[27](1921), Hardy (1922b), Lerch[28] (1892a) (reproduced in Landau 1909a, §72), van der Lune (1979), Mordell (1928,1929), Rademacher[29](1930), Rooney (1994) and Schnee (1930). J.Tate (1950) in his thesis proved functional equations for a large class of functions including Riemann's $\zeta(s)$ and also covering Dirichlet's L-functions and their various generalizations. His proof appears in Lang (1964) and Narkiewicz (1990). The theorem of Hamburger (Hamburger 1921a,c,1922a) asserts that the functional equation characterizes (up to a constant factor) Riemann's zeta-function among meromorphic functions $f(s)$ having only finitely many poles which are developable in an absolutely convergent Dirichlet series in the half-plane to the right of 1 and which for large $|s|$ satisfy
$$|f(s)| \leq \exp(|s|^c)$$
with a certain constant c. The first proof of this was later simplified in Hamburger (1922b) and Siegel (1922). See Hecke (1944), Kaczorowski, Perelli (1999) and Knopp[30](1994). Similar characterization of L-functions were given by H.Hamburger (1921b,1922a III), T.M.Apostol (1972,1975) and T.Funakura (1996). The problem of the determination of Dirichlet series by their functional equations turned out to be closely related to the theory of modular forms, as demonstrated by E.Hecke (1936). For generalizations see Bochner[31], Chandrasekharan (1956), Chandrasekharan, Mandelbrojt (1957,1959) and Ryavec (1973).

5. Riemann's memoir, besides the proofs of Theorems 4.1 and 4.2, does not contain any definitively proved results. Instead there are heuristic arguments leading to several assertions concerning the zeta-function and prime numbers. These were to occupy mathematicians for years to come. Note however that contrary to common belief Riemann did not make a definite assertion that all zeros of $\zeta(s)$ in the strip $0 < \text{Re } s \leq 1$ lie on the line $\text{Re } s = 1/2$, but stated only that this is very likely. He considered actually zeros of the function $\Xi(s) = \xi(1/2 + is)$ with ξ defined by (4.9). That all the roots of

[25]Siegel, Carl Ludwig (1896–1981), Professor in Frankfurt, Göttingen and Princeton.
[26]Denjoy, Arnaud (1884–1974), Professor in Utrecht and Paris.
[27]Hamburger, Hans Ludwig (1889–1956), Professor in Köln.
[28]Lerch, Matiaš (1860–1922), Professor in Prague and Brno.
[29]Rademacher, Hans(1892–1969), Professor in Hamburg, Breslau and at the University of Pennsylvania.
[30]Knopp, Konrad Hermann Theodor (1882–1957), Professor in Königsberg and Tübingen.
[31]Bochner, Salomon (1899–1982), Professor at Princeton University and Rice University.

$\Xi(s)$ are real is equivalent to the statement that all non-real roots of $\zeta(s)$ satisfy Re $s = 1/2$, the assertion known now as *Riemann's Hypothesis* (RH). Riemann wrote:

"... *es*[32] *ist sehr wahrscheinlich, dass alle Wurzeln reell sind. Hiervor wäre allerdings ein strenger Beweis zu wünschen; ich habe indess die Aufsuchung desselben nach einigen flüchtigen Versuchen vorläufig bei Seite gelassen, da er für den nächsten Zweck meiner Untersuchung entbehrlich schien.*"

This problem is still open although several claims for its solution can be found in the literature. They could not however withstand a closer scrutiny. For comments on some of them the reader may consult Grandjot,Landau (1924).

6. Apart from the statement given above Riemann's memoir contains also the following assertions:

I. *The number* $N(T)$ *of roots* α *of* $\Xi(t)$ *satisfying* $0 \le \operatorname{Re} \alpha \le T$ *equals asymptotically*

$$\frac{T}{2\pi} \log\left(\frac{T}{2\pi}\right) - \frac{T}{2\pi}$$

thus the same asymptotics holds for the number of non-real roots ρ *of* $\zeta(s)$ *satisfying* $|\operatorname{Im} \rho| \le T$.

The next statement was originally formulated in terms of roots of $\Xi(t)$ and contained a printing error in the last term, pointed out by A.Genocchi (1860).

II. (Riemann's explicit formula) *Define*

$$f(x) = \frac{1}{2k} + \sum_{n=1}^{\infty} \frac{1}{n}\pi(x^{1/n})$$

if $x = p^k$ *with prime* p *and*

$$\sum_{n=1}^{\infty} \frac{1}{n}\pi(x^{1/n})$$

if this is not the case.

If ρ *runs over all non-real roots of* $\zeta(s)$ *then*

$$f(x) = \operatorname{li}(x) - \sum_{\rho} \operatorname{li}(x^{\rho}) + \int_x^{\infty} \frac{dx}{(x^2-1)x\log x} - \log 2.$$

---

32 "... *it is very probable that all roots are real. One would like to have an exact proof of that; however after few hasty trials I put aside the search for it for the time being since this did not seem to be necessary for the current purposes of my research.*"

The integral logarithm $\mathrm{li}(x^\rho)$ occuring here is defined for $x > 1$ and real α, β by

$$\mathrm{li}(x^{\alpha+\beta i}) = \int_{-\infty+\beta i \log x}^{(\alpha+\beta i)\log x} \frac{e^z\,dz}{z+\epsilon\pi i},$$

with

$$\epsilon = \epsilon(\beta) = \begin{cases} 1 & \text{if } \beta > 0 \\ -1 & \text{otherwise.} \end{cases}$$

III. *One has*

$$\pi(x) = \sum_{n=1}^{\infty} \mu(n) \frac{\mathrm{li}(x^{1/n})}{n}.$$

From the three assertions listed above the first two were proved by H.von Mangoldt[33](1895). He established (I) with the error term $O(\log^2 T)$, giving a sketch of the argument in von Mangoldt (1894), and later improved the error term in (I) to $O(\log T)$ (von Mangoldt 1905). We shall present his proofs of I and II in Chap.5 (see Theorems 5.2, 5.3 and 5.4). Simpler proofs of both assertions were found by E.Landau (1908g,h) and we shall show them in Chap.6.

Although the series in the statement III converges the stated formula is not correct. H.L.Montgomery showed me the following way to establish the convergence which is based on the convergence of the series $\sum_{n=1}^{\infty} \mu(n)/n$ and $\sum_{n=1}^{\infty} \mu(n)\log n/n$ established by von Mangoldt (1897) and E.Landau (1899b), respectively.

First of all observe that if $0 < \epsilon < e-1$ then with a certain constant c one has

$$\mathrm{li}(1+\epsilon) = \log\epsilon + c + O(\epsilon).$$

If thus $j > \log x$ then $1 < x^{1/j} < e$ hence

$$\mathrm{li}(x^{1/j}) = \log\log x - \log j + c + O\left(\frac{\log x}{j}\right).$$

This shows that

$$\sum_{j>\log x} \mu(j)\frac{\mathrm{li}(x^{1/j})}{j}$$

$$= \sum_{j>\log x} \mu(j)\frac{\log\log x - \log j}{j} + c \sum_{j>\log x} \frac{\mu(n)}{n} + O\left(\log x \sum_{j>\log x} \frac{1}{j^2}\right)$$

and the convergence of the first two series is a consequence of the quoted results of von Mangoldt and E.Landau. One sees now easily that

---

[33]Mangoldt, Hans von (1854–1925), Professor in Hannover (1884–1886), Aachen (1886–1904) and Danzig.

$$\sum_{j>\log x} \mu(j)\frac{\mathrm{li}(x^{1/j})}{j} = o(\log x),$$

and so Riemann's formula, if true, would imply

$$\pi(x) - \mathrm{li}(x) = \sum_{2\le j\le \log x} \mu(j)\frac{\mathrm{li}(x^{1/j})}{j} + o(\log x),$$

but the first summand here equals

$$-\frac{1}{2}\mathrm{li}(\sqrt{x}) + O(\mathrm{li}(x^{1/3})\log x) = -\frac{1}{2}\mathrm{li}(\sqrt{x}) + O(x^{1/3})$$

and thus the difference $\pi(x) - \mathrm{li}(x)$ would be negative for all sufficiently large x. However J.E.Littlewood (1914) proved that this is not true. We shall present his result in Chap.6.

For a numerical comparison of the right-hand side of the formula in (III) with $\pi(x)$ see Riesel,Göhl (1970).

7. Unpublished notes of Riemann concerning the zeta-function were discovered around 1930 and C.L.Siegel (1932) presented an analysis of them. They contain essentially two results: the first is a kind of approximate functional equation for the zeta-function, which is a special case of the equation discovered later by G.H.Hardy and J.E.Littlewood (1922,1929) in the form

$$\zeta(s) = \sum_{n\le x}\frac{1}{n^s} + \frac{(2\pi)^s}{2\Gamma(s)\cos(\pi s/2)}\sum_{n\le y}\frac{1}{n^{1-s}} + R(n,s)$$

with $2\pi xy = |\mathrm{Im}\ s|$ sufficiently large, $x, y \ge 1$ and

$$R(x,y,s) = O(x^{-\sigma}) + O\left(\frac{(2\pi)^\sigma}{2|\Gamma(s)\cos(\pi s/2)|}y^{\sigma-1}\right).$$

Riemann's formula dealt only with the special case $x = y$ but the error term $R(x, s)$ was described with greater precision.

The second result concealed in Riemann's notes is an integral formula for $\zeta(s)$ which, as we have already mentioned, was used by Siegel to give a quick proof of Theorem 4.2.

4.2. Dirichlet's L-functions in the Complex Plane

1. Although Dirichlet defined his L-functions only for real arguments one realizes immediately that his definition extends also to complex arguments lying in the half-plane Re $s > 0$ with the exception of the case when χ is the principal character. Moreover Lemma 2.2 shows that the product formula

$$L(s,\chi) = \prod_p \left(1 - \frac{\chi(p)}{p^s}\right)^{-1} \tag{4.11}$$

(of which (2.1) is a special case) holds in the half-plane Re $s > 1$. It did not take long to observe that Theorems 4.1 and 4.2 have their analogues for Dirichlet L-functions. According to Landau (1908g,1909a,p.899) the first proof of the functional equation relating $L(s,\chi)$ to $L(1-s,\bar{\chi})$ was given by H.Kinkelin (1861/62). Kinkelin's paper gave full details only in the case when χ was a character mod k with k either a prime power or an odd square-free number. The case of a quadratic character was treated[34] by A.Hurwitz (1882) and the general case by R.Lipschitz (1889). Hurwitz based his proof on his discovery that the functions

$$F_{k,l}(s) = \sum_{n \equiv l \pmod{k}} \frac{1}{n^s}$$

(defined for Re $s > 1$) have meromorphic continuations with a unique pole at $s = 1$. We present now his argument:

Theorem 4.4. *If $k \geq 2$ and $0 \leq l \leq k-1$ then the function $F_{k,l}(s)$ can be continued to a meromorphic function having a single pole at $s = 1$ which is simple and has residue $1/k$.*

Sketch of Hurwitz's proof: The starting point is the equality

$$\Gamma(s)F_{k,l}(s) = \int_0^\infty \frac{e^{(k-l)x}}{e^{kx}-1} x^{s-1}\,dx,$$

valid for Re $s > 1$ which is a consequence of (4.3). Following the idea of Riemann's first proof of the functional equation for $\zeta(s)$ Hurwitz considers the integral

$$I_a(s) = \int_C e^{(s-1)\log(-x)} \frac{e^{(k-l)x}}{e^{kx}-1}\,dx$$

where C denotes the path taken first from $+\infty$ along the upper part of the real axis up to a point $a > 0$, then along the circle T_a around 0 with radius a and then back to $+\infty$ along the lower part of the real axis. Here a has to be

[34]This case was also considered by T.J.Stieltjes (See his letter to Hermite, dated September 1885; Baillaud,Bourget 1905, letter 84).

sufficiently close to 0 so that inside T_a the integrand has its only singularity at 0. Moreover $\log(-x)$ is considered to be positive for $x < 0$ thus its imaginary part equals $-\pi i$ for x on the positive part of the real axis and πi on the negative part. This leads to the equality

$$I_a(s) = -2i\sin(\pi s)\int_a^\infty x^{s-1}\frac{e^{(k-l)x}}{e^{kx}-1}dx + \int_{T_a} e^{(s-1)\log(-x)}\frac{e^{(k-l)x}}{e^{kx}-1}dx,$$

and if one notes that for Re $s > 1$ the second term tends to zero as $a \to 0$ then one gets

$$\lim_{a\to 0} I_a(s) = -2i\sin(\pi s)\int_0^\infty x^{s-1}\frac{e^{(k-l)x}}{e^{kx}-1}dx = -2i\sin(\pi s)\Gamma(s)F_{k,l}(s).$$

In view of $\pi/\sin(\pi s) = \Gamma(s)\Gamma(1-s)$ one arrives finally at the equality

$$F_{k,l}(s) = \frac{i}{2\pi}\Gamma(1-s)\lim_{a\to 0} I_a(s) \tag{4.12}$$

valid in the half-plane Re $s > 1$.

Now Hurwitz observes that $\lim_{a\to 0} I_a(s)$ exists and is finite for all s thus the singularities of the left-hand side of (4.12) lie at the poles of $\Gamma(1-s)$, i.e. at nonnegative integers. As for $s = 2, 3, 4, \ldots$ one has

$$\lim_{a\to 0} I_a(s) = 0$$

and

$$\lim_{a\to 0} I_a(1) = \frac{2\pi i}{k}$$

hence (4.12) gives the needed continuation and the remaining assertions follow immediately. □

Corollary 1. *All Dirichlet's L-functions which correspond to non-principal characters can be continued to entire functions. If χ_0 is the principal character* mod k *then $L(s,\chi_0)$ can be continued to a meromorphic function with a single pole at $s = 1$. This pole is simple with residue equal to $\varphi(k)/k$.*

Proof. If χ is a character mod k and Re $s > 1$ then

$$L(s,\chi) = \sum_{\substack{(k,l)=1\\ 1\le l\le k}} \chi(l)F_{k,l}(s). \tag{4.13}$$

This shows that for non-principal χ the assertion follows from the theorem and the equality

$$\sum_{\substack{(k,l)=1\\ 1\le l\le k}} \chi(l) = 0.$$

For the principal character χ_0 use again (4.13) and

$$\sum_{\substack{(k,l)=1 \\ 1\le l\le k}} \chi_0(l) = \varphi(k). \qquad \square$$

In the case of principal character one can also obtain the assertion directly from Theorem 4.1 since for Re $s > 1$ one has

$$L(s,\chi_0) = \prod_{p|k}\left(1-\frac{1}{p^s}\right)\zeta(s).$$

Corollary 2. *In the half-plane* Re $s < 0$ *one has*

$$F_{k,l}(s) = \frac{\Gamma(1-s)}{\pi}\left(\frac{2\pi}{k}\right)^s \sum_{n=1}^{\infty} n^{s-1}\cos\left(\frac{2nl\pi}{k}+\frac{(s-1)\pi}{2}\right).$$

Proof. The proof is based on the consideration of the function

$$\psi(z) = (-z)^{s-1}\frac{e^{(k-l)z}}{e^{kz}-1},$$

which appeared already in the proof of Theorem 4.4. Hurwitz considered it in a rectangle Λ with sides parallel to the axes and centered at 0 from which one deletes a region bounded by a loop encompassing 0 and starting at that point of Λ which lies on the positive part of the real line. Integrating over the boundary of that region, taking in account the poles at $z = 2r\pi i/k$ for $r = \pm 1, \pm 2, \ldots$ and finally extending Λ to infinity one gets the assertion. $\square$

Corollary 3. *In the half-plane* Re $s > 1$ *the following equality holds:*

$$F_{k,l}(1-s) = \frac{\Gamma(s)}{\pi}\left(\frac{2\pi}{k}\right)^{1-s}\sum_{r=1}^{k}\cos\left(\frac{2lr\pi}{k}-\frac{s\pi}{2}\right)F_{k,r}(s).$$

Proof. If Re $s < 0$ then in view of

$$\begin{aligned}
&\sum_{n=1}^{\infty} n^{s-1}\cos\left(\frac{2nl\pi}{k}+\frac{(s-1)\pi}{2}\right)\\
&=\sum_{r=1}^{k}\sum_{t=0}^{\infty}(kt+r)^{1-s}\cos\left(\frac{2rl\pi}{k}+\frac{(s-1)\pi}{2}\right)\\
&=\sum_{r=1}^{k}\cos\left(\frac{2rl\pi}{k}+\frac{(s-1)\pi}{2}\right)F_{k,r}(1-s)
\end{aligned}$$

it suffices to apply the preceding corollary and replace s by $1-s$. $\square$

If for $0 < u \le 1$ and Re $s > 1$ we put

$$\zeta(s,u) = \sum_{n=0}^{\infty} \frac{1}{(n+u)^s} \tag{4.14}$$

then clearly

$$F_{k,l}(s) = \frac{1}{k^s}\zeta(s, l/k), \tag{4.15}$$

and Theorem 4.4 and its Corollary 2 lead to the following assertion:

Theorem 4.5. *The function $\zeta(s,u)$ defined for $0 \le u < 1$ and* Re $s > 1$ *by (4.14) can be continued to a meromorphic function having a single pole at $s = 1$ which is simple with residue 1. Moreover in the half-plane* Re $s < 0$ *one has*

$$\zeta(s,u) = 2^s \pi^{s-1} \Gamma(1-s) \sum_{n=1}^{\infty} n^{s-1} \cos(2nu\pi + s\pi/2). \tag{4.16}$$

Proof. For rational u the assertions result from Theorem 4.4, its Corollary 2 and (4.15). In the general case it suffices to observe that $\zeta(s,u)$ is for fixed $s \ne 1$ continuous in u. □

The function $\zeta(s,u)$ is called the *Hurwitz zeta-function*, although Hurwitz himself considered only the case of rational u. For an early study of $\zeta(s,u)$ see Mellin[35] (1899).

For other proofs of Hurwitz's formula (4.16) see Apostol (1951a), Mellin (1899), M.Mikolás (1957), Mordell (1929) and Oberhettinger (1956). For proofs of the continuability of $\zeta(s,u)$ see Balasubramanian, Ramachandra (1987) and Fine (1951). The coefficients of the Laurent expansion of $\zeta(s,u)$ at $s = 1$ were, in case $0 < u \le 1$, given in Berndt (1972a). See also K.Verma (1991).

2. After obtaining the formula (4.16) Hurwitz proceeded to the proof of the functional equation for L-functions with real characters. For a square-free integer $D \not\equiv 1 \pmod 4$ he considered the function

$$F(s,D) = \frac{1}{1 - 2^{-s}\chi_8(D)} \sum_{(n,2D)=1} \left(\frac{D}{n}\right) \frac{1}{n^s}, \tag{4.17}$$

with χ_8 being the character mod 8 defined by

$$\chi_8(D) = \begin{cases} 1 & \text{if } D \equiv \pm 1 \pmod 8 \\ -1 & \text{if } D \equiv \pm 3 \pmod 8 \\ 0 & \text{if } 2|D \end{cases}$$

[35]Mellin, Robert Hjalmar (1854–1933), Professor in Helsinki, one of the founders of the Finnish Academy of Sciences.

and for $D \equiv 1 \pmod 4$ he put

$$F(s,D) = \sum_{(n,2D)=1} \left(\frac{D}{n}\right) \frac{1}{n^s}. \tag{4.18}$$

(Here $\left(\frac{a}{b}\right)$ denotes the Jacobi symbol.)

It is not difficult to realize, using the quadratic reciprocity law, that in the first case $F(s,D)$ coincides with the L-function corresponding to the character $\chi \bmod D$ defined by $\chi(n) = \left(\frac{n}{|D|}\right)$ whereas in the second case $F(s,D)$ is equal to the L-function corresponding to a quadratic character mod $4|D|$ (if $D \equiv 3 \pmod 4$) resp. mod $8|D|$ (if D is even).

Hurwitz stated his result in the following form:

Theorem 4.6. *Let $D > 1$ be a square-free integer and let the functions $F(s,D)$ be defined by (4.17) resp. (4.18). If*

$$\kappa = \begin{cases} 1 & \textit{if } D \equiv 1 \pmod 4 \\ 4 & \textit{otherwise,} \end{cases}$$

then for all complex s one has

$$\Gamma\left(\frac{1-s}{2}\right) F(1-s,D) = \left(\frac{\kappa D}{\pi}\right)^{s-1/2} \Gamma\left(\frac{s}{2}\right) F(s,D). \tag{4.19}$$

Proof of Hurwitz. The proof is based on the following explicit formula for Gaussian sums of quadratic characters, proved by Gauss (1811), which we have already encountered, in the special case of prime D, in Dirichlet's proof of the non-vanishing of $L(1,\chi)$ (see (2.7)). We state it without proof:

Lemma 4.7. *Let D be a positive integer of the form $2^a D_1$ with $a = 0,2,3$ and D_1 square-free and odd. Let also χ be a real non-principal primitive character mod D (which means that for no proper divisor d of D one has $\chi(n) = 1$ for every n satisfying $n \equiv 1 \pmod d$ and $(n,D) = 1$). If $\zeta_D = \exp(2\pi i/D)$ and we put*

$$\tau(\chi) = \sum_{m=1}^{D} \chi(m)\zeta_D^m$$

then

$$\tau(\chi) = \begin{cases} D^{1/2} & \textit{if } \chi(-1) = 1 \\ iD^{1/2} & \textit{if } \chi(-1) = -1. \end{cases}$$

The main difficulty in the proof lies in the case of prime D, the general assertion being easily deduced from this particular case using quadratic reciprocity and the elementary result that every character mod D can be represented as a product of characters mod p^k with prime p (for details see e.g. Hasse (1950), Narkiewicz (1990, Chap.5) or Borevič, Šafarevič (1964)).

In the case of $D = p$, a prime, there are several proofs of this lemma available (see Berndt,Evans (1981) for a survey). It seems that the simplest proof is that found by W.C.Waterhouse (1970) which is reproduced in Narkiewicz (1990).

Hurwitz's proof of the functional equation is divided into four cases, corresponding to the residue of D mod 8, however his idea becomes more transparent if we consider an arbitrary L-function with a real non-principal character χ mod D satisfying the assumptions of the preceding lemma. The L-function corresponding to χ can be written in the form

$$L(s,\chi) = \sum_{l=1}^{D} \chi(l) F_{D,l}(s),$$

hence applying Corollary 3 to Theorem 4.4 we get in the half-plane Re $s > 1$

$$\begin{aligned} L(1-s,\chi) &= \sum_{l=1}^{D} \chi(l) F_{D,l}(1-s) \\ &= \frac{\Gamma(s)}{\pi}\left(\frac{2\pi}{D}\right)^{1-s} \sum_{r=1}^{D} F_{D,r}(s) \sum_{l=1}^{D} \chi(l) \cos\left(\frac{2rl\pi}{D} - \frac{s\pi}{2}\right) \\ &= \frac{\Gamma(s)}{\pi}\left(\frac{2\pi}{D}\right)^{1-s} \left(\cos(s\pi/2) \sum_{r=1}^{D} F_{D,r}(s) \sum_{l=1}^{D} \chi(l) \cos\left(\frac{2rl\pi}{D}\right)\right) \\ &\quad + \sin(s\pi/2) \sum_{r=1}^{D} F_{D,r}(s) \sum_{l=1}^{D} \chi(l) \sin\left(\frac{2rl\pi}{D}\right). \end{aligned}$$

Since the inner sums are equal to real resp. imaginary parts of suitable Gaussian sums $\tau(\chi)$ Lemma 4.7 implies

$$L(1-s,\chi) = \frac{\Gamma(s)}{\pi}\left(\frac{2\pi}{D}\right)^{1-s} \cos(\pi s/2)\sqrt{D} L(s,\chi)$$

in case $\chi(-1) = 1$ and

$$L(1-s,\chi) = \frac{\Gamma(s)}{\pi}\left(\frac{2\pi}{D}\right)^{1-s} \sin(\pi s/2)\sqrt{D} L(s,\chi)$$

in case $\chi(-1) = -1$. It is now not difficult, using the relations between $F(s, D)$ and the corresponding L-functions, to obtain the assertion. □

Note that this method cannot be used for other characters because in the non-quadratic case the values of Gaussian sums are in general neither real nor purely imaginary.

Hurwitz's zeta-function $\zeta(s,u)$ satisfies a functional equation which is a limiting case ($x=0$) of the functional equation for *Lerch's zeta-function*

$$\mathfrak{K}_{u,x}(s) = \sum_{k=0}^{\infty} \frac{\exp(2k\pi x i)}{(u+k)^s}.$$

This definition is applicable if x lies in the upper half-plane and $0 < u < 1$ but as shown by M.Lerch (1887/88) (see also Lerch (1892a,1893,1894)) the function $\mathfrak{K}_{u,x}(s)$ can be continued to an entire function satisfying a functional equation. (See Apostol 1951b, Berndt 1972b, Fine 1951, Mikolás 1971, Oberhettinger 1956.) Lerch's zeta function was later utilized by T.M.Apostol (1951a) in his proof of (4.16). Actually this function can already be found in Malmstén (1849) and Lipschitz (1857, 1889). Lerch (1892a,1894) himself calls $\mathfrak{K}_{u,x}(s)$ the *Lipschitz function* and series of the form

$$\sum_{n=-\infty}^{\infty} \frac{\exp(2nv\pi i)}{((w+n)^2+u^2)^{s/2}}$$

are called by him *Malmstén series*. The last series in the special case $u=0$ was also considered by H.J.Mellin (1899) who showed that for $v \notin \mathbf{Z}$ it can be continued to an entire function vanishing at $s=0$ and for integral v it is meromorphic with a simple pole at $s=1$ and zeros at $s=-2,-4,-6,\dots$. In a later paper he (Mellin 1902) stated that if $F(X_1,\dots,X_k)$ is a rational function whose non-zero coefficients have positive real parts and the a_i's are fixed then the function

$$\sum_{n_1,\dots,n_k} \frac{1}{F(a_1+n_1,\dots,a_k+n_k)^s}$$

can be continued to a meromorphic function with real poles.

A deep study of functions defined by series of the form

$$Z(s,Q) = \sum_{\substack{m,n=-\infty \\ (m,n)\neq(0,0)}}^{\infty} \frac{1}{Q(m,n)^s}$$

where $Q(x,y)$ is a positive definite quadratic form was made by P.Epstein[36] (1903). These functions can be continued to meromorphic functions in the plane with a simple pole at $s=1$. From the vast literature concerning *Epstein zeta-functions* we quote here only the classical papers of H.S.A.Potter, E.C.Titchmarsh (1935) ($Z(s,Q)$ has infinitely many zeros on Re $s = 1/2$), H.Davenport, H.Heilbronn[37] (1936) ($Z(s,Q)$ may have zeros in the half-plane Re $s > 1$), S.Chowla, A.Selberg (1949) and A.Selberg, S.Chowla (1967) ($Z(s,Q)$ may have real zeros in the interval $(1/2,1)$) and H.M.Stark (1967a) (an asymptotic formula for the number of zeros of $Z(s,Q)$ in the rectangle $-1 < \text{Re } s < 2$, $0 \le \text{Im } s \le T$).

3. R.Lipschitz (1889) considered L-functions mod k with k either odd or divisible by 8. He used Dirichlet's description of characters (see Sect.2.3) and assumed that for $i \neq 0$ the exponents b_i occuring in that description are not divisible by the corresponding primes p_i, i.e. the characters are primitive. This

---

[36]Epstein, Paul (1871–1939), Professor in Frankfurt.
[37]Heilbronn, Hans (1908–1975), Professor in Bristol and Toronto.

restriction lies in the nature of the problem, since in other cases the function $L(s,\chi)$ does not have a functional equation in the usual sense (Apostol 1972).

In his argument Lipschitz essentially followed the way used by Riemann in the case of the zeta-function. He considered the series[38]

$$\sum_{m=-\infty}^{\infty} \frac{\exp(2\pi(v+m)ti)}{(\pi(v+m)^2)^s} \tag{4.20}$$

(with $t>0$ and $0<v<1$) and gave two proofs of the formula

$$\begin{aligned}\Gamma\left(\frac{1-s}{2}\right) & \sum_{m=-\infty}^{\infty} \frac{\exp(-2\pi(v+m)ti)}{(\pi(v+m)^2)^{s/2}} \\ &= \Gamma\left(\frac{s}{2}\right) \sum_{m=\infty}^{\infty} \frac{\exp(2\pi mvi)}{(\pi(t+m)^2)^{(1-s)/2}}\end{aligned} \tag{4.21}$$

(valid for $0<\operatorname{Re} s<1$) modelled after Riemann's proofs of the functional equation for $\zeta(s)$. With the use of Gaussian sums he used the last formula to deduce the functional equation[39] for L-functions in the form

$$\frac{L(s,\chi)}{L(1-s,\bar{\chi})} = \frac{\Gamma(1-s)}{(2\pi)^{1-s}} k^{1-s}(\exp(-\pi i(1-s)/2) + \theta\exp(\pi i(1-s)/2))\tau(\chi)^{-1},$$

with $\theta=\pm1$ and $\tau(\chi)$ being the Gaussian sum corresponding to χ.

We sketch now his reasoning in the simplest case when the modulus equals an odd prime p:

Let the character χ be defined by

$$\chi(n) = \begin{cases} \omega^{\gamma(n)} & \text{if } (p,n)=1 \\ 0 & \text{otherwise} \end{cases},$$

where $\omega=\exp(2\pi i\tau/(p-1)$ with $1\le\tau\le p-2$ and $\gamma(n)$ denotes the index of n with respect to a fixed primitive root g mod p, i.e. $n\equiv g^{\gamma(n)} \pmod p$ and $0\le\gamma(n)\le p-2$.

Consider moreover for $r=1,2,\dots,p-1$ the Gaussian sums

$$\sum_{j=1}^{p-1} \omega^{\gamma(j)}\zeta_p^{jr},$$

denoted by Lipschitz, as it was customary at his time, by (ω,ζ_p^r). (Nowadays one uses the notation $\tau_r(\chi)$ for it). He needs the formula (which in the case of real χ gives the equality (2.15))

[38]Earlier (Lipschitz 1857) he studied the series $\sum_{n=1}^{\infty}\exp(nui)n^{-\sigma}$ for real values of σ.

[39]We write this formula using modern notation. Lipschitz himself utilized in place of $\tau(\chi)$ the explicit form of it.

$$(\omega, \zeta_p^r) = \omega^{-\gamma(r)}(\omega, \zeta_p),$$

i.e.

$$\tau_r(\chi) = \overline{\chi(r)}\tau_1(\chi),$$

whose proof presents no difficulties. In fact, using modern notation we have

$$\tau_r(\chi) = \sum_{j=1}^{p-1} \chi(j)\zeta_p^{jr} = \overline{\chi(r)} \sum_{j=1}^{p-1} \chi(j)\chi(r)\zeta_p^{jr}$$

$$= \overline{\chi(r)} \sum_{j=1}^{p-1} \chi(jr)\zeta_p^{jr} = \overline{\chi(r)}\tau_1(\chi),$$

as asserted.

This implies

$$(\omega^{-1}, \zeta_p^{-1}) = \omega^{-\gamma(r)}(\omega^{-1}, \zeta_p^{-r})$$

and multiplying the equality (valid for Re $s > 1$)

$$L(s, \chi) = \sum_{n=1}^{\infty} \frac{\omega^{\gamma(n)}}{n^s}$$

by $(\omega^{-1}, \zeta_p^{-1})$ one obtains a function which can be expressed as a linear combination of functions given in the formula (4.20). Applying now (4.21) the assertion results.

Lipschitz's paper (see also Lipschitz (1890)) contains an interesting classification of primes and integers: the class P_1 consists of the single prime 2 and the class I_1 consists of all powers of 2. If the classes $P_1, I_1, P_2, I_2, \ldots, P_k, I_k$ are already defined then P_{k+1} consists of all primes p such that $p-1$ belongs to I_k and I_{k+1} is defined as the set of all integers not contained in $I_1 \cup I_2 \cup \ldots \cup I_k$ whose all prime factors lie in $P_1 \cup P_2 \cup \ldots \cup P_k$. Dirichlet's theorem on primes in progressions implies that there are infinitely many such classes. A similar classification of primes was proposed by J.L.Selfridge and P.Erdős (see Erdős 1976, Guy 1994,A18).

4. A simpler proof of the functional equation for L-functions was later given by C.de la Vallée-Poussin (1897) and in this version it was presented in Landau's book (Landau 1909a). In this approach the functional equation is proved for all primitive characters (without the restriction $8|k$ imposed by Lipschitz) and is given in a simpler form. It is not difficult to check, using the properties of the Γ-function, that both forms of the functional equation are equivalent.

Theorem 4.8. (C.de la Vallée-Poussin 1897) *Let* χ *be a primitive non-principal character* mod k *and put*

$$\mathfrak{a} = \begin{cases} 0 & \text{if } \chi(-1) = 1, \\ 1 & \text{if } \chi(-1) = -1. \end{cases}$$

If

$$\xi(s,\chi) = \Gamma((s+\mathfrak{a})/2)\left(\frac{\pi}{k}\right)^{-(s+\mathfrak{a})/2} L(s,\chi)$$

then for all complex s one has

$$\xi(s,\chi) = \epsilon(\chi)\xi(1-s,\bar{\chi})$$

with

$$\epsilon(\chi) = i^{-\mathfrak{a}}\frac{\tau(\chi)}{k^{1/2}}.$$

The reader may find the de la Vallée-Poussin's proof in the book of H.Davenport (1967, Sect.9). Also J.Hadamard (1897) provided a proof. A proof modelled after Riemann's proof of the functional equation for $\zeta(s)$ is given in Prachar (1957, Chap.7). For other proofs see Apostol (1970), Ayoub (1967), Berndt (1971, 1973), Schnee (1930) and Speiser[40](1939). A proof of the functional equation of the L-function corresponding to the non-principal character mod 4 was given by P.Lichtenbaum (1931).

4.3. Stieltjes, Cahen, Phragmén

1. The first papers dealing with Riemann's memoir were presumably those of A.Genocchi (1860) and W.Scheibner (1860a). They brought expositions of the contents of Riemann's paper. An analysis of it is also contained in Brocard (1880).

Riemann's extension of the zeta-function to complex arguments was unsuccessfully used by G.H.Halphen[41] (1883) to prove the asymptotic relation

$$\vartheta(x) = \sum_{p\le x}\log p = (1+o(1))x, \tag{4.22}$$

a result which can be easily shown to be equivalent to the Prime Number Theorem.

The same assertion appeared earlier in de Polignac[42] (1859a).

Halphen utilized the expression of the sum $\sum_{n\le x} a_n$ of coefficients of an arbitrary Dirichlet series $f(s) = \sum_{n=1}^{\infty} a_n n^{-s}$ by the integral

[40]Speiser, Andreas (1885–1970), Professor at Basel.

[41]Halphen, George (1844–1889), worked in Paris.

[42]He even asserted more generally, that if f is a monotonic continuous function, then $\sum_{p\le x} f(p)$ is asymptotically equal to $\int_2^x (f(t)/\log t)dt$. Under certain more stringent assumptions this was later deduced from the Prime Number Theorem by E.Landau (1900a).

$$\frac{1}{2\pi i}\int_{c-i\infty}^{c+i\infty} f(s)\frac{x^s}{s}ds,$$

which in the special case

$$a_n = \begin{cases} 1/k & \text{if } n = p^k \text{ with prime } p \\ 0 & \text{if } n \text{ is not a prime power} \end{cases}$$

appeared already in Riemann's memoir. (See Kronecker 1878.) A correct proof of this formula under rather weak assumptions was later given by O.Perron[43] (1908).

After giving an example dealing with the mean value of Euler's φ-function, Halphen wrote:

"*Dans*[44] *une communication ultérieure ... j'indiquerai d'autres applications, dont une fort importante. Je prouverai, en effet, que la fonction de M.Tchebycheff, somme des logarithmes des nombres premiers inférieurs à x, est asymptotique à x, ce qu'on n'avait pu établir jusquà present.*"

However after two years (Halphen 1885) he acknowledged his inability to establish this assertion:

"*J'avoue*[45] *que des obstacles inattendus m'ont arrété: une proposition, affirmée par Riemann dans son Mémoire sur les nombres premiers, sert de fondement au travail que j'ai préparé; mais cette proposition que naturellement j'admettais n'a pu être prouvée ...*".

2. At the same time T.J.Stieltjes (1885a) made the first attempt to deal with Riemann's Hypothesis. He asserted that the series

$$\sum_{n=1}^{\infty} \frac{\mu(n)}{n^s} \tag{4.23}$$

converges for all real $s > 1/2$. In view of Theorem 4.10 below this would imply the convergence also in the open half-plane Re $s > 1/2$. Since Theorem 1.15 (i) implies that for Re $s > 1$ the sum of this series equals $\zeta(s)^{-1}$, the assertion of Stieltjes would give the analytic continuation of $\zeta(s)^{-1}$ to a function regular in the half-plane Re $s > 1/2$ and this clearly implies Riemann's Hypothesis. In a letter to Hermite[46] Stieltjes wrote:

---

[43] Perron, Oskar (1880–1975), Professor in Tübingen, Heidelberg and München.

[44] "*In a later communication ... I will indicate further applications, among them a very important one. I will prove that the function of Mr. Tchebycheff, the sum of the logarithms of prime numbers less than x, is asymptotic to x, which one could not establish up to now.*"

[45] "*I confess that unexpected obstacles stopped me: a proposition asserted by Riemann in his memoir on prime numbers was at the basis of the work, which I prepared; however this proposition, which I naturally did assume, could not be proved ... .*"

[46] Baillaud,Bourget 1905; letter 77, undated but written possibly in the first half of July 1885.

"*J'ai*[47] *été ... assez heureux ... en demontrant cette propriété annoncée comme trés probable par Riemann, que toutes racines de $\xi(t) = 0$ sont réelles. ... Mais toutes ces recherches demanderont encore beaucoup de temps ... . Comme je ne puis pas pousser, en ce moment, activement ce travail à cause d'autres devoirs, je me propose de prendre un peu haleine et de laisser tout cela pendant quelques mois. Mais il n'y aura pas d'inconvénient, je l'espére, à publier dans les* Comptes Rendus *la Note ci-jointe*[48] *qui, me semble, doit intéresser les géometres qui ont étudie le Mémoire de Riemann*".

As Hermite[49] demanded details of the demonstration Stieltjes presented them in his letter[50] of 11th July 1885, the main point being the assertion that the series (4.23) converges for all $s > 1/2$. He based it on an upper bound for

$$M(x) = \sum_{n \leq x} \mu(n)$$

namely

$$|M(x)| \leq B\sqrt{x}, \tag{4.24}$$

with a suitable constant B and stated also that possibly one could take $B = 1$. No indication of the proof was given. Stieltjes wrote merely:

"*Ma*[51] *démonstration est bien pénible; je tâcherai, lorsque je reprendrai ces recherches, de la simplifier encore*" ,

and in a later letter[52] to M.G.Mittag-Leffler[53] he stated:

"*Mais*[54] *la démonstration de ce lemme est purement arithmétique et trés difficile et je ne l'obtiens que comme résultat de toute une série de proposition préliminaires. J'espére que cette démonstration pourra encore être simplifiée, mais en 1885 j'ai déjà fait mon mieux ...*".

---

[47] "*I have been quite lucky proving this property, announced as very probable by Riemann, that all roots of $\xi(t) = 0$ are real. ... But all these investigations need still much time ... . As I cannot actively continue this work at this instant because of other duties, I propose to take some breath and leave all this for a few months. I hope however that it will not be inconvenient to publish in Comptes Rendus the included note, which, it seems, will be of interest to geometers who did study Riemann's memoir*".

[48] I.e. Stieltjes (1885a), presented to the Paris Academy on July 13th, 1885.

[49] Baillaud,Bourget 1905; letter 78 of July 9th 1885.

[50] Baillaud,Bourget 1905, letter 79.

[51] "*My proof is rather painful; when I will return to these investigations I will try to simplify it.*"

[52] Baillaud, Bourget 1905, II, p.449, letter of April 15th 1887.

[53] Mittag-Leffler, Magnus Gösta (1846–1927), profesor in Helsinki (1877–1881) and in Stockholm. He founded in 1882 the journal *Acta Mathematica* and edited it for 45 years.

[54] "*But the proof of this lemma is purely arithmetic and very difficult. I obtained it as a result of a sequence of preliminary statements. I hope that this proof could be simplified but already in 1885 I did my best ... .*"

At the same time Stieltjes (1885b) sketched the deduction of the Prime Number Theorem from (4.24) in the very strong form[55]

$$\pi(x) = \mathrm{li}(x) + O(x^{\alpha})$$

for every $\alpha > 3/4$. He planned to present a proof of it in his thesis but finally he devoted his thesis to another subject and the alleged proof of this assertion was never published.

Even several years later Stieltjes seemed to be convinced of the correctness of his arguments since, in letters to Mittag-Leffler[56] and to Hermite,[57] he repeated his claim. According to S.C.van Ween (1978) no trace of a proof of (4.24) was found in the preserved notes of Stieltjes. See Kahane 1966).

The statement (4.24) with $B = 1$ is known as *Mertens' conjecture* since it was checked up to $x = 10000$ in Mertens (1897b). An even stronger conjecture, asserting (4.24) with $B = 1/2$, was formulated by R.D.von Sterneck (1901). The first counterexample (for $x = 7.76 \cdot 10^9$ one has $M(x) = 47465 \geq 0.538\sqrt{x}$) to von Sterneck's conjecture was produced by G.Neubauer (1963). In fact there are infinitely many such counterexamples (Jurkat 1972) (see Anderson,Stark 1981 and Jurkat, Peyerimhoff 1976). Also Mertens' conjecture was later disproved. This was done by A.Odlyzko and H.J.J.te Riele (1985) (see Saffari 1970a,b, Grosswald 1972, te Riele 1979a and Pintz 1980d,1983b) who showed that

$$\limsup_{x\to\infty} M(x)/\sqrt{x} > 1.06$$

and

$$\liminf_{x\to\infty} M(x)/\sqrt{x} < -1.009.$$

Assuming the linear independence over the rationals of the set of imaginary parts of imaginary zeros of $\zeta(s)$ as well as the Riemann Hypothesis A.E.Ingham (1942) showed that these limits are equal to ∞ and $-\infty$, respectively, however the truth of these assumptions is still hypothetical. Later J.Pintz (1987) proved that there is a counterexample to Mertens' conjecture which does not exceed $\exp(10^{65})$. Surveys of this topic were given by F.Dress (1984) and H.J.J.te Riele (1985).

H.von Mangoldt (1897) showed

$$M(x) = o(x)$$

and this was later improved by E.Landau to

$$M(x) = O(x\exp(-c\sqrt{\log\log x}))$$

with a certain positive c (Landau 1901b),

$$M(x) = O(x\exp(-(\log x)^{1/12})) \qquad \text{(Landau 1903c)}$$

[55] At first he even claimed to have a proof of the bound $O(\sqrt{x}\log x)$ for the error term (Baillaud,Bourget 1905, I, p.445, letter of July 23rd 1885 to Mittag-Leffler) but in the next letter (ibidem, I,446–448, letter of March 23rd 1887) he withdraw this claim.

[56] Baillaud,Bourget 1905, II, 449–452, letter of 15th April 1887.

[57] Baillaud, Bourget 1905, II, 154–158, letter of 4th March 1891.

and

$$M(x) = O(x \exp(-a(\log x)^{1/2}))$$

with a certain $a > 0$ (Landau 1908b). These evaluations depend on zero-free regions on the zeta-function. In fact it is known (see for example Ellison (1975, Th.4.6)) that if $\zeta(s)$ does not vanish in the region $\sigma > 1 - \eta(t)$ (with $s = \sigma + it$) and in that region the functions $\zeta'(s)/\zeta(s)$, $\zeta(s)$ and $\zeta^{-1}(s)$ are $O(t^\epsilon)$ (with positive ϵ) for t tending to infinity then one has

$$M(x) = O(xe^{-H(x)})$$

with $H(x)$ depending on λ and ϵ and defined by

$$H(x) = \inf_{t \geq 0} \frac{\eta(t) \log x + \lambda \log t}{1 + \lambda + \epsilon}.$$

Conversely, D.Allison (1970) proved that every non-trivial bound for $M(x)$ implies the existence of certain non-vanishing regions for the zeta-function. J.E.Littlewood (1912) proved that Riemann Hypothesis is equivalent to the evaluation

$$M(x) = O(x^{1/2+\epsilon})$$

for every positive ϵ. Later Landau (1924c) showed that Riemann Hypothesis implies

$$M(x) = O(x^{1/2} \exp\left(\frac{\log x \log\log\log x}{\log\log x}\right))$$

and Titchmarsh (1927) succeeded in removing the third iteration of the logarithm in this evaluation. A proof is given in Titchmarsh (1951, Th.14.26).

The first effective non-trivial upper bound for $M(x)$, namely $|M(x)| < x/9 + 8$, appears in von Sterneck (1898). For later bounds see MacLeod (1967), Schoenfeld (1969), Diamond, McCurley (1981), Costa Pereira (1989), Dress (1993), Dress, El Marraki (1993) and El Marraki (1995).

A related conjecture was proposed by G.Pólya (1919) who considered Liouville's function $\lambda(n) = (-1)^{\Omega(n)}$ (where $\Omega(n)$ is the number of prime factors of n, counted with their multiplicities), and conjectured that for all $x \geq 2$ one has

$$L(x) = \sum_{n \leq x} \lambda(n) \leq 0.$$

He noted that this inequality would imply Riemann's Hypothesis and checked it up to $x = 1500$ observing that in certain cases one has $L(x) = 0$, e.g. for $x = 586$. Later H.Gupta (1940) pursued Pólya's calculation up to $x = 20\,000$ (see Gupta 1943). Pólya's conjecture was finally disproved by C.B.Haselgrove (1958), using the first 1500 zeros of $\zeta(s)$ and R.S.Lehman (1960) confirmed his computations.

3. At some point Stieltjes became interested in the Taylor expansion of the difference $\zeta(s) - 1/(s-1)$ at $s = 1$ and in a letter[58] to Hermite proved the following result:

---

[58]He possibly communicated his formula to Hermite earlier, since the latter asked in his letter of June, 23rd 1885 (Baillaud, Bourget 1905, letter 74, p.150) for more explanations concerning it. Stieltjes answered it by a long undated letter(Baillaud, Bourget 1905, letter 75, pp.151–155) containing the proof which we present.

Theorem 4.9. *If*

$$\zeta(s) = \frac{1}{s-1} + \sum_{k=0}^{\infty} \frac{(-1)^k A_k}{k!}(s-1)^k$$

then for $k = 0, 1, 2, \dots$ *one has*

$$A_k = \lim_{x\to\infty} \left(\sum_{n\le x} \frac{\log^k n}{n} - \frac{\log^{k+1} x}{k+1} \right).$$

Proof. Denote by C_k the right-hand side of the asserted formula for A_k and observe that the existence of the limit in that formula is an easy consequence of Lemma 2.12. The starting point of Stieltjes's argument is the equality

$$\zeta(1+s) = \frac{1}{\Gamma(s+1)} \int_0^{\infty} \frac{x^s\,dx}{e^x - 1}$$

valid for positive s, which we have already encountered in Riemann's and Hermite's proofs of Theorem 4.1. It implies

$$\zeta(1+s) - \frac{1}{s} = \frac{1}{\Gamma(s+1)} \int_0^{\infty} \left(\frac{1}{e^x-1} - \frac{e^{-x}}{x} \right) x^s\,dx = \frac{1}{\Gamma(s+1)} \sum_{j=0}^{\infty} \frac{a_j}{j!} s^j$$

where

$$a_j = \int_0^{\infty} \left(\frac{1}{e^x-1} - \frac{e^{-x}}{x} \right) \log^j x\,dx.$$

For positive t and $n = 1, 2, 3, \dots$ put

$$F_n(t) = \int_0^{\infty} e^{-tx} \log^n x\,dx.$$

Since for positive t and n one has

$$\Gamma^{(n)}(1) = \int_0^{\infty} \log^n x e^{-x}\,dx = t \int_0^{\infty} (\log x + \log t)^n e^{-tx}\,dx$$

one gets

$$F_n(t) = \int_0^{\infty} (\log x + \log t - \log t)^n e^{-tx}\,dx$$
$$= \frac{1}{t} \sum_{j=0}^{n} (-1)^j \binom{n}{j} \Gamma^{(n-j)}(1) \log^j t. \tag{4.25}$$

Putting here $t = 1, 2, \dots r$ and adding one arrives at

$$\int_0^{\infty} \frac{e^{-x} - e^{-(r+1)x}}{1-e^{-x}} \log^n x\,dx = \sum_{j=0}^{n} (-1)^j \binom{n}{j} \Gamma^{(n-j)}(1) \sum_{t=1}^{r} \frac{\log^j t}{t}.$$

The left-hand side of the last equality can be written in the form

$$a_n + f_n(r) - g_n(r)$$

where

$$g_n(r) = \int_0^\infty \left(\frac{1}{1-e^{-x}} - \frac{1}{x}\right) e^{-(r+1)x} \log^n x dx = o(1) \quad (r \to \infty)$$

and

$$f_n(r) = \int_0^\infty \left(\frac{e^{-x}}{x} - \frac{e^{-(r+1)x}}{x}\right) \log^n x dx.$$

Treating r for a while as a continous variable one obtains, using (4.25), the equality

$$f_n'(r) = F_n(r+1) = \frac{1}{r+1}\sum_{j=0}^{n} \binom{n}{j} \Gamma^{(n-j)}(1) \frac{\log^j(r+1)}{r+1}$$

which in view of $f_n(0) = 0$ implies

$$f_n(r) = \sum_{j=0}^{n} (-1)^j \binom{n}{j} \frac{\log^{j+1}(r+1)}{j+1} \Gamma^{(n-j)}(1),$$

so finally

$$a_n - \int_0^\infty \left(\frac{1}{1-e^{-x}} - \frac{1}{x}\right) e^{-(r+1)x} \log^n x dx$$

$$= \sum_{j=0}^{n} (-1)^j \binom{n}{j} \Gamma^{(n-j)}(1) \left(\sum_{m=1}^{r} \frac{\log^j m}{m} - \frac{\log^{j+1} r}{j+1}\right)$$

and letting r tend to infinity we get for a_n the expression

$$a_n = \sum_{j=0}^{n} (-1)^j \binom{n}{j} C_j \Gamma^{(n-j)}(1).$$

This clearly implies the equality

$$\sum_{j=0}^{\infty} \frac{a_j}{j!} s^j = \left(\sum_{m=0}^{\infty} (-1)^m \frac{C_m}{m!} s^m\right) \left(\sum_{k=0}^{\infty} \frac{\Gamma^{(k)}(1)}{k!} s^k\right)$$

and since the second factor on the right-hand side equals $\Gamma(s+1)$ the theorem follows by simply substituting $s-1$ for s. □

An independent proof of this theorem was given by J.L.V.W.Jensen (1887), who used it to prove the continuability of $(s-1)\zeta(s)$ to an entire function. See Césaro (1893), Franel (1899), Gram (1895) and Lerch (1914). Stieltjes' result was later rediscovered by W.E.Briggs and S.Chowla (1955) and other proofs were given in

Andrica, Tóth (1991), Ferguson (1963) and Verma (1963). Similar formulas can be also obtained for the coefficients of the power series of any Dirichlet series having a unique simple pole at the boundary of its covergence half-plane (Buschman, Briggs 1961).

E.Lammel (1966) noted that Theorem 4.9 implies directly the non-vanishing of $\zeta(s)$ in the circle $|s-1| \le 1$ (see also Israilov 1979) and S.Wigert (1909) observed that one can express Riemann's Hypothesis in terms of the coefficients A_k.

Explicit formulas for A_k were given in Zhang (1981) and Zhang,Williams (1994). For numerical values of the coefficients A_k see Keiper (1992). See also Liang,Todd (1972).

Formulas for the coefficients of the power series of $\zeta(s)$ at s in the half-plane Re $s > 1$ were given by D.Mitrović (1957), at $s = 0$ by T.M.Apostol (1985b) and at arbitrary $s \neq 1$ by M.Nakajima (1987,1988). For a generalization see Dilcher (1992).

4. The next attempt to prove (4.22) was made by E.Cahen (1893,1894) in his thesis. His deduction however cannot be regarded as satisfactory since he handled non-absolutely convergent integrals in a rather careless way[59]. Cahen himself conceded few years later in his book (Cahen 1900, footnote on p.323) that his proof was not rigorous but would be such if one would succeed in proving Riemann's conjecture on roots of the zeta-function.

Notwithstanding its errors the thesis of Cahen contains the first more or less precise treatment of *general Dirichlet series*, i.e. series of the form

$$\sum_{n=1}^{\infty} a_n \exp(-\lambda_n s),$$

with λ_n being a sequence of positive reals tending to infinity. In the case $\lambda_n = \log n$ this gives the usual Dirichlet series

$$f(s) = \sum_{n=1}^{\infty} \frac{a_n}{n^s}, \tag{4.26}$$

and most of Cahen's results were at that time new even in this particular case. In the sequel we shall restrict ourselves essentially to the usual Dirichlet series.

In the first part of his work Cahen dealt with problems of convergence and uniform convergence of Dirichlet series, establishing the following result:

[59] C.de la Vallée-Poussin (1896a,p.250) pointed out a gap in Cahen's proof and wrote: "*M.Cahen affirme (au bas de la page 44) la convergence d'une série dont tous les termes s'obtiennent en intégrant tous les termes d'une série convergente entre des limites infinies. La convergence de cette série ne repose sur rien et serait beaucoup plus difficile à démontrer que la théorème que M.Cahen veut établir*".

("*Mr.Cahen asserts (on the bottom of p.44) the convergence of a series whose terms are obtained by integrating over an infinite interval all terms of another convergent series. The convergence of that series is based on nothing and its proof is much more difficult than the proof of the theorem which Mr.Cahen wants to establish.*")

Theorem 4.10. *If all partial sums of the series (4.26) are for $s = s_0$ uniformly bounded then this series converges in the open half-plane* Re $s >$ Re s_0. *This convergence is uniform in every region of the form* Re $s >$ Re $s_0 + \epsilon$, $|\text{Im } s| \le T$, *with positive ϵ and T.*

Proof. Cahen bases his proof on the following convergence criterion, extending that of P. du Bois-Reymond (1870) (see Lemma 2.4), and which results directly from the partial summation formula:

Lemma 4.11. *(i) Let α_n, β_n be two sequences of complex numbers satisfying the following conditions:*

(a) *For a suitable constant B and all $N \ge 1$ one has*

$$|\sum_{n=1}^{N} \alpha_n| \le B,$$

(b) $\lim_{n\to\infty} \beta_n = 0$,
(c) *The series $\sum_{n=2}^{\infty} |\beta_n - \beta_{n-1}|$ converges.*

Then the series

$$\sum_{n=1}^{\infty} \alpha_n \beta_n$$

is convergent.

(ii) If moreover $\beta_n = \beta_n(s)$ are functions of a variable s and the convergence in (b) and (c) is uniform then the series (4.27) also converges uniformly. □

Put $\alpha_n = a_n n^{-s_0}$, $\beta_n = n^{s_0 - s}$. The conditions (a) and (b) of the lemma being obviously satisfied with uniform convergence in (b), it remains to check that the series

$$\sum_{n=1}^{\infty} \left| \frac{1}{(1+n)^{s-s_0}} - \frac{1}{n^{s-s_0}} \right|$$

converges uniformly in the horizontal strip Re $s >$ Re $s_0 + \epsilon$, $|\text{Im } s| \le T$. To obtain that Cahen writes $s - s_0 = a + bi$ with $a > 0$ and utilizes the rather complicated but elementary inequality

$$\begin{aligned} \left| \frac{1}{(1+n)^{s-s_0}} - \frac{1}{n^{s-s_0}} \right| &\le \frac{1}{(n-1)^a} - \frac{1}{n^a} + b \log\left(\frac{n}{n-1}\right) \frac{1}{(n(n-1))^{a/2}} \\ &\le \frac{1}{(n-1)^a} - \frac{1}{n^a} + \frac{b}{(n-1)(n(n-1))^{a/2}} \end{aligned}$$

from which the assertion follows immediately.

To avoid the use of the last inequality one can proceed as follows:

If Re $(s - s_0) > 2$ then

$$\left|\frac{1}{(1+n)^{s-s_0}} - \frac{1}{n^{s-s_0}}\right| \leq \frac{2}{n^2}$$

and if $\epsilon < \mathrm{Re}\,(s - s_0) < 2$ then

$$\left|\frac{1}{(1+n)^{s-s_0}} - \frac{1}{n^{s-s_0}}\right| = \left|\int_n^{n+1} (s-s_0)x^{-(s-s_0)-1}dx\right| \leq \left|\frac{s-s_0}{n^{(s-s_0)+1}}\right| \leq \frac{2}{n^{1+\epsilon}}$$

and this again leads to our assertion. □

A slightly simpler proof was later given by O.Perron (1908). It led to uniform convergence of the considered series in the region defined by $|s - s_0| \leq R$, $\mathrm{Re}\,(s - s_0) \geq \delta$ for any R and $\delta > 0$. Nowadays one shows rather (see e.g. Prachar 1957, Tenenbaum 1990a) that the series (4.26) converges uniformly in every sector $|\mathrm{Im}\,(s - s_0)| \leq \theta < \frac{\pi}{2}$, $s \neq s_0$ (which had already been pointed by Cahen).

Corollary 1. *If the series (4.26) is neither divergent nor convergent everywhere then there exists a real number s_0 with the property that for* Re $s > s_0$ *the series converges whereas for* Re $s < s_0$ *it diverges.*

Proof. Take for s_0 the greatest lower bound of the set of all these s at which (4.26) converges. □

The number s_0 is called the *abscissa of convergence* of the series (4.26). We shall denote it by $c(f)$.

This result was earlier established by J.L.W.V.Jensen (1884).

Corollary 2. *If s_0 is the abscissa of convergence of the series (4.26) then the sum of that series defines a function $f(s)$, which is regular for* Re $s > s_0$ *and whose derivatives are given by the formula*

$$f^{(k)}(s) = \sum_{n=1}^{\infty} \frac{a_n \log^k n}{n^s} \qquad (k = 1, 2, \ldots).$$

Proof. It is sufficient to show that for Re $s > s_0$ the series

$$\sum_{n=1}^{\infty} \frac{a_n \log n}{n^s}$$

converges to a regular function and use recurrence in k. Cahen's proof follows the same line as his proof of the preceding theorem and so is reduced to showing the uniform convergence of the series

$$\sum_{n=1}^{\infty} \left|\frac{\log(n+1)}{(1+n)^{s-s_0}} - \frac{\log n}{n^{s-s_0}}\right|,$$

which is then established in the same way as the theorem. □

In his next result Cahen shows that if the abscissa of convergence $c(f)$ of the series (4.26) is positive then it equals

$$\limsup_{n\to\infty} \frac{\log(|\sum_{j=1}^{n} a_n|)}{\log n}.$$

His proof is reproduced in Landau (1909a, Chap.68, Th.5) who noted that an earlier partial result is contained in Jensen (1888).

This formula implies immediately that the *abscissa* $a(f)$ *of absolute convergence* of this series, i.e. such number a that for Re $s > a$ the series (4.26) converges absolutely, but this fails for any s with Re $s < a$ (the existence of such a already had been shown in Scheibner (1860b)) equals

$$\limsup_{n\to\infty} \frac{\log(\sum_{j=1}^{n} |a_n|)}{\log n},$$

provided this number is positive.

The case $c(f) < 0$ was not considered by Cahen; he merely noted that in that case one should shift the variable but did not explain how to do it without knowing the value of $c(f)$ beforehand. The same approach appears in Landau (1909a). For a treatment of this case see Pincherle[60](1908). A formula valid in all cases was given by K.Knopp (1910). For an analogous result in the case of general Dirichlet series see Kojima (1914) and Fujiwara[61](1914). (See also Fujiwara 1920, Kakeya 1917, Kojima 1916 and Malmrot 1920).

Similarly one introduces the *abscissa* $u(f)$ *of uniform convergence* of the series (4.26). According to Landau (1909a,p.887) the proof of its existence appears already in the book of W.Scheibner (1860b) and it was later considered (also for general Dirichlet series) by H.Bohr (1910a, 1913a) who gave also (Bohr 1913c), in case when it is positive, the formula

$$u(f) = \limsup_{n\to\infty} \frac{\log T_n}{\log n}$$

for it, with

$$T_n = \sup_{-\infty<t<\infty} \left| \sum_{k=1}^{n} \frac{a_k}{k^{it}} \right|.$$

A formula in the general case, valid also for general Dirichlet series

$$\sum_{n=1}^{\infty} a_n e^{-\lambda_n s},$$

was given by M.Kuniyeda (1916). H.Bohr (1913b) proved that the width of the vertical strip in which a Dirichlet series converges uniformly but not absolutely cannot exceed 1/2. He asked whether for every $0 \le d \le 1/2$ there exists a Dirichlet

[60]Pincherle, Salvatore (1853–1936), Professor in Naples and Bologna.
[61]Fujiwara, Matsusaburo (1881–1946), Professor at the Tôhoku University.

series with this width equal to d. This question found a positive answer in Bohnenblust,Hille (1931a,b). For an early weaker result see Toeplitz[62](1913).

For general Dirichlet series it was proved in Neder (1922) that the only relation between the three abscissas is the trivial relation $c(f) \leq u(f) \leq a(f)$. The first example of such series with $c(f) = 0$ and $a(f) = \infty$ appears in Jansson (1921).

Cahen's thesis also contains several other results dealing with Dirichlet series but not all his assertions turned out to be correct and the proofs of certain correct theorems were fallacious. Correct versions of several Cahen's results appear in Hadamard (1908), Landau (1907b), Perron (1908) and in the first part of the thesis of H.Bohr (1910b), which was devoted to the summability of Dirichlet series.

In particular Cahen's assertion that the necessary condition for a function to be represented by a Dirichlet series given by L.Kronecker (1878) is also sufficient turned out to be incorrect, as shown in the quoted paper of O.Perron. Perron also pointed out that Kronecker's argument leading to its necessity is incorrect and corrected it.

A discussion of Cahen's results was given by Landau (1909a, Chap.67), who wrote the following comment[63]:

"*Herr*[64] *Cahen hat* 1894 *zuerst die Dirichletschen Reihen als Funktionen komplexen Argumentes studiert und viele grundlegende Tatsachen bewiesen. Alsdann enthält aber der auf die allgemeine Theorie der Dirichletschen Reihen bezüglicher Teil seiner Arbeit eine Reihe von Fehlschlüssen verschiedenster Art und mit ihrer Hilfe eine so große Zahl tiefliegender und merkwürdiger Gesetze, daß vierzehn Jahre erforderlich waren, bis es möglich wurde bei jedem einzelnen der Cahenscher Resultate festzustellen, ob es richtig oder falsch ist.*"

5. In the second chapter Cahen applied his general theory to Riemann's zeta-function. After recalling the main results of Riemann's paper he proceeded to the alleged proof (sketched earlier in Cahen (1893)) of the asymptotic formula (4.22). His main idea is sound and consists in expressing the Čebyšev's function $\psi(x) = \sum_{p^k \leq x} \log p$ by the integral

$$\frac{1}{2\pi i} \int_{a-i\infty}^{a+i\infty} \frac{\zeta'(s)}{\zeta(s)} \frac{x^s}{s} ds,$$

with $a > 1$ and then evaluating that integral with the use of Cauchy's theorem, pushing the integration line to the left of the line Re $s = 1$. He assumed

[62]Toeplitz, Otto (1881–1940), Professor in Bonn and Jerusalem.

[63]Landau 1909a,p.724.

[64]"*Mr Cahen first studied in 1894 the Dirichlet series as functions of a complex variable and proved several fundamental facts. However that part of his work which deals with the general theory of Dirichlet series contains a series of false reasonings of a different kind; he obtained with their use such a large quantity of deep and remarkable laws that it took fourteen years until it was possible to determine whether a particular Cahen's result is true or false.*"

tacitly that the zeta-function has no zeros in a certain half-plane Re $s > c$ for some $c < 1$. We read on p.115:

"*Pour*[65] *évaluer cette intégrale par la méthode de Halphen, j'intègre la fonction sous la signe* $\int$ *le long du contour d'un rectangle ABCD: BC, AD étant à l'infini, CD étant à gauche de la droite* $x = 1, \ldots$ *tel que le rectangle ne renferme pas d'autre pôle que 1 de la fonction sous le signe* $\int$."

There is no word spent on explaining whether such choice of the rectangle is possible. Although one could infer from Hadamard (1896c) that Cahen's proof is based on the alleged Stieltjes' proof of Riemann's conjecture[66], the paper of Stieltjes is not quoted in Cahen's thesis.

To the remainder of the proof, consisting of the evaluation of the resulting integral, the critical comments of C.de la Vallée-Poussin quoted above apply: the calculations are performed in a purely formal way.

6. Other incomplete proofs of either the Prime Number Theorem or equivalent statements were presented by C.Ajello (1896) and E.Césaro (see Césaro 1883, p.345) At the end of Jensen's paper (Jensen 1899) containing the proof of the well-known Jensen's formula one finds a statement asserting the Prime Number Theorem in a rather strong form, viz.

$$\pi(x) = \sum_{m \le x} \frac{1}{\log m} + O(x^{1/2+\epsilon})$$

for every positive ϵ, which is easily seen to be equivalent with

$$\pi(x) = \mathrm{li}(x) + O(x^{1/2+\epsilon}).$$

It seems that Jensen did not know the results of Hadamard and de la Vallée-Poussin since he wrote there

"*... le*[67] *probléme relatif au nombre des nombres premiers inférieurs à une limite donnée attend encore sa solution rigoreuse, et cela, parce que l'on*

[65] "*For evaluating that integral with Halphen's method I integrate the function under the integral sign along the sides of a rectangle ABCD; BC, AD being at infinity, CD being to the left of the line* $x = 1, \ldots,$ *so that the rectangle does not contain other poles than 1 of the function under the sign* $\int$."

[66] Hadamard writes: "*Dans son Mémoire ... M.Cahen présente une démonstration du théorème énoncé par Halphen. ... Toutefois son raisonnement dépend de la proposition de Stieltjes sur la réalite des racines de* $\zeta(\frac{1}{2} + it) = 0$." ("*In his paper ... Mr.Cahen presents a proof of a theorem anounced by Halphen. ... However his proof depends on a statement of Stieltjes about the reality of roots of* $\zeta(\frac{1}{2} + it) = 0$.")

[67] "*The problem of the number of primes less than a given bound still awaits its rigorous solution, because one could not prove up to now that the zeros of the transcendental function* $\xi(t)$ *of Riemann are all real. But with the aid of subsequent research on Dirichlet series for which I proved already in 1884 two fundamental theorems ... I succeeded in proving rigorously that* $\xi(t)$ *does not have zeros inside the circle* $|t - ri| = r$, *and this, in other words, affirmes that all roots of the*

n'a pu jusqu'ici démontrer que les zéros de la fonction transcendante entiére $\xi(t)$ de Riemann sont tous réels. Or, à l'aide des recherches subséquentes sur les séries de Dirichlet, pour laquelles en 1884[68] *déja j'ai démontré deux théorèmes fondamenteux, et ... je suis parvenu à démontrer rigouresement, que $\xi(t)$ n'a pas des zéros a l'intérieur du cercle $|t-ri|=r$, c'est, en d'autres termes, affirmer que toutes les racines de l'équation $\xi(t)[=0]$ sont réelles. L'on a surmonté de la sorte le dernier et le plus difficile des obstacles qui s'opposaient à la solution du problème, et il est maintenant aisé de démontrer (comme je le ferai voir dans le mémoire que je prépare*[69]*) que le nombre des nombres premiers inférieurs à n est*

$$\vartheta(n)=\sum_{m=2}^{n}\frac{1}{\log m}+\rho_n,$$

où, pour $\lim n=\infty$, $\frac{|\rho_n|}{(\sqrt{n})^{1+\epsilon}}$ *tend vers zéro, ϵ désignant une quantité fixe positive aussi petite que l'on veut. D'autre part on démontre trés aisément que ρ_n n'est pas d'un ordre plus petite que $\sqrt{n}$, c'est à dire, pour parler d'une maniére plus précise, qu'on peut toujours trouver une suite infinie des nombres n pour lesquels*

$$|\rho_n|>(\sqrt{n})^{1-\epsilon}."$$

7. The only correct and fully proved result dealing with the growth of $\pi(x)$ and obtained in the period between Riemann's memoir and Hadamard's novel approach to the zeta-function, which we shall describe in Chap.5, was a theorem of E.Phragmén (1891,1901/1902). He obtained in a simple and elegant way a general result dealing with sign-changes of real functions and applied it to two questions about primes: he solved a problem of Čebyšev concerning the distribution of primes in residue classes mod 4 and showed that if we define

---

equation $\xi(t)[=0]$ are real. One has passed one of the last and most difficult obstacles preventing the solution of the problem and it is now easy to show (as I shall do in a memoir which I prepare) that the number of primes less than n is $\vartheta(n)=\sum_{m=2}^{n}\frac{1}{\log m}+\rho_n$, where, for $\lim n=\infty$, $\frac{|\rho_n|}{(\sqrt{n})^{1+\epsilon}}$ *tends to zero, ϵ being a fixed positive quantity as small as one wishes. On the other hand one proves easily that ρ_n is not of order smaller than $\sqrt{n}$, i.e.,speaking in a more precise manner, one can always find an infinite sequence of numbers n, for which*

$$|\rho_n|>(\sqrt{n})^{1-\epsilon}."$$

[68]See Jensen 1884, 1888.

[69]One knows from von Koch (1900) that the promised memoir was to appear in *Acta Mathematica* but it never did, hence it is not clear how Jensen tried to prove Riemann's conjecture.

$$f(x) = \sum_{p^k \le x} \frac{1}{k} = \sum_{n \ge 1} \frac{1}{n}\pi(x^{1/n}),$$

(this function differs inessentially from the function considered by Riemann in his assertion II) then $f(x) - \mathrm{li}(x) - \log 2$ changes sign infinitely often. Observe that Phragmén preferred to work with integrals rather than with Dirichlet series. This does not make any big difference because of the following easy result proved in Phragmén's paper in a particular case:

Lemma 4.12. *Let a sequence $a_1, a_2, \dots$ of complex numbers be given and put $A(x) = \sum_{n \le x} a_n$. If* $\mathrm{Re}\, s > 0$ *and one has*

$$\lim_{x \to \infty} \frac{A(x)}{x^s} = 0$$

then the equality

$$\sum_{n=1}^{\infty} \frac{a_n}{n^s} = s \int_1^{\infty} \frac{A(x)}{x^{1+s}} dx$$

will hold if either the series or the integral converges.

Proof. It suffices to note that for any integer N one has

$$s \int_1^{N+1} \frac{A(x)}{x^{1+s}} dx = s \sum_{j=1}^{N} \int_j^{j+1} \frac{A(x)}{x^{1+s}} dx = \sum_{j=1}^{N} A(j) \left(\frac{1}{j^s} - \frac{1}{(j+1)^s} \right)$$

$$= A(1) + \sum_{j=2}^{N} \frac{A(j) - A(j-1)}{j^s} - \frac{A(N)}{(N+1)^s} = \sum_{j=1}^{N} \frac{a_j}{j^s} - \frac{A(N)}{(N+1)^s}. \qquad \square$$

Theorem 4.13. *Let $\alpha \ge 1$ and let $\phi(x)$ be a real function defined for $x \ge \alpha$. Assume that for all $\beta \ge \alpha$ the integral*

$$F(s) = \int_\beta^\infty \phi(t) t^{-s-1} dt$$

represents a function regular in the open half-plane $\mathrm{Re}\, s > 1$, *whose power series at $s = 1$ has its radius of convergence $R > 1$. Then for any given $\delta > 0$ neither $\phi(x) > \delta$ nor $\phi(x) < -\delta$ can hold for all sufficiently large x. If moreover there exists an infinite sequence $x_1, x_2, \dots$ of points of discontinuity of the function $\phi(x)$ and there exist positive numbers h_j, m_j such that for $j = 1, 2, \dots$ one has*

$$h_j \le x_{j+1} - x_j$$

and

$$m_j < \inf_{0 < h < h_j} \{\phi(x_j + h) - \phi(x_j - 0)\}$$

then the divergence of the series

$$\sum_{j=1}^{\infty} \frac{x_j m_j}{x_j + h_j}$$

implies that the function $\phi(x)$ changes its sign infinitely often.

Proof. Assume that $\phi(x) > 0$ holds for all large x, say for $x \geq x_0 > 1$, and write for $\mathrm{Re}\, s > 1$

$$F(s) = \sum_{k=0}^{\infty} c_k (s-1)^k.$$

By assumption this series converges in the disc $\{s : |s-1| < R\}$ and thus F can be continued analytically into the disc $|s-1| \leq r$, with $1 < r < R$.

For $\mathrm{Re}\, s > 1$ we have formally

$$F^{(k)}(s) = (-1)^k \int_{x_0}^{\infty} \phi(t) \frac{\log^k t}{t^{s+1}} dt,$$

but in view of $\phi(t) \log t > 0$ the last integral converges uniformly in every half-plane $\mathrm{Re}\, s > 1 + \epsilon$ (provided $\epsilon > 0$) and hence this equality can be justified. Indeed, one has obviously

$$\left| \frac{\phi(x) \log^k x}{x^{1+s}} \right| \leq \frac{\phi(x)}{x^{2+\epsilon/2}} \frac{\log^k x}{x^{\epsilon/2}} = O\left(\frac{\phi(x)}{x^{2+\epsilon/2}} \right).$$

In view of $c_k = F^{(k)}(1)/k!$ one has

$$k! c_k = (-1)^k \int_{x_0}^{\infty} \phi(t) \frac{\log^k t}{t^{s+1}} dt + o(s-1),$$

thus

$$k! |c_k| = \limsup_{s \to 1+0} \int_{x_0}^{\infty} \phi(t) \frac{\log^k t}{t^{s+1}} dt \leq \int_{x_0}^{\infty} \phi(t) \frac{\log^k t}{t^2} dt.$$

On the other hand for $y > x_0$ and real $s > 1$ one has

$$\int_{x_0}^{y} \phi(t) \frac{\log^k t}{t^{1-s}} dt \leq k! |c_k| + o(s-1)$$

which implies

$$\int_{x_0}^{y} \phi(t) \frac{\log^k t}{t^2} dt \leq k! |c_k|$$

and shows that the integral

$$\int_{x_0}^{\infty} \phi(t) \frac{\log^k t}{t^2} dt$$

converges to $k!|c_k|$.

Now we return to the series $\sum_{k=0}^{\infty} c_k(s-1)^k$. By assumption it converges absolutely at $s=2$ hence $\sum_{k=0}^{\infty}|c_k|$ converges to S, say. Thus

$$S=\sum_{k=0}^{\infty}\frac{1}{k!}\int_{x_0}^{\infty}\phi(t)\frac{\log^k t}{t^2}dt$$

and as $\phi(x)$ is positive we may interchange the sum and the integral to get

$$S=\int_{x_0}^{\infty}\frac{\phi(t)}{t^2}\sum_{k=0}^{\infty}\frac{\log^k t}{k!}dt=\int_{x_0}^{\infty}\frac{\phi(t)}{t}dt.$$

This is clearly incompatible with the inequality $\phi(x)>\delta>0$ holding for all large x hence we get the first assertion of the theorem. To obtain the second assertion it suffices only to observe that its assumptions imply

$$\int_{x_0}^{\infty}\frac{\phi(t)}{t}dt \geq \sum_{j=1}^{\infty}\int_{x_j}^{x_j+h_j}\frac{\phi(t)}{t}dt$$

$$\geq \sum_{j=1}^{\infty} m_j\int_{x_j}^{x_j+h_j}\frac{dt}{t}\geq\sum_{j=1}^{\infty}\frac{m_j h_j}{h_j+x_j}=+\infty,$$

thus S is infinite, again a contradiction.

In the case of negative ϕ it suffices to apply the obtained result to the function $-\phi$. □

Corollary 1. *If*

$$f(x)=\sum_{n\geq 1}\frac{1}{n}\pi\left(x^{1/n}\right)=\sum_{p^k\leq x}\frac{1}{k}$$

then the function $g(x)=f(x)-\mathrm{li}(x)+\log 2$ *changes its sign infinitely often.*

Proof. The proof is based on the formula

$$\int_1^{\infty}\frac{f(t)}{t^{1+s}}dt=\frac{\log\zeta(s)}{s} \tag{4.27}$$

valid for Re $s>1$. This appears in Riemann's memoir and is a consequence of Lemma 4.12.

Now observe that for Re $s>1$ we have by partial integration

$$\int_2^{\infty}\frac{\mathrm{li}(t)}{t^{s+1}}dt=\frac{\mathrm{li}(2)}{s2^s}+\frac{1}{s}\int_2^{\infty}\frac{dt}{t^s\log t}$$

and the substitution $x=(s-1)\log t$ shows that the last integral equals

$$\int_{(s-1)\log 2}^{\infty}\frac{dx}{xe^x}=-\log(s-1)+h_1(s)$$

with an entire function $h_1(s)$. Since

$$\int_2^\infty \frac{\log 2}{t^{s+1}}dt = \frac{\log 2}{s2^s} = \frac{\log 2}{s} + h_2(s)$$

with entire h_2 we obtain finally for the function

$$F(s) = \int_2^\infty \frac{g(t)}{t^{1+s}}dt$$

the equality

$$F(s) = \frac{\log\zeta(s) + \log(s-1) + \log 2}{s} + h(s)$$

with a certain entire function $h(s)$. Now Phragmén uses the functional equation for $\zeta(s)$ as well as Riemann's formula (4.8) to deduce that $\zeta(s)$ does not vanish in $|s-1| \leq 1$ and this implies, using $\zeta(0) = -1/2$, that the function $F(s)$ is regular in a disc $|s-1| \leq R$ with a certain $R > 1$. It remains to take for $x_1, x_2, \ldots$ consecutive prime numbers starting with $x_1 = 11$ and observe that the assumptions of the second part of the theorem are satisfied with $h_j = 1$ and $m_j = 1/2$. □

The second corollary to Theorem 4.13 proves Čebyšev's first statement about the difference $\Delta(x;4,3,1) = \pi(x;4,3) - \pi(x;4,1)$, contained in his letter to Fuss (Čebyšev 1853), and which we mentioned in Chap.3:

Corollary 2. *The ratio*

$$\frac{\Delta(x;4,3,1)}{\sqrt{x}/\log x}$$

attains values arbitrarily close to 1.

Proof. Denote by Ω the family of all functions regular in the disc $|s-1| \leq 1$ and observe that for any finite subset of Ω there exists a number $R > 1$ such that all functions in this subset are regular in $|s-1| \leq R$.

If for $j = 1, 3$ we put[70]

$$f_j(x) = \sum_{\substack{p^k \leq x \\ p^k \equiv j \pmod 4}} \frac{1}{k}$$

then for Re $s > 1$ one gets by Lemma 4.12

$$\int_2^\infty \frac{f_j(x)}{x^{1+s}}dx = \frac{1}{s} \sum_{\substack{p^k \leq x \\ p^k \equiv j \pmod 4}} \frac{1}{kp^{ks}}.$$

[70] In Phragmén's paper the functions f_j (denoted there by ϕ and ψ) are defined as having jumps $1/kp^k$ at prime powers p^k and being constant between them. This seems to be merely a printing error.

An application of Dirichlet's Lemma 2.9 gives

$$\int_2^\infty \frac{f_1(x)}{x^{1+s}}dx = \frac{1}{2s}(\log L_0(s) + \log L(s))$$

and

$$\int_2^\infty \frac{f_3(x)}{x^{1+s}}dx = \frac{1}{2s}(\log L_0(s) - \log L(s)),$$

where $L_0(s) = L(s, \chi_0)$ with χ_0 being the principal character mod 4, i.e.

$$L_0(s) = \left(1 - \frac{1}{2^s}\right)\zeta(s)$$

and $L(s)$ is the L-function corresponding to the remaining character mod 4, i.e. to the character $\chi(n) = \left(\frac{-1}{n}\right)$. Using the functional equation (Theorem 4.6) Phragmén gets the non-vanishing of $L(s)$ in the disc $|s-1| \leq 1$ and the equality $L(0) = 1/2$. This shows that one can write

$$\log L(s) = -\log 2 + sh(s),$$

with $h \in \Omega$ and it follows that the function defined by the integral

$$\int_2^\infty \frac{f_1(x) - f_3(x) + \log 2}{x^{1+s}}dx$$

belongs to Ω. Let $f(x)$ be the function occuring in Corollary 1. Equality (4.27) implies

$$\int_1^\infty \frac{f(x^{1/2})}{x^{1+s}}dx = \frac{\log \zeta(2s)}{s}$$

but since the functional equation for $\zeta(s)$ implies $\zeta(0) = -1/2$ one has

$$\zeta(s) = \frac{1}{s-1} + \frac{1}{2} + sh_1(s)$$

with $h_1 \in \Omega$, hence (since $\zeta(2s)$ does not vanish in the disc $|s-1| \leq 1$)

$$\int_2^\infty \frac{f(x^{1/2})}{x^{1+s}}dx = -\frac{\log(2s-1) + \log 2}{s} + h_2(s)$$

holds with some $h_2 \in \Omega$ and moreover one sees without trouble that there exists $h_3 \in \Omega$ such that

$$\int_2^\infty \frac{\mathrm{li}(x^{1/2})}{x^{1+s}}dx = -\frac{\log(2s-1)}{s} + h_3(s).$$

This shows that if one puts

$$S(x) = f_1(x) - f_3(x) - \frac{1}{2}(f(x^{1/2}) - \mathrm{li}(x^{1/2}) - \log 2)$$

then the integral

$$\int_2^\infty \frac{S(x)}{x^{1+s}}dx$$

defines a function belonging to Ω. Similarly as in the proof of Corollary 1 one infers that the function $S(x)$ changes its sign infinitely often, i.e. there exist two sequences t_i and u_i tending to infinity and satisfying $t_1 < u_1 < t_2 < u_2 < \dots$ and $S(t_i) < 0$, $S(u_i) > 0$. One can freely assume that $u_i - t_i$ does not exceed 1/2 for all i. This implies $0 \le f_k(u_i) - f_k(t_i) \le 1$ for $k = 1, 3$, $0 \le f(u_i) - f(t_i) \le 1$ and $\mathrm{li}(u_i^{1/2}) - \mathrm{li}(t_i^{1/2}) = O(1)$ and the evaluation

$$S(u_i) = O(1) \tag{4.28}$$

results.

Now observe that

$$f_3(x) - f_1(x) = \pi(x;4,3) - \pi(x;4,1) + \sum_{k\ge 2}\frac{1}{k}\left(\sum_{\substack{p^k\le x\\ p^k\equiv 3 \pmod 4}} 1 - \sum_{\substack{p^k\le x\\ p^k\equiv 1 \pmod 4}} 1\right)$$

and since for odd p we have $p^2 \equiv 1 \pmod 4$ and

$$\sum_{k\ge 3}\sum_{p^k\le x} 1 = O(x^{1/3}\log x)$$

it follows that

$$f_3(x) - f_1(x) = \pi(x;4,3) - \pi(x;4,1) - \frac{1}{2}\pi(\sqrt{x}) + O(x^{1/3}\log x).$$

This implies

$$\pi(x;4,3) - \pi(x;4,1) - \frac{1}{2}\mathrm{li}(\sqrt{x}) = S(x) + O(x^{1/3}\log x)$$

and using (4.28) one arrives at

$$\pi(u_i;4,3) - \pi(u_i;4,1) - \frac{1}{2}\mathrm{li}(\sqrt{u_i}) = O(u_i^{1/3}\log u_i)$$

for $i \to \infty$. Dividing by $\frac{1}{2}\mathrm{li}(\sqrt{u_i})$ one is led to

$$\lim_{i\to\infty}\frac{2(\pi(u_i;4,3) - \pi(u_i;4,1))}{\mathrm{li}(\sqrt{u_i})} = 1$$

and the assertion follows. □

Later Landau (1905b) modified Phragmén's proof of Theorem 4.13, by noting that the assumption of regularity of the considered integrals in $|s-1| \le R$ for some

$R > 1$ may be replaced by the existence of analytic continuation along the interval $[0,1]$ (see Theorem 5.16).

It was stated by G.H.Hardy and J.E.Littlewood (1918,p.127) that the methods of their paper may be used to show that $\Delta(x;4,3,1)$ attains positive and negative values of arbitrary large modulus, more precisely one has

$$\limsup_{x\to\infty} \frac{\Delta(x;4,3,1)\log x}{\log\log x} > 0$$

and

$$\liminf_{x\to\infty} \frac{\Delta(x;4,3,1)\log x}{\log\log x} < 0.$$

This implies in particular that $\Delta(x;4,3,1)$ changes its sign infinitely often (the first negative value occuring for $x = 26861$ (Leech 1957)). (See Knapowski, Turán 1962,II).

A numerical study of $\pi(x;k,l)$ led A.J.C.Cunningham (1911,1913) to the conjecture

$$\pi(n;k,1) < \frac{\pi(n)}{\varphi(k)}.$$

The above result of Hardy and Littlewood shows that this conjecture already fails in the case $k = 4$, since if $\Delta(x;4,3,1)$ is negative then $\pi(x;4,3) \le \pi(x;4,1) - 1$ and the truth of Cunningham's conjecture would imply

$$\pi(x) - 1 = \pi(x;4,1) + \pi(x;4,3) \le 2\pi(x;4,1) - 1 < \pi(x) - 1,$$

a contradiction. Much later it turned out (Kaczorowski 1993,1994,1995,1996) that for every $k \ge 3$ this conjecture contradicts the *Extended Riemann Hypothesis* for L-functions corresponding to characters mod k, which asserts that all zeros of $L(s,\chi)$ in the half-plane Re $s > 0$ lie on the line Re $s = 1/2$. For a numerical study of $\Delta(x;4,3,1)$ see Shanks (1959).

The comparison of the behaviour of primes lying in various progressions is the subject of *comparative prime number theory* which was systematically built up in a series of papers of S.Knapowski and P.Turán (Knapowski 1958, 1959, 1960/61, 1961/62, 1962c, Knapowski,Turán 1962, 1963, 1964, 1965a,b, 1966, 1972, 1977). In the first part of Knapowski,Turán (1964) one finds a list of 44 problems concerning this subject. Most of them are still open. We mention here only some sample relevant results:

I.Kátai (1964) proved that if the L-functions corresponding to characters mod k do not have real zeros in $[0,1]$ and moreover the congruences $x^2 \equiv r \pmod{k}$ and $x^2 \equiv s \pmod{k}$ have the same number of solutions then

$$\limsup_{x\to\infty} \frac{(\pi(x;k,r) - \pi(x;k,s))\log x}{\sqrt{x}} > 0.$$

It was conjectured (the *Shanks–Rényi*[71] *race problem*) that if $l_1, l_2, \dots, l_r$ are distinct reduced residue classes mod k then there exist infinitely many integers x for which one has

$$\pi(x;k,l_1) > \pi(x;k,l_2) > \ldots > \pi(x;k,l_r).$$

M.Rubinstein and P.Sarnak (1994) showed that this assertion is a consequence of the Extended Riemann Hypothesis and the conjectured linear independence (over

[71]Rényi, Alfred (1921–1970), Professor in Budapest.

the rationals) of the imaginary parts of non-real zeros of all L-functions. See also Friedlander, Granville (1992), Kaczorowski (1993,1994,1995a,1996) and Stark (1971).

For an application of Phragmén's Theorem 4.13 to the study of the asymptotic behaviour of the function $\sum_{n\leq x}\sigma(n)/n$ see Wigert (1914).

Exercises

1. (E.L.Stark 1979) For $0 < r < 1$ put

$$p_r(x) = 1 + 2\sum_{k=1}^{\infty} r^k \cos(kx).$$

(i) Prove

$$p_r(x) = \frac{1-r^2}{1-2r\cos x + r^2}.$$

(ii) Prove

$$0 < p_r(x) \leq \frac{\pi^2(1-r)}{2rx^2}.$$

(iii) Put $M_r = \int_0^\pi x^2 p_r(x)dx$ and prove

$$0 < M_r = 4\pi\left(\frac{\pi^2}{12} - \sum_{k=1}^{\infty}(-1)^{k+1}\frac{r^k}{k^2}\right) < \frac{\pi^3(1-r)}{2r}.$$

(iv) Deduce from (iii) the equality

$$\sum_{k=1}^{\infty}\frac{(-1)^{k+1}}{k^2} = \frac{\pi^2}{12}$$

and obtain $\zeta(2) = \pi^2/6$.

2. (i) Prove that $\zeta(0) = -1/2$.

(ii) Show that for $n = 1, 2, \ldots$ one has

$$\zeta(1-2n) = (-1)^n\frac{B_{2n}}{2n}.$$

(iii) Prove

$$\lim_{x\to 0+0}\sum_{n=1}^{\infty}\frac{(-1)^{n-1}}{n^x} = \frac{1}{2}.$$

3. (i) Show that the real zeros of $\zeta(s)$ coincide with negative even integers.

(ii) (For a simple proof see Apostol 1985a) Show that if χ is a non-principal Dirichlet character then $L(s,\chi)$ vanishes at non-positive real integers in case $\chi(-1) = -1$ and at negative odd integers in case $\chi(-1) = 1$. Prove also that $L(s,\chi)$ does not have other negative real zeros.

4. (Riemann 1860, Cramér,Landau 1921) (i) Show that for Re $s = 1/2$ the function

$$\pi^{-s/2}\Gamma\left(\frac{s}{2}\right)\zeta(s)$$

is real and its logarithmic derivative is totally imaginary, provided $\zeta(s) \neq 0$.

(ii) Deduce from (i) that

$$\lim_{t\to\infty} \frac{\zeta'}{\zeta}(\frac{1}{2} + it) = +\infty.$$

(iii) Prove that if all zeros of $\zeta(s)$ on the line Re $s = 1/2$ are simple then this line contains only finitely many zeros of $\zeta'(s)$.

5. (Hadamard 1909, Ramaswami 1934) Establish the identity

$$(1 - 2^{1-s})\zeta(s) = \sum_{k=1}^{\infty} \frac{s(s+1)\cdots(s+k-1)\zeta(s+k)}{k!2^{s+k}}.$$

6. (Franel 1899) Prove that for Re $s > 0$ one has

$$\zeta(s) = \frac{1}{s-1} + \lim_{n\to\infty}\left(1 + \frac{1}{2^s} + \cdots + \frac{1}{n^s} - \frac{n^{1-s}-1}{1-s}\right).$$

7. (Landau,Walfisz[72] 1920) (i) Prove for Re $s > 1$ the identity

$$\sum_p \frac{1}{p^s} = \sum_{k=1}^{\infty} \frac{\mu(k)}{k}\zeta(ks).$$

(ii) Show that the series in (i) can be continued to a function regular in Re $s > 0$ apart for a discrete set of logarithmic singularities.

(iii) Prove that the function defined for Re $s > 1$ by the series

$$\sum_p \frac{\log p}{p^s}$$

can be continued to a function in Re $s > 0$ whose only singularities are simple poles.

8. (i) (Cahen 1894) Prove that if the convergence abscissa $c(f)$ of the Dirichlet series

$$f(s) = \sum_{n=1}^{\infty} \frac{a_n}{n^s}$$

is positive, then it is given by the formula

$$c(f) = \limsup_{n\to\infty} \frac{\log\left(\left|\sum_{j=1}^{n} a_j\right|\right)}{\log n}.$$

[72] Walfisz, Arnold (1892–1962), 1930–36 worked at Warsaw University and later in Tbilisi.

(ii) Let $a(f)$ be the abscissa of absolute convergence of the series. Prove that $a(f) - c(f) \leq 1$ and show that this inequality is best possible.

(iii) Give an example of a general Dirichlet series $\sum_{n=1}^{\infty} a_n e^{-\lambda_n s}$ with $0 < \lambda_1 < \lambda_2 < \ldots \longrightarrow \infty$ which has a half-plane of convergence but which does not converge absolutely anywhere.

9. (Schnee 1909b) Put

$$\epsilon_n = \begin{cases} 1 & \text{if } [\log^2 n] \text{ is even} \\ -1 & \text{otherwise} \end{cases}$$

and show that the series

$$\sum_{n=2}^{\infty} \frac{\epsilon_n}{n^s \log n}$$

converges absolutely for Re $s > 1$, converges nonabsolutely on the line Re $s = 1$ and diverges for Re $s < 1$.

10. (i) (Dirichlet 1857) Prove that if $\lim_{n\to\infty} a_n = A$, then for real s one has

$$\lim_{s\to 0+0} s \sum_{n=1}^{\infty} \frac{a_n}{n^{1+s}} = A.$$

(ii) (Schnee 1908) Show that the same assertion is valid if s is a complex variable tending to 0 inside an angle $-\theta \leq \arg s \leq \theta$ for some positive $\theta < \pi/2$.

11. (Schnee 1909a) If $f(s)$, $a(f)$ and $c(f)$ are as in Exercise 8 and $\sigma > (a(f) + c(f))/2$, then

$$\lim_{T\to\infty} \frac{1}{2T} \int_{-T}^{T} |f(\sigma + it)|^2 dt = \sum_{n=1}^{\infty} \frac{|a_n|^2}{n^{2\sigma}}.$$

12. (Apostol 1970,1972) For a non-principal Dirichlet character χ mod k define the Gaussian sum $\tau_n(\chi)$ by the formula

$$\tau_n(\chi) = \sum_{j=1}^{k} \chi(j)\zeta_k^{jn}$$

and

$$\tau(\chi) = \tau_1(\chi).$$

(i) Prove that the equality

$$\tau_n(\chi) = \bar{\chi}(n)\tau(\chi)$$

holds for all n if and only if χ is a primitive character mod k.

(ii) Prove that if for Re $s > 1$ one defines

$$F(x, s) = \sum_{n=1}^{\infty} \exp(2\pi i n x) n^{-s}$$

then the equality

$$\sum_{m=1}^{k-1} \chi(m)F(m/k,s) = \tau(\chi)L(s,\chi)$$

holds if and only if χ is a primitive character mod k.

(iii) Prove the equality

$$L(1-s,\chi) = (2\pi)^s \Gamma(s) k^{s-1} (\exp(-is\pi/2)$$
$$+\chi(-1)\exp(is\pi/2) \sum_{m=1}^{k-1} \chi(m)F(m/k,s).$$

(iv) Show that if $L(s,\chi)$ satisfies the functional equation given in Theorem 4.8 then χ is a primitive character mod k.

13. (Apostol 1954) Prove the identity

$$\sum_{\substack{1\leq m\leq n \\ (m,n)=1}} \zeta(s,\frac{m}{n}) = k^s \zeta(s) \sum_{d|n} \frac{\mu(d)}{d^s}.$$

14. (Apostol,Chowla 1959; Subbarao 1980) Prove that the function $F_{k,l}(s)$ does not have an Euler product when $k > 2$, $(k,l) = 1$.

15. (Salerno,Vitolo,Zannier 1991) Show that if the sequence of coefficients a_n of a Dirichlet series is periodic then it can be extended to a meromorphic function which is either entire or has a single simple pole at $s = 1$. Prove also that this function satisfies a functional equation.

16. (Stark 1993) Let $k > 1$ be given and let $f(n)$ be an arithmetic function, which is periodic mod k and satisfies $\sum_{j=1}^k f(j) = 0$ and $f(k-n) = f(n)$ for $n = 1,2,\ldots,k-1$. Moreover put

$$L(s,f) = \sum_{n=1}^{\infty} \frac{f(n)}{n^s}.$$

(i) Show that for Re $s > 1$ one has

$$L(s,f) + s\zeta(1+s) \sum_{n=1}^{k} nf(n)$$
$$= \sum_{n=1}^{k} \frac{f(n)}{n^s} + \sum_{m=1}^{\infty} f(n) \left(\frac{1}{(mk+n)^s} - \frac{1}{(mk)^s} + \frac{sn}{mk^{1+s}} \right).$$

(ii) Deduce from (i) that $L(s,f)$ has a continuation to a function regular in the half-plane Re $s > -1$.

(iii) Prove the equality

$$L(0,f) = -\frac{1}{k}\sum_{n=1}^{k} nf(n).$$

(iv) Use (iii) and the functional equation for Dirichlet's L-functions to obtain a formula for $L(1,\chi)$ in the case $\chi(-1) = 1$.

17. (Phragmén 1891) (i) Show that the functions

$$\sum_{p^k \le x} \frac{\log p}{p^k} - \log x + \gamma$$

and

$$\sum_{p^k \le x} \frac{1}{p^k} - \log\log x + \gamma$$

change their sign infinitely often.

18. (i) (Erdős 1946) Prove that if $f(n)$ is a non-decreasing multiplicative function (i.e. $f(mn) = f(m)f(n)$ holds when $(m,n) = 1$) then with a suitable c one has $f(n) = n^c$.

(ii) (Turán[73] 1959) Assume that the series $f(s) = \sum_{n=1}^{\infty} a_n n^{-s}$ with real monotonic coefficients is convergent in a half-plane and satisfies there the Euler product formula

$$f(s) = \prod_p \left(1 + \sum_{k=1}^{\infty} \frac{a_{p^k}}{p^{ks}}\right).$$

Prove that with a suitable real b one has $f(s) = \zeta(s+b)$.

19. (von Koch 1901) For $x, y > 1$ put

$$S(x,y) = \sum_{n=1}^{\infty} \frac{(-1)^{n-1}}{n!} x^{ny}\zeta(ny).$$

(i) Prove that the series

$$\sum_p \sum_{k=1}^{\infty} \frac{1}{k}(1 - \exp(-x^y p^{-ky}))$$

converges for real $y \ge 1 + \epsilon$ (with $\epsilon > 0$) uniformly to $S(x,y)$.

(ii) Prove that

$$\lim_{x\to\infty} S(x,y) = \alpha(x) + \sum_{k=1}^{\infty} \frac{1}{k}\pi(x^{1/k}),$$

where

$$\alpha(x) = \begin{cases} -\exp(-N) & \text{if } x = p^N \text{ for some prime } p \\ 0 & \text{otherwise.} \end{cases}$$

[73]For an analogue for L-functions see Turán (1965).

5. The Prime Number Theorem

5.1. Hadamard's First Paper on the Zeta-function and its Consequences

1. The last twenty years of the nineteenth century witnessed a rapid progress in the theory of complex functions, summed up in the monumental treatises of Émile Picard[1] (1891–1896) and Camille Jordan[2] (1893-1896). The development of the theory of integral functions, started by Karl Weierstrass (1876) and rounded up by Jacques Hadamard (1893), revived the interest in Riemann's memoir and forced attempts to use these new developments to solve questions left open by Riemann. This led to the first proofs of the Prime Number Theorem[3] in the form

$$\theta(x) = (1 + o(1))x$$

obtained independently by J.Hadamard (1896a,b,c) and C.de la Vallée-Poussin (1896a). They both started with establishing the non-vanishing of $\zeta(s)$ on the line Re $s = 1$ but obtained this result in completely different ways. Also the deduction of Prime Number Theorem from that result is done by them differently.

K.Weierstrass showed that every non–zero entire function f can be written in the form

$$f(s) = s^k e^{g(s)} \prod_{n=1}^{\infty} \left(1 - \frac{s}{\alpha_n}\right) e^{F_n(s)}, \tag{5.1}$$

where $k \geq 0$, $g(s)$ is entire, α_n is the sequence of all non-zero zeros of f arranged according to their absolute values, multiple zeros appearing according to their multiplicity and $F_1(s), F_2(s), \dots$ are polynomials, chosen to ensure the almost uniform convergence of the resulting product in the complex plane.

---

[1] Picard, Charles Émile (1856–1941), Professor in Toulouse (1879–1881) and in Paris.

[2] Jordan, Camille (1838–1922), Professor in Paris.

[3] According to E.Landau (1903a) this name of the equivalent theorems 5.8 and 5.13 appears first in its German form *Primzahlsatz* in the thesis of H.von Schaper (1898).

Hadamard in his 1893 paper made this formula more explicit, proving essentially that if $M(r)$ denotes the maximum of $|f(s)|$ in the disk $|s| \le r$ and the order $\mathrm{ord}(f)$ of f, defined by

$$\mathrm{ord}(f) = \limsup_{r\to\infty} \frac{\log\log M(r)}{\log r},$$

is finite then either f is a polynomial or the polynomials $F_1,F_2,\dots$ appearing in (5.1) can be written in the form

$$F_n(X) = \sum_{k=1}^{[\mathrm{ord}(f)]} \frac{1}{k\alpha_n^k} X^k.$$

In particular the degree of F_n does not exceed $\mathrm{ord}(f)$. Moreover the series

$$\sum_n \frac{1}{|\alpha_n|^{1+\mathrm{ord}(f)}}$$

converges. Hadamard showed further that if the coefficients A_n of the Maclaurin series of f satisfy, with some positive c, the inequality

$$|A_n| < \frac{1}{(n!)^c},$$

then $\mathrm{ord}(f) \le 1/c$.

Hadamard's original proof of these assertions was rather complicated since he did not have at his disposal many tools, like e.g. Jensen's theorem[4], which was proved a few years afterwards (Jensen 1899).

A very simple proof of Hadamard's results was found by E.Landau (1927b). See also Chandrasekharan (1941, 1970).

In the last part of his paper Hadamard applies his results to $\zeta(s)$ obtaining the following result:

Theorem 5.1. *If $\rho_1, \rho_2, \dots$ is the sequence of all non-real zeros of $\zeta(s)$ arranged according to their absolute values then*

$$A\frac{n}{\log n} \le |\rho_n| \le B\frac{n}{\log n},$$

holds with suitable constants $B > A > 0$ and the series

$$\sum_n \frac{1}{|\rho_n|^\mu}$$

[4] According to S.Mandelbrojt (1967) Hadamard was actually in posession of Jensen's formula before Jensen but did not not publish it, since he could not find for it any important application.

converges for $\mu > 1$. Moreover the function $\Xi(s) = \xi(1/2 + is)$ with $\xi(s)$ defined by (4.9) is of order not exceeding[5] 1.

Proof of Hadamard. Expanding the integral occuring in Riemann's formula for the function $\Xi(s)$ (cf. (4.10)) in the Taylor series one gets

$$\Xi(s) = \frac{1}{2} - \left(s^2 + \frac{1}{4}\right) \sum_{n=0}^{\infty} (-1)^n C_{2n} s^{2n},$$

where

$$C_n = \frac{1}{2^n n!} \int_1^{\infty} f(t) t^{-3/4} \log^n t \, dt$$

and

$$f(t) = \sum_{n=1}^{\infty} e^{-n^2 \pi t}.$$

This leads to

$$\Xi(s) = \sum_{n=0}^{\infty} A_n s^{2n}$$

where

$$A_n = (-1)^n (C_{2n}/4 - C_{2n-2})$$

for $n = 1, 2, \ldots$. In view of the rapid convergence of the series defining $f(t)$ one sees without great trouble that

$$\frac{(1 + o(1))^{2n}}{4^n} \frac{\log^{2n}(n)}{(2n)!} \leq |A_n| \leq \left(\frac{1}{2} + o(1)\right)^{2n} \frac{\log^{2n}(2n)}{(2n)!},$$

hence the result of Hadamard quoted above implies for α_n, the n-th zero of $\Xi(s)$, the bounds

$$A \frac{n}{\log n} \leq |\alpha = bb_n| \leq B \frac{n}{\log n}$$

with suitable constants $B > A > 0$.

The first two assertions follow now immediately from the equality $\rho_n = \frac{1}{2} + i\alpha_n$. To obtain the last note that $\Xi(\sqrt{s}) = \sum_{n=0}^{\infty} A_n s^n$ and thus the obtained bound for $|A_n|$ shows that

$$A_n = O\left(\frac{1}{(n!)^c}\right)$$

holds for every $c < 2$ and hence the order of $\Xi(\sqrt{s})$ is $\leq 1/2$ and the assertion follows. □

A simple proof of Theorem 5.1 based on Jensen's theorem can be found in Edwards (1974).

---

[5] In fact its order equals 1.

2. The following corollaries do not occur explicitly in Hadamard's paper, except the first identity in Corollary 1, but they follow easily from his results and in Hadamard (1896a) the first formula from the fourth corollary is explicitly quoted, being one of the main tools in his proof of the Prime Number Theorem.

Corollary 1. *For all $s \neq 1$ one has*

$$\Xi(s) = \Xi(0) \prod_{i=1}^{\infty} \left(1 - \frac{s^2}{\alpha_i^2}\right),$$

where α_i runs over all zeros of $\Xi(s)$ and

$$(s-1)\zeta(s) = \frac{1}{2} \frac{e^{as}}{\Gamma(\frac{s}{2}+1)} \prod_{\rho} \left(1 - \frac{s}{\rho}\right) e^{s/\rho}$$

with a suitable constant a, ρ running over all non-real zeros of $\zeta(s)$.

The product formula for $\Xi(s)$ already appears, in an equivalent form, in Riemann's memoir with a rather vague sketch of a proof. See Edwards (1974, Sect.1.11.)

The exact value of a is irrelevant for our purposes but it is not difficult to show that it equals $\log(2\pi) - 1 - \gamma/2$.

Proof. Observe first that Theorem 5.1 implies the equality

$$\Xi(\sqrt{s}) = \Xi(0) \prod_{i=1}^{\infty} \left(1 - \frac{s}{\beta_i}\right),$$

β_i running over all zeros of $\Xi(\sqrt{s})$. Putting $z = \sqrt{s}$ and noting that one has $\beta_i = \alpha_i^2$, α_i running over all zeros of $\Xi(s)$, we get the first assertion. The second is a consequence of the first and the relation between Ξ and ζ. Full details of the rather dull computation may be found in §76 of Landau's *Handbuch* (Landau 1909a). □

A simple proof of the second part of this corollary not utilizing Hadamard's theory of entire functions but based on Jensen's formula was given in Ananda-Rau (1924). It was further simplified in Landau (1927b).

Corollary 2. *Let $N(T)$ be the number of zeros of $\zeta(s)$ lying in the strip $0 < \operatorname{Im} s \leq T$, each counted according to its multiplicity. Then there exist positive constants c_1, c_2 such that for sufficiently large T one has*

$$c_1 T \log T \leq N(T) \leq c_2 T \log T.$$

Proof. Let $\rho_1, \rho_2, \ldots$ be the sequence of all zeros of $\zeta(s)$ lying in the upper half-plane and arranged according to their absolute values. If $\rho_n \leq T < \rho_{n+1}$, then $n = N(T)$ and Theorem 5.1 gives

$$A\frac{n}{\log n} \le T < B\frac{n+1}{\log(n+1)} = O(\frac{n}{\log n}).$$

This implies the assertion. □

From this one gets immediately the following result:

Corollary 3. *The zeta-function has infinitely many zeros in the strip* $0 \le \operatorname{Re} s \le 1$. □

A much simpler proof of this fact was given later by H.Bohr (1912c). A short proof of the existence of at least one root of the zeta-function in the strip $1/2 \le \operatorname{Re} s \le 1$ was provided by K.Ramachandra (1995).

Corollary 4. *If* $s \neq 1$ *is not a zero of* $\zeta(s)$, *then with suitable constants* C, C_1 *and* C_2 *one has*

$$\frac{\zeta'(s)}{\zeta(s)} = \frac{1}{1-s} + \sum_{\alpha}\left(\frac{1}{s-\alpha} + \frac{1}{\alpha}\right) + C$$

$$= \frac{1}{1-s} - \frac{1}{2}\frac{\Gamma'}{\Gamma}\left(\frac{s}{2}+1\right) + \sum_{\rho}\left(\frac{1}{s-\rho} + \frac{1}{\rho}\right) + C_1$$

$$= \frac{1}{1-s} - \frac{1}{2}\frac{\Gamma'}{\Gamma}\left(\frac{s}{2}+1\right) + \sum_{\rho}\frac{1}{s-\rho} + C_2,$$

α *and* ρ *running over all resp. all non-real zeros of* $\zeta(s)$. *The first two series occuring here converge absolutely and the last converges if the* ρ*'s are arranged according to their absolute values.*

Proof. The first two equalities follow from Corollary 1 and the product representation

$$\frac{1}{\Gamma(s)} = e^{As}s\prod_{m=1}^{\infty}\left(1+\frac{s}{m}\right)e^{-s/m}.$$

The last equality follows from the observation that the series

$$\sum_{\substack{\rho \\ \operatorname{Im}\rho>0}}\left(\frac{1}{\rho}+\frac{1}{\bar\rho}\right)$$

converges absolutely due to Corollary 2. □

5.2. von Mangoldt

1. Corollary 2 to Theorem 5.1 was the first step toward verification of the assertion of Riemann's memoir dealing with the asymptotics for the number $N(T)$ of roots ρ of $\zeta(s)$ with $0 < \text{Im } \rho \leq T$ for T tending to infinity (assertion I in Sect.4.1). In 1887 Stieltjes considered this question and obtained the formula

$$N(T) = \frac{1}{2\pi} T \log T - \frac{1+\log(2\pi)}{2\pi} T + \frac{1}{\pi} \Delta_{1/2+Ti,2+Ti} \arg \xi(s) + O(T^{-1}),$$

where $\Delta_{a,b} h(s)$ denotes the variation of the function h on the segment joining a and b and $\xi(s)$ is defined by (4.9). He communicated this in a letter to Mittag-Leffler[6] dated April 15th 1887.

The final step was made by H.von Mangoldt[7] (1895) who used Corollary 1 to Theorem 5.1 to establish the following result:

Theorem 5.2. *For the number $N(T)$ of zeros ρ of $\zeta(s)$ satisfying $0 < \text{Im } \rho \leq T$ and counted according to their multiplicities the following asymptotic equality holds for $T > 12$:*

$$N(T) = \frac{T}{2\pi} \log\left(\frac{T}{2\pi}\right) - \frac{T}{2\pi} + O(\log^2 T).$$

The error term does not exceed $0.34 \log^2 T + 1.35 \log T + 2.58$.

Sketch of the proof: Let $\alpha_1, \alpha_2, \ldots$ be the roots of Riemann's function $\Xi(t)$ lying in the upper half-plane, ordered according to the size of their real parts and each occuring according to its multiplicity. We obviously have

$$N(T) = \#\{\alpha_j : \text{ Re } \alpha_j \leq T\}$$

while Riemann's functional equation implies $|\text{Im } \alpha_j| \leq 1/2$. Choose $0 < a < b$, put $A = \{j : \ a \leq \text{Re } \alpha_j < b\}$, $N = \#A$ and let $t = x - 3i/2$ $(a \leq x < b)$. If now S denotes the sum of angles under which one sees the path of t from points belonging to A (counted according to the multiplicity of the corresponding zero) then an elementary geometric argument gives

$$N \arctan((b-a)/2) \leq S.$$

On the other hand the first formula of Corollary 1 to Theorem 5.1 implies that S does not exceed the increase Φ of the argument of $\Xi(t)$ for $a \leq x < b$.

---

[6] Baillaud,Bourget 1905,II,452–457.

[7] This paper of von Mangoldt is a good example of exactness and thoroughness in manipulating nonabsolutely convergent series and integrals. It largely exceeded the standard of his time, which can be observed by reading Cahen (1894) or Stieltjes (1885a, 1885b, 1889), and seems to be one of the first papers having these qualities.

To obtain an upper bound for Φ von Mangoldt puts first $t = \tau - 3i/2$ (with positive τ) in Riemann's formula

$$\Xi(t) = \Gamma\left(\frac{5}{4} + i\frac{t}{2}\right)(-1/2 + it)\pi^{-1/4 - it/2}\zeta\left(\frac{1}{2} + it\right)$$

and then uses the complex version of Stirling's formula, due to Stieltjes (1889):

$$\log \Gamma(1+z) = (z + \frac{1}{2})\log z - z + \frac{1}{2}\log(2\pi) + J(z),$$

with the error term satisfying

$$|J(z)| < \frac{1}{12}|z|^{-1}\left(\cos(\arg z/2)\right)^{-2}.$$

Denoting by $F(\tau)$ the imaginary part of $\log \Xi(\tau - 3i/2)$ and by $f(t)$ the imaginary part of $\log \zeta(2 + i\tau) + J(i\tau/2)$ one obtains after a short calculation which uses certain lemmas from Stieltjes' paper the formula

$$\begin{aligned} F(\tau) =& \frac{\tau}{2}\log(\tau/2) + \arctan(\tau/2) + \arctan(\tau) \\ &+ \pi/4 - \frac{\tau}{2}(1 + \log \pi) + f(\tau) \quad (\tau > 0) \end{aligned} \tag{5.2}$$

with

$$-\log\left(\frac{\pi^2}{6}\right) - \frac{1}{3\tau} < f(\tau) < \log\left(\frac{\pi^2}{6}\right).$$

The remaining part of the proof is founded on (5.2) and the equality

$$\Phi = F(b) - F(a). \tag{5.3}$$

First of all it is shown that there are no zeros of Ξ with real part in the interval $[-12, 12]$, hence non-trivial zeros of Riemann's zeta-function in the upper half-plane have imaginary parts larger than 12. In fact if there is a zero in that interval then by symmetry there would be at least two of them, hence, taking $a = -12$, $b = 12$ and using (5.2), (5.3) and the inequality

$$2\arctan\left(\frac{b-a}{2}\right) < \Phi,$$

one is led by a pure numerical computation to a contradiction.

It was noted by von Mangoldt that this approach does not work if 12 is replaced by 13. He also wrote that possibly there is a zero of the zeta-function with imaginary part between 12 and 13. This is however not true because the first zero equals $1/2 + i14.134\ldots$, as shown by J.P.Gram (1895), who computed the first three zeros. Later (Gram 1903) he extended his computation calculating the first 15 zeros and noting that they all lie on the line $\text{Re } s = 1/2$. At the same time E.Lindelöf (1903) utilized, for the computation of the first ten zeros, a formula for replacing sums by integrals, which was stated by G.Plana (1820) and N.H.Abel (1823,1825) without precise assumptions and then proved by L.Kronecker (1889) (see Weber 1903 and Gomes-Teixeira 1904).

In the next step it is shown that if k, h are real numbers satisfying $\tan 1 \leq k \leq h-4$ then the number $A = A(k,h)$ of α_i's with real parts belonging to the interval $[h-k, h+k]$ satisfies

$$A \leq k \log h. \tag{5.4}$$

This is achieved by observing that

$$A \leq A \arctan\left(\frac{b-a}{2}\right) < \Phi = F(b) - F(a)$$

and now a short calculation utilizing (5.2) and Lagrange's mean value theorem gives the desired bound.

Now let $T > 12$ and put in (5.3) $a = T$, $b = -T$. Since the function $\Xi(t)$ is even and attains real values on the lines Re $t = 0$ and Im $t = 0$ we have $F(-T) = -F(T)$ and $F(0) = 0$. Thus the equality (5.2) leads to

$$\Phi = F(T) - F(-T) = 2F(T) = T\log(T/2\pi) - T + 5\pi/2 + R(T)$$

with $|R| < 2.548$. To deduce from this formula the number $N = N(T)$ of zeros with real part in $[-T, T]$ one has to observe that if we write $\beta_i = \mathrm{Re}\, \alpha_i$ and $\gamma_i = \mathrm{Im}\, \alpha_i$ (recall that $|\gamma_n| \leq 1/2$) then

$$\Phi = 2\sum_{n=1}^{\infty}\left(\arctan\frac{T+\beta_n}{3/2+\gamma_n} + \arctan\frac{T-\beta_n}{3/2+\gamma_n}\right). \tag{5.5}$$

If we put

$$S_1 = \sum_{n=1}^{T}\left(\arctan\frac{3/2+\gamma_n}{T+\beta_n} + \arctan\frac{3/2+\gamma_n}{T-\beta_n}\right)$$

and

$$S_2 = \sum_{n=T+1}^{\infty}\left(\arctan\frac{3/2+\gamma_n}{\beta_n - T} - \arctan\frac{3/2+\gamma_n}{\beta_n+T}\right)$$

then one sees without great trouble that

$$\Phi = 2\pi N - 2S_1 + sS_2.$$

It remains now to evaluate the sums S_1 and S_2. The first of them clearly satisfies

$$0 \leq S_1 < \sum_{n=1}^{N}\frac{2N}{T} + \sum_{n=1}^{N}\arctan\frac{2}{T-\beta_n}$$

and to obtain an upper bound for the right-hand side put $\kappa = \tan 1$ and $\lambda = 1 + \left[\frac{T-12}{2}\kappa\right]$. The numbers $T - 2\lambda\kappa$, $T - 2(\lambda-1)\kappa, \ldots, T - 4\kappa$, $T - 2\kappa$

divide the interval $[0, T]$ into $1+\lambda$ intervals $I_1, I_2, \ldots, I_{1+\lambda}$. The first of them, viz. $I_1 = [0, T-2\lambda\kappa]$ lies to the left of the number 12, hence it does not contain any β_n. The equality (5.4) shows that if $\lambda > 1$ then the interval I_2 contains at most $\kappa \log(T - (2\lambda - 1)\kappa) \le \kappa \log T$ β_n's, and the corresponding term in S_1 satisfies

$$\arctan \frac{2}{T - \beta_n} < \frac{2}{T - \beta_n} \le \frac{1}{(\lambda - 1)\kappa}.$$

Hence the sum of all terms corresponding to β_n's from I_2 does not exceed $\log T/(\lambda-1)$. Similarly one deals with all the remaining intervals I_j save $I_{1+\lambda}$ and this leads to the bounds $\log T/(\lambda-2)$, $\log T/(\lambda-3), \ldots, \log T$. The same approach gives for the last interval the bound $\frac{1}{2}\kappa\pi \log T$. Putting all these bounds together one gets

$$0 \le S_1 \le \frac{2N}{T} + \log^2 T + \frac{1}{2}\kappa\pi \log T, \tag{5.6}$$

and a similar, but technically more complicated procedure, leads to

$$0 \le S_2 \le \log^2 T + C \log T + C_1 \tag{5.7}$$

with suitable constants C and C_1, which are given explicitly.

The assertion of the theorem follows now from (5.5), (5.6) and (5.7). □

2. In a later paper von Mangoldt (1905) was able to diminish the error term in his theorem and obtained the following evaluation:

Theorem 5.3. *For $T > 29$ one has*

$$N(T) = \frac{T}{2\pi} \log\left(\frac{T}{2\pi}\right) - \frac{T}{2\pi} + O(\log T)$$

and the error term does not exceed $0.432 \log T + 1.191662 \log\log T + 13.07873$.

To prove it von Mangoldt applied a technically more complicated version of the same method. His starting point is the formula

$$\Xi(T - ia) = \zeta(\frac{1}{2} + a + iT)\pi^{-1/4-a/2-iT/2}(a - \frac{1}{2} + iT)\Gamma\left(\frac{5}{4} + \frac{a}{2} + i\frac{T}{2}\right),$$

with $a > 1$ and T chosen so that no root of $\Xi(s)$ has its real part equal to T. This formula is a direct consequence of the formula (4.9) and Theorem 4.2 (i). Taking logarithms of both sides and using Stirling's formula this leads to an expression of the form

$$\begin{aligned}\log \Xi(T - ia) &= i\frac{T}{2} \log\left(\frac{1}{4} + \frac{a}{2} + i\frac{T}{2}\right) + i\frac{T}{2}(1 + \log \pi) \\ &+ \log \zeta\left(\frac{1}{2} + a + iT\right) + O(\log T),\end{aligned} \tag{5.8}$$

the branch of the logarithm being chosen so as to make $\log \Xi(-ia)$ real. (Actually von Mangoldt obtained a more precise result, needed to obtain the asserted numerical coefficients in the error term.)

Now note that $\pi N(t)$ equals the increase of the argument of $\log \Xi(t)$ when t runs first over the segment joining $-ia$ and $T-ia$ and then over the segment joining $T-ia$ and T. The increase on the first segment equals the imaginary part of $\log \Xi(T-ia)$ and one deduces from (5.8) that it equals

$$\frac{T}{2}\log\frac{T}{2\pi}-\frac{T}{2}+Z(a,T)+O(1),$$

where $Z(a,T)$ denotes the imaginary part of $\log\zeta(1/2+a+iT)$. Choosing now $a=3/2$ one gets easily $|Z(a,T)|\leq\log\zeta(2)=O(1)$.

The evaluation of the increase on the second segment is more difficult and it is done by von Mangoldt by comparing this increase with the increase of the same function on the segment joining $T-2ai$ and $T-ia$. From (5.8) one gets without problems that with the same choice of a the latter increase is bounded and so everything is reduced to the study of the difference of both increases. This is the most painstaking part of the paper; von Mangoldt uses here an extension of the geometrical method used in the previous paper and shows that this difference is of the order $O(\log T)$. □

We point out an immediate corollary, which is used in the proof of de la Vallée-Poussin's bound for the error term in the Prime Number Theorem (Theorem 5.19):

Corollary. *For $T\geq 2$ one has*

$$N(T+1)-N(T)=O(\log T).$$ □

Later the proof of Theorem 5.3 was simplified by E.Landau (1908b) and described in full detail in Landau (1909a, Chap.20). The simplification consists, essentially, in the replacement of the detailed geometrical analysis by a precise evaluation of the integral of $\zeta'(s)/\zeta(s)$ along a suitable contour. Landau (1908c) also provided a direct proof of the corollary. Both proofs will be presented in Chap.6, where a direct proof of the preceding corollary will also be given (see Lemma 6.9).

The constants in the bound for the error term were later improved by J.Grossmann (1913), who used von Mangoldt's approach, and Backlund (1914) whose proof will be sketched in the next chapter.

If one assumes Riemann's Hypothesis then one can obtain a sharper bound for the error term. First H.Bohr, E.Landau and J.E.Littlewood[8] (1913) proved the bound to be $o(\log T)$, then H.Cramér (1918b) improved it to

$$O(\log T\log\log\log^{1/2}T\log\log^{-1/2}T),$$

E.Landau (1920) got

[8] According to Landau (1920) this result is due to Bohr alone.

$$O(\log T \log\log\log^2 T \log\log^{-1} T)$$

and finally J.E.Littlewood (1924b) obtained the bound

$$O\left(\frac{\log T}{\log\log T}\right).$$

This is still the best known estimate. A simpler proof of Littlewood's result was given by E.C.Titchmarsh (1927).

It should be pointed out that no essential unconditional improvement of the error term has been made since 1905. The assertion made by J.Franel (1896) that the bound $O(\log T)$ for the error term in Theorem 5.3 can be replaced by $O(1)$ was shown by E.Landau (1911a) to be incompatible with Riemann's Hypothesis and the same applies already to the weaker evaluation $o(\log\log T)$. This is done by showing first that the asserted bound leads, under the assumption of Riemann Hypothesis, to the boundedness of the difference $\zeta(s) - 1/(s-1)$ in $\mathrm{Re}\, s > 1/2 + \delta$ for every $\delta > 0$. (Landau presented two proofs of this implication, the first of them being due to Franel and contained in his letter of February, 16th 1901 to von Koch, quoted in Landau (1911a, p.129).) However it was deduced in Bohr, Landau (1910) from Kronecker's approximation theorem that this difference is already unbounded in the half-plane $\mathrm{Re}\, s > 1$. A.Selberg (1946) proved that under Riemann Hypothesis even the evaluation $O((\log T/\log\log T)^{1/2})$ fails to be true. Without assuming Riemann Hypothesis he established the impossibility of the bound $O(\log^{1/3} T(\log\log T)^{-7/3})$ and later the exponent $-7/3$ was replaced by $-1/3$ by K.-M.Tsang (1986).

The integral over $[0,T]$ of the error term in Theorem 5.3 is equal to $\frac{7}{8}T + O(T^{\epsilon})$ for every positive ϵ (Cramér 1919b) and the error term in this last formula can be replaced by $O(\log T)$ (Littlewood 1922). On the other hand it cannot be $O(\log^{1/2} T(\log\log T)^{-9/4})$ (Tsang 1986) (previously A.Selberg (1946) had the exponent -4 in place of $-9/4$).

Theorem 5.3 implies immediately the evaluation

$$\int_0^T \frac{N(t)}{t}dt = \frac{T}{2\pi}\log\left(\frac{T}{2\pi}\right) - \frac{T}{\pi} + O(\log^2 T)$$

and F.Nevanlinna and R.Nevanlinna (1923) proved that the error term here may be replaced by

$$\frac{7}{8}\log T + c + O\left(\frac{\log T}{T}\right).$$

with a suitable constant c.

3. The first paper of von Mangoldt (1895) contains also a proof of the explicit formula stated by Riemann (assertion II):

Theorem 5.4. *The series*

$$S(x) = \sum_{\rho}(\mathrm{li}(x^{\rho}) + \mathrm{li}(x^{\bar\rho}))$$

(where ρ runs over all non-real roots of $\zeta(s)$ lying in the upper half-plane) converges for all $x \geq 2$ and if x is not a prime power then

$$\sum_{p^k \le x} \frac{1}{k} = \mathrm{li}(x) - S(x) + \int_x^\infty \frac{dt}{t(t^2-1)\log t} - \log 2,$$

the sum being taken over all prime powers not exceeding x.

Note that the left-hand side of this formula equals

$$\sum_{n=1}^{\infty} \frac{1}{n}\pi(x^{1/n}) = \pi(x) + O\left(\frac{\sqrt{x}\log\log x}{\log x}\right)$$

since

$$\begin{aligned}\sum_{n=2}^{\infty} \frac{1}{n}\pi(x^{1/n}) &= \sum_{2\le n\le \log_2 x} \frac{1}{n}\pi(x^{1/n}) \le \pi(\sqrt{x}) \sum_{2\le n\le \log_2 x} \frac{1}{n} \\ &= O\left(\frac{\sqrt{x}\log\log x}{\log x}\right).\end{aligned}$$

Proof. von Mangoldt gives a very careful argument providing every detail of justification for interchange of summation and integration in his formulas. He starts with a meticulous proof of the formula

$$\lim_{h\to\infty} \frac{1}{2\pi i}\int_{a-ih}^{a+ih} \frac{e^{us}}{s}ds = \begin{cases} 1 & \text{if } u > 0 \\ 0 & \text{if } u < 0 \, . \\ \frac{1}{2} & \text{if } u = 0 \end{cases}$$

Next the function[9]

$$\Lambda(n) = \begin{cases} \log p & \text{if } n \text{ is a power of the prime } p, \\ 0 & \text{otherwise} \end{cases}$$

(which is now associated with his name) is defined. Taking the logarithmic derivative of both sides of the Euler product formula (1.2) for $\zeta(s)$ one gets for Re $(s+r) > 1$ the equality

$$-\frac{\zeta'(s+r)}{\zeta(s+r)} = \sum_{n=1}^{\infty} \frac{\Lambda(n)}{n^{s+r}}.$$

Fix $a > 0$ and choose $r > 1 - a$. Integrating the last identity one obtains for $x > 1$, $h > 1$

$$\frac{1}{2\pi i}\int_{a-ih}^{a+ih} \frac{\zeta'(s+r)}{\zeta(s+r)}\frac{x^s}{s}ds = \sum_{n=1}^{\infty} \frac{\Lambda(n)}{n^r}\frac{1}{2\pi i}\int_{a-ih}^{a+ih} \frac{1}{s}\left(\frac{x}{n}\right)^s ds,$$

[9] von Mangoldt denoted this function by $L(n)$. The notation $\Lambda(n)$ and the name 'Mangoldt's function' was introduced later, possibly first in Landau's book (Landau 1909a), since previously Landau (e.g. in Landau 1903a) used the notation $L(n)$ for it. This function has been used earlier for other purposes by Bougaïeff (1888a,b) and Césaro (1888).

and putting

$$\Lambda(x,r) = \lim_{h\to\infty} \frac{1}{2\pi i} \int_{a-ih}^{a+ih} \frac{\zeta'(s+r)}{\zeta(s+r)} \frac{x^s}{s} ds$$

the equality

$$\Lambda(x,r) = \sum_{n\le x} \frac{\Lambda(n)}{n^r} - \frac{\epsilon_x}{2} \frac{\Lambda(x)}{x^r}$$

follows with

$$\epsilon_x = \begin{cases} 1 & \text{if } x \text{ is a prime power} \\ 0 & \text{otherwise.} \end{cases}$$

With the use of Hadamard's Corollary 1 to Theorem 5.1 von Mangoldt is led now to the equality

$$\lim_{h\to\infty} \sum_{n=1}^{\infty} \int_{a-ih}^{a+ih} \frac{2(s+r-1/2)}{(s+r-1/2)^2+\alpha_n^2} \frac{x^s}{s} ds$$
$$= -\Lambda(x,r) + \frac{1-x^{1-r}}{r-1} - \frac{\log\pi}{2} - \frac{1}{2}\gamma + \sum_{n=1}^{\infty} \frac{r+2nx^{-r-2n}}{2n(r+2n)},$$

with $\alpha_1, \alpha_2, \ldots$ being the roots of $\xi(s)$.

Every summand on the left-hand side of this equation tends for h tending to infinity to a finite limit, say $W_n(x,r)$, and the question arises whether one can interchange the summation over n with the limit $h \to \infty$. This is established by a long calculation, in which Theorem 5.2 plays an important role. If we now put

$$\Phi(x) = \sum_{n=1}^{\infty} W_n(x,r)$$

then the final formula takes the form

$$\Lambda(x,r) = -\Phi(x) + \frac{1-x^{1-r}}{r-1} - \log\sqrt{\pi} - \gamma/2 + \sum_{n=1}^{\infty} \frac{r+2nx^{-r-2n}}{2n(r+2n)} \,. \tag{5.9}$$

In the last part of the proof von Mangoldt considers the integral of $f(x,r) = \int_{x_0}^{x} \Lambda(x,r)dx$ for $-2 < r \le 1$ and uses (5.9) to deduce an equality which in particular case $r = 0$ is equivalent to assertion of the theorem. □

At the end of his paper von Mangoldt points out that if one assumes that all the roots of Ξ are real (i.e. the non-real roots of $\zeta(s)$ lie on the critical line) then from his result the relation

$$\sum_{p\le x} \log p = (1+o(1))x$$

can be deduced.

An elementary formula for the sum $\sum_{p^k\le x} 1/k$ was given in Brun (1928).

4. The integral formula for $A(x,r)$ occuring in the proof of Theorem 5.4 is a particular case of *Perron's formula* for the sum of coefficients of a general Dirichlet series

$$f(s) = \sum_{n=1}^{\infty} a_n e^{-\lambda_n s},$$

where λ_n is an increasing sequence of positive numbers tending to infinity. If this series converges for Re $s > c$ and a exceeds $\max\{0, c\}$ then for every $x \in (\lambda_k, \lambda_{k+1})$ one has

$$\sum_{n=1}^{k} a_n = \frac{1}{2\pi i} \int_{a-i\infty}^{a+i\infty} f(s) \frac{e^{xs}}{s} ds. \tag{5.10}$$

This formula was already used formally (without bothering about convergence) in special cases already by Riemann (1860), Halphen (1883) and Cahen (1893,1894) (see also Cantor (1880)) but the first exact proof in the general case (not assuming absolute convergence) appeared in the memoir of O.Perron (1908). This memoir forms a landmark in the development of the theory of Dirichlet series as it brought precise proofs of principal results of that theory which were already stated, mostly with unsatisfactory proofs, in Cahen's thesis (Cahen 1894). In particular Perron gave the first correct proof of the uniqueness theorem for general Dirichlet series

$$\sum_{n=1}^{\infty} a_n e^{-\lambda_n s}$$

critisizing earlier attempts to prove this result that had been made by de la Vallée-Poussin (1899), Bachmann (1894) and Kronecker (1901). The source of the errors in most of these proofs lay in the fact that these series may not have a half-plane of absolute convergence, as in the case of usual Dirichlet series. Perron presented also counterexamples both to a statement appearing in Kronecker (1901) which generalized the uniqueness theorem and to another statement of Cahen (1894) concerning a necessary and sufficient condition for a function to be the sum of a Dirichlet series. The error in Cahen's argument lied in an unjustified interchange of summation and integration and to repair this lacuna Perron was led to formulate and prove his result.

Perron pointed out that an important special case of formula (5.10) appears in von Mangoldt (1895) with a correct proof[10] which works in the case when the series for $f(s)$ converges absolutely on the line Re $s = a$. In applications to prime number theory only this case, in which the proof is straightforward, is of importance.

As shown in Perron (1908) the more general formula

[10]See the proof of Theorem 5.4.

$$\frac{1}{2\pi i}\int_{a-i\infty}^{a+i\infty}\frac{e^{xs}f(s)}{s^\rho}ds=\frac{1}{\Gamma(\rho)}\sum_{n\le x}a_n(x-\lambda_n)^{\rho-1}$$

for $x\neq\lambda_n$ and $\rho\geq 1$ remains true without the assumption of absolute convergence. See also Schnee (1910). Note that this formula may not be true for ρ smaller than 1, as shown by H.Bohr (1942) who constructed a Dirichlet series convergent in the half-plane Re $s>0$ and for which this formula fails for $x=1$ and every $\rho\in(0,1)$.

Another method of obtaining asymptotics for the sum of coefficients of Dirichlet series can be found in Mellin (1904). A version of it was considered by J.B.Steffensen (1912,1914a), who proved, for the usual Dirichlet series, the formula

$$2\pi i\sum_{n\le x}a_n=\lim_{\epsilon\to 0}\int_{a-i\infty}^{a+i\infty}\frac{(x+\frac{1}{2}-i\epsilon)^{z+1}-(x+\frac{1}{2}+i\epsilon)^{z+1}}{\exp(2\pi iz)-1}\frac{f(z+1)}{z+1}dz$$

and used it to give a new proof of the Prime Number Theorem.

5. The paper of von Mangoldt also contains the proof of another explicit formula, expressing this time the function $\psi(x)=\sum_{n\le x}\Lambda(n)$ in terms of roots of the zeta-function:

Theorem 5.5. *If $x>2$ is not a prime power, then one has*

$$\psi(x)=x-\frac{1}{2}\log\left(1-\frac{1}{x^2}\right)-\log(2\pi)-\lim_{T\to\infty}\sum_{|\rho|\le T}\frac{x^\rho}{\rho},$$

ρ running over all non-real zeros of $\zeta(s)$.

If x is a prime power, then one has to add the term $-\frac{1}{2}\Lambda(x)$ to the left-hand side of the above equality.

Proof. We sketch a modern proof, full details of which appear in Davenport (1967, Sect.17).

Due to the identity

$$\sum_{n=1}^{\infty}\frac{\Lambda(n)}{n^s}=-\frac{\zeta'}{\zeta}(s)\qquad(\text{Re }s>1)$$

and the formula

$$\frac{1}{2\pi i}\int_{a-i\infty}^{a+i\infty}\frac{y^s}{s}ds=\begin{cases}0 & \text{if } 0<y<1,\\ 1/2 & \text{if } y=1,\\ 1 & \text{if } y>1\end{cases}$$

valid for real $a>1$, one has

$$\psi(x)=\frac{1}{2\pi i}\int_{a-i\infty}^{a+i\infty}\frac{\zeta'}{\zeta}(s)\frac{x^s}{s}ds,$$

provided x is not a prime power, otherwise the term $-\frac{1}{2}\Lambda(x)$ should be added on the left-hand side. Replacing the infinite integral line by the vertical segment $[a-iT,a+iT]$, moving the line of integration to the left towards $-\infty$, letting T go to infinity and observing that the sum of the residues of the integrand coincides with the right-hand side of the asserted equality one gets the assertion after duly evaluating the error terms. □

A deduction of Theorem 5.5 from Theorem 5.4 was given by H.Cramér (1918a). Other proofs of Theorem 5.5 appear in Arwin (1923) and in the second volume of Landau (1927a).

5.3. Hadamard's Proof

1. The first proofs of the Prime Number Theorem were obtained in the year 1896 independently by J.Hadamard and C.de la Vallée-Poussin. Both authors based their arguments on the non-vanishing of the zeta-function at the line Re $s = 1$ but they achieved their goal in a completely different fashion. Hadamard's proof was simpler and this was acknowledged by Vallée-Poussin (1898) who wrote:

"J'ai[11] demontré, pour la première fois ... que la fonction $\zeta(s)$ n'a pas de racines de la forme $1+\beta i$. M.Hadamard a également, avant d'avoir en connaissance de mes recherches, trouvé le même théorème par une voie plus simple."

We shall now follow the steps of J.Hadamard (1896c) in his proof and start with establishing the crucial preliminary result:

Theorem 5.6. *The function $\zeta(s)$ does not have zeros on the line* Re $s = 1$.

Proof. Since Euler's product formula shows that $\zeta(s)$ has no zeros in the half-plane Re $s > 1$ Theorem 4.1 implies that in that half-plane one has

$$\log\zeta(s) = \log\frac{1}{s-1} + g(s),$$

where $g(s)$ is regular for Re $s > 1$ and in some neighbourhood of $s = 1$. Euler's product formula implies also that if we put $P(s) = \sum_p p^{-s}$ then

$$\log\zeta(s) = P(s) + g_1(s), \tag{5.11}$$

where

$$g_1(s) = \sum_{k=2}^{\infty}\sum_p \frac{1}{kp^{ks}}$$

[11] *"I proved for the first time ... that the function $\zeta(s)$ does not have roots of the form $1+\beta i$. Mr Hadamard, before knowing about my research, also found the same theorem in a simpler way."*

is regular for Re $s > 1/2$.

Assume now that $s_0 = 1 + it_0$ with real $t_0 \neq 0$ is a zero of $\zeta(s)$. Note first that this zero must be simple[12]. In fact, otherwise $1/\zeta(s)$ would have at least a double pole at s_0 but this is impossible because in that case for $s = \sigma + it_0$ ($\sigma > 1$) tending to s_0 we would have with a suitable constant $c > 0$

$$\frac{c}{(\sigma-1)^2} \leq \frac{1}{|\zeta(s)|} = \left|\sum_{n=1}^{\infty} \frac{\mu(n)}{n^s}\right| \leq \sum_{n=1}^{\infty} \frac{1}{n^\sigma} = \zeta(\sigma) = O\left(\frac{1}{\sigma-1}\right),$$

a contradiction.

Equality (5.11) shows that for $s = \sigma + it_0$ tending to s_0 one has

$$\log(|\zeta(s)|) = \mathrm{Re}(\log\zeta(s)) = \sum_p \mathrm{Re}\, p^{-s} + O(1),$$

(the constant implied by $O(\cdot)$ being dependent on t_0) and hence

$$\log(|\zeta(s)|) = S(s) + O(1),$$

with

$$S(s) = \sum_p \frac{\cos(t_0 \log p)}{p^\sigma}.$$

Now we use the assumption $\zeta(s_0) = 0$. Since s_0 is a simple zero we get

$$\log|\zeta(\sigma + it_0)| = \log(\sigma - 1) + O(1)$$

and using (5.11) as well as the equality just before it we arrive at

$$S(\sigma + it_0) = -P(\sigma) + O(1). \tag{5.12}$$

Fix now a positive and small number $\alpha < 1$ and denote by A the set of all primes satisfying with a suitable integer k the inequality

$$|t_0 \log p - (2k+1)\pi| < \alpha.$$

Moreover let $P_n(s)$ resp. $S_n(s)$ denote the n-th partial sum of the series $P(s)$ and $S(s)$ and put

$$P_n'(\sigma) = \sum_{\substack{p\leq n\\ p\in A}} \frac{1}{p^s},$$

$$S_n'(\sigma + it_0) = \sum_{\substack{p\leq n\\ p\in A}} \frac{\cos(t_0 \log p)}{p^\sigma},$$

$$P_n''(\sigma) = P_n(\sigma) - P_n'(\sigma), \qquad S_n''(\sigma + it_0) = S_n(\sigma + it_0) - S_n'(\sigma + it_0).$$

---

[12]Hadamard's original proof omits this observation; the proof is so easy that possibly he regarded it as evident.

Let now $\epsilon \in (1/(1+\cos(2\alpha)), 1)$ be given and put $\rho_n(x) = P_n'(x)/P_n(x)$. Since $\cos(t_0 \log p) \geq -1$ and for $p \notin A$ we have $\cos(t_0 \log p) \geq -\cos\alpha$ it follows that

$$S_n'(\sigma + it_0) \geq -P_n'(\sigma) = -\rho_n(\sigma)P_n(\sigma) \tag{5.13}$$

and

$$S_n''(\sigma + it_0) \geq -P_n''(\sigma)\cos\alpha = (\rho_n(\sigma) - 1)P_n(\sigma)\cos\alpha. \tag{5.14}$$

If for some n the inequality $\rho_n(\sigma) \leq 1 - \epsilon$ holds then

$$S_n(\sigma + it_0) = S_n'(\sigma + it_0) + S_n''(\sigma + it_0) \geq -\theta P_n(\sigma),$$

with $\theta = 1 - \epsilon + \epsilon\cos\alpha \neq 1$. If this was to happen for fixed σ and infinitely many n, then $S(\sigma + it_0) \geq -\theta P(\sigma)$ would result, which is not possible for σ sufficiently close to 1 due to (5.12), because $P(\sigma)$ is unbounded. This shows that for $\sigma > 1$ sufficiently close to 1 one has

$$\rho_n(\sigma) > 1 - \epsilon \tag{5.15}$$

for all large n.

In the same way as (5.13) and (5.14) one gets

$$S_n'(\sigma + 2t_0 i) \geq P_n'(\sigma)\cos(2\alpha) = \rho_n(\sigma)P_n(\sigma)\cos(2\alpha)$$

and

$$S_n''(\sigma + 2t_0 i) \geq -P_n''(\sigma) = (\rho_n(\sigma) - 1)P_n(\sigma),$$

and hence for x sufficiently close to 1 and large n (5.15) implies

$$S_n(\sigma + 2t_0 i) \geq \Theta P_n(\sigma),$$

with $\Theta = (1-\epsilon)\cos(2\alpha) - \epsilon$. Hence $S(\sigma + 2t_0 i) \geq \Theta P(\sigma)$ follows, showing that $\lim_{\sigma \to 1+0} S(\sigma + 2t_0 i) = +\infty$ and since the choice of ϵ implies $\Theta > 0$ we see that $\zeta(s)$ is unbounded in the neighbourhood of its regular point $1 + 2t_0 i$, which is obviously not possible. □

2. The argument given by Hadamard can be applied also to other functions. He himself noted that his proof utilizes only the following properties of the zeta-function:

(i) *The logarithm of* $\zeta(s)$ *equals the sum of a series of the form* $\sum a_n e^{-\lambda_n s}$, *with* λ_n *and* a_n *positive,*

(ii) $\zeta(s)$ *is regular at the line of convergence of this series, with the exception of a single simple pole.*

Hadamard applied this remark to prove the non-vanishing of Dirichlet L-functions:

Theorem 5.7. *If* χ *is a Dirichlet character* mod k *then for real* $t \neq 0$ *one has* $L(1 + it, \chi) \neq 0$.

Proof. Put

$$f(s) = \prod_{\chi} L(s,\chi)$$

with χ running over all characters mod k. Applying Dirichlet's Lemma 2.9 in the case $l = 1$ one gets for Re$s > 1$ the equality

$$\log f(s) = \varphi(k) \sum_{r=1}^{\infty} \frac{1}{r} \sum_{\substack{p \\ p^r \equiv 1 \pmod{k}}} \frac{1}{p^{rs}}.$$

Observe now that for principal χ the function $L(s,\chi)$ differs from $\zeta(s)$ by a factor regular in Re $s > 0$ and in remaining cases $L(s,\chi)$ is regular in that half-plane. This shows that both conditions (i) and (ii) above are satisfied. Hence $L(1+it,\chi) \neq 0$ for real non-zero t. □

3. The non-vanishing of the zeta-function on the line Re $s = 1$ is the main tool in Hadamard's proof of his principal result:

Theorem 5.8. (Prime Number Theorem) *For x tending to infinity one has*

$$\theta(x) = \sum_{p \le x} \log p = (1 + o(1))x.$$

Proof. Hadamard follows the ideas of G.H.Halphen (1883) and E.Cahen (1893,1894) and quoting Cahen's paper writes: "*Nous*[13] *allons voir qu'en modifiant légèrement l'analyse de l'auteur on peut établir le même résultat en toute rigeur.*" The novelty of Hadamard's argument lies first of all in replacing the integral

$$\frac{1}{2\pi i} \int_{a-i\infty}^{a+i\infty} \frac{\zeta'(s)}{\zeta(s)} \frac{x^s}{s} ds$$

utilized by Cahen by

$$\psi_\mu(x) = \frac{1}{2\pi i} \int_{a-i\infty}^{a+i\infty} \frac{\zeta'(s)}{\zeta(s)} \frac{x^s}{s^\mu} ds \tag{5.16}$$

for $\mu > 1$. This integral converges absolutely because on the line Re $s = a > 1$ we have

$$\left| \frac{\zeta'(s)}{\zeta(s)} \right| \le \sum_{m=1}^{\infty} \frac{\log m}{m^s} \sum_{n=1}^{\infty} \frac{|\mu(n)|}{n^s} \le \sum_{m=1}^{\infty} \frac{\log m}{m^a} \sum_{n=1}^{\infty} \frac{1}{n^a} = O(1)$$

and thus the integrand in (5.16) is $O(x^a |s|^{-\mu})$. This fact justifies the performed calculations.

[13] "*We shall see that by modifying slightly the author's analysis one can establish the same result rigorously.*"

First of all Hadamard gives a detailed proof of the following formula, valid for all positive a and μ:

$$J_\mu = \frac{1}{2\pi i}\int_{a-i\infty}^{a+i\infty} \frac{x^s}{s^\mu} ds = \begin{cases} 0 & \text{if } 0 < x < 1 \\ \frac{1}{\Gamma(\mu)}\log^{\mu-1} x & \text{if } x > 1. \end{cases}$$

For $\mu = 1, 2, \dots$ this is done directly, using the equality

$$\frac{1}{s^\mu} = (-1)^{\mu-1}\frac{1}{\Gamma(\mu)}\frac{d^{\mu-1}}{ds^{\mu-1}}\left(\frac{1}{s}\right).$$

If μ is non-integral and either $x < 1$ or $\mu < 1$ then one uses Cauchy's integral formula in a standard way and finally the remaining case can be reduced by partial integration to one of the cases already considered.

Applying the obtained formula in the case $\mu = 2$ to the function $f(s) = \zeta'(s)/\zeta(s)$ one gets in view of

$$\frac{\zeta'(s)}{\zeta(s)} = -\sum_n \frac{\Lambda(n)}{n^s} = -\sum_p \log p\left(\frac{1}{p^s} + \frac{1}{p^{2s}} + \cdots\right)$$

and $\Gamma(2) = 1$ the equality

$$\psi_2(x) = \sum_{p\le x} \log p \log\frac{x}{p} + \sum_{k=2}^{\infty}\sum_{p\le x^{1/k}} \log p\log\frac{x}{p^k}. \tag{5.17}$$

The second term in the sum on the right is $O(x^{1/2}\log^3 x)$. Indeed, since for k exceeding $\log x/\log 2$ the inner sum in the double sum in (5.17) is void, we can evaluate it by

$$\frac{\log x}{\log 2}\sum_{p\le\sqrt{x}} \log p\log\frac{x}{p^2} \le \sqrt{x}\frac{\log^3 x}{\log 2}.$$

Now it remains to evaluate the integral defining $\psi_2(x)$ in some other way to get asymptotics for the sum

$$\sum_{p\le x} \log p\log\frac{x}{p}.$$

Before doing that Hadamard shows elementarily how Theorem 5.8 would follow from such asymptotics:

Lemma 5.9. *If for x tending to infinity one has*

$$A(x) = \sum_{p\le x}\log p\log\frac{x}{p} = (1 + a(x))x$$

with $a(x) = o(1)$, then

$$\sum_{p\le x}\log p = (1+o(1))x.$$

Proof. For a given positive ϵ choose $0<h<1$ so that the inequalities

$$1-\epsilon/2 < \frac{h}{(1+h)\log(1+h)} < \frac{h}{\log(1+h)} < 1+\epsilon/2$$

hold and then choose x_0 sufficiently large so that for $x>x_0$ one would have $|a(x/2)|<\frac{1}{2}\epsilon\log(1+h)$. Then the assumptions imply

$$A(x+xh)-A(x) = hx + xR(x,h)$$

with

$$R(x,h) = ha(x+hx) - a(x) + a(x+hx).$$

Now for x sufficiently large one has

$$|R(x,h)| \le h\epsilon\log(1+h) + \frac{1}{2}\epsilon\log(1+h) \le \frac{3}{2}\epsilon\log(1+h)$$

and since

$$A(x+xh)-A(x) = \log(1+h)\theta(x) + \sum_{x<p\le x+xh}\log p\log\left(\frac{x(1+h)}{p}\right) \qquad (5.18)$$

and all terms here are positive, we arrive at

$$\theta(x) \le x\frac{h}{\log(1+h)} + \frac{3\epsilon}{2}x \le x + \frac{5\epsilon}{2}x.$$

But $\epsilon>0$ was arbitrary and thus the inequality

$$\limsup_{x\to\infty}\frac{\theta(x)}{x} \le 1 \qquad (5.19)$$

results.

On the other hand

$$\sum_{x<p\le x+xh}\log p\log\left(\frac{x(1+h)}{p}\right) \le \log(1+h)\sum_{x<p\le x(1+h)}\log p$$

and so the right-hand side of (5.18) does not exceed

$$\log(1+h)\theta(x) + \log(1+h)(\theta(x+xh)-\theta(x)) = \log(1+h)\theta(x+xh).$$

Thus

$$\theta(x(1+h)) \ge \frac{x(h+R(x,h))}{\log(1+h)} \ge \frac{hx}{\log(1+h)} - \frac{3}{2}\epsilon x$$

and replacing x by $x/(1+h)$ we get

$$\theta(x) \geq \frac{x(h + R(x/(1+h),h))}{(1+h)\log(1+h)} \geq \frac{hx}{(1+h)\log(1+h)} - \frac{3}{2}\epsilon x \geq x - 2\epsilon x.$$

This leads finally to

$$\liminf_{x\to\infty} \frac{\theta(x)}{x} \geq 1 \tag{5.20}$$

because $\epsilon > 0$ was arbitrary.

The assertion of the lemma results now from (5.19) and (5.20). □

Corollary. *If $\psi_2(x) = x + o(x)$ then $\theta(x) = x + o(x)$.* □

4. The final step of the proof consists in the evaluation of the integral defining $\psi_2(x)$. This is achieved by the use of Cauchy's integration formula. Since Theorem 5.1 implies the convergence of the series $\sum_\rho |\rho|^{-2}$, ρ running over all non-real zeros of $\zeta(s)$, one can, for a given $\epsilon > 0$, find a finite set $\mathcal{A}$ of zeros such that the inequality $\sum_{\rho\notin\mathcal{A}} |\rho|^{-2} < \epsilon$ is satisfied. Let ϑ, Θ be the maximal values of Re ρ resp. of Im ρ for $\rho \in \mathcal{A}$. The integration path is chosen in the following way: let $u > \Theta$ and put $A = 2 - ui$, $B = 2 + ui$. Moreover choose y in the open interval (Θ, x) so that there are no zeros of $\zeta(s)$ on the horizontal line Im $s = y$ and put $C = a + yi$, $D = a - yi$ with $a \in (\vartheta, 1)$. Choose also $t < 0$ and put $E = t + yi$, $F = t - yi$. Let G be the point of intersection of the horizontal line passing through B and the line traced from the origin and pasing through E. Finally let H be the point symmetric to G with respect to the real line. The integral will be taken along the polygon $\Gamma = ABGECDFHA$.

Lemma 5.10. *One can choose a sequence $u_1 < u_2 < \dots$ of real numbers tending to infinity such that on the segments BG and AH our integral tends to zero.*

Proof. It suffices to deal with the segment BG. We shall show that with a suitable choice of u_r one has

$$\left|\frac{\zeta'(s)}{\zeta(s)}\right| = O(u_r \log^2 u_r) \tag{5.21}$$

on it. Choose $A > 1$ and observe that Theorem 5.3 implies that the set of zeros $\rho = \alpha + \beta i$ of the zeta-function satisfying $A^{3r+1} \leq \beta \leq A^{3r+2}$ is for sufficiently large r non-empty and finite, having no more than BrA^{3r} elements with a suitable constant B. Moreover the same theorem implies that one can find two consecutive roots $\rho_1 = \alpha_1 + \beta_1 i$ and $\rho_2 = \alpha_2 + \beta_2 i$ satisfying

$$\beta_2 - \beta_1 > \frac{A^{3r+2} - A^{3r+1}}{BrA^{3r}}.$$

Put $u_r = \beta_1 + 1 + (\beta_2 - \beta_1)/2$. Then for every zero ρ we shall have

$$|u_r - \mathrm{Re}\,\rho| > \frac{A(A-1)}{2Br}.$$

Now one utilizes Corollary 4 to Theorem 5.1. The infinite series

$$\sum_\rho \left(\frac{1}{s-\rho} + \frac{1}{\rho}\right)$$

occuring there can be split into two sums S_1 and S_2, the first taken over zeros ρ satisfying

$$A^{3r} \le \mathrm{Im}\,\rho \le A^{3r+3}, \tag{5.22}$$

and the second taken over all other zeros. If s lies on BG then S_1 does not exceed

$$BrA^{3r}\frac{2BrA^3}{A(A-1)} = O(u_r \log^2 u_r).$$

To evaluate S_2 observe that for s on BG and ρ not satisfying (5.22) one has $|s-\rho| > A^{3r}(A-1)$ hence if we write $|S_2| = \Sigma_1 + \Sigma_2$, where the summation in Σ_1 is extended over ρ satisfying $|s-\rho| \ge |\rho/10|$, then $\Sigma_1 = O(u_r)$ and

$$\Sigma_2 = O\left(u_r \sum_{cu_r \le |s-\rho| < |\rho|/10} \left|\frac{1}{\rho(s-\rho)}\right|\right)$$

with a suitable $c > 0$. If one notes now that $|s-\rho| < |\rho|/10$ implies $|\rho| \le c_1 u_r$ with a certain positive c_1 and uses Theorem 5.1 to obtain that the number of such ρ's is $O(u_r \log u_r)$ then finally

$$\Sigma_2 = O\left(u_r \frac{1}{u_r} \sum_{|\rho| \le c_1 u_r} \left|\frac{1}{\rho}\right|\right) = O(u_r \log u_r)$$

results. In view of Corollary 4 to Theorem 5.1 this implies (5.21).

The assertion of the lemma follows now immediately, since (5.21) shows that the considered integral over BG is

$$O\left(\frac{\log^2 u_r}{u_r}\left|\int_{BG} x^s\,ds\right|\right) = O\left(\frac{\log^2 u_r}{u_r}\right) = o(1),$$

as r tends to infinity. □

The lemma which we have just proved implies that the line of integration AB can be moved to the polygon $\Gamma' = HFDCEG$ provided one adds the sum of residues corresponding to the pole $s = 1$ of $\zeta'(s)$ and to the zeros of $\zeta(s)$ lying inside Γ. The first sum equals $-x$ and the sum of the absolute values of all remaining residues is smaller than ϵx, since the corresponding zeros do not belong to A.

It thus remains to show that the integral over Γ' is $o(x)$. This is obvious for integrals over $FDCE$, and so we are left with integrals over EG and FH. It suffices to deal only with the first, and as on that segment the ratio $|(s-\alpha)/\alpha|$ exceeds a positive constant for every root α of $\zeta(s)$, it follows from the Corollary 4 to Theorem 5.1 that on EG we have $\frac{\zeta'}{\zeta}(s) = O(|s|)$ and this implies that the integral over that segment is $o(x)$. Thus finally we get $\psi_2(x) = x + o(x)$ and the theorem follows from Corollary to Lemma 5.9. □

Corollary. *For x tending to infinity one has*

$$\psi(x) = (1 + o(1))x.$$

Proof. The assertion follows immediately from the theorem and Lemma 3.3 (iii). □

5. In the final part of his paper Hadamard points out that the same approach leads to the proof of the following quantitative form of Dirichlet's Theorem 3.1:

Theorem 5.11. *If $(k,l) = 1$, then $\theta(x;k,l) = \sum_{p\equiv l \pmod k} \log p$ is asymptotically equal to $x/\varphi(k)$.*

Hadamard sketches the proof in few lines, saying that the argument is essentially the same as in the proof of Theorem 5.8 except that instead of the integral $\psi_\mu(x)$ one has to consider the integral

$$-\frac{1}{2\pi i}\int_{AB}\left[\sum_{\chi}\overline{\chi(m)}\frac{L'(s,\chi)}{L(s,\chi)}\right]\frac{x^s}{s^\mu}ds$$

(we have changed, slightly, Hadamard's notation) with $\mu > 1$. This leads to the formula

$$\frac{1}{\Gamma(\mu)}\sum_{\substack{p\leq x\\ p\equiv l \pmod k}} \log p \log^{\mu-1}\frac{x}{p} = (1+\epsilon(x))\frac{x}{\varphi(k)}$$

with $\lim_{x\to\infty}\epsilon(x) = 0$.

Hadamard concluded his memoir with a remark about a possible application of his method to the series considered by H.Weber (1882) and A.Meyer (1888) in their proofs of Dirichlet's assertion about primes represented by quadratic forms.

5.4. The Proof of de la Vallée-Poussin

1. Now we shall sketch the approach of C.de la Vallée-Poussin (1896a) which differs essentially from that of Hadamard although it is based on the same results, including Theorem 5.1 and its Corollary 1 as well as Čebyšev's upper bound for $\theta(x)$ (Corollary 2 to Theorem 3.5). The main tool in de la Vallée-Poussin's proof of Theorem 5.6 is the following simple auxiliary result from harmonic analysis, which essentially expresses the uniqueness of the Fourier expansion of an almost periodic function:

Lemma 5.12. *If a complex-valued function $f(x)$ defined for real $x > 1$ can be written in the form*

$$f(x) = \sum_{n=1}^{\infty} c_n e^{i\alpha_n \log x} + g(x)$$

where α_n is a sequence of distinct real numbers without a finite limit point, $c_n \neq 0$ for all $n \geq 1$, the series $\sum_{n=1}^{\infty} c_n$ converges absolutely and $g(x)$ tends to zero for $x \to \infty$, then both the exponents α_n and the coefficients c_n are uniquely determined by f.

Proof. (The proof presented here differs from the original argument only in using the exponential function in place of trigonometrical functions.) For real y and $N = 1, 2, \dots$ consider

$$A_N(y) = \int_1^N \frac{f(x)}{x} \exp(-iy \log x) dx$$

$$= \sum_{n=1}^{\infty} c_n \int_1^N \frac{\exp(i(\alpha_n - y)\log x)}{x} dx + o(\log N)$$

$$= \sum_{n=1}^{\infty} c_n I_n^{(N)}(y) + o(\log N),$$

where

$$I_n^{(N)}(y) = \begin{cases} -\frac{i}{\alpha_n - y}\left(\exp(i(\alpha_n - y)\log N) - 1\right) & \text{if } y \neq \alpha_n \\ \log N & \text{if } y = \alpha_n. \end{cases}$$

So, if $\epsilon(y)$ denotes the shortest positive distance from y to an element of the sequence α_j distinct from y, then for $y \neq \alpha_n$ we get

$$|I_n^{(N)}(y)| \leq \frac{2}{\epsilon(y)}.$$

This implies

$$|A_N(\alpha_n) - c_n \log N| \le \frac{2}{\epsilon(\alpha_n)} \sum_{k \ne n} |c_k| + o(\log N),$$

and for $y \notin \{\alpha_1, \alpha_2, \ldots\}$

$$|A_N(y)| \le \frac{2}{\epsilon(y)} \sum_{k=1}^{\infty} |c_k| + o(\log N).$$

Finally we arrive at

$$\lim_{N\to\infty} \frac{A_N(y)}{\log N} = \begin{cases} c_n & \text{if } y = \alpha_n \\ 0 & \text{otherwise} \end{cases}$$

and the assertion follows immediately. □

Corollary. *If a complex-valued function $f(x)$ defined for real $x > 1$ can be written in the form*

$$f(x) = \sum_{n=1}^{\infty} (c_n \cos(\alpha_n \log x) + d_n \sin(\alpha_n \log x)) + g(x)$$

where α_n is a sequence of distinct real numbers without a finite limit point, $c_n^2 + d_n^2 > 0$ for all $n \ge 1$, both series $\sum_{n=1}^{\infty} c_n$ and $\sum_{n=1}^{\infty} d_n$ are absolutely convergent and $\lim_{x\to\infty} g(x) = A$ then A, α_n, c_n and d_n are uniquely determined by f.

Proof. Use Euler's formulas

$$\cos t = \frac{e^{it} + e^{-it}}{2}, \qquad \sin t = \frac{e^{it} - e^{-it}}{2i},$$

and apply the lemma to the function $f(x) - A$. □

2. The main idea of the proof of Theorem 5.6 lies in showing that the existence of a root $1 + \beta i$ of $\zeta(s)$ with real β implies the existence of two representations of the function

$$\frac{1 + \cos(\beta \log x)}{x} \sum_{p<x} \log p$$

in the form given in the preceding corollary.

In the first step of his proof de la Vallée-Poussin continues the zeta-function into the half-plane Re $s > 0$ by means of the following simple argument:

In view of the formula

$$\int_0^1 \frac{dx}{(n+x)^s} = \frac{1}{(n+1)^s} + s \int_0^1 \frac{x\,dx}{(n+x)^{s+1}},$$

which is easily established by partial integration for Re $s > 1$, one obtains

$$\zeta(s) = 1 + \sum_{n=1}^{\infty} \left(\int_0^1 \frac{dx}{(n+x)^s} - s \int_0^1 \frac{x\,dx}{(n+x)^{s+1}} \right).$$

But

$$\sum_{n=1}^{\infty} \int_0^1 \frac{dx}{(n+x)^s} = \int_1^{\infty} \frac{dx}{x^s} = \frac{1}{s-1}$$

and hence

$$\zeta(s) = 1 + \frac{1}{s-1} - s \sum_{n=1}^{\infty} \int_0^1 \frac{x\,dx}{(n+x)^{s+1}}. \tag{5.23}$$

Now it suffices to observe that the series on the right-hand side converges absolutely in the half-plane Re $s > 0$ to obtain the continuation of $\zeta(s)$ to a function regular in that half-plane with the exception of a simple pole at $s = 1$ with residue equal to 1.

After recalling Hadamard's Theorem 5.1 and Corollary 1 to it de la Vallée-Poussin notes that on the line Re $s = 1$ there are no multiple roots of the zeta-function. Here is his argument:

From Euler's product formula one gets for Re $s > 1$

$$\frac{\zeta'(s)}{\zeta(s)} = -\sum_p \frac{\log p}{p^s - 1} = -\sum_p \frac{\log p}{p^s} - \sum_p \frac{\log p}{p^s(p^s-1)}$$

and hence the order of the possible root $\rho = 1 + \beta i$ equals

$$\begin{aligned} \lambda &= \lim_{s \to \rho} \left(-(s-\rho) \sum_p \frac{\log p}{p^s} \right) \\ &= -\lim_{\epsilon \to 0+0} \epsilon \sum_p \frac{\log p}{p^{1+\beta i+\epsilon}}. \end{aligned}$$

But λ is real, thus

$$\lambda = -\lim_{\epsilon \to 0+0} \epsilon \sum_p \frac{\log p \cos(\beta \log p)}{p^{1+\epsilon}},$$

which implies

$$|\lambda| \le \lim_{\epsilon \to 0+0} \epsilon \sum_p \frac{\log p}{p^{1+\epsilon}} = \lim_{\epsilon \to 0+0} \epsilon \zeta'(1+\epsilon)$$

and the last limit equals 1 because the pole of $\zeta(s)$ at $s = 1$ is simple. Thus $\lambda = 1$, and hence all roots of the zeta-function at the line Re $s = 1$ must be simple.

3. A very important role in de la Vallée'-Poussin's proofs of Theorems 5.6 and 5.8 is taken by the identity

$$I(y,u,v) = \frac{1}{2\pi i}\int_{a-i\infty}^{a+i\infty} \frac{\zeta'}{\zeta}(s)\frac{y^s\,ds}{(s-u)(s-v)}$$
$$= \frac{1}{u-v}\left(y^u\frac{\zeta'}{\zeta}(u) - y^v\frac{\zeta'}{\zeta}(v)\right) - \frac{y}{(u-1)(v-1)} + \sum_\rho \frac{y^\rho}{(u-\rho)(v-\rho)}$$
$$+ \sum_{m=1}^{\infty} \frac{y^{-2m}}{(2m+u)(2m+v)}\ . \tag{5.24}$$

where u and v are complex numbers which are neither zeros nor poles of $\zeta(s)$, $y > 1$ and a is a real number exceeding $\max\{1, \operatorname{Re} u, \operatorname{Re} v\}$. To prove this identity multiply the first formula in Corollary 4 to Theorem 5.1 by $y^s/(s-u)(s-v)$ and perform the integration.

As pointed out by Vallée-Poussin in a footnote, the same aim can be achieved by noting that the right-hand side of (5.24) equals the sum of residues of the integrand in the half-plane $\operatorname{Re} s < a$.

The integral $I(y,u,v)$ can be also evaluated in another way. Because for $\operatorname{Re} s > 1$ one has

$$\zeta'(s)/\zeta(s) = -\sum_{n=1}^{\infty} \frac{\Lambda(n)}{n^s}$$

an application of the formula

$$\lim_{h\to\infty} \frac{1}{2\pi i}\int_{a-ih}^{a+ih} \frac{e^{us}}{s}ds = \begin{cases} 1 & \text{if } u > 0 \\ 0 & \text{if } u < 0 \\ \frac{1}{2} & \text{if } u = 0 \end{cases}$$

leads, in the case when y is not a prime power, to the equality[14]

$$I(y,u,v) = -\frac{1}{u-v}\left(y^u\sum_{n<y}\frac{\Lambda(n)}{n^u} - y^v\sum_{n<y}\frac{\Lambda(n)}{n^v}\right).$$

Note that by continuity considerations one sees easily that the last equality holds for all complex u, v and $y > 1$. Put $v = 0$ in it, use (5.24), divide by y^u and let u tend to 1. Since

$$\lim_{u\to 1}\frac{y^{1-u}}{1-u} = \log y$$

we obtain the equality

---

[14] We have simplified here the notation used by Vallée-Poussin, who did not utilize the function $\Lambda(n)$.

$$\sum_{n<y}\frac{\Lambda(n)}{n}-\frac{1}{y}\sum_{n<y}\Lambda(n)=\log y-\lim_{u\to 1}\left(\frac{\zeta'}{\zeta}(u)+\frac{u}{u-1}\right)$$
$$+\frac{1}{y}\frac{\zeta'}{\zeta}(0)-\sum_{\rho}\frac{y^{\rho-1}}{\rho(\rho-1)}-\sum_{m=1}^{\infty}\frac{y^{-2m-1}}{2m(2m+1)}. \tag{5.25}$$

Assuming still $v=0$ let now u tend to a simple root ρ_1 of the zeta-function. In view of

$$\frac{uy^{\rho_1-u}}{\rho_1(\rho_1-u)}=\frac{1}{\rho_1-u}-\frac{1}{\rho_1}+\frac{u}{\rho_1}\frac{y^{\rho_1-u}-1}{\rho_1-u}=\frac{1}{\rho_1-u}-\frac{1}{\rho_1}+\log y+O(\rho_1-u)$$

we arrive at

$$\sum_{n<y}\frac{\Lambda(n)}{n^{\rho_1}}-\frac{1}{y^{\rho_1}}\sum_{n<y}\Lambda(n)=-\log y+\frac{1}{\rho_1}$$
$$-\lim_{u\to\rho_1}\left(\frac{\zeta'}{\zeta}(u)+\frac{1}{u-\rho_1}\right)+\frac{1}{y^{\rho_1}}\frac{\zeta'}{\zeta}(0)+\frac{\rho_1}{1-\rho_1}y^{1-\rho_1}$$
$$-\rho_1\sum_{\rho\neq\rho_1}\frac{y^{\rho-\rho_1}}{\rho(\rho-\rho_1)}-\rho_1\sum_{m=1}^{\infty}\frac{y^{-2m-\rho_1}}{2m(2m+\rho_1)}.$$

Assume now the existence of a root $\rho_1=1+\beta i$ of $\zeta(s)$ lying on the line $\operatorname{Re} s=1$. Adding the obtained equalities and noting that for y tending to infinity the expression

$$\left(\frac{1}{y}+\frac{1}{y^{\rho_1}}\right)\frac{\zeta'(0)}{\zeta(0)}-\sum_{m=1}^{\infty}\frac{y^{-2m-1}}{2m(2m+1)}-\rho_1\sum_{m=1}^{\infty}\frac{y^{-2m-1}}{2m(2m+\rho_1)}$$

tends to zero, one gets for $y\to\infty$

$$\sum_{n<y}\Lambda(n)(\frac{1}{n}+\frac{1}{n^{\rho_1}})-(\frac{1}{y}+\frac{1}{y^{\rho_1}})\sum_{n<y}\Lambda(n)$$
$$=\frac{\rho_1}{1-\rho_1}y^{\beta i}-\sum_{\rho}\frac{y^{\rho-1}}{\rho(\rho-1)}-\rho_1\sum_{\rho\neq\rho_1}\frac{y^{\rho-\rho_1}}{\rho(\rho-\rho_1)}+C+o(1)$$

with

$$C=\frac{1}{\rho_1}-1-\lim_{u\to 0}\left(\frac{\zeta'}{\zeta}(1+u)+\frac{\zeta'}{\zeta}(\rho_1+u)\right).$$

Observe now that both series

$$\sum_{\rho}\frac{y^{\rho-1}}{\rho(\rho-1)}$$

and

$$\sum_{\rho\neq\rho_1}\frac{y^{\rho-\rho_1}}{\rho(\rho-\rho_1)}$$

converge uniformly in $y > 1$ and, since their terms corresponding to ρ in the open half-plane Re $s < 1$ tend to zero when y grows indefinitely, the series

$$\sum_{\substack{\rho \\ Re\ \rho<1}} \frac{y^{\rho-1}}{\rho(\rho-1)}$$

and

$$\sum_{\substack{\rho \\ Re\ \rho<1}} \frac{y^{\rho-\rho_1}}{\rho(\rho-1)}$$

are both $o(1)$. Also since the series

$$\sum_{\substack{p^m<y \\ m\geq 2}} \frac{\log p}{p^m}$$

converges and

$$\sum_{\substack{p^m<y \\ m\geq 2}} \log p = o(y)$$

we can write the preceding formula in the simpler form

$$\sum_{p<y} \log p \left(\frac{1}{p} + \frac{1}{p^{\rho_1}}\right) - \left(\frac{1}{y} + \frac{1}{y^{\rho_1}}\right) \sum_{p<y} \log p = P(y) + C + o(1)$$

where

$$P(y) = \frac{\rho_1}{1-\rho_1} y^{-\beta i} - \sum_{\rho}{}^{*} \frac{y^{\rho-1}}{\rho(\rho-1)} - \rho_1 \sum_{\rho\neq\rho_1}{}^{*} \frac{y^{\rho-\rho_1}}{\rho(\rho-\rho_1)},$$

the asterisk in $\sum^*$ denoting that the sum is to be taken over ρ's satisfying Re $\rho = 1$.

By considering separately the real and imaginary parts of both sides of this formula one gets

$$\sum_{p<y} \frac{\log p}{p}(1 + \cos(\beta \log p)) - \frac{1 + \cos(\beta \log y)}{y} \sum_{p<y} \log p = \text{Re } P(y) + g_1(y),$$

and

$$-\sum_{p<y} \frac{\log p}{p} \sin(\beta \log p) + \frac{\sin(\beta \log y)}{y} \sum_{p<y} \log p = \text{Im } P(y) + g_2(y)$$

where $g_1(y)$ and $g_2(y)$ tend to zero for y tending to infinity.

Corollary 2 to Theorem 3.5 is now invoked to obtain the existence of

$$\lim_{y\to\infty} \sum_{p<y} \frac{\log p}{p}(1 + \cos(\beta \log p)),$$

from which the following easy corollaries are then deduced:

$$\lim_{y\to\infty} \frac{1}{y} \sum_{p<y} \log p |\sin(\beta \log p)| = 0,$$

and

$$\lim_{y\to\infty} \frac{1}{\log y} \sum_{p<y} \frac{\log p}{p} |\sin(\beta \log p)| = 0.$$

After that a more difficult step occurs: the proof of convergence of the series

$$\sum_{p} \frac{\log p}{p} \sin(\beta \log p),$$

which involves rather delicate analysis, since the convergence is not absolute. Putting all this together one arrives at the formula

$$\left(\frac{1}{y} + \frac{1}{y^{\rho_1}}\right) \sum_{p<y} \log p = -P(y) + C_1 + o(1),$$

with a constant C_1. Putting $A(y) = -\text{Re } P(y)$, $B(y) = -\text{Im } P(y)$ one gets

$$A(y) = \frac{1 + \cos(\beta \log y)}{y} \sum_{p<y} \log p + C_1 + o(1),$$

and

$$B(y) = -\frac{\sin(\beta \log y)}{y} \sum_{p<y} \log p + C_1 + o(1).$$

From the two last equalities one gets

$$A(y) = \frac{\beta}{y} \int_1^y B(t)dt + C_1 + o(1)$$

with a suitable constant C_1, thus we obtain two expressions for $A(y)$, namely

$$A(y) = \beta \text{Im} \Big(-\frac{\rho_1}{(1-\rho_1)(2-\rho_1)} y^{1-\rho_1}$$

$$+\rho_1 \sum_{\rho \neq \rho_1}{}^{*} \frac{y^{\rho-\rho_1}}{\rho(\rho-\rho_1)(\rho-\rho_1+1)} + C_2 + o(1)\Big)$$

with a certain constant C_2 and

$$A(y) = \text{Re } \Big(-\frac{\rho_1}{1-\rho_1} y^{1-\rho_1} + \sum_{\frac{y^{\rho-1}}{\rho(\rho-1)}}{}^{*}$$

$$+\rho_1 \sum_{\rho \neq \rho_1}{}^{*} \frac{y^{\rho-\rho_1}}{\rho(\rho-\rho_1)} + C_2 + o(1)\Big).$$

Now it is a simple but rather dull task to transform these expressions to the form considered in Lemma 5.12 and to observe that, contrary to that lemma, the coefficients of the cosine terms $\cos(\beta \log y)$ are different. This contradiction shows that no root of $\zeta(s)$ can have its real part equal to 1.

The original proof of de la Vallée Poussin avoids computations at this stage and contains a rather shaky argument instead in which the Corollary to Lemma 5.12 is applied to a possibly divergent series.

Another proof of the formula (5.25) was given a few years later by H.von Koch (1903/04). He based his result on Riemann's explicit formula and quoted M.Lerch (1903) and E.Phragmén (1904) for the result asserting that if f is a function continuous in a real interval $[a,b]$ and having the property that for $n = 0,1,2,\dots$ all integrals $\int_a^b f(x)x^n dx$ vanish, then f must vanish in that interval (see Lerch 1892b). This was first established by J.Liouville (1837) for continuous functions having only a finite number of sign changes in $[a,b]$. (If $x_1, x_2, \dots, x_m$ are these points then the integral of $f(x)(x-x_1)(x-x_2)\cdots(x-x_m)$ vanishes but the integrand is of constant sign). Actually Lerch's theorem was obtained earlier by Stieltjes who presented two proofs of it in his letter to Hermite of September, 12th 1893 (see Baillaud,B.,Bourget,H. (1905),II,337–339. See also Landau (1908a)).

4. The final step of the proof of Theorem 5.8 is based on the formula

$$\sum_{p<y} \frac{\log p}{p-1} - \frac{1}{y}\sum_{p<y} \log p = \log y - 1 - C + o(1) \tag{5.26}$$

valid for $y \to \infty$ with

$$C = \lim_{s\to 1}\left(\frac{\zeta'(s)}{\zeta(s)} + \frac{1}{s-1}\right).$$

This formula follows from (5.25) since all roots of $\zeta(s)$ have their real part smaller than 1, the series

$$\sum_{\rho} \frac{y^{\rho-1}}{\rho(\rho-1)}$$

converges uniformly on the half-line $y \geq 1$ and moreover one has

$$\sum_{p^m<y} \log p - \sum_{p<y} \log p = o(y)$$

and

$$\sum_{p<y} \frac{\log p}{p-1} - \sum_{p^m<y} \frac{\log p}{p^m} = o(1).$$

To prove Theorem 5.8 integrate the equality (5.26) from 1 to x. The result is

$$\int_1^x \sum_{p<y} \frac{\log p}{p-1} dy - \int_1^x \frac{1}{y}\sum_{p<y} \log p \, dy = x\log x - (2+C)x + o(x). \tag{5.27}$$

The first integral is now transformed to a finite sum:

$$\int_1^x \sum_{p<y} \frac{\log p}{p-1} dy = \sum_{j=1}^{[x]} \int_j^{j+1} \sum_{p<y} \frac{\log p}{p-1} dy + O(\log x)$$
$$= \sum_{j\le x} \sum_{p<j} \frac{\log p}{p-1} + O(\log x).$$

Applying Lemma 2.5 with $a_j = 1$ and $b_j = \sum_{p<j} \log p/(p-1)$ we get

$$\int_1^x \sum_{p<y} \frac{\log p}{p-1} dy = \sum_{p<x} \frac{\log p}{p-1}([x]-p) + O(\log x)$$
$$= (x+1)\sum_{p<x} \frac{\log p}{p-1} - \sum_{p<x} \log p + O(\log x).$$

Now using (5.26) and observing that, due to Theorem 3.8, we have

$$\sum_{p<x} \frac{\log p}{p-1} = O(\log x)$$

(for $x \to \infty$) we obtain

$$\int_1^x \sum_{p<y} \frac{\log p}{p-1} dy = x\left(\sum_{p<x} \frac{\log p}{p-1} - \frac{1}{x}\sum_{p<x} \log p\right) + O(\log x).$$

The last equality, jointly with (5.26) and (5.27) leads to

$$\int_1^x \frac{1}{y} \sum_{p<y} \log p \, dy = (1+o(1))x \tag{5.28}$$

and it remains to deduce from this that the integrand tends to unity.

Denoting the integral occuring in (5.28) by $T(x)$ take a positive ϵ and observe that

$$T((1+\epsilon)x) - T(x) = (\epsilon + o(1))x.$$

Yet

$$T((1+\epsilon)x) - T(x) \ge \log(1+\epsilon)\theta(x)$$

and

$$T((1+\epsilon)x) - T(x) \le \log(1+\epsilon)\theta((1+\epsilon)x)$$

hold and these lead to

$$\limsup_{x\to\infty} \frac{\theta(x)}{x} \le \frac{\epsilon}{\log(1+\epsilon)}$$

and

$$\frac{\epsilon}{\log(1+\epsilon)} \leq \liminf_{x\to\infty} \frac{\theta(x(1+\epsilon))}{x}.$$

Replacing in the last relation x by $x/(1+\epsilon)$ and noting that

$$\lim_{\epsilon\to 0} \frac{\epsilon}{\log(1+\epsilon)} = 1$$

we obtain finally

$$\lim_{x\to\infty} \frac{\theta(x)}{x} = 1. \qquad \square$$

This concludes de la Vallée-Poussin's proof of the Prime Number Theorem.

5. The constant C appearing in (5.26) equals the Euler's constant γ. For this fact de la Vallée-Poussin gives the following elegant proof, avoiding the use of the integral representation of $\zeta(s)$ utilized in the standard proof (see e.g.Titchmarsh (1951, Chap.2)):

Since for $x > 1$ we have

$$\zeta(x) = (1 - 2^{1-x})^{-1} \sum_{n=1}^{\infty} (-1)^{n+1} \frac{1}{n^x},$$

one obtains

$$\frac{\zeta'(x)}{\zeta(x)} = \frac{M(x)}{R(x)} - \frac{2\log 2}{2^x - 2}$$

with

$$M(x) = \sum_{n=2}^{\infty} (-1)^n \frac{\log n}{n^x}$$

and

$$R(x) = \sum_{n=1}^{\infty} \frac{(-1)^{n+1}}{n^x}.$$

Now de l'Hôpital's rule gives

$$\lim_{x\to 1} \left(\frac{1}{x-1} - \frac{2\log 2}{2^x - 2} \right) = \frac{\log 2}{2}$$

and since by the continuity of Dirichlet series we have $\lim_{x\to 1} R(x) = \log 2$ thus

$$C = \frac{\log 2}{2} + \frac{M(1)}{\log 2}$$

and it remains to calculate $M(1)$. To this end choose a positive integer N and differentiate the identity

$$\sum_{n=1}^{2N} \frac{(-1)^{n+1}}{n^x} = (1 - 2^{1-x}) \sum_{n=1}^{N} \frac{1}{n^x} + \sum_{n=N+1}^{2N} \frac{1}{n^x}$$

to get

$$\sum_{n=2}^{2N} \frac{(-1)^n \log n}{n^x} = \frac{\log 2}{2^{x-1}} \sum_{n=1}^{N} \frac{1}{n^x} - \left(1 - 2^{1-x}\right) \sum_{n=1}^{N} \frac{\log n}{n^x} - \sum_{n=N+1}^{2N} \frac{\log n}{n^x}.$$

Substituting $x = 1$, observing that

$$\sum_{n=N+1}^{2N} \frac{\log n}{n} = \log 2 \log N + \frac{\log^2 2}{2} + o(1)$$

and taking into account the definition of γ (see footnote 20 in Chap.3) one gets

$$\sum_{n=2}^{2N} \frac{(-1)^n \log n}{n} = \gamma \log 2 - \frac{\log^2 2}{2} + o(1),$$

hence

$$M(1) = \gamma \log 2 - \frac{\log^2 2}{2}.$$

Another deduction of Theorem 5.8 from (5.26) was presented by E.Landau (1910) who applied a modification of the following tauberian result of G.H.Hardy (1910):

If a_n is a sequence satisfying $na_n = O(1)$,

$$s_m = a_1 + a_2 + \cdots + a_m$$

and the limit

$$\lim_{n\to\infty} \frac{1}{n} \sum_{m=1}^{n} s_m$$

exists and equals s, then the series $\sum_{n=1}^{\infty} a_n$ converges and its sum equals s.

Landau's modification consisted in replacing the assumption $na_n = O(1)$ by $na_n \geq c$ for a fixed c (which implies of course that the a_n's are real). He also showed that this result may be used to modify the last step of Hadamard's proof of Theorem 5.8.

6. Neither Hadamard nor de la Vallée-Poussin pointed out in their papers the following corollary[15] to Theorem 5.8, which is the usual form of the Prime Number Theorem:

Theorem 5.13. *For x tending to infinity one has*

$$\pi(x) = (1 + o(1))\mathrm{li}(x) \tag{5.29}$$

and

[15] It seems that it was J.J.Sylvester (1892, footnote) who first observed that (5.30) is a consequence of the asymptotic equality of $\theta(x)$ and x.

$$\pi(x) = (1 + o(1))\frac{x}{\log x}. \tag{5.30}$$

Note that since partial integration gives

$$\mathrm{li}(x) = \int_2^x \frac{dt}{\log t} + O(1) = \frac{x}{\log x} + \int_2^x \frac{dt}{\log^2 t} + O(1) = \frac{x}{\log x} + O(\frac{x}{\log^2 x}),$$

the formulas (5.29) and (5.30) are equivalent.

H.von Mangoldt (1898) presented a rather complicated deduction of (5.29) from Theorem 5.8 but also quoted a letter of de la Vallée-Poussin with the following very short deduction of (5.30) from that theorem:

Proof of de la Vallée-Poussin. Theorem 5.8 implies

$$\sum_{x/\log x<p<x} \log p = (1 + o(1))x,$$

but for primes in the interval $(x/\log x, x)$ one has $\log p = \log x + O(\log\log x)$, hence

$$(\log x + O(\log\log x))(\pi(x) - \pi(x/\log x)) = (1 + o(1))x,$$

and thus

$$\pi(x) = (1 + o(1))\frac{x}{\log x + O(\log\log x)} + \pi(x/\log x) = (1 + o(1))\frac{x}{\log x},$$

the last step being justified either by Čebyšev's Theorem (see Corollary 4 to Theorem 3.5) or by the observation that $\pi(x) = o(x)$ (see Corollary 2 to Theorem 1.4). □

Essentially the same argument may be used to deduce from Theorem 5.11 the following analogue of Theorem 5.13 for primes in progressions:

Theorem 5.14. *If* $(k, l) = 1$ *and* $\pi(x; k, l)$ *denotes the number of primes* $p \le x$ *satisfying* $p \equiv l \pmod k$, *then*

$$\pi(x; k, l) = \left(\frac{1}{\varphi(k)} + o(1)\right)\frac{x}{\log x}.$$ □

7. In the second part of his paper de la Vallée-Poussin gave a proof of Theorem 5.11, based on arguments, analogous to these used in his proof of Theorem 5.8 which we presented. Later (de la Vallée-Poussin 1896a III, 1897) he used a similar, but technically more complicated, approach to obtain a quantitative version of the Dirichlet-Weber theorem, concerning representation of primes by binary quadratic forms. He showed that if an irreducible form $f(x, y) = ax^2 + bxy + cy^2$ has integral coefficients satisfying $(a, 2b, c) = 1$

and is either indefinite or positive definite then the number $\pi_f(T)$ of primes not exceeding T which are represented by f is asymptotically equal to a constant multiple of $\mathrm{li}(T)$. His argument was later slightly simplified by E.Landau (1915b).

This problem can be stated equivalently as a question of distribution of prime ideals in ideal classes in the ring of integers of a quadratic number field. A very general result of Landau (1907a) implies, in particular, the relation

$$\pi_f(T) = c\,\mathrm{li}(T) + O(T\exp(-c\log^\lambda T))$$

with certain positive c and λ, both depending on T. Later Landau (1908f) proved the existence of a λ which is independent of f. The details in the quadratic case were worked out in the thesis of P.Bernays[16] (1912), who obtained Landau's result with $\lambda = 1/8$. Further progress was achieved by Landau (1914a) who showed that one can take $\lambda = 1/2$.

5.5. Other Proofs of the Non-vanishing of $\zeta(1+it)$ and $L(1+it,\chi)$

1. The first simplification of the rather complicated proofs given by Hadamard and de la Vallée-Poussin for Theorem 5.6, which we presented in the preceding section, appears at the end of the second part of de la Vallée-Poussins paper (de la Vallée-Poussin 1896a,II,pp.395–397):

De la Vallée-Poussin's second proof of Theorem 5.6. Assume that $1+it$ (with real t) is a zero of the zeta-function. Since the argument given at the beginning of the first proof implies its simplicity and moreover we have

$$\sum_p \frac{\log p}{p^s} = -\frac{\zeta'}{\zeta}(s) + g(s) \tag{5.31}$$

with $g(s)$ regular in $\mathrm{Re}\, s > 1/2$, the equalities

$$\lim_{\epsilon\to 0} \epsilon \sum_p \frac{\log p}{p^{1+\epsilon+ti}} = -1$$

and

$$\lim_{\epsilon\to 0} \epsilon \sum_p \frac{\log p}{p^{1+\epsilon}} = 1 \tag{5.32}$$

follow. Adding these and taking real parts we get

$$\lim_{\epsilon\to 0} \epsilon \sum_p \frac{\log p}{p^{1+\epsilon}}(1+\cos(t\log p)) = 0$$

---

[16]Bernays, Isaak Paul (1888–1977), Professor in Göttingen and Zürich.

with all terms of the last series being non-negative. In view of the inequality

$$3+4\cos\alpha+\cos(2\alpha)=2(1+\cos\alpha)^2\geq 0, \tag{5.33}$$

we get

$$1-\cos(2t\log p)\leq 4(1+\cos(t\log p))$$

and this shows that

$$\lim_{\epsilon\to 0}\epsilon\sum_p \frac{\log p}{p^{1+\epsilon}}(1-\cos(2t\log p))=0. \tag{5.34}$$

Subtracting now (5.32) from (5.34) we get

$$\lim_{\epsilon\to 0}\epsilon\sum_p \frac{\log p}{p^{1+\epsilon}}\cos(2t\log p)=1,$$

i.e.

$$\lim_{\epsilon\to 0}\epsilon\ \mathrm{Re}\left(\sum_p \frac{\log p}{p^{1+\epsilon+2ti}}\right)=1,$$

but in view of (5.31) this contradicts the regularity of $\zeta(s)$ at $s=1+2ti$, which implies that the last limit is non-positive. □

This proof can be adapted (Landau 1902,1903a) to the case of Dedekind zeta-functions

$$\zeta_K(s)=\sum_I N(I)^{-s},$$

I ranging over all non-zero ideals of $\mathbf{Z}_K$, the ring of integers of the field K and $N(I)$ being the norm of I, i.e. the cardinality of the factor ring $\mathbf{Z}_K/I$.

2. The by now standard proof of Theorem 5.6 found in most text-books uses the same idea but usually one starts, following Mertens (1898), with the equality

$$4\log|\zeta(1+\epsilon+it)|+3\log|\zeta(1+\epsilon)|+\log|\zeta(1+\epsilon+2ti)|$$
$$=\sum_p\sum_{k=1}^{\infty}\frac{1}{k}\,\frac{3+4\cos(kt\log p)+\cos(2kt\log p)}{p^{k(1+\epsilon)}}$$

and uses (5.33) to deduce

$$|\zeta(1+\epsilon)^3\zeta(1+\epsilon+ti)^4\zeta(1+\epsilon+2ti)|\geq 1. \tag{5.35}$$

Were now $1+ti$ a root of $\zeta(s)$, then in view of Corollary 2 to Lemma 2.3 the left-hand side of this inequality would tend to 0 for ϵ tending to 0, a clear contradiction. □

The same method may be applied to the proof of the non-vanishing at $s=1$ of L-functions associated with non-real characters (Landau 1903b):

Proof. For Re $s > 1$ one gets from (4.11)

$$L(s,\chi) = \exp\left(\sum_p \sum_{k=1}^{\infty} \frac{\chi(p)^k}{kp^{ks}}\right),$$

$$L(s,\chi^2) = \exp\left(\sum_p \sum_{k=1}^{\infty} \frac{\chi^2(p)^k}{kp^{ks}}\right)$$

and

$$\zeta(s) = \exp\left(\sum_p \sum_{k=1}^{\infty} \frac{1}{kp^{ks}}\right),$$

which for real $x > 1$ gives

$$\zeta^3(x)L(x,\chi)^4 L(x,\chi^2) = \exp\left(\sum_p \sum_{k=1}^{\infty} \frac{3 + 4\chi^k(p) + (\chi^k(p))^2}{kp^{kx}}\right).$$

If $\chi(p)$ does not vanish then $|\chi(p)| = 1$, thus we can write

$$\chi^k(p) = \cos\alpha + i\sin\alpha$$

with suitable real $\alpha = \alpha(k,p)$. This leads to

$$|\zeta^3(x)L(x,\chi)^4 L(x,\chi^2)|$$

$$= \exp\left(\sum_{\substack{p \\ \chi(p)=0}} \sum_{k=1}^{\infty} \frac{3}{kp^{kx}} + \sum_{\substack{p \\ \chi(p)\neq 0}} \sum_{k=1}^{\infty} \frac{3 + 4\cos\alpha + \cos(2\alpha)}{kp^{ku}}\right)$$

and now (5.33) implies

$$|\zeta^3(x)L(x,\chi)^4 L(x,\chi^2)| \geq 1. \tag{5.36}$$

Now assume that $L(1,\chi)$ vanishes. The functions $L(s,\chi)$ and $L(s,\chi^2)$ are regular at $s = 1$ and so for x close to unity we have

$$L(x,\chi) = O(x-1),\ L(x,\chi^2) = O(1).$$

Since Corollary 2 to Lemma 2.3 implies $\zeta(x) = O((x-1)^{-1})$ thus (5.36) leads to

$$1 = O(x-1)$$

which for x tending to unity is not possible. □

A simple proof of non-vanishing at $s = 1$ of a Dirichlet L-function corresponding to a non-real character χ can be also obtained by the following argument, which translates into the language of complex functions the proof given by Dirichlet (1837c) in the case of a prime modulus which we presented at the end of Sect.2.2:

Proof. Assume that ψ is a non-real character mod k with $L(1,\psi) = 0$. Then evidently $L(1,\overline{\psi})$ also vanishes, and since both $L(s,\psi)$ and $L(s,\overline{\psi})$ are regular at $s = 1$ the limits

$$\lim_{s\to 1} \frac{L(s,\psi)}{s-1}$$

and

$$\lim_{s\to 1} \frac{L(s,\overline{\psi})}{s-1}$$

exist and this shows the existence of

$$A = \lim_{s\to 1} \frac{\prod_{\chi\neq\chi_0} L(s,\chi)}{(s-1)^2}.$$

Since the function $L(s,\chi_0)$, corresponding to the principal character has at $s = 1$ a simple pole, we obtain

$$\lim_{s\to 1} \prod_{\chi \bmod k} L(s,\chi) = 0. \tag{5.37}$$

However for real $x > 1$ we have

$$\log \prod_{\chi \bmod k} L(x,\chi) = \sum_{\chi \bmod k} \log L(x,\chi)$$

$$= \sum_{\chi \bmod k} \sum_p \sum_{n=1}^{\infty} \frac{\chi(p^n)}{np^{nx}} = \sum_p \sum_{n=1}^{\infty} \frac{1}{np^{nx}} \sum_{\chi \bmod k} \chi(p^n)$$

$$= \varphi(k) \sum_p \sum_{\substack{n \\ p^n\equiv 1 \pmod{k}}} \frac{1}{np^{nx}} \geq 0$$

and hence

$$\prod_{\chi \bmod k} L(x,\chi) \geq 1,$$

contradicting (5.37). □

A similar idea was utilized by de la Vallée-Poussin (1896a,II). His proof seems to be the shortest available:

Proof. Taking the logarithmic derivative of $L(s,\chi)$ for Re $s > 1$ one gets

$$\frac{L'}{L}(s,\chi) = -\sum_p \frac{\chi(p)\log p}{p^s} + g(s)$$

with $g(s)$ being regular in the closed half-plane Re $s \geq 1$. If we define

$$\lambda_\chi = -\lim_{s\to 1}(s-1)\frac{L'}{L}(s,\chi)$$

and denote by $r(\chi)$ the order of the possible zero of $L(s,\chi)$ at $s=1$ then

$$\lambda_\chi = \begin{cases} 1 & \text{if } \chi \text{ is the principal character} \\ -r_\chi & \text{otherwise.} \end{cases}$$

Then for real $x > 1$

$$1 - \sum_{\chi\neq\chi_0} r(\chi) = \lim_{x\to 1+0}(x-1)\sum_\chi\sum_p \frac{\chi(p)\log p}{p^x}$$

$$= \lim_{x\to 1+0}(s-1)\sum_p\sum_\chi \frac{\chi(p)\log p}{p^x} = \varphi(k)\lim_{x\to 1+0}(x-1)\sum_{p\equiv 1 \pmod k} \frac{\log p}{p^x} \geq 0,$$

which implies $\sum_{\chi\neq\chi_0} r(\chi) \leq 1$ and so we see that for at most one character χ mod k can the corresponding L-function vanish. However the numbers $L(1,\chi)$ and $L(1,\overline{\chi})$ are conjugated, so either they both vanish or none of them and this eliminates the possibility of $L(1,\chi)=0$ for a non-real character. □

Still another proof is given in Landau (1911c).

3. The case of real characters is more difficult. We present first a proof of de la Vallée-Poussin (1896a,II):

Proof. For a real character χ put

$$f(s) = \frac{L(s,\chi)L(s,\chi_0)}{L(2s,\chi_0)} = \prod_{\substack{p \\ \chi(p)=1}} \frac{p^s+1}{p^s-1}.$$

By Corollary 1 to Theorem 4.4 the function $f(s)$ is meromorphic and the assumption $L(1,\chi)=0$ implies its regularity in the half-plane Re $s > 1/2$. Moreover we have $f(1/2)=0$. For positive ϵ the function $f(1+\epsilon+s)$ is thus regular in the half-plane Re $s > -1/2-\epsilon$ and we may write

$$f(1+\epsilon+s) = f(1+\epsilon) + \sum_{j=1}^{\infty} \frac{f^{(j)}(1+\epsilon)}{j!} s^j,$$

the power series having a radius of convergence at least $1/2+\epsilon$. Since for Re $s > 1$ one has

$$f(s) = \sum_{n=1}^{\infty} \frac{\alpha_n}{n^s}$$

with certain non-negative coefficients α_n, we obtain by differentiation

$$f^{(j)}(1+\epsilon) = (-1)^j A_j$$

where

$$A_j = \sum_{n=1}^{\infty} \frac{a_n \log^j n}{n^{1+\epsilon}} > 0.$$

This implies

$$f(1+\epsilon+s) = f(1+\epsilon) + \sum_{j=1}^{\infty} (-1)^j \frac{A_j}{j!} s^j$$

which for $s = -1/2 - \epsilon$ gives

$$0 = f(1/2) = f(1+\epsilon) + \sum_{j=1}^{\infty} \frac{A_j}{j!} (1/2+\epsilon)^j > 0,$$

a clear contradiction. □

A similar idea was used by Landau (1906a) who based his proof on a slightly earlier theorem (Landau 1905b) concerning singularities of Dirichlet series with nonnegative coefficients[17], which we now present. Note that this proof is actually a modification of E.Phragmén's proof of Theorem 4.13 (Phragmén 1891).

Theorem 5.15. (Landau 1905b) *The abscissa of absolute convergence of a Dirichlet series with non-negative coefficients cannot be a regular point for the sum of that series.*

Proof. Let x be the abscissa of absolute convergence of the series

$$f(s) = \sum_{n=1}^{\infty} \frac{a_n}{n^s}$$

with $a_n \geq 0$ and assume that x is a regular point of f. Then f can be prolonged to a function holomorphic in an open disk containing x, thus the Taylor expansion of f at a point $x_0 > x$ converges at some point $x_1 < x$. Writing down this expansion we get

$$f(x_1) = \sum_{j=0}^{\infty} (-1)^j \frac{(x_1 - x_0)^j}{j!} \sum_{k=1}^{\infty} a_k \frac{\log^j k}{k^{x_0}}, \tag{5.38}$$

because

$$f^{(j)}(s) = \sum_{n=1}^{\infty} (-1)^j \frac{a_n \log^j n}{n^s}.$$

[17] This theorem is an analogue for Dirichlet series of the Vivanti-Pringsheim theorem (Pringsheim 1894, Vivanti 1893), stating that any function whose Maclaurin series has positive coefficients must have a singularity at the intersection of the positive real half-line with its circle of convergence.

Since all terms of the series (5.38) are non-negative we may change the order of summation which leads to

$$f(x_1) = \sum_{k=1}^{\infty} \frac{a_k}{k^{x_0}} \sum_{j=0}^{\infty} \frac{(x_0 - x_1)^j}{j!} \log^j k = \sum_{k=1}^{\infty} \frac{a_k}{k^{x_1}},$$

but our assumption implies that the series on the right-hand side is divergent, giving a contradiction. □

A stronger version of Landau's result, in which the nonnegativity of coefficients is replaced by the assumption that all coefficients lie in an angle smaller than π was proved by M.Fekete[18] (1910).

Essentially the same argument leads to the following integral analogue of Theorem 5.15:

Theorem 5.16. (Landau 1905b) *Let $\varphi(x)$ be a function which is nonnegative for $x \geq \alpha \geq 1$ and assume that there exists a real number β such that the integral*

$$\int_\alpha^\infty \frac{\varphi(x)}{x^{s+1}} dx$$

converges for $s > \beta$ and diverges for $s < \beta$. Then β is a singular point of the function $F(s)$ defined by this integral in the half-plane Re $s > \beta$.

Proof. Assume that $F(s)$ is regular at β and choose $\gamma > \beta$. Observing that the integral defining $F(s)$ converges uniformly in every half-plane Re $s \geq \beta + \delta$ ($\delta > 0$) due to

$$|\varphi(x)x^{-s-1}| \leq \varphi(x)x^{-\mathrm{Re}s-1} \leq \varphi(x)x^{-\delta-1}$$

we may write for Re $s > \beta$

$$F(s) = \sum_{n=0}^{\infty} \frac{F^{(n)}(\gamma)}{n!}(s-\gamma)^n = \sum_{n=0}^{\infty} \left(\frac{(-1)^n}{n!} \int_\alpha^\infty \varphi(x) \frac{\log^n x}{x^{\gamma+1}} dx \right) (s-\gamma)^n.$$

By assumption this series converges at a certain real point $\xi < \beta$, i.e.

$$\sum_{n=0}^{\infty} \left(\frac{1}{n!} \int_\alpha^\infty \varphi(x) \frac{\log^n x}{x^{\gamma+1}} dx \right) (\gamma - \xi)^n$$

is convergent, but all its terms are nonnegative, so we may interchange the integration and summation to get the convergence of

$$\int_\alpha^\infty \varphi(x) x^{-\gamma-1} \sum_{n=0}^{\infty} \left(\frac{1}{n!} \log^n x (\gamma - \xi)^n \right) dx = \int_\alpha^\infty \varphi(x) x^{-\xi-1} dx.$$

[18]Fekete, Michael (1886–1957), Professor in Budapest and Jerusalem.

This contradicts the choice of β. □

Before we continue with the main topic of this section we would like to point out a simple corollary of the last theorem:

Corollary. *The difference $\psi(x) - x$ changes its sign infinitely often.*

Proof. If we put

$$\varphi(x) = \sum_{n \le x} (\Lambda(n) - 1) = \psi(x) - [x]$$

and

$$I(s) = \int_1^\infty \frac{\varphi(x)}{x^{1+s}} dx$$

then an application of Lemma 4.12 leads to the equality

$$I(s) = \frac{1}{s} \sum_{n=1}^{\infty} \frac{\Lambda(n) - 1}{n^s} = -\frac{1}{s} \left(\frac{\zeta'}{\zeta}(s) + \zeta(s) \right)$$

valid for Re $s > 1$. This gives (in view of the non-vanishing of $\zeta(s)$ on Re $s = 1$) the analytic continuation of $I(s)$ to the closed half-plane Re $s \ge 1$. If the difference $\psi(x) - x$ would have a constant sign for sufficiently large x then the abscissa β of convergence of the integral $I(s)$ would be less than 1. In particular $I(1)$ would converge. Since $\psi(x) = \sum_{n \le x} \Lambda(n)$ it follows from the Corollary to Theorem 5.8 that Lemma 4.12 is applicable and thus we obtain the convergence of the series

$$\sum_{j=1}^{\infty} \frac{\Lambda(j) - 1}{j}.$$

This implies the convergence of its subseries

$$\sum_p \frac{\log p - 1}{p}$$

contradicting Theorem 1.4. □

Theorems 5.15 and 5.16 may be used to prove also other oscillation theorems. See e.g. Dress (1984) and Exercise 18.

4. We give now two proofs of $L(1, \chi) \ne 0$ for real χ, both due to Landau (1906a,1908b).

Landau's first proof. Let χ be a real non-principal Dirichlet character, assume that $L(1, \chi) = 0$ and put

$$f(s) = \zeta(s) L(s, \chi). \tag{5.39}$$

It follows from Theorem 4.1 that $f(s)$ is regular for Re $s>0$ and we shall now show that this leads to a contradiction.

·Obviously for Re $s>1$ we have the equality

$$f(s)=\sum_{n=1}^{\infty}\frac{a_n}{n^s},$$

where

$$a_n=\sum_{d|n}\chi(d). \tag{5.40}$$

and Lemma 2.11 shows that the coefficients a_n are all nonnegative and that the inequality $a_{n^2}\geq 1$ holds. Thus the Dirichlet series of $f(s)$ diverges at $s=1/2$ and hence, by Theorem 5.15, this function must have a singularity in the interval $[1/2,1]$, contradiction. □

Another simple proof of $L(1,\chi)\neq 0$ for real characters can be found in the last section of Landau (1908b). It has certain affinity with the elementary proofs of Mertens, which we presented in Chap.2.

Landau's second proof. If $f(s)$ and a_n are defined by (5.39) and (5.40), respectively, then

$$\sum_{n\leq x}a_n=\sum_{mn\leq x}\chi(n)=\sum_{1\leq n\leq\sqrt{x}}\chi(n)\left[\frac{x}{n}\right]+\sum_{1\leq m\leq\sqrt{x}}\ \sum_{\sqrt{x}<n\leq x/m}\chi(n).$$

The Corollary to Lemma 2.10 implies that the second summand here is $O(\sqrt{x})$ we obtain

$$\sum_{n\leq x}a_n=\sum_{1\leq n\leq\sqrt{x}}\chi(n)\frac{x}{n}+O(\sqrt{x})=\ L(1,\chi)x+O(x\sum_{n>\sqrt{x}}\frac{\chi(n)}{n})+O(\sqrt{x})$$
$$=L(1,\chi)x+O(\sqrt{x})$$

the last equality being obtained by partial summation. Assume now that $L(s,\chi)$ vanishes at $s=1$. Then $f(s)$ is regular for Re $s>0$ and moreover

$$\sum_{n\leq x}a_n=O(\sqrt{x})$$

hence the series for $f(s)$ converges for Re $s>1/2$. Using Lemma 2.12 we deduce

$$\lim_{s\to 1/2}|f(s)|=\infty$$

and this contradicts the regularity of $f(s)$ at $s=1/2$. □

5. A.E.Ingham (1930) (see also Bateman 1997) extended this approach to functions defined by Dirichlet series

$$\Phi(s) = \sum_{n=1}^{\infty} \frac{f(n)}{n^s}$$

with *completely multiplicative* $f(n)$ (i.e. non-zero functions satisfying the equality $f(mn) = f(m)f(n)$ for all integers m, n) and $|f(n)| = 0$ or 1, provided $\Phi(s)$ has a good behaviour on the interval $[1/2, 1]$.

Theorem 5.17. (Ingham 1930) *Let $f(n)$ be a complex-valued completely multiplicative function whose non-zero values lie on the circumference of the unit circle. If the function defined for* Re $s > 1$ *by the series*

$$\Phi(s) = \sum_{n=1}^{\infty} \frac{f(n)}{n^s}$$

can be continued to a function regular in a domain containing the interval $[1/2, 1]$ *then* $\Phi(1) \neq 0$.

Proof. Put

$$F(n) = \sum_{d|n} f(d), \quad \Psi(s) = \sum_{n=1}^{\infty} \frac{\overline{f(n)}}{n^s},$$

$$A(s) = \sum_{n=1}^{\infty} \frac{|f(n)|^2}{n^s} = \sum_{f(n)\neq 0} \frac{1}{n^s}$$

and

$$B(s) = \sum_{n=1}^{\infty} \frac{|F(n)|^2}{n^s}.$$

The function F is multiplicative by Lemma 1.16 and the series defining $\Psi(s)$, $A(s)$ as well as $B(s)$ all converge absolutely for Re $s > 1$, the last because of the inequality

$$|F(n)| \leq d(n) = O(n^\epsilon)$$

valid for every $\epsilon > 0$. Euler's product formula gives now the equality

$$B(s) = \prod_p \sum_{m=0}^{\infty} \frac{|F^2(p^m)|}{p^{ms}},$$

valid for Re $s > 1$.

The crucial point of the proof lies in the identity

$$B(s) = \frac{\zeta(s)\Phi(s)\Psi(s)A(s)}{A(2s)}$$

due essentially to S.Ramanujan (1916) and B.M.Wilson (1922), and which is proved by putting $p^{-s}=x$ in the factor corresponding to the prime p of the left-hand side and splitting it into partial fractions (see e.g.Titchmarsh 1951, Chap.1).

Now observe that

$$A(s)=\prod_{\substack{p\\ f(p)\neq 0}}\left(1-\frac{1}{p^s}\right)^{-1}$$

and thus

$$\frac{A(s)}{A(2s)}=\frac{\zeta(s)}{\zeta(2s)}\prod_{\substack{p\\ f(p)=0}}\frac{1}{1+p^{-s}}.$$

Put

$$C(s)=\frac{\zeta^2(s)\Phi(s)\Psi(s)}{\zeta(2s)}.$$

For Re $s>1$ the equality

$$C(s)=\prod_{\substack{p\\ f(p)=0}}\left(1+\frac{1}{p^s}\right)B(s)$$

holds, hence $C(s)$ is representable by a Dirichlet series with non-negative coefficients. By assumption $\Phi(s)$ can be continued to a regular function in $[1/2,1]$ and the same applies to $\Psi(s)$ due to $\Psi(s)=\overline{\Phi(\overline{s})}$ and we obtain that $C(s)$ is also regular there, with the possible exception of a pole at $s=1$ of order not exceeding 2. If now $\Phi(1)=0$, then $\Psi(1)=0$ and thus $C(s)$ cannot have a pole at $s=1$, hence Theorem 5.15 shows that the Dirichlet series of $C(s)$ converges for Re $s>1/2$. Write for real $x>1/2$

$$C(x)=\sum_{n=1}^{\infty}\frac{a_n}{n^x}$$

with $a_n\geq 0$. Since $a_1=|F(1)|^2=1$ we have $C(x)\geq 1$ and this gives for $x>1/2$, in view of

$$\Psi(x)=\overline{\Phi(x)},$$

the inequality

$$|\zeta(x)\Phi(x)|^2\geq\zeta(2x)$$

but this leads to a contradiction, since for x tending to $1/2$ the left-hand side of this inequality remains bounded and the right-hand side tends to infinity. □

As a corollary one gets another proof of Theorem 5.7:

Proof. Let t be a real number and put $f(n)=\chi(n)n^{-it}$. Then

$$\Phi(s) = \sum_{n=1}^{\infty} \frac{\chi(n)}{n^{s+it}} = L(s+it, \chi)$$

is regular for Re $s > 0$ and Theorem 5.17 implies $L(1+it) \neq 0$. □

Ingham's method can be utilized to obtain similar results but subject to weaker assumptions. See Rankin (1939), Wintner[19](1946) and G.Shapiro (1949). Wintner considered real-valued functions $f(n)$ satisfying $-1 \leq f(n) \leq 1$ and Shapiro, following the lines of Wintner's argument, extended this to arbitrary bounded completely multiplicative complex-valued functions.

A proof of Theorem 5.6 on similar lines was given by R.Narasimhan (1968) and his idea was later extended to the proof of Theorem 5.7 by D.J.Newman (1993).

H.J.Besenfelder (1977) gave still another proof of Theorem 5.6 utilizing rather heavy analytic tools, including a kind of explicit formula obtained in Besenfelder, Palm (1977). His method can be also applied to prove Theorem 5.7.

5.6. Bounding the Error Term

1. The first evaluation of the error term in the Prime Number Theorem was obtained by C.de la Vallée-Poussin (1899) who deduced it from the existence of an explicit region to the left of the line Re $s = 1$ in which Riemann's zeta-function does not have any zeros:

Theorem 5.18. *There exist positive constants c_0, c_1 and c_2 such that in the region*

$$\sigma > \begin{cases} 1 - c_1/\log(t - c_2) & \text{if } |t| > c_0 \\ 1 - c_1/(\log c_0 - c_2) & \text{if } |t| \leq c_0 \end{cases}$$

one has $\zeta(\sigma + it) \neq 0$.

De la Vallée-Poussin obtained the following values for the constants occuring in this theorem: $c_0 = 574$, $c_1 = 0.0328214$, $c_2 = 3.806$ and at the end of his memoir he showed that the values $c_0 = 705$, $c_1 = 0.034666$ and $c_2 = \log 47.886 = 3.8688\ldots$ are also admissible. Later it turned out that for large τ one can take $c_1 = 0.106998\ldots$ (Rosser,Schoenfeld 1975, see also Stečkin 1970a,b and Kondrat'ev 1977).

The principal tool in the proof is the formula from Corollary 4 to Theorem 5.1, which for Re $s > 1$ can be written in the form

$$\sum_{\rho} \frac{1}{s-\rho} = \frac{1}{s-1} - C + \frac{1}{2}\frac{\Gamma'}{\Gamma}\left(1 + \frac{s}{2}\right) - \sum_{p}\sum_{k=1}^{\infty} \frac{\log p}{p^{ks}}. \tag{5.41}$$

Actually one has $C = \log \sqrt{\pi}$ but the precise value of this constant is not important here.

[19]Wintner, Aurel Friedrich (1903-1964).

For $s = u + it$ with $u > 1$ put

$$A(s) = \sum_p \sum_{k=1}^{\infty} \left(\frac{\log p}{p^{ku}} + \frac{\log p}{p^{ks}} \right).$$

Thus

$$\text{Re } A(s) = \sum_p \sum_{k=1}^{\infty} \left(\frac{\log p(1 + \cos(kt \log p))}{p^{ku}} \right)$$

and in view of

$$1 + \cos \alpha = \frac{1}{2} \frac{1 - \cos(2\alpha)}{1 - \cos \alpha} \geq \frac{1 - \cos(2\alpha)}{4}$$

one gets

$$\text{Re } A(s) \geq \frac{1}{4} \sum_p \sum_{k=1}^{\infty} \left(\frac{\log p(1 - \cos(2kt \log p))}{p^{ku}} \right). \tag{5.42}$$

Now put $s' = u + 2it$. Using (5.41) and (5.42) one arrives after a short calculation at the inequality

$$\begin{aligned} &\text{Re} \left(\sum_\rho \frac{1}{s - \rho} + \frac{1}{4} \sum_\rho \frac{1}{s' - \rho} \right) \\ &< \text{Re} \left(\frac{1}{s-1} + \frac{1}{2} \frac{\Gamma'}{\Gamma}(1 + \frac{s}{2}) \right) + \frac{3}{4} \sum_p \sum_{k=1}^{\infty} \frac{\log p}{p^{ku}} + \frac{1}{4} \text{Re } \frac{1}{s' - 1} \\ &\quad + \frac{1}{8} \text{Re} \left(\frac{\Gamma'}{\Gamma}(1 + \frac{s'}{2}) \right) + O(1), \end{aligned}$$

which for u tending to 1 can be written in the form

$$\text{Re} \left(\sum_\rho \frac{1}{s - \rho} + \frac{1}{4} \sum_\rho \frac{1}{s' - \rho} \right) < \frac{3}{4} \sum_p \sum_{k=1}^{\infty} \frac{\log p}{p^{ku}} + O_t(1). \tag{5.43}$$

Since for u tending to 1 we have

$$\sum_p \sum_{k=1}^{\infty} \frac{\log p}{p^{ku}} = -\frac{\zeta'(u)}{\zeta(u)} = \frac{1}{u - 1} + O(1)$$

and the real parts of both $1/(s - \rho)$ and $1/(s' - \rho)$ are positive for $u > 1$ then if $\rho = \alpha + \beta i$ is an arbitrary root of $\zeta(s)$ satisfying $0 < \alpha < 1$ the left-hand side of (5.43) can be minorized by $1/(u - \alpha)$ which leads to the inequality

$$\frac{1}{u - \alpha} < \frac{3}{4} \frac{1}{u - 1} + O_t(1).$$

This gives a clear contradiction if α and u are close to 1. Using explicit bounds for the error terms in the formula (5.43) Vallée-Poussin obtained after an elementary numerical computation the assertion of the theorem. □

2. The existence of zero-free region established in the preceding theorem enabled Vallée-Poussin to obtain a bound for the error term in the Prime Number Theorem:

Theorem 5.19. *There exist positive constants C and D such that for $x \geq 2$ one has*[20]

$$\psi(x) = x + O(x\exp(-C\sqrt{\log x})), \quad \theta(x) = x + O(x\exp(-C\sqrt{\log x})) \quad (5.44)$$

and

$$\pi(x) = \mathrm{li}(x) + O(x\exp(-D\sqrt{\log x})). \quad (5.45)$$

In the following sketch of de la Vallée-Poussin's proof we introduced certain inessential simplifications, resulting from using Landau's symbol $O(\cdot)$ in places where the original proof contained careful estimations of the implied constants.

Proof. The first main tool is a bound for the sum

$$S_b(y) = \sum_{\substack{\rho=\alpha+\beta i \\ \beta>b}} \frac{y^{\alpha-1}}{\beta^2}$$

for positive y and a fixed b (about which one assumes that it exceeds the imaginary part of the first non-trivial zero of the zeta-function) ρ running, as before, over all non-real zeros of the zeta-function. Theorem 5.18 implies the bound

$$y^{\alpha-1} < \exp\left(-\frac{c_1 \log y}{\log\beta - c_2}\right)$$

and it is an easy exercise in calculating derivatives to obtain the evaluation

$$\frac{y^{\alpha-1}}{\beta^{1/2}} \leq \exp(-B\sqrt{\log y})$$

with a certain positive constant B. Since by Theorem 5.1 the series

$$\sum_{\rho} \frac{1}{\beta^{3/2}}$$

converges we obtain

$$S_b(y) = O(\exp(-B\log y)\sum_{\substack{\rho \\ \beta>b}} \frac{1}{\beta^{3/2}}) = O(\exp(-B\log y)) \quad (5.46)$$

with the implied constant depending on b.

[20]De la Vallée-Poussin wrote the error terms with the extra factor $\sqrt{c\log x}$. Its omission diminishes slightly the value of the constant C.

The same theorem shows that the part of the sum

$$\sigma = \sum_\rho \frac{y^{\alpha-1}}{|\rho|^2}$$

corresponding to ρ's in the half-plane Re $s < 1/2$ is $O(1/\sqrt{y})$. In the remaining part of σ the sum of terms with $|\beta| \geq b$ is majorized by $S_b(y)$ and the remaining terms add to $O(1)$ since b is fixed. Therefore we get, using (5.46), the evaluation

$$\sigma = O(\log y \exp(-2\sqrt{c_1 \log y})). \tag{5.47}$$

Put

$$A(y) = \sum_{n<y} \frac{\Lambda(n)}{n} - \frac{1}{y}\psi(y)$$

and

$$B(y) = \log y - \lim_{u\to 1}\left(\frac{\zeta'}{\zeta}(u) + \frac{u}{u-1}\right) + \frac{1}{y}\frac{\zeta'}{\zeta}(0)$$
$$-\sum_\rho \frac{y^{\rho-1}}{\rho(\rho-1)} - \sum_{m=1}^{\infty} \frac{y^{-2m-1}}{2m(2m+1)}.$$

Since (5.25) implies $A(y) = B(y)$ (for real $y > 1$) we have

$$xA(x) - \int_1^x A(y)dy = xB(x) - \int_1^x B(y)dy, \tag{5.48}$$

thus putting

$$C(x) = \int_1^x \frac{\psi(y)}{y} dy$$

we obtain after a short calculation[21]

$$C(x) = x - \sum_\rho \frac{x^\rho}{\rho^2} + c_1 \log x + c_2 + O\left(\frac{1}{x^2}\right) \tag{5.49}$$

with certain constants c_1 and c_2.

Consider now, for positive ϵ,

$$D(x,\epsilon) = \frac{C((1+\epsilon)x) - C(x)}{\epsilon} = \frac{1}{\epsilon}\int_x^{x+h} \frac{\psi(y)}{y} dy$$
$$= x - \sum_\rho \frac{(1+\epsilon)^\rho - 1}{\epsilon} \frac{x^\rho}{\rho^2} + O(1).$$

Since $\psi(y)$ is nondecreasing one obtains

[21] Here we simplify slightly, without changing the main idea, the original argument of de la Vallée-Poussin who performed his computation conserving all terms in the expressions considered.

$$\frac{\log(1+\epsilon)}{\epsilon}\psi(x) \le D(x,\epsilon) \le \frac{\log(1+\epsilon)}{\epsilon}\psi((1+\epsilon)x). \tag{5.50}$$

Using (5.49), (5.50) and the bound Re $\rho < 1$ and noting that all terms of $C(x)$ are real, one gets finally

$$\psi(x) \le \frac{\epsilon}{\log(1+\epsilon)}x - \sum_\rho \frac{(1+\epsilon)^\rho - 1}{\log(1+\epsilon)}\frac{x^\rho}{\rho^2} + O(1) \le \frac{\epsilon}{\log(1+\epsilon)}x$$

$$+\sum_\rho \frac{3}{\log(1+\epsilon)}\frac{x^{\mathrm{Re}\rho}}{|\rho|^2} + c + 1 + O\left(\frac{1}{x^2\log(1+\epsilon)}\right) \tag{5.51}$$

and (changing x to $x/(1+\epsilon)$ in (5.50))

$$\psi(x) \ge \frac{\epsilon}{(1+\epsilon)\log(1+\epsilon)}x - \sum_\rho \frac{1-(1+\epsilon)^{-\rho}}{\log(1+\epsilon)}\frac{x^\rho}{\rho^2} + O(1)$$

$$\ge \frac{\epsilon}{(1+\epsilon)\log(1+\epsilon)}x - \sum_\rho \frac{3}{\log(1+\epsilon)}\frac{x^{\mathrm{Re}\rho}}{|\rho|^2} + O(1). \tag{5.52}$$

The first terms on the right-hand sides of (5.51) and (5.52) are both equal $x + O(\epsilon x)$ and in view of

$$\frac{(1+\epsilon)^\rho + 1}{\log(1+\epsilon)} = O(\epsilon^{-1})$$

and

$$\frac{(1+\epsilon)^{-\rho} + 1}{\log(1+\epsilon)} = O(\epsilon^{-1})$$

one gets

$$\psi(x) = x + O(\epsilon x) + O\left(\frac{1}{\epsilon}\sum_\rho \frac{x^{\mathrm{Re}\rho}}{|\rho|^2}\right) + O\left(\frac{1}{\epsilon x^2}\right).$$

If now $\epsilon > 0$ satisfies

$$\epsilon^2 = \sum_\rho \frac{x^{\mathrm{Re}\rho - 1}}{|\rho|^2}$$

then using (5.47) one arrives at

$$\psi(x) = x + O(x\exp(-c\sqrt{\log x}))$$

with a certain constant $c > 0$. This proves the first equality in (5.44) and the second results from Lemma 3.3 (iii).

It remains to deduce (5.45). This was done by de la Vallée-Poussin in a very complicated way, using a variant of (5.24) as well as the functional equation of the zeta-function and performing a lot of integrations. Later E.Landau (1900a) noted that this deduction can be made using only elementary tools. We present now his argument:

It is sufficient to establish (5.45) for integral values of $x > 2$. Observe that if we write $r(x) = \theta(x)/x - 1$ then

$$1 + nr(n) - (n-1)r(n-1) = \begin{cases} \log n & \text{if } n \text{ is prime,} \\ 0 & \text{otherwise} \end{cases}$$

hence

$$\pi(x) = \sum_{n=2}^{x} \frac{1 + nr(n) - (n-1)r(n-1)}{\log n}$$

and partial summation gives

$$\pi(x) = \sum_{n=2}^{x} \frac{1}{\log n} + \sum_{n=2}^{x-1} nr(n) \left(\frac{1}{\log n} - \frac{1}{\log(n+1)} \right) + \frac{xr(x)}{\log x}.$$

Lemma 2.12 implies the equality

$$\sum_{n=2}^{x} \frac{1}{\log n} = \mathrm{li}(x) + O(1)$$

and hence

$$\pi(x) = \mathrm{li}(x) + O\left(\sum_{n=2}^{x-1} \frac{nr(n)}{\log^2 n} \right) + O\left(\frac{xr(x)}{\log x} \right) + O(1).$$

As (5.44) implies $r(x) = O(\exp(-C\sqrt{\log x}))$ and

$$\sum_{n=2}^{x-1} \frac{nr(n)}{\log^2 n} = O(\int_2^x \frac{t\exp(-C\sqrt{\log t})}{\log^2 t} dt) = O(\int_2^x \exp(-C\sqrt{\log t}) dt)$$

$$= O(\int_{\sqrt{\log 2}}^{\sqrt{\log x}} u\exp(-Cu + u^2)du)) = O(x\sqrt{\log x} \int_{\sqrt{\log 2}}^{\sqrt{\log x}} e^{-Cu} du)$$

$$= O(x\sqrt{\log x}\exp(-C\sqrt{\log x}))$$

the evaluation (5.45) follows with an arbitrary $D < C$. □

This proof shows that the size of the error term in the Prime Number Theorem depends on the region in which the zeta-function does not vanish. This fact was made explicit by A.E.Ingham (1932), who proved the following result:

Let $\eta(t)$ be function having a continuous derivative and decreasing for $t \geq 0$. Assume further that the following three conditions are satisfied: $0 < \eta \leq 1/2$, $\lim_{t\to\infty} \eta'(t) = 0$ and $1/\eta(t) = O(\log t)$ for t tending to infinity.

If $\zeta(s)$ does not vanish in the region

$$\{\sigma + it : \ \sigma > 1 - \eta(|t|)\}$$

then the differences $\pi(x) - \mathrm{li}(x)$, $\theta(x) - x$ *and* $\psi(x) - x$ *are all* $O(x\exp(-c\omega(x)))$ *with a suitable* $c > 0$ *and*

$$\omega(x) = \min_{t\geq 1}(\eta(t)\log x + \log t).$$

Ingham's result implies in particular that if one can take

$$\eta(t) = \frac{c}{\log^\alpha(t)}$$

then one gets for the error term the bound $O(x\exp(-c_1\log^\beta x))$ with $\beta = 1/(1+\alpha)$ and a suitable constant c_1. Later investigations showed that this can be done with any $\alpha > 3/4$ (Čudakov[22] 1936a), any $\alpha > 5/7$ (Korobov 1958a) and any $a > 2/3$ (Korobov 1958c, Vinogradov[23] 1958). The two last authors actually proved that the zeta-function does not vanish in the region

$$\mathrm{Re}\ s > 1 - \frac{c}{\log^{2/3} t(\log\log(t+c_1))^{1/3}}, \qquad \mathrm{Im}\ s = t > 0$$

(with suitable positive c and c_1) which implies the error term of the form

$$O(x\exp(-c_2\log^{3/5} x(\log\log x)^{-1/5})).$$

The first complete proof (due to H.-E.Richert) appeared in Walfisz (1963). See also Ellison (1975). Two proofs were given by Y.Motohashi (1978a,1981), the second being based on Selberg's sieve, and O.V.Popov (1994) made explicit the constant $c = 0.00006888$.

The converse of Ingham's theorem in the particular case $\eta(t) = c\log^{-\alpha}(t)$ was proved by P.Turán (1950,II) and in the general case a partial converse was obtained by W.Staś (1961). Finally J.Pintz (1980c) (see Pintz 1983a) first strengthened Ingham's result by replacing the factor 1/2 in the assertion by 1 and then showed the truth of an almost complete converse theorem, which we now quote:

Assume that $\zeta(s)$ *has infinitely many zeros in the region* $\sigma \geq 1-\eta(\log|t|)$, *where* $0 < \eta(t) \leq 1/2$ *is continuous and decreases for* $t \geq 0$. *Moreover assume that its derivative increases for large* t *and tends to* 0 *when* t *tends to infinity.*

If now $0 < \epsilon < 1$ *is fixed,*

$$\omega(x) = \min_{t\geq 0}\{\eta(t)\log x + t\}$$

and

$$H(x) = \exp((1+\epsilon)\omega(x))$$

then $\pi(x) - \mathrm{li}(x)$ *is not* $o(x/(H(x)\log x))$ *and* $\theta(x) - x$ *is not* $o(x/H(x))$

The existence of a zero-free region for the zeta-function also follows from any non-trivial upper bound for the sum $\sum_{n\leq x}\mu(n)$ as proved in Allison (1970).

For analogues for arithmetic progressions of the results of Ingham and Turán see Wiertelak (1971,1972) and Staś, Wiertelak (1974).

---

[22] Proofs may be found in Titchmarsh (1951), Chandrasekharan (1970).

[23] Vinogradov, Ivan Matveevich (1891-1983), Professor in Moscow.

The argument used at the end of the proof presented above was utilized by Landau (1900a) to obtain asymptotic formulas for sums of the form $\sum_{p\le x} F(p,x)$, where the function $F(n,x)$ is assumed to satisfy the following conditions:

a) For every fixed n one has

$$0 \le F(n,x) = o\left(\int_2^x \frac{F(t,x)}{\log t}\,dt\right),$$

b) For $2 \le m \le n \le x$ one has

$$\frac{F(m,x)}{\log n} \ge \frac{F(n,x)}{\log n}.$$

Under these assumptions Landau proved (see Exercise 3) that

$$\sum_{p\le x} F(p,x)$$

and

$$\int_2^x \frac{F(t,x)}{\log t}\,dt$$

are asymptotically equal, thus justifying a series of earlier statements of de Polignac (1859a,b).

3. As a simple corollary to Theorem 5.19 one can solve a problem proposed by F.J.E.Lionnet (1872):

Corollary 1. (Landau 1901c) *For sufficiently large x one has*

$$\pi(2x) < 2\pi(x),$$

i.e. in the interval $[2,x]$ there are more prime numbers than in $(x,2x]$.

Proof. Observe that partial integration gives

$$\mathrm{li}(x) = \frac{x}{\log x} + \frac{x}{\log^2 x} + 2\int_2^x \frac{dt}{\log^3 t} + O(1),$$

where last integral is $O(x\log^{-3} x)$. We have also

$$\frac{1}{\log(2x)} = \frac{1}{\log x} - \frac{\log 2}{\log^2 x} + O\left(\frac{1}{\log^3 x}\right)$$

and

$$\frac{1}{\log^2(2x)} = \frac{1}{\log^2 x} + O\left(\frac{1}{\log^3 x}\right).$$

Thus Theorem 5.19 implies

$$\pi(2x) = \frac{2x}{\log x} + 2(1-\log 2)\frac{x}{\log^2 x} + O\left(\frac{x}{\log^3 x}\right)$$

and

$$2\pi(x) = \frac{2x}{\log x} + \frac{2x}{\log^2 x} + O\left(\frac{x}{\log^3 x}\right)$$

which makes the assertion evident. □

It was later announced by J.B.Rosser, L.Schoenfeld (1975) that the assertion of this corollary holds for all $x \geq 11$ and a proof can be found in Kopetzky,Schwarz (1979). The related conjecture (Hardy, Littlewood 1923a) asserting that the inequality $\pi(x+y) \leq \pi(x) + \pi(y)$ holds for all $x, y \geq 2$ is still unsettled, though D.Hensley and I.Richards (1973, 1973/74) (see also Richards 1974) showed that it is incompatible with another conjecture going back to L.E.Dickson (1904):

If $a_1, a_2, \dots, a_m$ are integers such that for every prime p there exists an integer n_p such that the product $\prod_{i=1}^{m}(n_p + a_i)$ is not divisible by p, then there exist infinitely many integers n such that all numbers $n + a_1, n + a_2, \dots, n + a_m$ are primes.

It was shown by C.Karanikolov (1971) that for $x \geq 347$ and $k \geq \sqrt{e}$ one has $\pi(kx) < k\pi(x)$ (see Martić,Vasić 1973).

4. The inequalities (5.51) and (5.52) may be used to show that under the assumption of Riemann's Hypothesis about the roots of the zeta-function one can essentially improve the bound for the error term in both parts of Theorem 5.19 and obtain the assertion contained in the paper of Stieltjes (1885b). This observation does not appear in de la Vallée-Poussin's paper and the first proof of it, in a much stronger form, was given by H.von Koch (1901) (see Theorem 5.21 below).

Corollary 2. *Assume that all non-real roots of $\zeta(s)$ lie on the line* Re $s =$ 1/2. *Then for $x \geq 2$ one has*

$$\psi(x) = x + O(x^{3/4}),$$
$$\theta(x) = x + O(x^{3/4}),$$
$$\pi(x) = \mathrm{li}(x) + O(x^{3/4}\log x).$$

Proof. To prove the first assertion it suffices to repeat the argument following the formula (5.52) in the proof of Theorem 5.19, observing that the assumption implies

$$\sum_{\rho} \frac{x^{\mathrm{Re}\rho}}{|\rho|^2} = O(x^{1/2})$$

and choosing ϵ to be equal $x^{-1/4}$.

The second assertion follows from the first and Lemma 3.3 (iii) and to obtain the last one should proceed as at the end of the proof of Theorem 5.19 using $r(x) = O(x^{3/4})$ and the evaluation

$$\sum_{n\le x} \frac{1}{n^{1/4}\log^2 n} = O\left(\frac{x^{3/4}}{\log^2 x}\right). \qquad \square$$

5. In his paper of 1898 de la Vallée-Poussin utilized Theorem 5.19 to provide a short proof of the following result:

Theorem 5.20. *For $x \ge 2$ one has*

$$\sum_{n=1}^{\infty} \frac{\mu(n)}{n} = 0,$$

more precisely

$$\sum_{n\le x} \frac{\mu(n)}{n} = O\left(\frac{1}{\log x}\right).$$

The first assertion was already stated by Euler (1748, Chap.15, §277) and proved by H.von Mangoldt (1897). We used this fact in Chap.4 to show the convergence of a series considered by Riemann. In the thesis of Landau (1899a) one finds a simpler proof (see Exercise 7). Later Landau (1901b) showed that one can, in the above theorem, replace $O(1/\log x)$ by $o(1/\log x)$. A similar assertion of Euler (1737) about the sum $\sum_{n=1}^{\infty} \lambda(n)n^{-1}$, where $\lambda(n) = (-1)^{\Omega(n)}$ and $\Omega(n)$ denotes the number of prime factors of n counted according to their multiplicities, follows from von Mangoldt's result with the use of the identity

$$\lambda(n) = \sum_{d^2|n} \mu\left(\frac{n}{d^2}\right).$$

This identity is a consequence of the equality

$$\sum_{n=1}^{\infty} \frac{\lambda(n)}{n^s} = \frac{\zeta(2s)}{\zeta(s)},$$

valid for Re $s > 1$, but it can also be easily proved in an elementary way.

Proof. De la Vallée-Poussin established first the following equality, which is an immediate consequence of his formula (5.26) and Theorem 5.19, although he preferred to give another proof:

$$\sum_{p^m\le x} \frac{\log p}{p^m} = \log x + A + r(x), \tag{5.53}$$

with

$$r(x) = O(\exp(-B\sqrt{\log x})) \tag{5.54}$$

where A, B are suitable positive constants.

Differentiating the formula

$$\frac{1}{\zeta(s)} = \sum_{n=1}^{\infty} \frac{\mu(n)}{n^s}$$

one gets for Re $s > 1$

$$\frac{\zeta'(s)}{\zeta^2(s)} = \sum_{n=1}^{\infty} \frac{\mu(n)\log n}{n^s}.$$

On the other hand, multiplying the Dirichlet series for $1/\zeta(s)$ and the logarithmic derivative of $\zeta(s)$ one is led to

$$\frac{\zeta'(s)}{\zeta^2(s)} = -\sum_{p^m} \frac{\log p}{p^{ms}} \sum_{k=1}^{\infty} \frac{\mu(k)}{k^s}$$

and thus, using the uniqueness theorem for Dirichlet series, one obtains for $x \geq 2$ the identity

$$\sum_{k\leq x} \frac{\mu(k)}{k^s} \sum_{p^m \leq x/k} \frac{\log p}{p^{ms}} = -\sum_{n\leq x} \frac{\mu(n)\log n}{n^s}. \tag{5.55}$$

Another deduction of this identity avoiding the uniqueness theorem but using integration instead may be found in Sect.5.6 of the book of H.M.Edwards (1974). The following completely elementary argument does the same:

It suffices to establish for $n = 1, 2, \ldots$ the identity

$$\mu(n)\log n = -\sum_{d|n} \Lambda(d)\mu(n/d).$$

For $n = 1$ both sides vanish. If $n > 1$ is squarefree, say $n = p_1 p_2 \cdots p_s$, then the assertion amounts to the obvious equality

$$(-1)^s \log n = -\sum_{i=1}^{s} (-1)^{s-1} \log p_i.$$

If however n is not square-free and d is a divisor of n with $\Lambda(d) \neq 0$ and $\mu(n/d) \neq 0$ then d must be a prime-power, say $d = p_1^k$ and we get $n = p_1^k p_2 \cdots p_s$ with $p_1, p_2, \ldots, p_s$ being distinct primes and $k \geq 2$. One sees now easily that for a divisor δ of n the product $\Lambda(\delta)\mu(n/\delta)$ does not vanish if and only if either $\delta = p_1^{k-1}$ or $\delta = p_1^k$. Thus the right hand-side of the assserted formula equals $(-1)^s \log p_1 + (-1)^{s-1} \log p_1 = 0$ and as the left-hand side also vanishes we are finished.

Putting $s = 1$ in (5.55) we get

$$\sum_{k\leq x} \frac{\mu(k)}{k} \sum_{p^m \leq x/k} \frac{\log p}{p^m} = -\sum_{n\leq x} \frac{\mu(n)\log n}{n}.$$

Using (5.53) one transforms this to the form

$$(\log x + A) \sum_{n \leq x} \frac{\mu(n)}{n} + \sum_{n \leq x} \frac{\mu(n)}{n} r(x/n) = 0$$

and it remains to evaluate the second summand on the right-hand side. In view of (5.54) it is of the order

$$O\left(\sum_{n \leq x} \frac{1}{n} \exp(-B\sqrt{\log x - \log n}) \right) = O(1)$$

and the theorem follows immediately. □

6. A few years after the appearance of de la Vallée-Poussin's evaluation of the error term in Prime Number Theorem an improvement of it was deduced from Riemann's Hypothesis by H.von Koch (1901) (for a slightly weaker result see von Koch (1900)):

Theorem 5.21. *If all non-real roots of* $\zeta(s)$ *lie on the line* Re $s = 1/2$ *then one has*

$$\psi(x) = x + O(\sqrt{x} \log^2 x),$$

and

$$\pi(x) = \mathrm{li}(x) + O(\sqrt{x} \log x).$$

Proof. The argument is based on a new formula for the function

$$f(x) = \sum_{p^k \leq x} \frac{1}{k} = \sum_{k=1}^{\infty} \frac{1}{k} \pi(x^{1/k}),$$

namely

$$f(x) = \lim_{s \to \infty} S(x, s),$$

where

$$S(x, s) = \sum_{k=1}^{\infty} \frac{(-1)^{k-1}}{k!} x^{ks} \log \zeta(ks) + \epsilon(x), \tag{5.56}$$

with

$$\epsilon(x) = \begin{cases} 1/ke & \text{if } x \text{ is a } k\text{-th power of a prime} \\ 0 & \text{otherwise.} \end{cases}$$

To obtain this note that for real $s > 1$ one has

$$\log \zeta(s) = \sum_{k=1}^{\infty} \frac{1}{k} \sum_{p} \frac{1}{p^{ks}},$$

thus

$$S(x, s) = \sum_{k=1}^{\infty} \sum_{p} \sum_{j=1}^{\infty} \frac{(-1)^{k-1}}{k!} \frac{1}{j} \left(\frac{x}{p^j} \right)^{ks}$$

and since the series on the right-hand side is absolutely convergent for $s > 1$ one can interchange the order of summation and obtain

$$S(x,s) = \sum_p \sum_{j=1}^{\infty} \frac{1}{j}(1 - \exp(-x^s p^{-sj})).$$

The last series being uniformly convergent for $s > 1 + \epsilon$ for every positive ϵ we may let s tend to infinity under the sum sign and this leads to (5.56).

Similarly, starting with the identity

$$\frac{\zeta'}{\zeta}(s) = -\sum_{k=1}^{\infty} \sum_p \frac{\log p}{p^{ks}}$$

one is led to

$$\psi(x) = \omega(x) + \lim_{s\to\infty} \Psi(x,s)$$

where

$$\omega(x) = \begin{cases} \frac{\log p}{e} & \text{if } x \text{ is a power of a prime } p, \\ 0 & \text{otherwise} \end{cases}$$

and

$$\Psi(x,s) = \sum_{k=1}^{\infty} \frac{(-1)^{k-1}}{k!} \frac{\zeta'}{\zeta}(ks)x^{ks} = -\sum_{p^k} (1 - \exp(-x^s p^{-ks})) \log p\,.$$

Corollary 4 to Theorem 5.1 implies now

$$\frac{\zeta'(ns)}{\zeta(ns)} = C + \frac{1}{1-ns}$$
$$+ \sum_{m=1}^{\infty} \left(\frac{1}{ns+2m} - \frac{1}{2m} \right) + \sum_\rho \left(\frac{1}{ns-\rho} + \frac{1}{ns-\bar\rho} \right),$$

ρ running over all non-real roots of $\zeta(s)$ lying in the upper half-plane. With the use of this formula von Koch achieves after six pages of calculations, using the explicit value of C, the equality

$$\Psi(x,s) = x - \log(2\pi) - 2s^2 \mathrm{Re} \left(\int_0^x \sum_\rho \frac{(xy^{-1})^\rho}{\rho(s-\rho)} y^{2s-1} \exp(-y^s)dy \right)$$
$$+ O(sx^s \exp(-x^s)) + O(\frac{x}{s}) + O(\log s) \tag{5.57}$$

valid for $s > 1$ and every $\epsilon > 0$.

Now choose x of the form $n + 1/2$ with $0 < n \in \mathbf{Z}$ and split the sum defining $\Psi(x,s)$ in two parts S_1, S_2, the first corresponding to $p^k < x$. Since in our case the inequality $p^k < x$ implies $p^k \le x - 1/2$ we get

$$\exp(-x^s p^{-ks}) \le x^{-s} p^{ks} < \left(1 - \frac{1}{2x}\right)^s,$$

hence

$$\psi(x) \ge S_1 = \psi(x) - \sum_{p^k < x} \exp(-x^s p^{-ks}) \log p \ge \psi(x) \left(1 + O\left(\left(1 - \frac{1}{2x}\right)^s\right)\right)$$

In the second sum we have $p^k > x$, thus $1 - \exp(-x^s p^{-ks}) < (xp^{-k})^s$ and, if s exceeds $x + 2$, then

$$S_2 \le \sum_{p^k > x} x^s p^{-sk} \log p < x^s \sum_{n > x} \frac{\log n}{n^s}$$
$$< \left(\frac{\log(x + 1/2)}{(x + 1/2)^s} + \int_{x+1/2}^{\infty} \frac{\log t}{t^s} dt\right) x^s$$

and calculating the last integral we are led to

$$S_2 = O\left(\log x \left(\frac{x}{x + 1/2}\right)^s\right).$$

Finally

$$\psi(x) = \Psi(x, s) + O\left(\psi(x)\left(1 - \frac{1}{2x}\right)^s\right) + O\left(\log x \left(\frac{x}{x + 1/2}\right)^s\right)$$

and using Theorem 3.5 (i) one arrives at

$$\psi(x) = \Psi(x, s) + O(x \exp(-s/2x)) \tag{5.58}$$

provided $s > x + 2$.

Now take $s = x^2$. The integral occuring in (5.57) does not exceed

$$\int_0^1 \left(\sum_\rho \left|\frac{x^\rho y^{2s-\rho-1}}{\rho(s-\rho)}\right|\right) dy + \sum_\rho \left|\frac{x^\rho}{\rho(s-\rho)}\right| \left|\int_1^x y^{2s-1} \exp(-y^s) dy\right.$$
$$\le \sum_\rho \left|\frac{x^\rho}{\rho(s-\rho)(2s - \operatorname{Re}\rho)}\right| + \frac{1}{s} \sum_\rho \left|\frac{x^\rho}{\rho(s-\rho)}\right|,$$

and since $|2s - \rho| > s$ and we assume the truth of Riemann's Hypothesis, i.e. $\operatorname{Re}\rho = 1/2$, this expression is

$$O\left(\frac{x^{1/2}}{s} \sum_\rho \frac{1}{|\rho(s-\rho)|}\right).$$

A short elementary computation leads now to

$$\Psi(x,x^2) = x + O(x^{1/2+2\epsilon}\sum_\rho \frac{1}{|\rho|^{1+\epsilon}}) = x + O(x^{1/2+2\epsilon}) \tag{5.59}$$

for every positive ϵ.

The equalities (5.58) and (5.59) imply immediately the first assertion of the theorem with the error term $O(x^{1/2+2\epsilon})$ for any $\epsilon > 0$, and this appears in the main text of von Koch's memoir as well as in von Koch (1900). In a note added in proof von Koch remarks, that using von Mangoldt's Theorem 5.2, one can improve this evaluation. Indeed, partial summation utilizing that result gives

$$\sum_\rho \frac{1}{|\rho|^{1+\epsilon}} = O\left(\frac{1}{\epsilon^2}\right),$$

hence (5.58) and (5.59) give

$$\psi(x) = x + O\left(\frac{x^{1/2+2\epsilon}}{\epsilon^2}\right)$$

and choosing $\epsilon = 1/\log x$ one obtains

$$\psi(x) = x + O(\sqrt{x}\log^2 x).$$

This gives the first assertion of Theorem 5.21. The second is deduced by von Koch using the following formula obtained by de la Vallée-Poussin (1899):

$$f(x) = \operatorname{li}(x) - \log 2 + \frac{\psi(x) - x - \log(2\pi)}{\log x} - \frac{1}{\log x}\left(\sum_{m=1}^\infty \int_0^\infty \frac{x^{-2m-u}du}{(2m+u)^2}\right)$$
$$+\frac{1}{\log x}\left(\sum_\rho \int_0^\infty \frac{x^{\rho-u}du}{(\rho-u)^2}\right),$$

and noting that $\pi(x) = f(x) + O(\sqrt{x})$, but it is also obtainable directly by observing that

$$\theta(x) = \psi(x) + O(\sqrt{x}\log x),$$
$$\pi(x) = \sum_{n\le x} \frac{\theta(n) - \theta(n-1)}{\log n}$$

and applying partial summation (see e.g. Landau (1903a,§7).). □

Later von Koch gave another proof of his theorem (von Koch 1902), based on an explicit formula of the form $f(x) = A(x) + B(x)$, where $A(x)$ is a sum of a series converging both uniformly and absolutely and $B(x)$ is uniformly bounded. He pursued this topic in von Koch (1910) and applied his method to obtain the first approximative explicit formula[24] for $\psi(x)$, expressing it by roots of the zeta-function from a bounded region:

$$\psi(x) = x - \sum_{\substack{\rho=\alpha+\beta i \\ |\beta|<x^c}} \frac{x^\rho}{\rho} \Gamma\left(1 - \frac{c\rho \log x}{kx^c}\right) + O(x^{1-c} \log^2 x), \tag{5.60}$$

for any $0 < c < 1$ and arbitrary fixed $k < \pi/2$. (See Sect.6.3 for further development in this topic).

Other proofs of Theorem 5.21 were given by E.Holmgren (1902/03) (his method is explained in Sect.5.5 of Edwards (1974)), and J.J.Benedetto (1979/80).

For most values of x the bound in the last theorem was essentially improved by P.X.Gallagher (1980) who showed that under Riemann's Hypothesis one has

$$\psi(x) - x = O(\sqrt{x}(\log \log x)^2)$$

provided x does not lie in a certain fixed set E of finite logarithmic measure, i.e. satisfying

$$\int_E \frac{dt}{t} < \infty.$$

If the Riemann's Hypothesis is not true, i.e.

$$\Theta = \sup\{\rho : \ \zeta(\rho) = 0, 0 < \mathrm{Re}\ \rho < 1\} > 1/2$$

then, as shown by E.Grosswald (1965),

$$\psi(x) - x = O(x^\Theta)$$

and

$$\pi(x) - \mathrm{li}(x) = o\left(\frac{x^\Theta}{\log x}\right)$$

and if, moreover, there exists a root of the zeta-function on the line Re $s = \Theta$ then these bounds cannot be reduced.

Corollary. *Assume that all non-real roots of $\zeta(s)$ lie on the line* Re $s = 1/2$.

(i) If C is a sufficiently large constant then with a suitable $A = A(C) > 0$ there are at least $A\sqrt{x}\log x$ primes in the interval $(x, x + C\sqrt{x}\log^2 x)$.

(ii) If p_n denotes the nth prime and we put $d_n = p_{n+1} - p_n$ then

$$d_n = O(p_n^{1/2} \log^2 p_n).$$

(iii) If $f(x)$ is a nondecreasing function such that $\sqrt{x}\log^2 x = o(f(x))$ and $f(x) = o(x)$ then

$$\pi(x + f(x)) - \pi(x) = (1 + o(1))\frac{f(x)}{\log x}.$$

(iv) There is at least one prime between two sufficiently large cubes of integers.

[24] The proof of this result contains a small oversight, a correction of which can be found in a footnote on p. 272 of Landau (1912a).

Proof. (i) The theorem implies

$$\pi(x + C\sqrt{x}\log^2 x) - \pi(x) = \int_x^{x+C\sqrt{x}\log^2 x} \frac{dt}{\log t} + O(\sqrt{x}\log x)$$

and since this integral equals $(C + o(1))\sqrt{x}\log x$ and the error term is less than $B\sqrt{x}\log x$ with a fixed B, the assertion follows with $A(C) = C - B$, provided $C > B$.

(ii) Follows immediately from (i).

(iii) The same argument as in (i) is applicable here: we have

$$\pi(x + f(x)) - \pi(x) = (1 + o(1))\frac{f(x)}{\log x}$$

$$= \int_x^{x+f(x)} \frac{dt}{\log t} + O(\sqrt{x}\log x)$$

and the assumptions imply that the right-hand side of this equality equals

$$(1 + o(1))f(x)/\log x.$$

(iv) It suffices to observe that for every C and sufficiently large n one has

$$(n+1)^3 - n^3 \geq 3n^2 > C\sqrt{n^3}\log^2(n^3)$$

and apply (i). □

Under the same assumption it was later shown by H.Cramér (1920) that the assertion in (ii) can be improved to

$$d_n = O(p_n^{1/2}\log p_n).$$

He obtained also the following analogue of (i):

If C is sufficiently large then

$$\pi(x + C\sqrt{x}\log x) - \pi(x) > \sqrt{x}.$$

In the same paper Cramér touched on a conjecture put forward by E.Landau in his talk at the International Congress of Mathematicians held in Cambridge (Landau 1912c): does every interval between two squares of integers contain at least one prime? Cramér was unable to answer this question, even under the assumption of Riemann Hypothesis. This problem is still open, but Cramér succeeded in showing that under Riemann Hypothesis almost every interval $(n^2, (n+1)^2)$ contains primes, more precisely, the number of integers $n \leq x$ for which this fails is $O(x^a)$ for every $a > 2/3$. These results were based on a study of the function

$$F(z) = \sum_{\mathrm{Im}\rho > 0} \exp(\rho z),$$

(ρ denoting, as before, the non-real roots of the zeta-function) which he started in Cramér (1919a,b). Later (Cramér 1936) he conjectured that $d_n = O(\log^2 p_n)$ and gave heuristical arguments to support this assertion. A still stronger conjecture, viz. $d_n \le \log^2 p_n$ for $p_n \ge 11$ was made by A.Schinzel (1961) (see Gjeddebaek 1966).

The existence of a prime in the interval $I = I_x = (x, x + \sqrt{x}]$ is equivalent to the inequality $p_I(x) > x$, where $p_A(x)$ denotes the greatest prime factor of the product of all integers contained in an interval A. Several bounds of the form $p_I(x) \ge x^\alpha$, valid for sufficiently large x, are known. The first, with $\alpha = 15/26 = 0.5769\dots$, was proved by K.Ramachandra (1969) who later noted (Ramachandra 1970) that from a result of H.-E.Richert (1969) one can deduce this bound with $\alpha = 0.625$. Later the values $\alpha = 0.66$ (Graham 1981), $\alpha = 0.692\dots$ (Jia Chaohua 1986), $\alpha = 0.7$ (Baker 1986), $\alpha = 0.723$ (Liu Hong Quang 1993), $\alpha = 0.732$ (Baker, Harman 1995) and $\alpha = 0.728$ (Jia Chaohua 1996b) were obtained. For intervals $J = J_\epsilon = (x, x + x^{1/2+\epsilon})$ (with an arbitrary positive ϵ) stronger results are available: M.Jutila (1973) proved $p_J(x) > x^\beta$ with β being any number smaller than 2/3, and this was improved consecutively to $\beta = 0.73$ (Balog 1980), $\beta = 0.772$ (Balog 1984), $\beta = 0.82$ (Balog, Harman, Pintz 1983), $\beta = 11/12 - \epsilon$ (Heath-Brown 1996) and $\beta = 17/18 - 2\epsilon$ (Heath-Brown,Jia Chaohua 1998).

Although the existence of primes in intervals I_x is still unsettled, the related question of the existence of numbers with at most two prime factors in $[x - \sqrt{x}, x]$ found, for large x, a positive answer in a paper of Chen Jing Run (1975) (earlier this was established for numbers with at most eleven (Brun 1920a), nine (Mientka 1961) resp. four (Uchiyama 1963) prime factors). Later several authors obtained similar assertions for smaller intervals of the form $[x - x^\alpha, x]$ with a certain $a = 0.4911$ (Heath-Brown 1978), $\alpha = 0.477$ (Chen Jing Run 1979a), $\alpha = 0.455$ (Halberstam, Heath-Brown, Richert 1979), $\alpha = 0.45$ (Iwaniec, Laborde 1981), $\alpha = 0.4436$ (Fouvry 1990), $\alpha = 0.44$ (Wu (1992), $\alpha = 0.4386$ (Li Hong Ze 1994) and the current record-holder is Hong Quan Liu (1996) who reached $\alpha = 0.436$.

Riemann's Hypothesis implies (Selberg 1943) that if $f(n)$ is any function tending to infinity then for almost all integers n the interval $[n, n + f(n)\log^2 n]$ contains a prime number. This cannot be true without any exception, as shown by H.Maier (1985). Without any hypothesis it is known that for almost all n there are products of two primes in $[n, n + f(n)\log^5 n]$, provided $f(n)$ tends to infinity (Mikawa 1989). For previous results see Heath-Brown (1978), Motohashi (1979a), Wolke (1979) and Harman (1981,1982).

The best known upper bound obtained for d_n without the assumption of Riemann's Hypothesis is due to R.C.Baker, G.Harman (1996b) who obtained

$$d_n = O(p_n^\alpha)$$

for any $\alpha > 107/200 = 0.535$. For a survey of earlier results concerning d_n see Heath-Brown (1988b).

Existing lower bounds for large values of d_n are very far from the upper bounds quoted above. It follows immediately from the Prime Number Theorem that

$$\lambda = \limsup \frac{d_n}{\log p_n} \ge 1$$

and by elementary arguments this was improved in Backlund (1929) to $\lambda \ge 2$ and in Brauer,Zeitz (1930) to $\lambda \ge 4$. Finally E.Westzynthius (1931) showed that $\lambda = \infty$, more precisely

$$\limsup_{n\to\infty} \frac{d_n \log\log\log\log p_n}{\log p_n \log\log\log p_n} \ge 2\exp(\gamma).$$

This was improved in Erdős (1935a) to

$$\limsup_{n\to\infty} \frac{d_n (\log\log\log p_n)^2}{\log p_n \log\log p_n} > 0$$

and in Rankin (1938) to

$$\limsup_{n\to\infty} \frac{d_n (\log\log\log p_n)^2}{\log p_n \log\log p_n \log\log\log\log p_n} \geq 1/3.$$

The value 1/3 was replaced first by $\exp(\gamma/2)$ (Schönhage 1963), then by $\exp(\gamma)$ (Ricci 1952, Rankin 1962/63) and $1.3126\dots\exp(\gamma)$ (Maier,Pomerance 1990) and finally by $2\exp(\gamma)$ (Pintz 1997), which is the best known result.

One conjectures that the equality $d_n = 2$ holds infinitely often (the *twin prime conjecture*) and this forms a part of a conjecture put forward in Hardy, Littlewood (1923a) which implies that for every even integer $2k$ one has $d_n = 2k$ infinitely often (see Sect.6.7).

Tables of the first appearance of a particular value of d_n appear in Glaisher (1876/77), (covering the range $p_n \leq 2 \cdot 10^6$, the largest difference found being equal to 147), Western (1934) ($p_n < 10^7$, $\max d_n = 154$), Lander,Parkin (1967) ($\max d_n = 314$), Brent (1973,1980) and Young, Potter (1989) ($\max d_n = 674$). A heuristical argument led D.Shanks (1964) to conjecture that if k is an even integer and $p_m = p_m(k)$ is the smallest prime with $d_m = k$, then $\log p_m(k)$ is asymptotically equal to $\sqrt{k}$. The largest difference of consecutive primes actually known seems to be equal to 916, found by T.Nicely in February 1998.

The assertion (iii) of Cramér was strengthened by A.Selberg (1943). Working under the Riemann Hypothesis he obtained the asymptotic equality in (iii) under the weaker assumption: $f(x) \leq x$ and $\sqrt{x}\log x = o(f(x))$. (For other proofs see Goldston (1982), where also another proof of Cramér's bound for d_n is given, and Kueh (1990)). Still assuming the Riemann Hypothesis Selberg proved (iii) for almost all integral x under weaker assumptions: $f(x)$ increases, $f(x)/x$ decreases to zero and $\log^2 x = o(f(x))$. If the last condition is replaced by

$$\liminf_{x\to\infty} \frac{\log f(x)}{\log x} > \frac{19}{77}$$

then the same assertion holds independently of the Riemann Hypothesis. (Earlier this has been known with 19/77 replaced by 48/77 (Ingham 1937).) Analogues of Selberg's results for primes in progressions were obtained by K.Prachar (1976) under the assumption of the Extended Riemann Hypothesis.

7. The error term in Theorem 5.21 is nearly best possible. It was stated by Jensen (1899) without proof that $|\pi(x_j) - \mathrm{li}(x_j)|$ exceeds $x_j^{1/2-\epsilon}$ for every positive ϵ and a sequence x_j tending to infinity and E.Schmidt[25] (1903) provided a proof of the following, more precise, assertion:

Theorem 5.22. *Put*

$$\Delta(x) = \pi(x) - \mathrm{li}(x) + \frac{1}{2}\mathrm{li}(x^{1/2}).$$

If θ is the upper bound of the real parts of zeros of $\zeta(s)$ and $\epsilon > 0$ is given then there exist arbitrarily large x, y with

[25] Schmidt, Erhard (1876-1959), Professor in Berlin.

$$\Delta(x) > x^{\theta-\epsilon}$$

and

$$\Delta(y) < -y^{\theta-\epsilon}.$$

Proof. Put

$$L(x) = \sum_{2 \le m \le x} \frac{1}{\log m}$$

and

$$L_1(x) = \sum_{2 \le m \le x} \frac{1}{\sqrt{m} \log m}.$$

One sees easily that for $x > 2$ we have

$$L_1(x) - c_1 - c_2 < \mathrm{li}(\sqrt{x}) < L_1(x) - c_2,$$

where $c_1 = 1/(\sqrt{2}\log 2)$ and $c_2 = \int_{\sqrt{2}}^{2} \log^{-1} t dt$, hence if $\Phi(x)$ is defined by

$$\pi(x) = L(x) - \frac{1}{2} L_1(x) + \Phi(x)$$

then it suffices to establish the assertion of the theorem with $\Delta(x)$ replaced by $\Phi(x)$.

We have

$$\sum_p \frac{1}{p^s} = \sum_{n=2}^{\infty} \frac{\pi(n) - \pi(n-1)}{n^s}$$

$$= \sum_{n=2}^{\infty} \frac{L(n) - L(n-1)}{n^s} - \frac{1}{2} \sum_{n=2}^{\infty} \frac{L_1(n) - L_1(n-1)}{n^s} + \sum_{n=2}^{\infty} \frac{\Phi(n) - \Phi(n-1)}{n^s}$$

$$= \sum_{n=2}^{\infty} \frac{n^{-s}}{\log n} - \frac{1}{2} \sum_{n=2}^{\infty} \frac{n^{-s-1/2}}{\log n} + \sum_{n=2}^{\infty} \Phi(n) \left(\frac{1}{n^s} - \frac{1}{(n+1)^s} \right).$$

Observe now that for for any fixed $a > 1$ and $\mathrm{Re}\, s > 1$ one has

$$\sum_{n=2}^{\infty} \frac{n^{-s}}{\log n} = \sum_{n=2}^{\infty} \frac{n^{-a}}{\log n} - \int_a^s \zeta(w) dw + s - a$$

in which case the preceding equality implies

$$\sum_p \frac{1}{p^s} = \frac{s-a}{2} + \sum_{n=2}^{\infty} \frac{2 - n^{-1/2}}{2n^a \log n} - \int_a^s \zeta(w) dw$$

$$+ \frac{1}{2} \int_a^s \zeta(w + \frac{1}{2}) dw + s \sum_{n=2}^{\infty} \frac{\Phi(n)}{n^{s+1}} - \sum_{n=2}^{\infty} \Phi(n) R_n(s),$$

where

$$R_n(s) = \frac{1}{(n+1)^s} - \frac{1}{n^s} + \frac{s}{n^{s+1}}.$$

Since

$$\sum_p \frac{1}{p^s} = \log\zeta(s) - \frac{1}{2}\log\zeta(2s) + h_1(s)$$

with $h_1(s)$ regular for Re $s > 1/3$ we obtain finally for Re $s > 1$ the equality

$$\sum_{n=2}^{\infty} \frac{\Phi(n)}{n^{s+1}} = F(s),$$

with

$$F(s) = \frac{1}{s}\int_a^s \left(\frac{\zeta'}{\zeta}(w) - \frac{\zeta'}{\zeta}(2w) + \zeta(w) - \frac{1}{2}\zeta(w+\frac{1}{2})\right) dw + h_2(s),$$

where $h_2(s)$ is regular for Re $s > 1/3$. Since the integrand is regular at every point in Re $s \geq 1/2$ which is not a zero of $\zeta(s)\zeta(2s)$ we have that $F(s)$ is regular in the half-plane Re $s > \theta$.

Let N_1, N_2 be the sets of positive integers at which Φ is nonnegative resp. nonpositive and assume that for all large $n \in N_2$ one has $\Phi(n) > -n^{\theta-\epsilon}$. Then the series

$$\sum_{n\in N_2} \frac{\Phi(n)}{n^{s+1}}$$

converges absolutely and uniformly in the half-plane Re $s > \theta - \epsilon$ and hence is regular there. The equality

$$\sum_{n\in N_1} \frac{\Phi(n)}{n^{s+1}} = -\sum_{n\in N_2} \frac{\Phi(n)}{n^{s+1}} + F(s)$$

shows that the radius ρ_a of convergence of the Taylor series of the function

$$G(s) = \sum_{n\in N_1} \frac{\Phi(n)}{n^{s+1}}$$

at $s = a$ satisfies $r_a > a - \theta$. Write $r_a = a - \theta + \eta$ with $\eta > 0$ and choose a root $\rho = \alpha + \beta i$ of $\zeta(s)$ with $\alpha > \max\{\theta - \eta, \theta - \epsilon\}$. We see that the radius of convergence of the Taylor series of $G(s)$ at ρ is smaller than r_a, but this is easily shown to be impossible by comparing the magnitudes of Taylor coefficients.

The assumption $\Phi(n) < n^{\theta-\epsilon}$ for all large $n \in N_1$ can be refuted in exactly the same way. □

Under the assumption of the Riemann Hypothesis, i.e. $\theta = 1/2$, a slightly more complicated argument utilizing von Koch's Theorem 5.21 led Schmidt to the assertion that there are arbitrarily large numbers x, y with

$$\Delta(x) > c\frac{\sqrt{x}}{\log x}$$

and

$$\Delta(y) < -c\frac{\sqrt{y}}{\log x}$$

for a suitable positive c.

Corollary 1. *One does not have*

$$\pi(x) = \mathrm{li}(x) + O\left(\frac{x^{1/2}}{\log x}\right).$$ □

Later J.E.Littlewood (1914) (see Theorem 6.20 (ii)) proved that even the weaker evaluation of the error term

$$\pi(x) = \mathrm{li}(x) + O\left(\frac{x^{1/2}\log\log\log x}{\log x}\right)$$

cannot hold. In a later paper (Littlewood 1927) he found a simple proof of Schmidt's result based on repeated integrations of von Mangoldt's explicit formula for $\psi(x)$. An explicit version of the second part of Theorem 5.22 appears in Pintz (1980b).

At the end of his paper Schmidt noted that his result shows that if the inequality $\pi(x) < \mathrm{li}(x)$ (which was verified by Gauss and Goldschmidt up to $x = 3\cdot 10^6$) holds generally, then Riemann's conjecture is true. We give here his deduction, pointing out that it is now known (Littlewood 1914) that the difference $\pi(x) - \mathrm{li}(x)$ changes its sign infinitely often (see Theorem 6.16.).

Corollary 2. *If for all sufficiently large x one has $\pi(x) < \mathrm{li}(x)$, then all non-real roots of $\zeta(s)$ lie on the line* Re $s = 1/2$.

Proof. If the number θ in the theorem does exceed $1/2$ then for sufficiently small positive ϵ and all large x the difference

$$x^{\theta-\epsilon} - \frac{1}{2}\mathrm{li}(x^{1/2})$$

would be positive, hence from the second part of the theorem we could infer $\pi(x_n) - \mathrm{li}(x_n) > 0$ for a sequence x_n tending to infinity. □

In his paper Schmidt also obtained, with the same type of argument, a similar assertion for the difference $f(x) - \mathrm{li}(x)$, where

$$f(x) = \sum_{n=1}^{\infty}\frac{1}{n}\pi\left(x^{1/n}\right) = \sum_{p^k\le x}\frac{1}{k}.$$

Exercises

1. Show that the constant a occurring in Corollary 1 to Theorem 5.1 equals

$$\log(2\pi) - 1 - \gamma/2,$$

γ being Euler's constant.

2. (Obláth 1936) Prove that for any given α if $n < \log^\alpha x$ and x is sufficiently large then each of the intervals

$$(0, x), (x, 2x), \ldots, ((n-1)x, nx)$$

contains more primes than the preceding.

3. (Landau 1900a) Let $f(t, x)$ be a nonnegative function defined for $1 \leq t \leq x$ and satisfying the following two conditions:

(a) The function

$$\frac{f(t,x)}{\log t}$$

is for fixed x nonincreasing in t,

(b) The integral

$$I_f(x) = \int_2^x \frac{f(t,x)}{\log t} dt$$

is unbounded and for fixed t one has

$$\lim_{x \to \infty} \frac{f(t,x)}{I_f(x)} = 0.$$

Prove that

$$\sum_{p \leq x} f(p, x) = (1 + o(1)) I_f(x).$$

4. Let $k \geq 2$ be a given integer.

(i) (Posed in Hargreave (1849), proved in Landau (1900a).) Prove that the number of integers $n \leq x$ which are divisible by k distinct primes is asymptotically equal to

$$\frac{1}{(k-1)!} \frac{x \log\log^{k-1} x}{\log x}.$$

(ii) (Landau 1900a). Obtain the same asymptotics for the number of integers $n \leq x$ which are products of k primes, not necessarily distinct.

5. (de la Vallée-Poussin 1898) Prove

$$\sum_{p \leq x} \frac{\log p}{p-1} = \log x + A + O(\exp(-B\sqrt{\log x}))$$

with some constants A and $B > 0$.

6. (von Mangoldt 1897) (i) Let $M(x) = \sum_{n \leq x} \mu(n)$. Prove the evaluation

$$M(x) = O(x \exp(-C\sqrt{\log x}))$$

with a suitable $C > 0$.

(ii) Prove

$$|\sum_{n \le x} \frac{\mu(n)}{n}| \le 1.$$

7. (Landau 1899a) Put

$$f(x) = \sum_{n \le x} \frac{\mu(n) \log n}{n},$$

$$g(x) = \sum_{n \le x} \frac{\mu(n)}{n}$$

and

$$h(x) = \sum_{p} \frac{\log p}{p} g(x/p).$$

(i) Show

$$f(x) = -h(x) + O(1).$$

(ii)Define $\epsilon(x)$ by $\theta(x) = (1 + \epsilon(x))x$ and prove

$$h(x) = 1 + \sum_{n \le x} \left(\epsilon(n) - \epsilon(n-1) + \frac{\epsilon(n-1)}{n} \right) g(x/n).$$

(iii) Deduce $h(x) = o(\log x)$.

(iv) (von Mangoldt 1897) Prove that $g(x) = o(1)$.

8. (Stated in Möbius (1832), proved in Landau (1899b)) Prove the equality

$$\sum_{n=1}^{\infty} \frac{\mu(k) \log k}{k} = -1.$$

9. (i) (Stieltjes 1887) Prove that if the series $\sum_{n=1}^{\infty} a_n$ converges absolutely and $\sum_{n=1}^{\infty} b_n$ converges and we put $c_n = \sum_{d|n} a_d b_{n/d}$ then the series $\sum_{n=1}^{\infty} c_n$ is also convergent.

(ii) (von Mangoldt 1897) Show that

$$\sum_{n=1}^{\infty} \frac{\lambda(n)}{n} = 0,$$

where $\lambda(n) = (-1)^{\Omega(n)}$ is the function of Liouville.

10. (i) (Hall 1972 (for $p = 2$), Apostol 1973b) Prove that if p is a prime then for Re $s \ge 1$ one has

$$\left(1 + \frac{1}{p^s}\right) \sum_{n=1}^{\infty} \frac{\mu(n)}{n^s} = \left(1 - \frac{1}{p^s}\right) \sum_{n=1}^{\infty} \frac{\mu(n)\mu((p,n))}{n^s}.$$

(ii) (Kluyver 1904, Landau 1904a) Let p be a prime. Show that

$$\sum_{\substack{n \\ p|n}} \frac{\mu(n)}{n} = 0.$$

11. (Cipolla 1902) Show that if p_n denotes the nth prime number then

$$p_n = n \log n + n \log\log n + O(n).$$

12. (Wigert 1906/07) Let $d(n)$ be the number of positive divisors of n. Prove

$$\limsup_{n\to\infty} \frac{\log d(n) \log\log n}{\log n} = \log 2.$$

13. (Phragmén; cf. von Koch 1900) Prove that for positive ϵ one cannot have

$$\pi(x) - \mathrm{li}(x) = O(x^{1/2-\epsilon}).$$

14. (i) Assume that the series

$$f(s) = \sum_{n=2}^{\infty} \frac{c_n}{n^s}$$

converges absolutely in the half-plane $\mathrm{Re}\, s > 1$. Prove that the function $F(s) = \exp f(s)$ can be represented by a Dirichlet series

$$\sum_{n=1}^{\infty} \frac{b_n}{n^s},$$

converging in a certain half-plane and show that if $a_n \geq 0$ then $b_n \geq a_n \geq 0$.

(ii) (Narasimhan 1968) Use (i) and Theorem 5.15 to deduce the non-vanishing of $\zeta(s)$ on the line $\mathrm{Re}\, s = 1$. (Hint: If $\zeta(s)$ vanishes at $s = 1 + i\alpha$ then consider $F(s) = \zeta^2(s)\zeta(s+ia)\zeta(s-ia)$.)

15. (Hobby, Silberberger 1993) Show that the set of all ratios of prime numbers is dense in the half-line $(0, \infty)$.

16. (Landau 1910) Let $a_1, a_2, \ldots$ be a sequence of real numbers and put

$$s_n = a_1 + a_2 + \cdots + a_n.$$

(i) Prove that if $a_n \geq 0$ and

$$\lim_{n\to\infty} \frac{1}{n} \sum_{m=1}^{n} \frac{s_m}{m} = g$$

then

$$\lim_{n\to\infty} \frac{s_n}{n} = g.$$

(ii) Prove that in (i) one can replace the assumption $a_n \geq 0$ by $a_n \geq -c$ for some fixed c.

(iii) Prove that if

$$\lim_{n\to\infty} \frac{1}{n} \sum_{m=1}^{n} s_m = g$$

and with a suitable constant c we have $na_n \geq -c$ then

$$\lim_{n\to\infty} s_n = g.$$

(iv) Use (iii) to deduce $\theta(x) = (1 + o(1))x$ from the formula (5.26).
Hint: consider

$$a_n = \begin{cases} \dfrac{\log n - 1}{n} & \text{if } n \text{ is prime,} \\ -\dfrac{1}{n} & \text{otherwise.} \end{cases}$$

17. (i) (Axer[26] 1910) Let $f(n)$ be an arithmetical function and put $S(x) = \sum_{n\leq x} f(n)$ and $T(x) = \sum_{n\leq x} |f(n)|$. Prove that if $S(x) = o(x)$ and $T(x) = O(x)$ then

$$\sum_{n\leq x} f(n) \left(\frac{x}{n} - \left[\frac{x}{n}\right] \right) = o(x).$$

(ii) (Landau 1910) Use (i) to deduce the formula

$$\sum_{p\leq x} \frac{\log p}{p} = \log x + c + o(1)$$

from $\theta(x) = (1 + o(1))x$.

18. Prove that the difference $\theta(x) - x$ changes its sign infinitely often.

19. (Mertens 1874b) Show that if $(k, l) = 1$ then

$$\sum_{\substack{p\leq x \\ p\equiv l \pmod k}} \frac{1}{p} = \frac{1}{\varphi(k)} \log\log x + c(k,l) + o(1)$$

holds with certain constant $c(k, l)$.

20. Let $F(n)$ be a completely multiplicative complex-valued arithmetical function (i.e. $F(mn) = F(m)F(n)$ holds for all positive integers m, n), and let $|F(n)| = 0$ or 1 for all n. Assume also that both functions

$$f_1(s) = \sum_{n=1}^{\infty} \frac{F(n)}{n^s} \qquad f_2(s) = \sum_{n=1}^{\infty} \frac{F^2(n)}{n^s}$$

can be continued to functions regular in Re $s \geq 1$, $s \neq 1$ and $s = 1$ is either a regular point or a simple pole.

(i) Prove that $f_1(1 + it) \neq 0$ holds for all real $t \neq 0$.

(ii) Prove that if both functions f_1, f_2 are regular at $s = 1$, then $f_1(1) \neq 0$.

---

[26] A simple proof is given in Landau (1910).

6. The Turn of the Century

6.1. Progress in Function Theory

1. The end of the XIXth century had seen a rapid progress in the theory of complex functions. In the preceding chapter one could notice the impact of Hadamard's theory of entire functions on the determination of the asymptotic behaviour of functions associated with prime numbers, like $\pi(x)$ and $\theta(x)$. For further development of this theory three results in function theory, all obtained between 1896 and 1910, were of particular importance: Hadamard's three circle theorem (1896), Jensen's formula (1899) and the theorem of Phragmén–Lindelöf (1908). With the use of them the main results of Hadamard, von Mangoldt and de la Vallée-Poussin were given simpler proofs, particularly through the activity of Edmund Landau.

We start with the *three circle theorem* stated without proof in Hadamard (1896d):

Theorem 6.1. *Let* $0 < r_1 < r_2 < r_3$ *and let* f *be a function regular in the disk* $|z - z_0| \leq r_3$. *If* M_j *denotes for* $j = 1,2,3$ *the maximum of* $|f(z)|$ *on the circle* $|z - z_0| = r_j$ *and we put* $m_j = \log(r_j/r_1)$ $(j = 2,3)$, *then*

$$M_2^{m_3} \leq M_1^{m_3 - m_2} M_3^{m_2}.$$

Proof of Hadamard[1] . We may assume $M_1 > 0$. Observe that for positive α every branch of $(s - s_0)^\alpha$ is regular at every point of the annulus $r_1 < |s - s_0| < r_3$ hence by the maximum principle we infer that the function $g(s) = f(s)(s - s_0)^{-\alpha}$ satisfies there the inequality

$$|g(s)| \leq \max(M_1 r_1^{-\alpha}, M_3 r_3^{-\alpha}).$$

Putting

$$\alpha = \frac{\log(M_3/M_1)}{\log(r_3/r_1)}$$

[1] This proof was published with Hadamard's permission in Bohr,Landau (1913). Thus the statement in the commentaries to Littlewood's paper (1912) in his *Collected Papers* that " ... *no proof by Hadamard appears ever to have been published* ..." is inexact.

we get

$$M_2 \le M_1 \left(\frac{r_2}{r_1}\right)^{\alpha} = M_1 \left(\frac{M_3}{M_1}\right)^{m_2/m_3}$$

which is equivalent with the assertion. □

2. The next theorem is named after J.L.W.V.Jensen, who in 1899 stated and proved it. It is perhaps not generally known that the essence of this result can be found already in an older paper of C.G.J.Jacobi (1827), albeit only in the case when f is a polynomial with real coefficients. However, as pointed out by E.Landau (1914b), Jacobi's proof works without essential changes also in the general case[2].

Theorem 6.2. (Jacobi 1827, Jensen 1899) *Let f be a function regular in the disk $D = \{z : |z - z_0| \le R\}$ and assume that it does not have zeros either on the circumference of D or at its center. Let $z_1, z_2, \dots, z_n$ be the zeros of $f(z)$ in D, each appearing in this sequence in accordance with its multiplicity, and put $Z = |(z_1 - z_0)(z_2 - z_0) \cdots (z_n - z_0)|$. Then*

$$\frac{1}{2\pi} \int_0^{2\pi} \log(|f(z_0 + Re^{it})|)dt = \log\left(|f(z_0)|\frac{R^n}{Z}\right). \tag{6.1}$$

This theorem may be stated also in the following equivalent form:

Theorem 6.3. *Let f be a function satisfying the assumptions of Theorem 6.2 and moreover let $f(z_0) = 1$. For $0 < r \le R$ denote by $n(r)$ the number of zeros of f in the disk $|z - z_0| \le r$. Then*

$$\frac{1}{2\pi} \int_0^{2\pi} \log(|f(z_0 + Re^{it})|)dt = \int_0^R \frac{n(r)}{r} dr.$$

To show that these two assertions are equivalent first observe that in Theorem 6.2 it suffices to consider the case $z_0 = 0$ and $f(0) = 1$, replacing $f(z)$ by $g(z) = f(z_0 + z)/f(z_0)$ and noting that to every zero z of f there corresponds the zero $z - z_0$ of g and conversely. Denote by $0 < r_1 < r_2 < \dots < r_{s-1} < R$ the distinct absolute values of the z_i's and let λ_j denote the number of z_i's satisfying $z_i = r_j$. Let also $r_s = R$ and $n = n(R) = n(r_{s-1})$. Putting

$$I = \int_0^R \frac{n(r)}{r} dr$$

we obtain

[2] Landau quotes also two other old papers dealing with Jacobi's result: Gauger (1894) and Richert (1882), which however I was unable to consult.

$$I = \sum_{j=1}^{s-1} \int_{r_j}^{r_{j+1}} \frac{n(r)}{r} dr = \sum_{j=1}^{s-1} n(r_j) \int_{r_j}^{r_{j+1}} \frac{dr}{r} = \sum_{j=1}^{s-1} n(r_j)(\log r_{j+1} - \log r_j).$$

In view of $n(r_j) = \lambda_1 + \lambda_2 + \cdots + \lambda_j$ and

$$Z = |z_1 z_2 \cdots z_n| = \prod_{j=1}^{s-1} r_j^{\lambda_j}$$

we get

$$\log\left(\frac{R^n}{Z}\right) = n \log R - \sum_{j=1}^{s-1} \lambda_j \log r_j$$

hence the equality

$$I = \log\left(\frac{R^n}{Z}\right)$$

results by partial summation and the equivalence of both versions of Jensen's equality follows.

3. Jensen stated his result in a more general form, admitting poles of f in the disk D. We present now his original proof, but only in the case when f is regular.

Jensen's proof of (6.1). The starting point is the following computation of an integral:

Lemma 6.4. *If* $|w| < r$ *then*

$$\frac{1}{2\pi} \int_0^{2\pi} \log\left(\left|1 - \frac{re^{it}}{w}\right|\right) dt = \log\left(\frac{r}{|w|}\right).$$

Proof. Let $z = re^{it}$. In view of $|1 - z/w| = |z/w||1 - w/z|$ and

$$\log\left(1 - \frac{w}{z}\right) = -\sum_{k=1}^{\infty} \frac{1}{k}\left(\frac{w}{z}\right)^k$$

we get

$$\begin{aligned} \log(|1 - \frac{z}{w}|) &= \log\frac{r}{|w|} - \operatorname{Re} \sum_{k=1}^{\infty} \frac{1}{k}\left(\frac{w}{z}\right)^k \\ &= \log\frac{r}{|w|} - \frac{1}{2}\sum_{k=1}^{\infty} \frac{1}{kr^k}\left(w^k e^{-itk} + \overline{w}^k e^{itk}\right) \end{aligned}$$

and the assertion results immediately by integrating the last equality over the interval $[0, 2\pi]$. □

Assuming $z_0 = 0$, which does not restrict the generality, Jensen writes

$$f(z) = f(0) \prod_{j=1}^{n} \left(1 - \frac{z}{z_j}\right) f_1(z)$$

where f_1 is regular in D, does not vanish there and satisfies $f_1(0) = 1$. Thus $f_1(z) = \exp g(z)$ with g regular in D and vanishing at $z = 0$. If now $|z| = r$ then taking real parts of the logarithms of both sides of the last equality, dividing by 2π, integrating over $[0, 2\pi]$ and using the lemma one arrives at the assertion. □

4. Jensen's formula immediately implies the following useful corollary:

Corollary 1. *If f satisfies the assumptions of Theorem 6.2, M denotes the maximum of $|f(z)|$ on $|z - z_0| = R$ and $n(r)$ denotes, as in Theorem 6.3, the number of zeros of f in the disk $|z - z_0| \le r \le R$, then*

$$|f(z_0)| \left(\frac{R}{r}\right)^{n(r)} \le M.$$

Proof. If the zeros $z_1, z_2, \ldots$ of f in $|z - z_0| \le R$ are arranged according to their absolute values and every zero appears in this sequence as many times as its order, then (6.1) implies

$$\log(|f(z_0)|) + \sum_{j=1}^{n(R)} \log\left(\frac{R}{|z_j - z_0|}\right) \le \log M.$$

Since all summands here are nonnegative this leads to

$$\log(|f(z_0)|) + n(r) \log\left(\frac{R}{r}\right) \le \log M$$

and the assertion follows. □

The following corollary, proved in Bohr,Landau (1914a), turned out to be useful in studying the zeros of $\zeta(s)$ and $L(s, \chi)$. We shall utilize it later in the proof of Theorem 6.18.

Corollary 2. *If $r < R$ and A are positive numbers then there exists an explicit constant $C = C(r, R, A)$ with the property that if f is a function regular in the disc $K_R = \{z : \ |z| \le R\}$ and $|f(0)| > A$, then the number $n = n(r)$ of zeros of f lying in the disc $|z| \le r$ counted with their multiplicities does not exceed*

$$C \int\!\!\int_{K_R} |f(x + iy)|^2 dx dy.$$

Proof. We shall use the following elementary inequality, valid for all nonnegative and integrable functions $f(x)$, defined on an interval $[a, b]$:

$$\log\left(\frac{1}{b-a}\int_a^b f(t)dt\right) \leq \frac{1}{b-a}\int_a^b \log(f(t))dt.$$

A six-line proof of it, utilizing only the inequality $\log t \leq t-1$ $(t>0)$, can be found in Hardy, Littlewood, Pólya (1934;Th.184).

Let $r < (r+R)/2 \leq \rho \leq R$ and denote by Z_ρ, Z the absolute value of the product of all zeros of f in $|z| \leq \rho$ and $|z| \leq r$, respectively. Jensen's formula (6.1) applied to the function $f^2(z)$ leads, in view of $Z \leq r^n$ to

$$\log\left(\frac{1}{2\pi}\int_0^{2\pi} |f(\rho e^{it})|^2 dt\right) \geq \frac{1}{2\pi}\int_0^{2\pi} \log(|f(\rho e^{it})|^2)dt$$
$$\geq \log\left(A^2\frac{\rho^{2n(\rho)}}{Z_\rho^2}\right) \geq \log\left(A^2\frac{\rho^{2n}}{Z^2}\right) \geq \log\left(A^2(\rho r^{-1})^{2n}\right)$$

hence

$$\left(\frac{\rho}{r}\right)^{2n} A^2 \leq \frac{1}{2\pi}\int_0^{2\pi} |f(\rho e^{it})|^2 dt < \int_0^{2\pi} |f(\rho e^{it})|^2 dt.$$

Observe now that with a suitable positive $C = C(R, r, A)$ we have

$$\frac{A^2}{r^{2n}}\int_{\frac{R+r}{2}}^R \rho^{2n+1}d\rho \geq \frac{A^2}{r^{2n}}\left(\frac{R+r}{2}\right)^{2n+1}\left(R - \frac{R+r}{2}\right) \geq Cn$$

and thus

$$n \leq \frac{1}{C}\frac{A^2}{r^{2n}}\int_{(R+r/2)}^R \rho^{2n+1}d\rho \leq \frac{1}{C}\int_{\frac{R+r}{2}}^R \rho d\rho \int_0^{2\pi} |f(\rho e^{it})|^2 dt$$
$$< \iint_{|u+iv|\leq R} |f(u+iv)|^2 dudv$$

□

5. A very simple proof of Theorem 6.2 was given later by R.Backlund[3] (1918):

Backlund's proof of (6.1). We may assume $z_0 = 0$ and $R = 1$, replacing $f(z)$ by $f(R(z - z_0))$. Assume first that $n = 0$, i.e. f does not vanish in D. The function $g(z) = \log|f(z)|$ equals the real part of $\log f(z)$, which is regular in D, hence g is harmonic in D as a function of two real variables Re z and Im z. This implies

$$\frac{1}{2\pi}\int_0^{2\pi} \log(|f(e^{it})|)dt = \log(|f(0)|), \tag{6.2}$$

[3] Backlund's proof can be found also in Ahlfors (1953) and Edwards (1974).

which in our case coincides with (6.1).

In the general case observe that the function

$$F(z) = f(z) \prod_{i=1}^{n} \frac{1 - z\bar{z}_i}{z - z_i}$$

is regular and non-vanishing in D and satisfies

$$|F(0)| = \frac{|f(0)|}{|z_1 z_2 \cdots z_n|} \neq 0.$$

Applying to it the formula (6.2) we obtain

$$\frac{1}{2\pi} \int_0^{2\pi} \log(|F(e^{it})|)dt = \log(|F(0)|) = \log|f(0)| - \log(|z_1 z_2 \cdots z_n|). \quad (6.3)$$

Since for z on the unit circle T we have

$$\left| \frac{1 - z\bar{z}_i}{z - z_i} \right| = \left| \frac{\bar{z} - \bar{z}z\bar{z}_i}{z - z_i} \right| = \left| \frac{\bar{z} - \bar{z}_i}{z - z_i} \right| = 1,$$

we get $|F(z)| = |f(z)|$ on T and now the equality (6.3) gives the assertion.

□

Among numerous applications of Jensen's formula one should point out one made by Landau (1927b) who applied it in his simple proof of Corollary 1 to Theorem 5.1.

6. The next important result was the Phragmén–Lindelöf theorem (Phragmén, Lindelöf 1908), which gives upper bounds for a function in an unbounded region Ω, provided a bound for it on the lines bordering Ω is available and inside Ω the function does not grow too quickly. The paper of Phragmén and Lindelöf[4] contains several versions of this result of different generality. An earlier, weaker, version occurs in Phragmén (1904). We present here the classical version of this theorem with its original proof:

Theorem 6.5. (Phragmén–Lindelöf 1908, sect.3) *Let Ω be an angle of size π/α (with α positive) and vertex w and let f be a function continuous in Ω and regular in its interior. Assume that on the boundary $\partial\Omega$ of Ω one has*

$$|f(z)| \leq C$$

with some constant C and denote by $M(r)$ the maximum of the absolute value of f on that part of the circle $|z| = r$ which lies in Ω. If

$$\beta = \limsup_{r \to \infty} \frac{\log \log M(r)}{\log r} < \alpha \quad (6.4)$$

[4] Lindelöf, Ernst (1870–1946), Professor in Helsinki.

then for all z in Ω one has

$$|f(z)| \le C.$$

Proof. Without restricting the generality we may assume that $w = 0$ and

$$\Omega = \{z : -\frac{\pi}{2\alpha} \le \arg z < \frac{\pi}{2\alpha}\}.$$

Choose a number γ in the interval (β, α) and consider the function $\omega(z) = \exp(-z^\gamma) = \exp(-\exp(\gamma \log z))$, where, as usual, we take this branch of the logarithm which is rel for positive arguments. Then for $z = re^{i\phi}$ one has

$$|\omega(z)| = \exp(-r^\gamma \cos(\gamma\phi)) < \exp(-\delta r^\gamma) \le 1$$

with $0 < \delta = \cos(\gamma\pi/2\alpha)$ and moreover for $0 < \epsilon < 1$ the function $F_\epsilon(z) = \omega^\epsilon(z) f(z)$ satisfies uniformly the relation $\lim_{z\to\infty} F_\epsilon(z) = 0$, provided the variable z remains in the interior of Ω. Since

$$|F_\epsilon(z)| \le |f(z)| \le C$$

holds on $\partial\Omega$, and the function F_ϵ is regular and vanishes at the point in infinity, the maximum modulus principle applied to the intersection of Ω with the sequence of discs $D_N = \{z : |z| \le N\}$ implies $|F_\epsilon(z)| \le C$ for $z \in \Omega$ and thus

$$|f(z)| \le C|\omega^\epsilon(z)|.$$

Since C is fixed and $\epsilon > 0$ can be arbitrarily small the assertion follows. □

The same proof works also in a more general case:

Corollary 1. (Phragmén,Lindelöf 1908, sect.5) *The assertion of the theorem holds also in the case when Ω is a domain contained in an angle satisfying the conditions of the theorem.* □

Corollary 2. (Phragmén,Lindelöf 1908, sect.6) *Let Ω be a connected domain contained in the vertical strip $\{z : a \le \mathrm{Re}\, z \le b\}$ and let f be continuous in Ω and regular in its interior. If $|f(z)| \le C$ holds on the boundary of Ω and in its interior f satisfies (6.4) with $\alpha = \pi/(b-a)$ then $|f(z)| \le C$ holds in Ω.*

Proof. The mapping $w = \exp(z)$ maps Ω onto a domain to which the preceding corollary may be applied. □

In Prachar (1957, Appendix, §8) one finds a direct proof of this corollary.

7. An important application of Theorem 6.5 was found by Lindelöf (1908), who used it to improve the upper bound for the absolute value of $\zeta(s)$ in the critical strip $0 < \text{Re } s < 1$. The knowledge of such bounds turned out to be crucial in the study of the error term in the Prime Number Theorem. Earlier weaker bounds were obtained by H.Mellin and E.Landau.

Theorem 6.6. *For positive δ put*

$$\Omega_\delta = \{\sigma + it : \ 0 < \delta \leq \sigma \leq 1 - \delta < 1, \quad |t| \geq \delta.\}$$

(i) (Mellin 1900) *For every fixed $0 < \sigma < 1$ one has*

$$\zeta(\sigma + it) = O\left(|t|^{1-\sigma}\right)$$

and this evaluation is for every fixed $\delta > 0$ uniform in Ω_δ.

(ii) (Mellin 1900) *For real t bounded away from 0 one has*

$$\zeta(1 + it) = O(\log |t|)$$

and

$$\zeta(it) = O(|t|^{1/2} \log |t|).$$

(iii) (Landau 1905a) *For every σ in $(0, 1)$ one has*

$$\zeta'(\sigma + it) = O(|t|^{1-\sigma} \log |t|)$$

and this bound is uniform in Ω_δ.

(iv) (Landau 1905a) *One has*

$$\zeta(s) = O(|t|^\alpha \sqrt{\log |t|})$$

with

$$\alpha = \begin{cases} \frac{3}{4}(1 - \sigma) & \text{if } 1/2 \leq \sigma < 1 \\ \frac{1}{4}(2 - \sigma) & \text{if } 0 < \sigma \leq 1/2 \end{cases}$$

the bounds being uniform for $s \in \Omega_\delta$

(v) (Lindelöf 1908) *For any fixed $t_1 > 1$ one has in the region $0 \leq \sigma \leq 1$, $t \geq t_1 > 1$ the estimate*

$$\zeta(\sigma + it) = O\left(t^{(1-\sigma)/2} \log t\right).$$

Proof. Obviously it suffices to consider the case of positive t.

(i) A simple proof of Mellin's result was presented by E.Landau (1905a) who used de la Vallée Poussin's formula (5.23) written in the form

$$\zeta(s) = 1 + \frac{1}{s-1} + \sum_{n=1}^{m} \left(\frac{1}{s-1} \left(\frac{1}{(n+1)^{s-1}} - \frac{1}{n^{s-1}} \right) + \frac{1}{(n+1)^s} \right) - s \sum_{n=m+1}^{\infty} \int_0^1 \frac{x\, dx}{(n+x)^{1+s}}, \tag{6.5}$$

with m being an arbitrary integer. Choosing now $m = [t] \geq \delta > 0$ and evaluating trivially all terms one gets

$$\zeta(s) = O\left(\frac{1}{t}\right) + O\left(\frac{1}{(m+1)^{\sigma-1}}\right) + O\left(\int_1^{[t]} \frac{dx}{x^\sigma}\right)$$

$$+ O\left(t \sum_{n>[t]} \int_0^1 \frac{x dx}{(n+x)^{1+\sigma}}\right) = O\left(\frac{t^{1-\sigma}}{1-\sigma}\right) + O\left(\frac{t^{1-\sigma}}{\sigma}\right)$$

and (i) follows.

One finds an even simpler proof of (i) in Landau (1907b): If $s = \sigma + it$ and $m = [t]$ then the first sum in the identity

$$(1 - 2^{1-s})\zeta(s) = \sum_{n=1}^{m} \frac{(-1)^{n+1}}{n^s} + \sum_{n=m+1}^{\infty} \frac{(-1)^{n+1}}{n^s}$$

is $O(t^{1-\sigma})$. The second sum can be written in the form

$$(-1)^m \sum_{k=1}^{\infty} \left(\frac{1}{(m+2k-1)^s} - \frac{1}{(m+2k)^s}\right) = (-1)^m s \sum_{k=1}^{\infty} \int_{m+2k-1}^{m+2k} \frac{dx}{x^{1+s}}$$

and hence it is

$$O\left(t \int_{m+1}^{\infty} \frac{dx}{x^{1+\sigma}}\right) = O(t^{1-\sigma}).$$

In view of $|1 - 2^{1-s}| \geq 2^{1-\sigma} - 1$ the assertion (i) follows.

(ii) One applies Lemma 2.12, the simplest case of the Euler-Maclaurin formula, to the sum $\sum_{k=n}^{\infty} k^{-s}$, to get for Re $s > 1$ the equality

$$\zeta(s) = \sum_{k=1}^{n-1} \frac{1}{k^s} + \frac{1}{2n^s} + \frac{n^{1-s}}{s-1} - s \int_n^{\infty} \frac{x - [x] - 1/2}{x^{1+s}} dx \tag{6.6}$$

which analytically[5] continues the zeta-function to the half-plane Re $s > 0$. Putting $s = 1 + it$, $n = [t]$ and using the bound

$$\sum_{k=1}^{n-1} \frac{1}{k} = O(\log n)$$

one obtains the first assertion. The second follows by the functional equation.

(iii) Differentiating the equality (6.5), putting $m = [t]$ and evaluating trivially the obtained summands one arrives at the assertion.

(iv) To prove (iv) Landau (1905a) utilized Voronoi's[6] (1903) evaluation in Dirichlet's divisor problem:

---

[5] The integral converges actually for Re $s > -1$.

[6] Voronoi, Georgii Feodosevič (1868–1908), Professor in Warsaw.

If $d(n)$ denotes the number of positive divisors of n then

$$D(x) = \sum_{n \le x} d(n) = x \log x + (2\gamma - 1)x + \Delta(x) \tag{6.7}$$

with $\Delta(x) = O(x^{1/3} \log x)$ and γ being Euler's constant.

Voronoi's result essentially improved upon Dirichlet's bound $\Delta(x) = O(x^{1/2})$ (Dirichlet 1849). There is a long story of later improvements and the best known result is due to M.N.Huxley (1993b) who obtained

$$\Delta(x) = O(x^{23/73} \log^{461/146} x),$$

the previous record being held by H.Iwaniec and C.J.Mozzochi (1988) who had the bound $O(x^{7/22+\epsilon})$ for every $\epsilon > 0$. If θ denotes the least upper bound of numbers ϑ for which $\Delta(x) = O(x^{\vartheta})$ holds then it is conjectured that $\theta = 1/4$ and it is known (Hardy 1916) that $\theta \ge 1/4$. It is remarkable that in 90 years since Voronoi's achievement the value of θ has been improved only by $1/3 - 23/73 = 4/219 = 0.01826\ldots$, i.e. by less than six percent.

We sketch now Landau's argument. The function

$$F(s) = \zeta^2(s) + \zeta'(s) - 2\gamma\zeta(s)$$

can be written for Re $s > 1$ as a Dirichlet series

$$F(s) = \sum_{n=1}^{\infty} \frac{a_n}{n^s}$$

with $a_n = d(n) - \log n - 2\gamma$. Stirling's formula and (6.7) imply

$$S(x) = \sum_{n \le x} a_n = O(x^{1/3} \log x)$$

and using partial summation one obtains that $F(s)$ converges for Re $s > 1/3$ and thus gives a continuation of $F(s)$ into that half-plane.

To get an upper bound for $|F(s)|$ in the strip $1/3 <$ Re $s < 1$ one writes with a fixed m

$$F(s) = \sum_{n=1}^{m} \frac{a_n}{n^s} + \sum_{n=m+1}^{\infty} \frac{S(n) - S(n-1)}{n^s}$$

which after partial summation leads to

$$F(s) = \sum_{n=1}^{m} \frac{a_n}{n^s} + s \sum_{n=m+1}^{\infty} S(n) \int_0^1 \frac{du}{(n+u)^{1+s}} - \frac{S(m)}{(m+1)^s}.$$

Choosing now $m = \left[t^{3/2}\right]$, noting that

$$\sum_{n\le m} \frac{1}{n^\sigma} = O\left(\frac{m^{1-\sigma}}{1-\sigma}\right),$$

$$\sum_{n\le m} \frac{d(n)}{n^\sigma} = O\left(\sum_{n\le m} \frac{\log n}{n^\sigma}\right) = O\left(\frac{m^{1-\sigma}\log m}{1-\sigma}\right)$$

and

$$\left|\sum_{n=m+1}^{\infty} S(n)\int_0^1 \frac{du}{(n+u)^{1+s}}\right| \le \sum_{n=m+1}^{\infty} \frac{n^{1/3}\log n}{n^{1+\sigma}} = O\left(\frac{\log m}{m^{\sigma-1/3}}\right)$$

one gets

$$F(s) = O\left(t^{(3/2)(1-\sigma)}\log t\right).$$

Thus for $1/3 < \sigma < 1$ one has

$$|\zeta^2(s)| \le |\zeta'(s)| + 2\gamma|\zeta(s)| + O\left(t^{(3/2)(1-\sigma)}\log t\right).$$

Using now (i) and (iii) one gets the assertion (iv) for $1/3 < \mathrm{Re}\ s < 1$ and the functional equation takes care of the strip $0 < \mathrm{Re}\ s \le 1/3$.

This proof was later simplified in Landau (1907b). Note that if one repeats Landau's argument replacing Voronoi's exponent $1/3$ by the conjectured value $\theta = 1/4 + \epsilon$ (with $\epsilon > 0$) then putting $m = [t^\alpha]$ with $\alpha = 1/(1-\theta)$ one arrives at the bound

$$\zeta(s) = O(t^{(2/3+\epsilon)(1-\sigma)})$$

which is weaker than Lindelöf's bound (v).

(v) The first step in Lindelöf's proof of (v) consists in the observation that from the functional equation of the zeta-function and its boundedness in every half-plane $\mathrm{Re}\ s > 1+\epsilon > 1$ one deduces that if $0 < \epsilon < \sigma_0$ are given, then in the region $-\sigma_0 \le \sigma \le -\epsilon$, $t \ge \delta > 0$ the inequalities

$$At^{1/2-\sigma} \le |\zeta(\sigma+it)| \le Bt^{1/2-\sigma} \tag{6.8}$$

hold with suitable positive constants A and B, depending only on ϵ and δ.

Indeed for $\mathrm{Re}\ s \ge 1+\epsilon > 1$ we have $\zeta(s) = O(1)$ and evaluating the Γ-factors in the functional equation of $\zeta(s)$ with the aid of the asymptotic formula

$$|\Gamma(\sigma+it)| = \exp(-\frac{1}{2}\pi|t|)|t|^{\sigma-1/2}(\sqrt{2\pi}+o(1))$$

(valid for $|t| \to \infty$ uniformly for σ in any fixed finite interval) we obtain (6.8).

Applying the formula (6.6) for $n = 1$ one gets for $\mathrm{Re}\ s > 0$ the equality

$$\zeta(s) = \frac{1}{2} + \frac{1}{s-1} - s\int_1^\infty \frac{x-[x]-1/2}{x^{1+s}}dx$$

which implies that for any positive δ the bound $\zeta(s) = O(t)$ holds uniformly in the region $\mathrm{Re}\ s \ge \delta$, $|t| \ge \delta$.

Integrating by parts the last integral one is led to

$$\zeta(s) = \frac{1}{2} + \frac{1}{s-1} - \frac{s^2+s}{2}\int_1^\infty \frac{\phi(x)}{x^{s-2}}dx,$$

where $\phi(x)$ denotes the function of period 1 equal to $x^2 - x + 1/6$ in the interval $[0,1]$. Since the last integral converges for Re $s > -1$ we get the evaluation $\zeta(s) = O(t^2)$ uniformly in every region Re $s \geq -1+\delta$, $|t| \geq \delta > 0$.

Note that the last bound jointly with (6.8) shows that in every fixed strip $a \leq$ Re $s \leq b$ one has

$$\zeta(s) = O(|t|^k) \tag{6.9}$$

with a certain $k = k(a,b)$.

Consider now the function

$$F(s) = s^{-3/4}\zeta(s)^{1/2}\log s.$$

It is regular in the region Ω defined by the inequalities $-2 \leq$ Re $s \leq -1$ and Im $s > 0$ because $\zeta(s)$ does not vanish there. Using (6.8) one sees that for $s = \sigma + it$ in Ω we have

$$A_1 t^{-(1+\sigma)/2}\log t \leq |F(s)| \leq B_1 t^{-(1+\sigma)/2}\log t$$

with certain positive A_1 and B_1 and this shows that the function $G(s) = F(s-2)$, which is regular and non-vanishing in the region defined by $0 \leq \sigma \leq 1$, $t > 0$ satisfies there the inequalities

$$A_2 t^{(1-\sigma)/2}\log t \leq |G(s)| \leq B_2 t^{(1-\sigma)/2}\log t$$

with $0 < A_2 \leq B_2$.

If now

$$f(s) = \frac{\zeta(s)}{G(s)}$$

then the obtained inequalities jointly with (ii) imply that $f(s)$ remains bounded at lines Re $s = 0$ and Re $s = 1$, provided $|\text{Im } s| \geq \delta > 0$. Now (6.9) allows us to apply the Phragmén–Lindelöf theorem in the form given in Corollary 2 to Theorem 6.5 to obtain $|\zeta(s)| \leq |f(s)|$ in $0 \leq \sigma \leq 1$, $t \geq 1$ and thus (v) follows. □

Another proof of the last theorem is given in Izumi,Izumi (1976). The implied constant was evaluated in Backlund (1918).

Littlewood (1912) utilized the three circle theorem (Theorem 6.1) to obtain still better upper bounds[7] for $\zeta(s)$ and $1/\zeta(s)$ under the assumption of Riemann's Hypothesis. He established in particular that in every half-plane Re $s \geq 1/2 + \delta$ with $\delta > 0$ it implies both

$$|\zeta(s)| = O(t^\epsilon)$$

[7] See e.g. Titchmarsh (1951, Ch.14).

and

$$|\zeta(s)|^{-1} = O(t^{\epsilon})$$

for every $\epsilon > 0$. Littlewood's result shows that under the Riemann Hypothesis one has for every fixed $\sigma \in [0,1]$ and $t \to \infty$

$$\log(|\zeta(\sigma + it)|) = o(\log t)$$

and on the other hand it was established by E.C.Titchmarsh (1928) that this cannot be improved to

$$\log(|\zeta(\sigma + it)|) = o(\log^{a} t)$$

with $a < 1 - \sigma$ (for another proof of the last assertion see Ramachandra (1974)). See Levinson[8](1972) and Montgomery (1977) for a further strengthening.

H.-E.Richert (1967) proved the bound

$$\zeta(\sigma + it) = O(t^{100(1-\sigma)^{3/2}} \log^{2/3} t)$$

and the number 100 in the exponent has been replaced first by 50 (Edgorov 1972) and then by 21 (Panteleeva 1988) and 20 (Bartz 1990).

8. In his 1908 paper Lindelöf introduced the function $\mu(\sigma)$ defined as the lower bound of exponents μ for which

$$\zeta(\sigma + it) = O(|t|^{\mu})$$

holds for all large $|t|$. He noted that $\mu(\sigma)$ is continuous, convex and satisfies the relation

$$\mu(\sigma) = \mu(1 - \sigma) + \frac{1}{2} - \sigma,$$

as a consequence of the functional equation. Moreover one has $\mu(\sigma) = 0$ for $\sigma \geq 1$ and $\mu(\sigma) = 1/2 - \sigma$ for $\sigma \leq 0$. He expressed also the belief that for any positive ϵ the function $\zeta(\sigma + it)$ remains bounded when $\sigma \geq 1/2 + \epsilon$ and t is sufficiently large. This would imply

$$\mu(\sigma) = \begin{cases} 0 & \text{if } \sigma \geq 1/2 \\ \frac{1}{2} - \sigma & \text{if } \sigma < 1/2. \end{cases}$$

The last statement is known as *Lindelöf's Hypothesis* and remains still unsettled though the boundedness conjecture turned out to be false. In fact H.Bohr proved (see Bohr,Landau 1910) that the difference $\zeta(s) - 1/(s-1)$ is unbounded in the open half-plane Re $s > 1$ and even the relation $\zeta(1 + it) = o(\log\log t)$ fails to be true. This result cannot be improved, at least if one assumes Riemann's Hypothesis. Indeed, J.E.Littlewood (1928) showed that this conjecture implies

$$|\zeta(1 + it)| \leq (2e^{\gamma} + o(1)) \log\log t,$$

[8] Levinson, Norman (1912–1975), Professor at the Massachusetts Institute of Technology.

with γ denoting, as usual, the Euler's constant. Earlier he showed (Littlewood 1912) that Lindelöf's Hypothesis is a consequence of Riemann's Hypothesis and deduced from Lindelöf's Hypothesis the bound

$$\zeta(1+it) = O(\log\log t \log\log\log t).$$

A few years later R.J.Backlund (1918/19) proved that Lindelöf's Hypothesis is equivalent to the assertion that for every positive δ the number of zeros ρ of the zeta-function, lying in the rectangle

$$\operatorname{Re}\rho \geq \frac{1}{2} + \delta, \qquad T \leq \operatorname{Im}\rho \leq T+1$$

is $o(\log T)$. For several other statements equivalent with that hypothesis see Hardy, Littlewood (1923b) and Titchmarsh (1951, Chap.13). In particular Hardy and Littlewood proved that Lindelöf's Hypothesis is equivalent with the evaluation

$$J_k(T) = \frac{1}{T}\int_1^T |\zeta(\frac{1}{2}+it)|^{2k} dt = O(T^\epsilon)$$

for all positive integers k and arbitrary $\epsilon > 0$. This bound is known to be true for $k=1$ and $k=2$. In fact $J_1(T)$ is asymptotically equal to $\log T$, as shown in Hardy, Littlewood (1918,1922) (see Titchmarsh 1951, Theorem 7.3) and later developments led to the formula

$$J_1(T) = \log T + 2\gamma - 1 - \log(2\pi) + O(T^{-\alpha})$$

with any $\alpha < 155/227 = 0.6828\ldots$ (Heath-Brown 1993). Previously A.Ivić (1985) had this for all $\alpha < 73/108$. In the case $k=2$ Hardy and Littlewood (1922) established $J_2(T) = O(\log^4 T)$, Ingham (1927) proved

$$J_2(T) = \frac{\log^4 T}{2\pi^2} + O(\log^3 T)$$

and much later this was made more precise by D.R.Heath-Brown (1979c) who showed that with suitable constants $c_0, c_1, \ldots, c_4$ one has

$$J_2(T) = \sum_{j=0}^{4} c_j \log^j T + O(T^{-\alpha})$$

for any $\alpha < 1/8$. Later A.Ivić and Y.Motohashi (1995) showed that the error term in this formula is $O(T^{-1/3}\log^C T)$ with some C and is not $o(T^{-1/2})$. Cf. Ivić, Motohashi (1994).

If one denotes by $k(\sigma)$ the upper bound of real numbers k for which

$$\int_1^T |\zeta(\sigma+it)|^{2k} dt = c_k T + o(T)$$

holds with

$$c_k = \sum_{n=1}^{\infty} \frac{d_k^2(n)}{n^{2\sigma}},$$

$d_k(n)$ being the number of positive integral solutions of the equation

$$n = x_1 x_2 \cdots x_k,$$

then one has

$$\mu(\sigma) \le \frac{1}{2k(\sigma)},$$

as shown by C.B.Haselgrove (1949).

The convexity of the function $\mu(\sigma)$ (which follows from the general theory of Dirichlet series) implies the inequality

$$\mu(\sigma) \le 2\mu(1/2)(1-\sigma),$$

thus Lindelöf's hypothesis is equivalent to the equality $\mu(1/2) = 0$. There is still a long way to go towards this goal. A list of all previous bounds for $\mu(1/2)$ proved in Walfisz (1924), Titchmarsh (1931,II; 1942), Phillips (1933), Min (1949), Haneke (1962/63) and Kolesnik (1973, 1982[9], 1985) can be found[10] on p.118 of the second edition of Titchmarsh (1951), the best bound given there being $\mu(1/2) \le 139/858 = 0.162004\ldots$ (Kolesnik 1985). Later E.Bombieri, H.Iwaniec (1986) obtained $9/56 = 0.16071\ldots$ (cf. Huxley, Watt 1988), N.Watt (1989) obtained an exponent below 0.16 by proving $\mu(1/2) \le 89/560 = 0.15892\ldots$, M.N.Huxley, G.Kolesnik (1991) achieved $17/108 = 0.1574$ and the current record is due to M.N.Huxley (1993a):

$$\mu(1/2) \le 89/570 = 0.15614\ldots .$$

Under the Riemann Hypothesis it was proved (see e.g. the book of Titchmarsh (1951, Th.14.14)) that

$$\zeta(1/2+it) = O(\exp(A \log t/ \log\log t))$$

with a certain constant $A < 1/2$ (another proof is contained in Ramachandra, Sankaranarayanan (1993)) and on the other hand it is known (Balasubramanian, Ramachandra 1977) that for a sequence t_n tending to infinity one has

$$|\zeta(1/2+it_n)| \ge c\exp\left(\frac{3}{4}\left(\frac{\log t_n}{\log\log t_n}\right)^{1/2}\right)$$

for some positive constant c. (Cf. Montgomery 1977).

[9] The bound $\mu(1/2) \le 35/216$ proved there was obtained a little earlier by A.Ivić (1980).

[10] To this list one should add the inequality $\mu(1/2) \le 1/2$ proved in van der Corput, Koksma (1930).

6.2. Landau's Approach to the Prime Number Theorem

1. Prime number theory in the first years of the twentieth century was dominated by the person of Edmund Landau. His first work on this subject (Landau 1899a), which brought him the doctorate of the Berlin University, contained a simple proof[11] of the equality

$$\sum_{n=1}^{\infty} \frac{\mu(n)}{n} = 0$$

established two years earlier by von Mangoldt (1897). This was the starting point of a series of more than fifty papers devoted to various aspects of prime number theory and Riemann's zeta-function. He wrote also the first monograph presenting analytic aspects of the prime number theory (Landau 1909a). Its two volumes served as a guide for a whole generation of mathematicians.

2. Landau's first important result was an essential simplification of the proof of the Prime Number Theorem (Landau 1903a). His first approach to this subject led however to a worse bound for the error term than that obtained by C.de la Vallée-Poussin (1899) (see Theorem 5.19), who gave the bound $O(x\exp(-c\log^{\alpha} x))$ with $\alpha = 1/2$, whereas Landau obtained only $\alpha = 1/12$ in (5.44) and $\alpha = 1/13$ in (5.45). However Landau did not utilize the Weierstrass-Hadamard theory of entire functions nor the fact that $\zeta(s)$ can be continued across the imaginary axis. In fact, he did not use any deeper results about the distribution of the zeros of $\zeta(s)$, restricting himself to rather elementary tools from function theory, which he applied in his characteristic and clever way.

Sketch of Landau's first proof. The main ingredients of the proof are summarized in the following auxiliary result:

Lemma 6.7. *(i) For $t \geq 10$ and $1 \leq \sigma \leq 2$ one has*

$$|\zeta(\sigma + it)| \leq 2\log t, \tag{6.10}$$

(ii) For $t \geq 10$ and $1 - 1/\log t \leq \sigma \leq 1$ one has

$$|\zeta'(\sigma + it)| \leq 6\log^2 t. \tag{6.11}$$

(iii) For $t \geq 10$ and $1 - b/\log^9 t \leq \sigma \leq 2$ (with a suitable $b > 0$) one has

$$\frac{\zeta'}{\zeta}(\sigma + it) = O(\log^9 t). \tag{6.12}$$

[11] See Exercise 7 in the preceding chapter.

(iv) The function $\zeta(s)$ does not vanish in the region

$$\text{Re } s > 1 - \frac{b}{\log^9 t}, \qquad t = \text{Im } s \geq 10 \tag{6.13}$$

for a certain $b > 0$.

Note that the assertion (iv) is much weaker than Theorem 5.18 proved a few years earlier by de la Vallée-Poussin but, as we already pointed out, Landau tried to use only very simple tools.

Proof. (i) From the equality (5.23) (or (6.5)) one gets the identity

$$\zeta(s) = \frac{1}{s-1}\frac{1}{(m+1)^{s-1}} + \sum_{n=1}^{m+1}\frac{1}{n^s} - s\sum_{n=m+1}^{\infty}\int_0^1 \frac{udu}{(n+u)^{s+1}}, \tag{6.14}$$

valid for Re $s > 0$ and $m = 0, 1, 2, \dots$ and after putting here $m = [t]$ and $s = 1 + \epsilon + 2it$ $(0 \leq \epsilon \leq 1)$ one obtains the bound

$$|\zeta(1 + \epsilon + 2it)| < 2\log t$$

for $t > 10$, implying (6.10).

(ii) Differentiation of (6.14) gives

$$\zeta'(s) = -\frac{1}{(s-1)^2}\frac{1}{(m+1)^{s-1}} - \frac{1}{s-1}\frac{\log(m+1)}{(m+1)^{s-1}} - \sum_{n=1}^{m+1}\frac{\log n}{n^s}$$
$$- \sum_{n=m+1}^{\infty}\int_0^1 \frac{xdx}{(n+x)^{s+1}} + s\sum_{n=m+1}^{\infty}\int_0^1 \frac{x\log(n+x)}{(n+x)^{s+1}}dx.$$

Put now $s = 1+\epsilon+ti$ with $-1/\log t \leq \epsilon \leq 1$, $t \geq 10$ and choose $m = [t]-1$. Then clearly $4t/5 < m \leq t-1$ and thus

$$\left|\frac{1}{(s-1)^2}\frac{1}{(m+1)^{s-1}}\right| \leq \frac{1}{t^2(m+1)^\epsilon} \leq \frac{5}{4t^{2+\epsilon}} \leq \frac{5e}{4t^2} \leq \frac{5e}{400} < 0.034,$$

$$\left|\frac{1}{s-1}\frac{\log(m+1)}{(m+1)^{s-1}}\right| \leq \frac{\log t}{t(m+1)^\epsilon} \leq \frac{5\log t}{4t^{1+\epsilon}} \leq \frac{5e\log t}{4t} \leq \frac{5e\log 10}{40} < 0.783$$

and

$$\left|\sum_{n=1}^{m+1}\frac{\log n}{n^s}\right| \leq \log t\sum_{n\leq t}\frac{n^{1/\log t}}{n} \leq e\log t\sum_{n\leq t}\frac{1}{n}$$
$$\leq e(\log^2 t + \log t) \leq (\log^2 t + \frac{\log^2 t}{\log 10}) < 3.899\log^2 t.$$

This shows that the contribution of the first three summands in our formula for $\zeta'(s)$ in the case $t \geq 10$, $\epsilon \geq -1/\log t$ does not exceed $0.817+3.899\log^2 t$. For the remaining summands we get

$$\left|\sum_{n=m+1}^{\infty}\int_0^1 \frac{x\,dx}{(n+x)^{s+1}}\right| \leq \sum_{n\geq t}\frac{1}{n^{2+\epsilon}}\int_0^1 x\,dx \leq \frac{1}{2}\sum_{n\geq t}\frac{1}{n^{3/2}} \leq \frac{1}{\sqrt{t}} \leq \frac{1}{3}$$

and

$$\left|s\sum_{n=m+1}^{\infty}\int_0^1 \frac{x\log(n+x)}{(n+x)^{s+1}}dx\right| \leq (2+t)\sum_{n\geq t}\frac{\log(n+1)}{n^{2+\epsilon}}\int_0^1 x\,dx$$
$$\leq (1+\frac{t}{2})\sum_{n\geq t}\frac{\log(n+1)}{n^{2+\epsilon}} \leq (1+\frac{t}{2})\sum_{n\geq t}\frac{\log n + 1/n}{n^{2+\epsilon}}$$
$$\leq (1+\frac{t}{2})\int_t^{\infty}\frac{\log x + 1/x}{x^{2+\epsilon}}dx \leq e(1+\frac{t}{2})\left(\int_t^{\infty}\frac{\log x}{x^2}dx + \int_t^{\infty}\frac{dx}{x^3}\right)$$
$$\leq (1+\frac{t}{2})\left(\frac{\log t}{2t}+\frac{1}{2t}+\frac{1}{2t^2}\right) \leq 0.68\log t + 1.210 \leq 0.296\log^2 t + 1.210.$$

Hence the inequality

$$|\zeta'(1+\epsilon+it)| \leq 1.447 + 5.109\log^2 t \leq \left(1.447\log^2 10 + 5.109\right)\log^2 t < 6\log^2 t$$

results and this gives (6.11).

(iii) Using (i) and (ii) one infers

$$|\zeta(1+\epsilon+it) - \zeta(1+it)| = \left|\int_{1+it}^{1+\epsilon+it}\zeta'(z)dz\right| \leq 6|\epsilon|\log^2 t. \tag{6.15}$$

Since for positive ϵ one has

$$\zeta(1+\epsilon) = \sum_{n=1}^{\infty}\frac{1}{n^{1+\epsilon}} < 1 + \int_1^{\infty}\frac{du}{u^{1+\epsilon}} = \frac{1+\epsilon}{\epsilon}$$

the inequality (5.35) implies

$$|\zeta(1+\epsilon+it)| \geq \left(\frac{\epsilon}{1+\epsilon}\right)^{3/4}\frac{1}{|\zeta(1+\epsilon+2ti)|^{1/4}}\,. \tag{6.16}$$

Now choose $\epsilon = c_1\log^{-9} t$ with $0 < c_1 < 12^{-4}$ and note that (6.10) and (6.16) imply

$$|\zeta(1+\epsilon+it)| \geq c_2\log^{-7} t$$

with $c_2 = c_1^{3/4}/2$. The inequality (6.15) gives now

$$|\zeta(1+it)| \geq |\zeta(1+\epsilon+it)| - 6\epsilon\log^2 t \geq \frac{c_2 - 6c_1}{\log^7 t}$$

and since our choice of c_1 shows that $c_2 > 6c_1$ we obtain

$$|\zeta(1+it)| > \frac{c_3}{\log^7 t} \qquad (t \geq 10) \tag{6.17}$$

with[12] some positive c_3.

Using again (6.15) one gets in the region (6.13) with a sufficiently small positive b the bound

$$|\zeta(\sigma+it)| \geq \frac{c_4}{\log^7 t}$$

and now (ii) implies the evaluation (6.12) for $1 - b/\log^9 t < \sigma < 1$. Using upper bounds for $\zeta'(s)$ to the right of the line Re $s = 1$, based on trivial evaluations of its Dirichlet series, Landau also obtained the assertion (iii) for $1 \leq \text{Re } s \leq 2$, $t \geq 10$.

Finally note that the assertion (iv) follows immediately from (iii). □

The bound (6.10) for $\sigma = 1$ makes effective Mellin's bound in Theorem 6.6 (ii). In Backlund (1918) one finds the more precise bound $|\zeta(1+it)| < \log t$ for $t > 50$. Later H.Weyl[13](1921) using his method of estimation of trigonometric sums, which became later known as Weyl's sums, improved this to

$$\zeta(1+it) = O\left(\frac{\log t}{\log\log t}\right)$$

and the best known bound is

$$\zeta(1+it) = O(\log^{2/3} t)$$

proved independently by I.M. Vinogradov (1958) and N.M.Korobov (1958b,c). The previous records were: $O(\log^{3/4} t \log\log^{1/2+\epsilon} t)$ (Flett[14]1950), $O(\log^{3/4} t \log\log^{1/2} t)$ (Walfisz 1958) and $O(\log^{5/7+\epsilon} t)$ (Korobov 1958a).

On the other hand J.E.Littlewood (1925) proved

$$\limsup_{t\to\infty} \frac{|\zeta(1+it)|}{\log\log t} \geq e^\gamma$$

and showed that under the assumption of Riemann's Hypothesis almost the full converse is true, namely this upper limit does not exceed $2e^\gamma$.

The bound (6.12) was improved in Gronwall[15](1913a,b) to

$$\frac{\zeta'}{\zeta}(\sigma+it) = O(\log t)$$

for $\sigma \geq 1 - c/\log t$ (with positive c) and t bounded away from 0. Note that, as shown in Landau (1911a), the evaluation

---

[12]Landau (1909a,p.888) regarded the inequality (6.17) as one of his most important discoveries.

[13]Weyl, Hermann (1885–1955), Professor in Zürich, Göttingen and Princeton.

[14]Flett, Thomas Muirhead (1923–1977).

[15]Gronwall, Thomas Hakon (1877–1932), worked in the Department of Physics of Columbia University.

$$\frac{\zeta'}{\zeta}(\sigma + it) = O(\log\log t)$$

fails already in the half-plane Re $s > 1$.

3. In the final part of his proof of the Prime Number Theorem Landau evaluates the same integral as did Hadamard, namely

$$I(x) = \frac{1}{2\pi i}\int_{2-i\infty}^{2+i\infty} \frac{x^s\zeta'(s)}{s^2\zeta(s)}ds$$

but proceeds in another way. He observes first, using easy bounds resulting from the consideration of the Dirichlet series for $\zeta'(s)/\zeta(s)$, that truncating the integral $I(x)$ at the level x^2 one gets a good approximation for Hadamard's function $A(x)$:

$$A(x) = \sum_{p\le x}\log p\log\frac{x}{p} = -\frac{1}{2\pi i}\int_{2-ix^2}^{2+ix^2} \frac{x^s\zeta'(s)}{s^2\zeta(s)}ds + O(x^{1/2}\log^2 x).$$

He then moves the integration path across the line Re $s = 1$ so that everything is reduced to the integration over the polygon Λ with the following vertices: $2 - x^2 i$, $2 + x^2 i$, $1 - a/\log^9(x^2) + x^2 i$, $1 - a/\log^9(10) + 10i$, $1 - a/\log^9(10) - 10i$, $1 - a/\log^9(x^2) - x^2 i$. Here $0 < a < 10^{-6}$ is any number smaller than the minimal distance from the line Re $s = 1$ of root of $\zeta(s)$ with imaginary parts not exceeding[16] 10. The integrand has a unique pole at $s = 1$ with residue -1 and the use of (6.12) leads without any difficulty to

$$A(x) = x + O(x\exp(-\log^c x))$$

with $c = 1/11$. The deduction of both versions of the Prime Number Theorem in the forms

$$\theta(x) = x + O(x\exp(-\log^{1/12} x))$$

and

$$\pi(x) = \mathrm{li}(x) + O(x\exp(-\log^{1/13} x))$$

is now made in the same manner as done by Hadamard, the only difference being the necessity of taking care of the error term[17]. □

The inequality (6.17) was subsequently improved (Landau 1906d) to

$$|\zeta(1+it)| \ge \frac{C}{\log^6 t}$$

[16] In fact there are no such roots, as observed already by J.P.Gram (1895), who showed that the imaginary parts of the first three roots of $\zeta(s)$ in the upper half-plane are approximatively equal to 14.135, 20.82 and 25.1 (cf. Gram 1903).

[17] Actually this part of Landau's argument contains a small oversight, repaired in Landau (1908b, footnote 64 on p.70).

(with some $C > 0$) with the use of a result of Hadamard (1904). In Landau (1908b, §2) a simpler proof appears, differing from the given above only in the respect that instead of the inequality $4(1+\cos\alpha) \geq 1-\cos(2\alpha)$ it utilizes

$$17 + 24\cos\alpha + 8\cos(2\alpha) \geq 0.$$

In the same paper (§3) the inequality

$$5 + 8\cos\alpha + 4\cos(2\alpha) + \cos(3\alpha) \geq 0$$

is used to get a stronger bound

$$|\zeta(1+it)| \geq \frac{C}{\log^5 t}.$$

Landau noted that one cannot in this way, using more complicated trigonometrical inequalities, push the exponent of the logarithm in this inequality below 3.587 (further results concerning this question are contained in Landau (1933), Tschakaloff[18](1940), van der Waerden[19],Rosser, Schoenfeld (1962), French (1966) and Stečkin (1970a)). With more sophisticated tools Landau (1908b,§8) was able to establish finally

$$|\zeta(1+it)| \geq \frac{C}{\log t \log\log t}$$

with a certain positive C and J.E.Littlewood showed[20] that Riemann's Hypothesis implies

$$|\zeta(1+it)| \geq \frac{C_1}{\log\log t}$$

(with $C_1 > 0$) while, on the other hand, one finds in Bohr, Landau (1923) the proof that one does not have

$$|\zeta(1+it)| \geq \frac{C\log\log\log t}{\log\log t}$$

with an arbitrarily large C. Later they showed (Bohr,Landau 1924) that even

$$|\zeta(1+it)| \geq \frac{M}{\log\log t}$$

is false for sufficiently large M (thus Littlewood's result cannot be improved) and E.C.Titchmarsh (1933) made this more explicit by establishing

$$\liminf_{t\to\infty} |\zeta(1+it)| \log\log t \leq \frac{\pi^2}{6}e^{-\gamma},$$

γ denoting the Euler's constant. For a simple proof see Levinson (1972), where one also finds a proof that the inequality

$$\frac{1}{|\zeta(1+it)|} \geq \frac{6e^{\gamma}}{\pi^2}(\log\log t - \log\log\log t) + O(1)$$

holds for suitable arbitrarily large values of t.

---

[18]According to E.Landau (1933, corr.) the results of this paper were published as early as in 1923 in the yearbook of Sofia University.

[19]Waerden, Bartel Leendert van der (1903–1996), Professor in Leipzig, Amsterdam and Zürich.

[20]For a proof see a footnote in Bohr, Landau (1923).

4. Usually a proof of a result based on a new idea lends itself to various generalizations. This also happened in this case and, in the second part of Landau's paper, the same method is applied to give the first proof of the Prime Ideal Theorem, describing the asymptotic distribution of prime ideals in the ring $\mathbf{Z}_K$ of integers of an algebraic number field K. The place of Riemann's $\zeta(s)$ is taken here by *Dedekind's zeta-function* $\zeta_K(s)$, which we already encountered in subsection 5.5.1 of the preceding chapter. At that time one knew only that $\zeta_K(s)$ could be extended to the open half-plane Re $s > 1 - 1/n$, where n is the degree of the field in question (Landau 1902) but later E.Hecke (1917a) extended $\zeta_K(s)$ to a meromorphic function (see Tate (1950) for a modern proof). For certain particular progressions (e.g. $1 \bmod n$) the Prime Ideal Theorem implies the quantitative form of Dirichlet's Theorem 2.1. The case of an arbitrary progression was attacked by Landau in his next paper (Landau 1903b) where the following improvement of Theorems 5.11 and 5.14 was established:

Theorem 6.8. *If* $(k,l) = 1$ *then*

$$\theta(x; k.l) = \sum_{\substack{p \leq x \\ p \equiv l \pmod{k}}} \log p = \frac{x}{\varphi(k)} + O(x \exp(-\log^\beta x)) \tag{6.18}$$

and

$$\pi(x; k, l) = \frac{1}{\varphi(k)} \mathrm{li}(x) + O(x \exp(-\log^\gamma x)) \tag{6.19}$$

with β and γ being positive constants, depending on k.

Proof. The reasoning runs parallel to the above proof of the Prime Number Theorem. First an analogue of the inequality (6.16) is deduced for the product

$$L(s) = \prod_\chi L(s, \chi)$$

of all L-functions corresponding to characters mod k. Since Euler's product formula for L-functions implies for Re $s > 1$ the equality

$$L(s, \chi) = \exp\left(\sum_{m=1}^\infty \frac{1}{m} \sum_p \frac{\chi(p^m)}{p^{ms}}\right)$$

one gets

$$L(s) = \exp\left(\varphi(k) \sum_{m=1}^\infty \frac{1}{m} \sum_{\substack{p \\ p^m \equiv 1 \pmod{k}}} \frac{1}{p^{ms}}\right),$$

which leads (for positive ϵ and real t) to

$$|L^4(1+\epsilon+it)L^4(1+\epsilon)|$$
$$= \exp\left(\varphi(k)\sum_{m=1}^{\infty}\frac{1}{m}\sum_{\substack{p \\ p^m\equiv 1 \pmod k}}\frac{4(1+\cos(mt\log p))}{p^{m(1+\epsilon)}}\right).$$

In view of $4(1+\cos\alpha) \geq 1-\cos(2\alpha)$ this implies

$$|L^4(1+\epsilon+it)L^4(1+\epsilon)| \geq \frac{L(1+\epsilon)}{|L(1+\epsilon+2it)|}$$

and thus using the fact that $L(s)$ has a simple pole at $s=1$ (coming from the factor $L(s,\chi_0)$ corresponding to the principal character mod k) one gets

$$|L(1+\epsilon+it)| \geq \frac{c\epsilon^{3/4}}{|L^{1/4}(1+\epsilon+2it)|} \tag{6.20}$$

with a suitable positive constant c.

In the next step Landau deduces the bounds

$$|L(s,\chi)| = O(\log t) \tag{6.21}$$
$$|L'(s,\chi)| = O(\log^2 t) \tag{6.22}$$

in the range $s=\sigma+it\in\Omega$, where Ω is determined by the inequalities $1-1/\log t\leq\sigma\leq 2$ and $t\geq 3$. For the L-function associated with the principal character this is a simple consequence of the corresponding result for the zeta-function (see Lemma 6.7) and for the remaining characters this is achieved essentially by partial summation, using the trivial bound $|\sum_{n=1}^{N}\chi(n)|<k$. From (6.21) and (6.22) the bounds

$$|L(s)| = O(\log^{c_1} t)$$
$$|L'(s)| = O(\log^{c_2} t)$$

follow with suitable constants c_1, c_2. The remainder of the proof proceeds as in the case of the Prime Number Theorem: upper bounds for $L'(s)/L(s)$ and $L'(s,\chi)/L(s,\chi)$ are deduced as well as the lower bound

$$|L(s,\chi)| > \frac{1}{\log^{c_3} t},$$

all bounds valid in Ω. Finally Landau defines for fixed l, coprime to k,

$$K_l(s) = \sum_{p\equiv l \pmod k}\frac{\log p}{p^s},$$

notes that $K_l(s)$ is regular in Ω with a simple pole at $s=1$ with residue $1/\varphi(k)$ and integrates the function $x^sK(s)/s$ over essentially the same path as in the previous paper. The established bounds are sufficient to obtain in this way the formula

$$\sum_{\substack{p\le x \\ p\equiv l \pmod k}} \log p \log\frac{x}{p} = \frac{x}{\varphi(k)} + O(x\exp(-\log^c x))$$

from which both assertions of the theorem follow as in the case of Prime Number Theorem. □

The dependence of the exponent γ on k in this theorem was eliminated in Landau (1908f,h) (with any $\gamma < 1/8$ resp. $\gamma < 1/2$) and in §§131–132 of Landau (1909a) the evaluation

$$\pi(x;k,l) = \frac{1}{\varphi(k)}\mathrm{li}(x) + O(x\exp(-c\sqrt{\log x}))$$

with some absolute constant $c > 0$ (e.g. $c = 0.2$) was obtained. Note that this assertion can already be tracked back to C.de la Vallée-Poussin (1899), who after providing a non-trivial error term in the Prime Number Theorem (Theorems 5.9 and 5.12) stated that his method can also be applied to the case of an arithmetic progression and promised to return to this subject, but never did. Landau (1909a, p.899) writes that Vallée-Poussin in a letter confirmed that he had exactly the last formula in mind.

It is known (von Koch 1901; cf. Prachar 1957, Theorem 7.5.1) that if all non-real zeros ρ of L-functions corresponding to characters mod k satisfy Re $\rho \le c$ then for $x \ge k$

$$\pi(x;k,l) = \frac{1}{\varphi(k)}\mathrm{li}(x) + O(x^c\log x)$$

hence, in particular, the Extended Riemann Hypothesis implies

$$\pi(x;k,l) = \frac{1}{\varphi(k)}\mathrm{li}(x) + O(x^{1/2}\log x)$$

for $x \ge k$ (cf. Titchmarsh 1930a). Note that the constant implied by the $O(\cdot)$-symbol occuring there must depend on k since it follows from results of J.Friedlander and A.Granville (1989, 1992b) that even the estimate

$$\pi(x;k,l) = \frac{1+o(1)}{\varphi(k)}\mathrm{li}(x)$$

cannot hold uniformly for all $k < x/\log^B x$ and $(k,l)=1$ with a fixed positive B.

The next step was taken by A.Page (1935) who utilized results of E.C.Titchmarsh (1930a) to obtain the evaluation

$$\pi(x;k,l) = \frac{1}{\varphi(k)}\mathrm{li}(x) + O(x\exp(-c\sqrt{\log x})) + O\left(\frac{x^\rho}{\varphi(k)\log x}\right),$$

ρ being the greatest real zero (if it exists) of L-functions corresponding to real characters mod k and all implied constants being effective. There is an old conjecture that such zeros ρ do not exist and for L-functions with small conductors this was confirmed in Rosser (1949,1950) and Low (1967/68). In his paper Page established the inequality

$$\rho \le 1 - \frac{c}{\sqrt{k}\log^2 k}$$

(with some $c > 0$) for any such hypothetical zero.

For L-functions corresponding to non-real characters one always has

$$\rho \leq 1 - \frac{c}{\log k}$$

(with certain positive c) but for real characters no such evaluation is known. The best result in that case is a consequence of *Siegel's theorem* (Siegel 1936) stating that for every positive ϵ there exists $C = C(\epsilon) > 0$ such that

$$L(1,\chi) \geq \frac{C}{k^{\epsilon}}.$$

A.Walfisz (1936b) used this theorem to deduce the inequality

$$\rho \leq 1 - \frac{c}{k^{\epsilon}}$$

for the real zeros of L-functions and this lead to the bound $O(x\exp(-c\sqrt{\log x}))$ for the error term in the formula for $\pi(x;k,l)$ uniformly for $k \leq \log^a x$ for any fixed a (the *Siegel–Walfisz theorem*).

Unfortunately Siegel's proof is not effective, i.e. it does not allow to compute the values of $c(\epsilon)$ for small ϵ and the same applies to all other known proofs. There are many of them: see Chowla (1950), Čudakov (1942), Estermann (1948), Goldfeld (1974), Heilbronn (1938), Knapowski (1968), Linnik (1950), Motohashi (1978b), Pintz (1973/74,1977d), Ramachandra (1975b,1980,1990), Rodosskij (1956) and Tatuzawa (1951,1985).

The first improvement of the effective result of Page was obtained by H.Davenport (1966) who proved

$$\rho \leq 1 - \frac{C}{\sqrt{k}\log\log k}$$

with an effective positive constant C and the factor $\log\log k$ in the denominator was later removed by W.Haneke (1973). J.Pintz (1976b) pointed out a gap in Haneke's argument, filled up by Haneke in a corrigendum. This paper of Pintz contains another proof of Haneke's assertion with the implied constant being $12/\pi + o(1)$. A.Schinzel pointed out (see footnote 3 in Pintz (1976b)) that using the results of Baker, Schinzel (1971) one can improve this constant to $16/\pi + o(1)$. (Cf. Goldfeld, Schinzel 1975, Mallik 1981).

In any case it is known that there is an explicit constant $c > 0$ such that in the interval $[1 - c/\log k, 1)$ there can be at most one zero of an $L(s,\chi)$ for real χ mod k (such a zero is called a *Siegel zero*). This was first shown by E.Landau (1918b) with an unspecified constant c and J.Pintz (1977c) proved that one can take $c = 1/454$, whereas for large k one can even have $c = 2 + o(1)$. (Cf. Heath-Brown 1992, Lepistö 1974, Metsänkylä 1970, Miech 1968a and Pintz 1976b).

E.C.Titchmarsh (1930a) obtained, with the use of Brun's sieve, for $k < x^a$ (with fixed $0 < a < 1$) the inequality (the *Brun–Titchmarsh theorem*)

$$\pi(x;k,l) \leq \frac{cx}{\varphi(x)\log x}$$

with a constant $c = c(a)$. Actually his method may be used to prove for $k < x$ the inequality

$$\pi(x;k,l) \leq \frac{cx}{\varphi(x)\log(x/k)}$$

with an absolute constant c and nowadays this inequality is usually called the Brun-Titchmarsh theorem. S.Chowla (1934b) observed that if $x > \exp(\log^5 k)$ then one can take for c any number exceeding 2 and later research showed that this is true also for smaller values of x (Montgomery, Vaughan 1973, Klimov 1980). Y.Motohashi (1975) showed that for $k \leq x^{\beta}$ with $0 < \beta < 2/5$ one can even take $c = 2 - \delta$ with some positive δ, depending on β. Later the range of admissible β in this result was extended to $0 < \beta < 2/3$ (Iwaniec 1982). Cf. also Baker (1996), Bombieri, Davenport (1968), Deshouillers, Iwaniec (1984), Fouvry (1984), Goldfeld (1975), Hooley (1972,1975b), van Lint, Richert (1965), Montgomery (1971) and Motohashi (1974,1975).

Y.Motohashi (1979a) proved that if there exists a constant C such that for $x \geq k^C$ the Brun-Titchmarsh theorem holds with $c = 2 - \epsilon$ (with a positive ϵ) then the L-functions corresponding to characters mod k do not have real zeros in the interval

$$(1 - \frac{C_1 \epsilon}{\log k}, 1)$$

with a certain positive C_1, depending on C.

For explicit estimates of the functions

$$\theta(x; k, l) = \sum_{\substack{p \leq x \\ p \equiv l \pmod{k}}} \log p$$

and

$$\psi(x; k, l) = \sum_{\substack{p^r \leq x \\ p^r \equiv l \pmod{k}}} \log p$$

see McCurley (1984a,d) and Ramaré, Rumely (1996).

If for relatively prime k and l we put

$$E(x, k, l) = \sum_{\substack{n \leq x \\ n \equiv l \pmod{k}}} \Lambda(n) - \frac{x}{\varphi(k)}$$

then Theorem 6.8 implies the evaluation

$$E(x, k, l) = O(\frac{x}{\log^a x})$$

for every positive a. A very important result concerning $E(x, k, l)$ was obtained independently by E.Bombieri (1965) and (in a slightly weaker version) A.I.Vinogradov (1965) (cf. Bombieri 1974 and Vaughan 1975, 1980) using the large sieve:

For any given $A > 0$ and a suitable $B = B(A) > 0$ one has

$$\sum_{k \leq x^{1/2} \log^{-B} x} \max_{y \leq x} \max_{(k,l)=1} |E(y, k, l)| = O\left(\frac{x}{\log^A x}\right).$$

Similar results dealing with the square of $E(x, k, l)$ were obtained by M.B.Barban (1963), H.Davenport, H.Halberstam (1966) and H.L.Montgomery (1970) (cf. Hooley 1975a).

5. In the year 1908 Landau gave (Landau 1908d) two new proofs of the Prime Number Theorem in its simplest form, i.e. without giving any evaluation of the error term. The first is similar to the proof presented above but instead of dealing with the the function $\zeta'(s)/\zeta(s)$ Landau considers $Z(s) = \log\zeta(s)$. This necessitates a small modification of the integration path to take care of the essential singularity of the integrand at $s = 1$ but the remainder of the argument is carried out along the lines of the preceding proof, using bounds obtained in Lemma 6.7. At the final stage one gets the equality

$$\sum_{p\le x} \log\left(\frac{x}{p}\right) = \frac{x}{\log x} + O\left(\frac{x}{\log^2 x}\right).$$

Replacing in it x by $(1+\delta)x$ and subtracting one obtains

$$\pi(x)\log(1+\delta) + \sum_{x<p\le(1+\delta)x} \log\left(\frac{x+\delta x}{p}\right) = \frac{\delta x}{\log x} + O\left(\frac{x}{\log^2 x}\right)$$

and since the left-hand side of this equality lies between $\pi(x)\log(1+\delta)$ and $\pi(x+\delta x)\log(1+\delta)$ the assertion $\pi(x) = (1+o(1))x/\log x$ is easily deduced.

The second proof is more interesting. In it Landau considers the function

$$\Phi(s) = \frac{\zeta'}{\zeta}(s) - \zeta(s) = \sum_{n=1}^{\infty} \frac{\Lambda(n) - 1}{n^s}$$

which is regular in Re $s \ge 1$ and computes the integral

$$\frac{1}{2\pi i}\int_{2-i\infty}^{2+i\infty} \frac{x^s}{s^8}\Phi(s)ds = \frac{1}{7!}\sum_{n\le x}(\Lambda(n)-1)\log^7\left(\frac{x}{n}\right).$$

Moving the line of integration to the line Re $s = 1$ and observing that due to Theorem 5.6 the integral converges absolutely one deduces that for x tending to infinity this integral is $o(x)$. This implies

$$\lim_{x\to\infty}\frac{1}{x}\sum_{n\le x}(\Lambda(n)-1)\log^7\left(\frac{x}{n}\right) = 0$$

and since a short computation leads to

$$\sum_{n\le x}(\Lambda(n)-1)\log^7\left(\frac{x}{n}\right) = \sum_{p\le x}\log p \log^7(x/p) - x\int_1^x \frac{\log^7 u}{u^2}du + O(\sqrt{x}\log^9 x)$$

one obtains the relation

$$\lim_{x\to\infty}\frac{1}{x}\frac{1}{7!}\sum_{p\le x}\log p\log^7(x/p) = 1.$$

Now Landau uses a very general lemma permitting him to deduce from the last equality the assertion $\sum_{p\leq x}\log p = (1+o(1))x$. We shall present his argument in a more general setting in section 4, devoted to tauberian approach to the Prime Number Theorem.

In the same 1908 year a big paper of Landau (1908b) appeared, in which he deduced the de le Vallée-Poussin's bound $O(x\exp(-c\sqrt{\log x}))$ for the error term in the Prime Number Theorem directly from Theorem 5.18 (for a further variant see Landau (1909b)). Another proof was given by J.B.Steffensen (1914b).

One should also point out another paper of Landau (1911d) on the same subject. In it he returned to the failed approaches to the Prime Number Theorem made by G.H.Halphen (1883) and E.Cahen (1893) who tried to use the identity

$$\psi(x) = -\frac{1}{2\pi i}\int_{a-i\infty}^{a+\infty}\frac{\zeta'}{\zeta}(s)\frac{x^s}{s}ds$$

valid for $a > 1$ and all positive x which are not prime powers. Because the integral here is not absolutely convergent a direct approach to evaluate it could not succeed. Landau combined efficiently the existence of zero-free regions to the left of the line Re $s = 1$ with an approximate formula for the sum of coefficients of a Dirichlet series $\sum_{n=1}^{\infty} a_n n^{-s}$ under the assumption $a_n = O(\log n)$ (he used later a variant of it in his proof of an approximate explicit formula for $\psi(x)$ which we shall present in Theorem 6.10 below) to get around this difficulty.

6. Both de la Vallée-Poussin's and Landau's proofs of the Prime Number Theorem with an error term show the strong influence of the size of the region to the left of the line Re $s = 1$ in which the zeta-function is zero-free on the growth of the error term. This was pointed out by H.Bohr and H.Cramér (1923, footnote on p.789) who noted that from Littlewood's zero-free region

$$\{\sigma + it : \ \sigma > 1 - c\log\log t/\log t\}$$

(with a certain constant $c > 0$ (Littlewood 1922)) one can deduce the bound

$$O(x\exp(-C\sqrt{\log x \log\log x}))$$

for the error term with a certain positive C. In A.E.Ingham's book (Ingham 1932) one finds a general result describing this influence. We quoted it in the preceding chapter.

In later years several ways of proving the Prime Number Theorem were given. In sect.4 we shall describe a tauberian approach to this problem, initiated by a result in Landau (1908b). Other analytic methods were given in Daboussi[21] (1984, 1989) (here only real analysis was used), Gerig (1976) (this proof uses harmonic analysis and provides a rather weak error term),

[21] For an exposition see Schwarz,Spilker (1994).

Grosswald (1964), Newman (1980) (a really simple proof[22]based on an integral formula going back essentially to Ingham (1935)), Byrnes,Giroux,Shisha (1984), Wolke (1985), Selvaray (1991). A proof based on the large sieve was given by A.Hildebrand (1986).

7. More than fifty years after first proofs of the Prime Number Theorem P.Erdős (1949a,b) and A.Selberg (1949b) found a method of proving it without the use of analytic tools. This came as a big surprise since it was commonly believed that this is not possible and a large literature arose even around the question concerning which statements are 'elementarily equivalent[23]' to the Prime Number Theorem, this equivalence meaning that one can derive one assertion from the other with elementary[24] means. In particular the following assertions turned out to be elementarily equivalent to Theorems 5.8 and 5.13:

$$(A) \qquad \sum_{n=1}^{\infty} \frac{\mu(n)}{n} = 0,$$

$$(B) \qquad M(x) = \sum_{n\leq x} \mu(n) = o(x),$$

$$(C) \qquad \sum_{n=1}^{\infty} \frac{\mu(n)\log n}{n} = -1,$$

and

$$(D) \qquad \sum_{p\leq x} \frac{\log p}{p} = \log x + c + o(1)$$

H.von Mangoldt (1897) deduced (A) from the Prime Number Theorem and obtained also the implication (A)→(B) (see also Landau (1899a)). The implication (C)→(A) was pointed out in Landau (1906c). The fact that (C) follows from Theorem 5.8 was proved in Landau (1899b) and the converse implication appeared in Landau (1906c), where also a direct proof of (B) based on Theorem 5.8 was presented (another proof of that fact was given in Helson (1973) and reproduced in Ellison (1975,p.69)).

[22]Variants of Newman's argument are presented in Korevaar (1982) and Zagier (1997).

[23]This notion has of course nothing to do with the modern notion of arithmetical equivalence of algebraic structures.

[24] The word 'elementary' was understood sometimes in different ways. Usually it meant 'without the use of complex functions', hence the use of real analysis was allowed, but other people were more strict and forbade the utilization of all limit processes.

In the year 1910 the equivalence of the Prime Number Theorem and (D) was established by A.Axer (1910) (see also Landau (1910), Karamata[25] (1954)) and one year later Landau (1911b) showed that Theorem 5.8 is implied by (A), thus Prime Number Theorem and (A) are equivalent. On the way he gave an elementary deduction of Prime Number Theorem from (B) so finally it became known that all statements (A), (B), (C), (D) are equivalent to the Prime Number Theorem in an elementary way. For a proof of their equivalence based on a tauberian theorem see Kienast (1924). A related result was obtained by A.Kienast (1926), who proved that the equality

$$\psi(x) = x + O\left(\frac{x}{\log^2 x}\right)$$

is equivalent to the convergence of the series

$$\sum_{n=1}^{\infty} \frac{\mu(n) \log^k n}{n}$$

for all $k \geq 1$.

Needless to say, all these assertions lost their meaning after elementary proofs of the Prime Number Theorem had been found but the replacement of the word 'elementary' by the word 'easy' may save something. The interested reader should consult Guerin, Buschman (1976), Nevanlinna (1964) and Specht (1964) on this topic. For results which are 'elementarily equivalent' to the quantitative version of Dirichlet's theorem see the paper of H.N.Shapiro (1949).

8. The method used by Erdős and Selberg did not lead to any evaluation of the error term which was also the case with several later elementary proofs given in Amitsur (1956/57,1962), Bang (1964), Breusch (1954), Eda (1953), Fogels (1950), Guerin, Buschman (1975), Kalecki (1969), Levinson (1966/67,1969), Nevanlinna (1962,1964) and Wright (1952). The first elementary proofs with an evaluation of the error term were independently given by P.Kuhn (1955) and J.G.van der Corput (1956). The first had

$$\theta(x) = x + O\left(\frac{x}{\log^c x}\right)$$

with $c = 1/10$ whereas the second had here $c = 1/200$. This has been subsequently improved to $c = 1/6 - \epsilon$ (for every $\epsilon > 0$) (Breusch 1960) and $c = 3/4$ (Wirsing 1962) before finally E.Bombieri (1962a,b) showed that one can take for c any positive number. The same result was announced in the quoted paper of Wirsing and a proof appeared in Wirsing (1964). Note that both proofs use real analysis, as do the papers quoted below. Bombieri's argument was

[25]Karamata, Jovan (1902–1967), Professor in Belgrade and Geneva.

simplified in Inojatova,Sobirov (1972), Sobirov,Inojatova (1972) and Sobirov (1973) by eliminating the use of integral transforms. (Cf. Dusumbetov 1962).

An important step was taken by H.G.Diamond and J.Steinig (1970) who succeeded in obtaining the relation

$$\psi(x) = x + O(x\exp(-\log^\alpha x(\log\log x)^\beta))$$

with $\alpha = 1/7$ and $\beta = -2$ without the use of complex analysis (a variant of this proof was given in Novák (1975)). Later, better values for the constants α and β were obtained: $\alpha = 1/6$, $\beta = -3$ (Lavrik, Sobirov 1973) (cf. Sobirov 1974), $\alpha = 1/6$, $\beta = 0$ (Srinivasan, Sampath 1988) and $\alpha = 1/5$, $\beta = 4$ (Nuriddinov 1991). See also Srinivasan, Sampath (1983). For a survey of elementary methods in prime number theory see Diamond (1982). The paper of A.F.Lavrik (1984) contains a survey of various approaches, both elementary and analytic, to the Prime Number Theorem.

Elementary proofs of the assertion (B) from the preceding subsection were given in Postnikov,Romanov (1955) (cf. Postnikov 1969), Kalecki (1967).

In Chapter 2 we pointed out the existence of elementary proofs of Dirichlet's theorem about primes in progression given by H.Zassenhaus (1949), A.Selberg (1949a) and A.O.Gelfond (1956). Elementary proofs of its quantitative form were provided by A.Selberg (1950), H.N.Shapiro (1950b), K.Yamamoto (1955) and S.A.Amitsur (1956/57). These proofs did not give any evaluation of the error term and the first such evaluation, in the form

$$\pi(x;k,l) = \frac{1}{\varphi(k)}\frac{x}{\log x} + O\left(\frac{x}{\log^{1.14} x}\right),$$

appears in Levin (1961a) and was slightly improved in Dusumbetov (1963). See Granville (1992) for a discussion what information about the distribution of primes in progressions one can get without using Dirichlet's characters.

6.3. von Mangoldt's Theorems Revisited

1. Without utilizing essentially new tools Landau simplified, in 1908, the proofs of Theorems 5.3 and 5.4, the first giving asymptotics for the number of zeros of $\zeta(s)$ in rectangles and the second being Riemann's explicit formula. First he did this (Landau 1908c,g) for the explicit formula stated in the following form:

$$f(x) = \operatorname{li}(x) - \sum_{\substack{\rho \\ \operatorname{Im}\rho>0}} (\operatorname{li}(x^\rho) + \operatorname{li}(x^{\bar\rho})) + \int_x^\infty \frac{dy}{(y^2-1)y\log y} - \log 2,$$

where the non-real roots ρ of $\zeta(s)$ are arranged according to their absolute values, the roots ρ and $1-\rho$ are paired together and the function f is defined by

$$f(x) = \sum_{p^k < x} \frac{1}{k} = \sum_{n=1}^{\infty} \frac{\pi(x^{1/n})}{n}$$

in case when x is not a prime power and by

$$f(x) = \lim_{h \to 0+0} \frac{f(x+h) - f(x-h)}{2}$$

otherwise.

The details of Landau's reasoning being rather complicated we give only a short sketch of it.

Sketch of Landau's proof of Theorem 5.4. Landau applied Cauchy's integral formula to the function $\log\zeta(s)x^s/s$ and his argument contained two new ingredients: he avoided the use of the decomposition of $\zeta'(s)/\zeta(s)$ into simple fractions, which occured in von Mangoldt's proof and decided to choose the integration contour in another way. Its construction utilized a family of rectangles $R_{z,g}$ with vertices $2 \pm T_g i$, $-z \pm T_g i$ where $g \geq 2$, $3 \leq z \equiv 1 \pmod 2$ were integers and T_g was a real number from the interval $[g, g+1]$ chosen in such a way that the strip

$$|\text{Im } s - T_g| \leq \frac{1}{c \log T_g}$$

was, with a suitable constant c not depending on g, without any zeros of the zeta-function. From that rectangle Landau deleted the segment $[-2, 1]$ of the real line as well as the segment joining $-2 + i\text{Im } \rho$ and ρ for every zero ρ of $\zeta(s)$ lying in the critical strip $0 < \text{Re } s < 1$.

The possibility of such a choice of the numbers T_g follows from Theorem 5.2 but Landau proved it directly using the Corollary to Theorem 5.3 which we state now as a separate lemma and give Landau's very simple proof:

Lemma 6.9. *For all positive T one has*

$$N(T+1) - N(T) = O(\log T).$$

Proof. We start with recalling the formula

$$\frac{\zeta'(s)}{\zeta(s)} = \frac{1}{1-s} - \frac{1}{2}\frac{\Gamma'}{\Gamma}\left(1 + \frac{s}{2}\right) + \sum_{\rho}\left(\frac{1}{s-\rho} + \frac{1}{\rho}\right) + C \qquad (6.23)$$

established in Corollary 4 to Theorem 5.1.

The equality

$$\frac{\Gamma'}{\Gamma}(s) = -\gamma - \frac{1}{s} + s\sum_{n=1}^{\infty} \frac{1}{n(s+n)}$$

which is uniform for $b_1 \leq x \leq b_2$, ($b_1 < b_2$ being any fixed positive reals) implies the bound

$$\frac{\Gamma'}{\Gamma}(x+iy) = O(\log(|y|)). \tag{6.24}$$

Moreover the identity

$$\frac{\zeta'}{\zeta}(s) = -\sum_{p}\sum_{m=1}^{\infty} \frac{\log p}{p^{ms}}$$

implies for Re $s > 1$ the estimate

$$\frac{\zeta'}{\zeta}(2+iT) = O(1).$$

This shows that for $s = 2 + iT$ one has

$$\sum_{\rho} \left(\frac{1}{2+iT-\rho} + \frac{1}{\rho} \right) = O(\log T)$$

and

$$\sum_{\rho} \operatorname{Re} \left(\frac{1}{2+Ti-\rho} + \frac{1}{\rho} \right) = O(\log T). \tag{6.25}$$

If we write $\rho = \beta + \gamma i$ then, in view of

$$\operatorname{Re} \left(\frac{1}{s-\rho} + \frac{1}{\rho} \right) = \frac{2-\beta}{(2-\beta)^2 + (T-\gamma)^2} + \frac{\beta}{\beta^2+\gamma^2} \geq \frac{1}{4} \frac{1}{1+(T-\gamma)^2},$$

(6.25) implies

$$\sum_{\rho} \frac{1}{1+(T-\gamma)^2} \leq 4 \sum_{\rho} \left(\frac{2-\beta}{(2-\alpha)^2+(T-\gamma)^2} \right) = O(\log T).$$

But, on the other hand,

$$\begin{aligned} \sum_{\rho} \frac{1}{1+(T-\gamma)^2} &\geq \sum_{\substack{\rho \\ T<\gamma\leq T+1}} \frac{1}{1+(T-\gamma)^2} \\ &\geq \sum_{\substack{\rho \\ T<\gamma\leq T+1}} \frac{1}{2} = \frac{1}{2}(N(T+1) - N(T)) \end{aligned}$$

and the assertion results. □

Corollary. *One has*

$$\sum_{\substack{\rho \\ |T-\rho|\geq 1}} \frac{1}{(T-\rho)^2} = O(\log T)$$

Proof. This is a consequence of (6.25). □

The lemma shows that it is possible to choose the numbers T_g in the manner described above. Using now the fact that if ρ is a non-real zero of $\zeta(s)$ of order λ then the difference $\log\zeta(s) - \lambda\log(s-\rho)$ has an inessential singularity at $s = \rho$ and utilizing the observation that if ρ is a zero of the zeta-function lying in $R_{z,g}$ and we integrate along the upper and lower part of the line joining $-z + \text{Im }\rho$ and ρ then

$$\lim_{z\to\infty} \int_{-z+i\text{Im }\rho}^{\rho} \log\zeta(s)\frac{x^s}{s}ds = \text{li}(x^\rho) - \text{sgn}(\text{Im }\rho)\pi i$$

Landau obtains the equality

$$\text{li}(x) - \sum_{\substack{\rho \\ T_g > \text{Im }\rho > 0}} \left(\text{li}(x^\rho) + \text{li}(x^{\bar\rho})\right) + \int_x^\infty \frac{dy}{y(y^2-1)\log y} - \log 2$$

$$= \frac{1}{2\pi i}\left(\int_{-\infty-T_g i}^{2-T_g i} + \int_{2-T_g i}^{2+T_g i} + \int_{2+T_g i}^{-\infty+T_g i}\right) \log\zeta(s)\frac{x^s}{s}ds.$$

The assertion results now after letting g tend to infinity, since the first and the third integral in brackets tend to zero. □

2. A very important explicit formula, simplifying essentially von Koch's (1902) formula (5.60) was proved in Landau (1912a):

Theorem 6.10. *For any $T > 2$ one has*

$$\psi(x) = x - \sum_{\substack{\rho=\beta+\gamma i \\ |\gamma|\le T}} \frac{x^\rho}{\rho} + O\left(\frac{x\log^2 x}{T} + \frac{x\log T}{T} + \log x\right).$$

Proof. We present only the main idea of the proof. All details may be found e.g. in Chandrasekharan (1970, Sect.5.2).

Landau starts with the observation that if the coefficients of the Dirichlet series $f(s) = \sum_{n=1}^\infty a_n n^{-s}$ satisfy $a_n = O(\log n)$ for $n \ge 2$ and for $1 < \eta < 2$ one has

$$\sum_{n=1}^\infty \frac{|a_n|}{n^\eta} = O\left(\frac{1}{\eta - 1}\right)$$

then for $1 < \eta < 2$ $x > 2$ and positive T the equality

$$\frac{1}{2\pi i}\int_{\eta-Ti}^{\eta+Ti} \frac{f(s)x^s}{s}ds = \sum_{n\le x} a_n + O\left(\frac{x^\eta}{T(\eta-1)} + \frac{x\log^2 x}{T} + \log x\right)$$

results. He proves it by a straightforward although somewhat tedious computation. This formula is a special case of the by now standard formula expressing the sum of coefficients of a Dirichlet series, a detailed proof of which appears e.g. in Prachar (1957, Theorem 3.1 of the Appendix).

Applying Cauchy's integration to the third formula of Corollary 4 to Theorem 5.1 over the rectangle with vertices at $\eta \pm Ti$, $-(2n+1) \pm Ti$ (with integral $n \geq 0$ and T distinct from the imaginary part of any zero of the zeta-function) and letting n tend to infinity one is lead to

$$\frac{1}{2\pi i}\int_{\eta-Ti}^{\eta+Ti} \frac{\zeta'}{\zeta}(s)\frac{x^s}{s}ds = -x + \frac{1}{2}\log(1-x^{-2})$$

$$+\sum_{\substack{\rho=\beta+\gamma i\\ |\gamma|\leq T}} \frac{x^\rho}{\rho} - \frac{1}{2\pi i}\int_{-\infty-Ti}^{\eta+Ti} \frac{\zeta'}{\zeta}(s)\frac{x^s}{s}ds + \frac{1}{2\pi i}\int_{-\infty+Ti}^{\eta+Ti} \frac{\zeta'}{\zeta}(s)\frac{x^s}{s}ds + O(1)$$

and it remains to bound the two last integrals. Because of the evident symmetry it suffices to deal with the second of them.

To do this Landau uses (6.24), Lemma 6.9 and the following two evaluations:

$$(a) \qquad \frac{\zeta'}{\zeta}(s) = O(\log(|s|)) \quad (\sigma \leq -1,\ |t| \geq 2),$$

$$(b) \qquad \sum_{\substack{\rho\\ |\rho-t|\geq 1}} \left(\frac{1}{s-\rho} + \frac{1}{\rho}\right) = O(\log(|t|)) \quad (-1 \leq \sigma \leq 2,\ |t| \geq 2).$$

The bound (a) is a consequence of the trivial bound for $\zeta'(s)/\zeta(s)$ in the half-plane Re $s \geq 2$ and the functional equation for $\zeta(s)$, while, to obtain (b), one uses Lemma 6.9 and its Corollary and Corollary 4 to Theorem 5.1.

Splitting the integration line into two parts separated by the point $-1+Ti$ we get

$$I = \int_{-\infty+Ti}^{\eta+Ti} \frac{\zeta'}{\zeta}(s)\frac{x^s}{s}ds = I_1 + I_2,$$

with I_1 being the integral over $(-\infty + Ti, -1 + Ti)$. Now (a) implies the evaluation

$$|I_1| = O\left(\frac{\log T}{T}\frac{1}{x\log x}\right)$$

and to evaluate I_2 observe that, due to Corollary 4 to Theorem 5.1, (6.25), Lemma 6.9 and the bound given in (b), we have

$$\left|\frac{\zeta'}{\zeta}(\sigma+iT) - \sum_{\substack{\rho\\ |\mathrm{Im}\ \rho-T|<1}} \frac{1}{s-\rho}\right| = O(\log t)$$

provided there is no zero of the zeta-function with imaginary part equal to T, which we shall assume for a while. This observation leads to

$$|I_2| \le \sum_{\substack{\rho \\ |\mathrm{Im}\,\rho - T|<1}} \left|\int_{-1+Ti}^{\eta+Ti} \frac{x^s}{s}\frac{ds}{s-\rho}\right| + O\left(\frac{\log T}{T}\frac{x^\eta}{\log x}\right)$$

and to compute the resulting integrals one deforms slightly each of the integration segments to avoid the possible zeros of $\zeta(s)$. This is done by bending the path along small semi-circles of radius $\eta - 1$ around every such zero. The integral over every such semi-circle is $O(x^\eta/T)$ and over the remaining part of the segment one gets the bound $O(x^\eta/(T(\eta-1)\log x))$. Using the bound for the number of zeros given by Lemma 6.9 one gets for the integral I_2 the bound

$$O\left(\frac{x^\eta \log T}{T}\left(1 + \frac{1}{(\eta-1)\log x}\right)\right).$$

Choosing finally $\eta = 1 + 1/\log x$ one gets the assertion in the case when there are no zeros of $\zeta(s)$ with imaginary part equal to T. But if there are such zeros then it suffices to apply the obtained result for $T - \epsilon$ with $\epsilon > 0$ and let ϵ tend to zero. □

Corollary. *One has*

$$\psi(x) = x - \sum_{\substack{\rho=\beta+\gamma i \\ |\gamma| \le x^{1/2}}} \frac{x^\rho}{\rho} + O(x^{1/2}\log^2 x).$$ □

The last corollary leads to a simple proof of von Koch's Theorem 5.21. In fact, it implies under the Riemann Hypothesis

$$|\psi(x) - x| \le \sum_{\substack{\rho=\beta+\gamma i \\ |\gamma| \le \sqrt{x}}} \frac{x^{1/2}}{|\rho|} + O(x^{1/2}\log^2 x)$$

and it suffices to apply Theorem 5.1 to obtain

$$\sum_{\substack{\rho=\beta+\gamma i \\ |\gamma| \le \sqrt{x}}} \frac{x^{1/2}}{|\rho|} = O(x^{1/2}\log^2 x).$$

Assuming the Riemann J.E.Littlewood (1924a) was able to reduce the error term in the last corollary to $O(x^{1/2}\log x)$. This however does not have any influence on the error term in Theorem 5.21. Littlewood's result was later made more precise by D.A.Goldston (1982) who proved (still under Riemann's Hypothesis)

$$\left|\psi(x) - x + \sum_{\substack{\rho=\beta+\gamma i \\ |\gamma| \le T}} \frac{x^\rho}{\rho}\right| < \frac{x}{2T} + 2x^{1/2}\log^2 T$$

for $x \ge 3$ and all sufficiently large T. Without any assumption he showed later (Goldston 1983) that for $x, T \ge 3$ one may write

$$\psi(x) = x - \sum_{\substack{\rho=\beta+\gamma i \\ |\gamma|\le T}} \frac{x^\rho}{\rho} + O\left(\frac{x\log x\log\log x}{T} + \frac{x\log T}{T} + \log x\right)$$

which improves Theorem 6.10 and in the case $3 \le T \le x$ shows that the error term is

$$O\left(\frac{x}{T}\log x\log\log x\right).$$

He achieved this by applying the inequality

$$\pi(x) - \pi(x-y) \le \frac{2y}{\log y} \quad (1 < y \le x)$$

(proved with the use of the large sieve in Montgomery, Vaughan (1973)) to obtain a better evaluation of the sum

$$\sum_{x/2\le n\le x-1} \frac{\Lambda(n)}{\log(n/x)}$$

occuring in the evaluation of one of the integrals in Landau's proof. Cf. also Wolke (1980,1983) and Perelli, Puglisi (1985).

Theorem 6.10 turned out to be very useful in studying the existence of primes in short intervals. The first such application was given by G.Hoheisel (1930) who established the existence of a constant $c < 1$ (he obtained the value $c = 1-1/33000$) such that for sufficiently large x one has

$$\pi(x + x^c) - \pi(x) = (1 + o(1))\frac{x^c}{\log x},$$

and thus the interval $[x - x^c, x]$ contains at least one prime number. Apart from Theorem 6.10 he utilized the non-vanishing of the zeta-function in the region

$$\Omega_C = \{\sigma + it :\ \sigma > 1 - C\log\log t/\log t\}$$

for $t \ge t_0 > 0$ and certain positive C (Littlewood 1922) as well as upper bounds for $N(\sigma, T)$ $(1/2 \le \sigma \le 1)$.

(Note that L.J.Moreno (1973) showed that one can obtain Hoheisel's result, even in a more general setting, using the smaller zero-free region obtained by de la Vallée-Poussin.)

There are actually three related problems there. The first consists in evaluation of c_0, the greatest lower bound of the exponent c in Hoheisel's result, the second deals with the optimal value of the exponent c_1 with the property that for all sufficiently large x there exists at least one prime in the interval $[x - x^{c_1}, x]$ and the third asks for the greatest lower bound c_2 of exponents c for which

$$\pi(x + x^c) - \pi(x) \ge B\frac{x^c}{\log x}$$

holds with a suitable constant $B = B(c)$ for all suficiently large x.

The value of c_0 was diminished to 249/250 by H.Heilbronn (1933) and shortly afterwards Hoheisel's approach obtained a more general form due to A.E.Ingham (1937), who proved that if $\zeta(s)$ does not vanish in Ω_C and the evaluation

$$N(\sigma, T) = O(T^{a(1-\sigma)}\log^b T)$$

(for T tending to infinity) holds uniformly in $\sigma \in [1/2, 1]$ with certain positive a and non-negative b then

$$c_0 \leq 1 - \frac{1}{a + b/C}.$$

Using the bounds for $N(\sigma, T)$ available at that time Ingham obtained in this way the inequality $c_0 \leq 5/8$. A proof of Ingham's theorem can be found in Chandrasekharan (1970,Sect.5.3) and, in a stronger version dealing with primes in progressions, in Tatuzawa (1950) (see Prachar (1957), Chap.9, Theorem 3.1).

If the Riemann Hypothesis is true then one can take for C an arbitrary large number and $a = 2$, thus Ingham's result shows that in that case for every positive ϵ and sufficiently large x there is a prime in the interval $[x, x + x^{1/2+\epsilon}]$. If one denotes by p_n the nth prime and $d_n = p_{n+1} - p_n$, then this shows that Riemann Hypothesis implies

$$d_n = O(p_n^{1/2+\epsilon})$$

for every $\epsilon > 0$, but this is weaker than the evaluation given in part (ii) of the Corollary to Theorem 5.21 and its improvement by H.Cramér quoted there.

It was proved by N.G.Čudakov (1936a,b) that for C one can take an arbitrary large number and this leads to $c_0 \leq 3/4$. Later research led to a substantial reduction of the value of c_0. The result of S.H.Min (1949) implied $c_0 \leq 38/61 = 0.62295\ldots$ and G.A.Kolesnik (1973) obtained a rather tiny improvement by showing $c_0 \leq 1759/2826 = 0.62243\ldots$. This is the best bound obtained with the use of Ingham's theorem. A new method of obtaining bounds for $N(\sigma, T)$ developed by H.L.Montgomery (1969b) led to $c_0 \leq 3/5 = 0.6$ and a modification of his method permitted M.N.Huxlcy (1972b) to get $c_0 \leq 7/12 = 0.58333\ldots$ (cf. Heath-Brown 1982,1988a).

Recently S.M.Gonek (1993) conjectured that the sum $\sum x^{i\mathrm{Im}\,\rho}$ extended over roots ρ of the zeta-function with $0 < \mathrm{Im}\,\rho \leq T$ is $O(Tx^{1/2+\epsilon}) + O(T^{1/2+\epsilon})$ for all x, $T \geq 2$ and $\epsilon > 0$ and proved that this conjecture implies $d_n = O(p_n^{\epsilon})$ for every $\epsilon > 0$ and thus $c_1 = 0$.

One has obviously $c_1 \leq c_2 \leq c_0$, hence the bounds for c_0 given above apply also to the remaining two constants.

Consecutive improvements of upper bounds for c_2 were obtained in Iwaniec, Jutila (1979) ($c_2 \leq 13/23 = 0.5652\ldots$), Heath-Brown, Iwaniec (1979) ($11/20 = 0.55$), Pintz (1981a,1984) ($17/31 = 0.548\ldots$), Iwaniec, Pintz (1984) ($23/42 = 0.5476\ldots$), Mozzochi (1986) ($11/20 - 1/384 = 0.5473\ldots$), Lou, Yao (1992,1993) ($6/11 = 0.545\ldots$ and $7/13 = 0.538\ldots$) and Baker, Harman (1996b) ($107/200 = 0.535$) which is the best result at this time.

It was shown by A.Selberg (1943) that for $\theta > 19/77 = 0.2467\ldots$ the inequality

$$\pi(x + x^\theta) - \pi(x) \geq B\frac{x^\theta}{\log x}$$

holds for almost all integers x with a suitable $C > 0$. This was later shown to hold for $\theta > 1/10$ (Harman 1982), $\theta > 1/14$ (Jia 1995, Watt 1995), $\theta > 1/15$ (Li Hong Ze 1997) and $\theta > 1/20$ (Jia 1996a).

3. We saw above (Lemma 6.9) that the Corollary to Theorem 5.3 can be established directly. Moreover E.Landau (1908h) noticed that Theorem 5.3 itself can be deduced from this corollary in a rather simple way:

Landau's proof of Theorem 5.3: The main tool here is complex integration. Choose $T > 1$ so that there are no roots of $\zeta(s)$ with imaginary part equal to

T and consider the rectangle R with vertices $-1+i$, $2+i$, $2+Ti$ and $-1+Ti$. Then no root of $\zeta(s)$ lies on the border of R and for the number $N(T)$ of roots ρ of the zeta-function satisfying $0 < \mathrm{Im}\ \rho \leq T$ one has the formula

$$N(T) = \frac{1}{2\pi i} \oint_R \frac{\zeta'}{\zeta}(s)ds,$$

the integral being taking anticlockwise. One sees immediately that the integrals taken over the lower border and the right border of R are bounded and hence, taking into account that $N(T)$ is real, one gets

$$N(T) = \frac{1}{2\pi}(J_1 + J_2) + O(1),$$

with

$$J_1 = \mathrm{Im} \int_{2+Ti}^{-1+Ti} \frac{\zeta'}{\zeta}(s)ds$$

and

$$J_2 = \mathrm{Im} \int_{-1+Ti}^{-1+i} \frac{\zeta'}{\zeta}(s)ds.$$

To evaluate the second integral one uses the equality

$$\frac{\zeta'}{\zeta}(s) = -\frac{1}{2}\frac{\Gamma'}{\Gamma}\left(\frac{1-s}{2}\right) - \frac{1}{2}\frac{\Gamma'}{\Gamma}\left(\frac{s}{2}\right) - \frac{\zeta'}{\zeta}(1-s) + \log \pi$$

resulting from the functional equation of $\zeta(s)$. As

$$\mathrm{Im}\ J_2 = -\int_0^T \mathrm{Re}\ \frac{\zeta'}{\zeta}(-1+it)dt$$

this equality leads to

$$\mathrm{Im}\ J_2 = \frac{1}{2}\int_1^T \mathrm{Re}\ \frac{\Gamma'}{\Gamma}(1 - \frac{t}{2}i)dt + \frac{1}{2}\int_1^T \mathrm{Re}\ \frac{\Gamma'}{\Gamma}(-\frac{1}{2} - \frac{t}{2}i)dt$$
$$- T\log\pi - \mathrm{Im} \int_{-1+Ti}^{-1+i} \frac{\zeta'}{\zeta}(1-s)ds.$$

The last integral here equals to $\int_{2-i}^{2-Ti} \zeta'(s)/\zeta(s)ds$ hence is bounded and, in view of the formula

$$\int_1^z \mathrm{Re}\ \frac{\Gamma'}{\Gamma}(a \pm i\frac{t}{2})dt = z\log(\frac{z}{2}) - z + O(\log z)$$

valid for any real a, one is led to

$$\mathrm{Im}\ J_2 = T\log T - (1 + \log(2\pi))T + O(\log T).$$

So

$$N(T) = \frac{T}{2\pi}\log T - \frac{1+\log(2\pi)}{2\pi}T + O(\log T) + \frac{1}{2\pi}\text{Im} \int_{2+Ti}^{-1+Ti} \frac{\zeta'}{\zeta}(s)ds.$$

and it remains to show that the integral J_1 is $O(\log T)$.

To obtain this one uses the equality (6.23), which leads to

$$\begin{aligned}\text{Im } J = &\sum_{\substack{\rho \\ |T-\text{Im}\rho|<1}} \text{Im} \int_{2+Ti}^{-1+Ti} \left(\frac{1}{s-\rho} + \frac{1}{\rho}\right) ds \\ &+ \sum_{\substack{\rho \\ |T-\text{Im}\rho|\geq 1}} \text{Im} \int_{2+Ti}^{-1+Ti} \left(\frac{1}{s-\rho} + \frac{1}{\rho}\right) ds \\ &+ O(1) + O\left(\int_{2+Ti}^{-1+Ti} \frac{\Gamma'}{\Gamma}(1+\frac{s}{2})ds\right).\end{aligned} \tag{6.26}$$

The last term here is $O(\log T)$ due to (6.24) and as the number of terms in the first sum does not exceed $N(T+1) - N(T-1) = O(\log T)$ by Lemma 6.9, and on the segment $[-1+Ti, 2+Ti]$ we have

$$\left|\frac{1}{s-\rho} + \frac{1}{\rho}\right| = O(1)$$

the first sum is $O(\log T)$.

To evaluate the second term in (6.26) consider the sum

$$S = \sum_{\rho} \left(\frac{1}{s+3-\rho} + \frac{1}{\rho}\right) = \frac{\zeta'}{\zeta}(s+3) - C + \frac{1}{s+2} + \frac{1}{2}\frac{\Gamma'}{\Gamma}\left(\frac{s+5}{2}\right) = O(\log T).$$

As the part of S corresponding to ρ's satisfying $|T - \text{Im } \rho| \leq 1$ is $O(\log T)$, which is established by the same argument as above, hence the same bound applies also to the remaining part of this sum, i.e.

$$S = \sum_{\substack{\rho \\ |T-\text{Im } \rho|\geq 1}} \left(\frac{1}{s+3-\rho} + \frac{1}{\rho}\right) = O(\log T). \tag{6.27}$$

Finally due to Corollary to Lemma 6.9 we get

$$\left|\sum_{\substack{\rho \\ |T-\text{Im } \rho|\geq 1}} \left(\frac{1}{s+3-\rho} + \frac{1}{\rho}\right) - \left(\frac{1}{s-\rho} + \frac{1}{\rho}\right)\right|$$

$$= \left|\sum_{\substack{\rho \\ |T-\text{Im } \rho|\geq 1}} \frac{3}{(s-\rho)(s+3-\rho)}\right| \leq 3 \left|\sum_{\substack{\rho \\ |T-\text{Im } \rho|\geq 1}} \frac{1}{\text{Im } (s-\rho)\text{Im } (s+3-\rho)}\right|$$

$$= 3 \sum_{\substack{\rho \\ |T-\text{Im } \rho|\geq 1}} \frac{1}{(T - \text{Im } \rho)^2} = O(\log T)$$

and now (6.27) implies

$$\sum_{\substack{\rho \\ |T-\mathrm{Im}\,\rho|\geq 1}} \left(\frac{1}{s-\rho}+\frac{1}{\rho}\right) = O(\log T).$$

Integrating this inequality we see that the second term in (6.26) is $O(\log T)$ and this implies the assertion. □

4. A few years later Backlund (1914,1916) obtained an even simpler proof of Theorem 5.3, based on Jensen's theorem. He started his proof in the same way as von Mangoldt did, and quickly arrived at the formula

$$N(T) = \frac{T}{2\pi}\log\left(\frac{T}{2\pi}\right) - \frac{T}{2\pi} + \frac{7}{8} + P(T) + O\left(\frac{1}{T}\right),$$

where $\pi P(T)$ is equal to the increase of the argument of $\zeta(s)$ on the polygon ABC with $A = 2$, $B = 2 + iT$, $C = 1/2 + iT$. To evaluate this increase observe that it does not exceed $(1+n)\pi$ where n is the number of zeros of $\mathrm{Re}\,\zeta(s)$ on the sides of ABC, thus $P(T) \leq 1 + n$. If we put

$$f(s) = \frac{1}{2}(\zeta(s+iT) + \zeta(s-iT))$$

then n is at most equal to the number L of zeros of the function f in the disk $|s-2| \leq 3/2$.

If M denotes the maximum of $|f(2+2e^{i\varphi})|$ on $[0, 2\pi]$ and $m = |f(2)| = |\mathrm{Re}\,\zeta(2+iT))|$ then an application of Jensen's formula to f gives

$$P(T) \leq 1 + n \leq 1 + L \leq 1 + \frac{\log(M/m)}{\log(4/3)}.$$

The final step utilizes the inequality (6.9) which gives $\log M = O(\log T)$ and to obtain $P(T) = O(\log T)$ it remains to observe that m is bounded away from zero, which is a consequence of

$$m = \left|\sum_{j=1}^{\infty} \frac{\cos(T\log j)}{j^2}\right| \geq 1 - \sum_{j=1}^{\infty} \frac{1}{j^2} > 0.3 \qquad \square$$

Commenting upon his proof Backlund points out correctly that contrary to previous approaches it does not use any deeper tools than the product formula and the functional equation for $\zeta(s)$.

6.4. Tauberian Methods

1. The *Rendiconti* paper of Landau (1908b), which we have, already, quoted, contains the first result from a long sequence of tauberian theorems for Dirichlet series. S.Bochner (1933) regarded it as the first deep tauberian theorem.

The name *tauberian theorem*[26] stems from a result of A.Tauber[27](1897) who showed that if the Abelian method of summation applied to a series $\sum_{n=1}^{\infty} a_n$ gives the value s, i.e.

$$\lim_{x\to 1-0} \sum_{n=1}^{\infty} a_n x^n = s$$

and moreover one has

$$\lim_{n\to\infty} n a_n = 0$$

then the series $\sum_{n=1}^{\infty} a_n$ converges and its sum equals s. For a proof see e.g. Hardy (1949, Theorem 85) where also a discussion of further developments is given.

Theorem 6.11. (Landau 1908b,Theorem VIII) *Assume that the series*

$$\sum_{n=1}^{\infty} a_n n^{-s}$$

has real coefficients satisfying $a_n = O(n^c)$ *with some* c *and moreover for a certain positive* A *one has*

$$a_n \geq -A.$$

Denote by $f(s)$ *the function defined by that series in its half-plane of convergence and assume that it can be continued to a function regular in the closed half-plane* Re $s \geq 1$ *and satisfies, with a suitable* $k > 0$, *the inequality* $f(\sigma + it) = O(t^k)$ *uniformly for* $\sigma \geq 1$, $|t| \geq 1$. *Then*

$$\lim_{x\to\infty} \frac{1}{x} \sum_{n\leq x} a_n \log\left(\frac{x}{n}\right) = 0.$$

Actually Landau proved a stronger result, stating that

$$\frac{1}{x} \sum_{n\leq x} a_n = 0$$

utilizing a more complicated version of Lemma 6.12 (i) below, but for the proof of the Prime Number Theorem this is irrelevant, since in this case one can use in the last step Hadamard's Lemma 5.9.

[26]It seems that it was first used in Hardy, Littlewood (1913a).
[27]Tauber, Alfred (1866–1942), Professor in Wien.

Proof. The assumptions imply the absolute convergence of the series defining f in the half-plane $\operatorname{Re} s > 1 + c$. If now one chooses a vertical line γ lying in that half-plane and an integer μ exceeding $k + 1$ then for every positive x the integral

$$I(x) = \int_\gamma f(s) \frac{x^s}{s^\mu} ds$$

converges absolutely and using Hadamard's formula

$$\frac{1}{2\pi i} \int_{a-i\infty}^{a+i\infty} \frac{x^s}{s^\mu} ds = \begin{cases} 0 & 0 < x \leq 1 \\ \dfrac{\log^{\mu-1} x}{(\mu-1)!} & x > 1 \end{cases}$$

(see Sect.5.3.3) one arrives at

$$I(x) = \frac{2\pi i}{(\mu-1)!} \sum_{n=1}^{\infty} a_n \log^{\mu-1}(x/n).$$

Shifting the line of integration to $\operatorname{Re} s = 1$ one gets

$$I(x) = x \int_{-\infty}^{\infty} \frac{f(1+it)}{(1+it)^\mu} \exp(it \log x) dt.$$

Now note that the integral on the right hand-side tends to zero for x tending to infinity. Landau proves this by calculating the integral, but this can also be seen directly, as our assumptions imply that the function

$$\frac{f(1+it)}{(1+it)^\mu}$$

is integrable on the real line and hence its Fourier transform tends to zero at infinity. This implies

$$\lim_{x\to\infty} \frac{1}{x} \sum_{n \leq x} a_n \log^{\mu-1}\left(\frac{x}{n}\right) = 0. \tag{6.28}$$

The assertion of the theorem is an immediate consequence of the last of the following sequence of elementary results:

Lemma 6.12. *(i)* (Landau 1908b, Satz III) *Let $f(x)$ be a real function having a derivative for $x > 0$. Assume that $xf'(x)$ is non-decreasing and the ratio $f(x)/x$ tends to 1 when x tends to infinity. Then*

$$\lim_{x\to\infty} f'(x) = 1.$$

(ii) (Landau 1908b, Satz V) *If c_n is a sequence of nonnegative real numbers and for some integer $k \geq 2$ one has*

$$\lim_{x\to\infty} \frac{1}{k!x} \sum_{n\le x} c_n \log^k\left(\frac{x}{n}\right) = 1,$$

then

$$\lim_{x\to\infty} \frac{1}{x} \sum_{n\le x} c_n \log\left(\frac{x}{n}\right) = 1.$$

(iii) (Landau 1908b, Satz VI) *If b_n is a sequence of real numbers bounded from below, i.e. $b_n \ge -A$ holds with a certain positive A and there exists a positive integer k such that*

$$\lim_{x\to\infty} \frac{1}{x} \sum_{n\le x} b_n \log^k\left(\frac{x}{n}\right) = 0,$$

then

$$\lim_{x\to\infty} \frac{1}{x} \sum_{n\le x} b_n \log\left(\frac{x}{n}\right) = 0.$$

Proof. (i) We give a proof obtained by G.H.Hardy and J.E.Littlewood (1913b) who presented original Landau's idea in a transparent form:

For any positive δ one has

$$\frac{f(x+\delta x)-f(x)}{\delta x} = \frac{1}{\delta x}\int_x^{x+\delta x} f'(t)dt \ge \frac{xf'(x)}{\delta x}\int_x^{x+\delta x}\frac{dt}{t} = \frac{f'(x)}{\delta}\log(1+\delta)$$

and since the left-hand side tends to 1 as x tends to infinity this leads to

$$\limsup_{x\to\infty} f'(x) \le \frac{\delta}{\log(1+\delta)}.$$

Similarly

$$\frac{f(x+\delta x)-f(x)}{\delta x} = \frac{1}{\delta x}\int_x^{x+\delta x} f'(t)dt$$
$$\le \frac{(x+\delta x)f'(x+\delta x)}{\delta x}\int_x^{x+\delta x}\frac{dt}{t} = \frac{(1+\delta)\log(1+\delta)f'((1+\delta)x)}{\delta}$$

and putting $y=(1+\delta)x$ one gets

$$\liminf_{y\to\infty} f'(y) \ge \frac{\delta}{(1+\delta)\log(1+\delta)}.$$

Since δ can be arbitrarily small, the assertion follows.

(ii) For $j\ge 2$ consider the function

$$f_j(x) = \frac{1}{j!}\sum_{n\le x} c_n \log^j\left(\frac{x}{n}\right)$$

and observe that for all positive x it has a derivative equal to $f_{j-1}(x)/x$. Indeed, if x is not an integer and $|h|$ is sufficiently small so that the interval $(x, x+h)$ does not contain any integer then

$$\frac{1}{h}(f_j(x+h) - f_j(x)) = \frac{1}{j!}\sum_{n\le x} c_n \frac{1}{h}\left(\log^j\left(\frac{x+h}{n}\right) - \log^j\left(\frac{x}{n}\right)\right)$$

and, for h tending to zero, the right hand-side tends to $f_{j-1}(x)/x$. If x is an integer then the same argument shows the existence of the right derivative of f_j at x, equal to $f_{j-1}(x)/x$ so it remains to note that due to $\log(x/x) = 0$ the left derivative at x equals

$$\lim_{h\to 0+0} \frac{1}{-h}(f(x-h) - f(x))$$
$$= \frac{1}{j!}\sum_{n\le x-1} c_n \lim_{h\to 0+0} \frac{1}{-h}(\log^j\left(\frac{x-h}{n}\right) - \log^j\left(\frac{x}{n}\right)) = \frac{f_j(x)}{x}.$$

Applying (i) to the function f_j consecutively for $j = k, k-1, \dots, 2$ one obtains (ii).

(iii) Observe first that Lemma 2.12 implies

$$\sum_{n\le x} \log^k(x/n) = \int_1^x \log^k\left(\frac{x}{t}\right) dt + O(\log^k x) = x\int_1^x \frac{\log^k u}{u^2} du + O(\log^k x)$$

for integral $k \ge 1$ and this, in view of

$$\int_1^x \frac{\log^k u}{u^2} du = \int_0^\infty e^{-w} w^k dw = \Gamma(1+k) = k!,$$

leads to

$$\lim_{x\to\infty} \frac{1}{x}\sum_{n\le x} \log^k\left(\frac{x}{n}\right) = k! \qquad (6.29)$$

If we now put $c_n = 1 + b_n/A \ge 0$ then

$$\lim_{x\to\infty} \frac{1}{x}\frac{1}{k!}\sum_{n\le x} c_n \log^k\left(\frac{x}{n}\right) = 1$$

thus (ii) implies

$$\lim_{x\to\infty} \frac{1}{x}\sum_{n\le x} c_n \log\left(\frac{x}{n}\right) = 1$$

and applying (6.29) for $k = 1$ we arrive at (iii). □

As we already noted, the theorem follows now immediately from (6.28) and part (iii) of the lemma. □

2. The Prime Number Theorem in the form $\theta(x) = (1+o(1))x$ follows rather easily from Theorem 6.11 although this approach does not lead to any evaluation of the error term. Consider the Dirichlet series

$$F(s) = \sum_{n=1}^{\infty} \frac{a_n}{n^s}$$

with

$$a_n = \begin{cases} \log n - 1 & \text{if } n \text{ is prime,} \\ -1 & \text{otherwise.} \end{cases}$$

One sees easily that this series converges absolutely for Re $s > 1$ and one has there the equality

$$F(s) = \frac{\zeta'(s)}{\zeta(s)} - \zeta(s) + g(s),$$

where

$$g(s) = -\sum_p \frac{\log p}{p^s(p^s - 1)}$$

is regular and bounded in the half-plane Re $s > 3/4$. The function $F(s)$ is regular on the line Re $s = 1$ and from Lemma 6.7 one infers that it satisfies the assumptions of Theorem 6.11. Applying that theorem we get

$$\lim_{x\to\infty} \sum_{p\leq x} \log p \log \frac{x}{p} - \sum_{n\leq x} \log \frac{x}{n} = 0$$

and Lemma 5.9 jointly with the case $k = 1$ of (6.29) give the assertion. □

A variant of this proof appears in Landau (1908d).

3. Another way of deducing the Prime Number Theorem from a similar tauberian result was shown by Hardy and Littlewood (1915,1918). They based their reasoning on the following result, which is a special case of a tauberian theorem proved earlier by them (Hardy, Littlewood 1913b,1914):

Theorem 6.13. *Let a_n be a sequence of non-negative real numbers and assume that the series*

$$f(x) = \sum_{n=1}^{\infty} a_n e^{-nx}$$

converges for all positive values of x. If

$$\lim_{x\to 0+0} x f(x) = A > 0$$

then for T tending to infinity one has

$$S(T) = \sum_{n\leq T} a_n = (A + o(1))T.$$

We shall obtain this theorem as a corollary to the following result of J.Karamata (1930) :

Theorem 6.14. *Let the coefficients* a_n *be non-negative and assume that*

$$\lim_{x\to 1-0}(1-x)\sum_{n=0}^{\infty} a_n x^n = C > 0.$$

If g *is a function bounded in in the interval* $(0,1)$ *and Riemann-integrable there, then*

$$\lim_{x\to 1-0}(1-x)\sum_{n=0}^{\infty} a_n g(x^n)x^n = C\int_0^1 g(t)dt.$$

Proof. The assumption implies that for positive m one has

$$\lim_{x\to 1-0}(1-x)\sum_{n=0}^{\infty} a_n x^{mn}x^n = \frac{C}{1+m} = C\int_0^1 t^m dt.$$

Indeed, it suffices to put $y = x^{1+m}$ and observe that

$$\lim_{x\to 1}\frac{1-x}{1-x^{1+m}} = \frac{1}{1+m}.$$

It follows that for every polynomial P one has

$$\lim_{x\to 1-0}(1-x)\sum_{n=0}^{\infty} a_n P(x^n)x^n = C\int_0^1 P(t)dt.$$

Now note that from a variant of the Weierstrass approximation theorem (which has its origin in Weierstrass 1885) for every positive ϵ one can find two polynomials p and P with $p(x) \le g(x) \le P(x)$ for $0 \le x \le 1$ and

$$\int_0^1 (P(t)-p(t))dt \le \epsilon$$

and since in view of $a_n \ge 0$ we have

$$C\int_0^1 p(t)dt = \lim_{x\to 1-0}(1-x)\sum_{n=0}^{\infty} a_n p(x^n)x^n \le \lim_{x\to 1-0}(1-x)\sum_{n=0}^{\infty} a_n g(x^n)x^n$$
$$\le \lim_{x\to 1-0}(1-x)\sum_{n=0}^{\infty} a_n P(x^n)x^n = C\int_0^1 P(t)dt$$

the assertion results. □

Karamata also gave another proof, presented in Hardy (1949, Theorem 98). Cf. Karamata 1931.

Corollary. *If*

$$f(x) = \sum_{n=0}^{\infty} a_n x^n$$

converges for $|x| < 1$, *all coefficients* a_n *are non-negative and*

$$\lim_{x \to 1-0} (1-x) f(x) = C > 0$$

then for N *tending to infinity one has*

$$\sum_{n \le N} a_n = (C + o(1)) N.$$

Proof. Apply Theorem 6.14 with

$$g(t) = \begin{cases} 0 & \text{if } 0 \le t \le e^{-1} \\ \frac{1}{t} & \text{if } e^{-1} \le t \le 1. \end{cases}$$

Then

$$\int_0^1 g(t) dt = 1$$

and since $g(e^{\ n/N})$ vanishes for $n > N$ and one has

$$\lim_{N \to \infty} N(1 - e^{-1/N}) = 1$$

we get

$$\begin{aligned} C &= \lim_{N \to \infty} (1 - e^{-1/N}) \sum_{n=0}^{\infty} a_n g(e^{-n/N}) e^{-n/N} \\ &= \lim_{N \to \infty} (1 - e^{-1/N}) \sum_{n \le N} a_n = \lim_{N \to \infty} \frac{1}{N} \sum_{n \le N} a_n, \end{aligned}$$

as asserted. □

Proof of Theorem 6.13. Put $x = e^{-y}$ in the last Corollary. □

Corollary 1. (Hardy, Littlewood 1914, Theorem D) *In Theorem 6.13 one can replace the assumption* $a_n \ge 0$ *by* $a_n \ge -B$ *with a positive constant* B *and the condition* $A > 0$ *by* $A \ge 0$.

Proof. Applying the theorem to the sequence $b_n = a_n + B \ge 0$ and utilizing

$$\lim_{x \to 0+0} x \sum_{n=0}^{\infty} e^{-nx} = 1$$

we obtain

$$\sum_{n\leq T} b_n = (A + B + o(1))T$$

which implies immediately the assertion. □

Corollary 2. *In the preceding corollary one can replace the assumption* $a_n \geq -B$ *by* $a_n \leq B$ *with a positive constant* B.

Proof. Apply the preceding corollary with a_n replaced by $-a_n$. □

These results imply the following theorem, proved in Hardy, Littlewood (1915), which is the basis of their proof of the Prime Number Theorem:

Theorem 6.15. *Assume that the series*

$$F(s) = \sum_{n=1}^{\infty} \frac{a_n}{n^s}$$

converges absolutely for $\operatorname{Re} s > 1$.

If $F(s)$ *can be extended to a function continuous in the closed half-plane* $\operatorname{Re} s \geq 1$ *and in that half-plane one has*

$$|F(s)| = O\left(e^{c|\operatorname{Im} s|}\right)$$

with a certain constant $c < \pi/2$ *then the series*

$$f(y) = \sum_{n=1}^{\infty} a_n e^{-ny}$$

converges for all positive values of y *and satisfies*

$$\lim_{y\to 0} y f(y) = 0.$$

If moreover $F(s)$ *can be extended to a function having at* $s = 1$ *a simple pole with residue* A, *then for* y *tending to* 0 *one has*

$$f(y) = (A + o(1))y^{-1}.$$

Proof. Since for positive y and $a > 1$ the equality

$$e^{-ny} = \frac{1}{2\pi i} \int_{a-i\infty}^{a+i\infty} \frac{\Gamma(s)}{(ny)^s} ds \tag{6.30}$$

holds, the use of Cauchy's integral theorem leads to

$$f(y)=\frac{1}{2\pi i}\int_{a-i\infty}^{a+i\infty}F(s)\frac{\Gamma(s)}{y^s}ds$$

$$=\frac{1}{2\pi}\int_{1-i\infty}^{1+i\infty}F(s)\frac{\Gamma(s)}{y^s}ds=y^{-1}\int_{-\infty}^{\infty}F(1+it)\Gamma(1+it)e^{-it\log y}dt.$$

Our assumptions imply that the function $F(1+it)\Gamma(1+it)$ is absolutely integrable on the real line hence

$$\lim_{|\log y|\to\infty}\int_{-\infty}^{\infty}F(1+it)\Gamma(1+it)\exp(it\log y)dt=0$$

and thus $f(y)=o(y^{-1})$ as y tends to zero. This gives the first assertion. To prove the second assume that $F(s)$ has a simple pole at $s=1$ and let A be its residue there. Since $F(s)$ is regular inside a certain annulus, say $0<|s-1|\le\delta$ we obtain, as in the previous case

$$f(y)=\frac{A}{y}+\frac{1}{2\pi i}\int_\gamma F(s)\frac{\Gamma(s)}{y^s}ds$$

where γ denotes the contour consisting of the two half-lines $[1-i\infty,1-\delta i]$, $[1+\delta i,1+i\infty]$ and the left half of the circle $|s-1|=\delta$. The same argument as in the first part shows that the integrals over the half-lines tend to zero for $y\to 0$ while the integral over the semi-circle equals

$$\Gamma(1+i\delta)F(1+i\delta)\frac{y^{1+i\delta}}{\log(1/y)}-\Gamma(1-i\delta)F(1-i\delta)\frac{y^{1-i\delta}}{\log(1/y)}$$
$$-\frac{1}{\log(1/y)}\int_\gamma y^{-s}\frac{d}{ds}(\Gamma(s)F(s))ds=O\left(\frac{1}{\log(1/y)}\right)$$

and hence tends to zero with y. This establishes

$$f(y)=(A+o(1))y^{-1},$$

as asserted. □

Corollary. *If the assumptions of the theorem are satisfied and moreover the coefficients a_n are either non-negative or bounded, then one has*

$$\sum_{n\le T}a_n=\begin{cases}(A+o(1))T & \text{if } F(s) \text{ has a simple pole at } s=1\\ o(T) & \text{otherwise.}\end{cases}$$

Proof. Apply either Theorem 6.13 or its Corollary 2. □

Now it is easy to prove the Prime Number Theorem. Apply the preceding Corollary 1 to $a_n=\Lambda(n)$. Since in this case $F(s)=-\zeta'/\zeta(s)$ we have $A=1$ and the needed upper bound for $F(s)$ results from Lemma 6.7 (iii). Thus

$$\psi(T) = \sum_{n \le T} \Lambda(n) = (1 + o(1))T$$

and in view of $\theta(T) = (1 + o(1))\psi(T)$ we obtain the assertion of Theorem 5.8. □

4. Many years later J.E.Littlewood (1971) presented a simpler deduction of the Prime Number Theorem from Theorem 6.13.

One needs a simple lemma first:

Lemma 6.16. *For all sufficiently large M the interval $(3M/4, 5M/4)$ contains a number T for which*

$$\frac{\zeta'}{\zeta}(\sigma \pm iT) = O(T^3)$$

holds for all $\sigma \in [-1, 2]$.

Proof. It suffices to consider $\sigma + iT$, the other case resulting by symmetry.

We use the formula

$$\frac{\zeta'}{\zeta}(s) = \frac{1}{1-s} - \frac{1}{2}\frac{\Gamma'}{\Gamma}\left(\frac{s}{2} + 1\right) + \sum_{\rho}\left(\frac{1}{s-\rho} + \frac{1}{\rho}\right) + O(1)$$

established in Corollary 4 to Theorem 5.1, ρ running over all non-real zeros of the zeta-function. For large M and $T \in [c_1 M, c_2 M]$ (with arbitrary $0 < c_1 < c_2$) the first two terms of the right hand-side of this formula are $O(\log T)$ due to (6.24), so we may concentrate our attention on the last term, equal to

$$\sum_{\rho} \frac{s}{\rho(s-\rho)}.$$

Let us split all zeros ρ into three classes:

$$R_1 = \{\rho :\ M/2 \le \gamma \le 2M\},$$

$$R_2 = \{\rho :\ \gamma \ge 2M \text{ or } \gamma < 0\}$$

and

$$R_3 = \{\rho :\ 0 < \gamma \le M/2\}.$$

Since the set of all zeros $\rho = \beta + \gamma i$ satisfying $3M/4 \le \gamma \le 5M/4$ has by Theorem 5.2 $O(M^2)$ elements, there exists a number T in the interval $[3M/4, 5M/4]$ such that for all ρ in R_1 and $s \in I = \{\sigma + iT,\ -1 \le \sigma \le 2\}$ the inequality

$$|s - \rho| \ge |\gamma - T| > \frac{1}{T^2}$$

holds. Evaluating the sum

$$S_1 = \sum_{\rho \in R_1} \frac{s}{\rho(s-\rho)}$$

trivially by using $|s| = O(T)$, $1/|\rho| = O(1/M)$ and $\#R_1 = O(M^2)$ we get

$$S_1 = O(M^3).$$

If $\rho \in R_2$ then for $s \in I$ we have $|s-\rho| > c|\rho|$ with a certain positive c, thus

$$S_2 = \sum_{\rho \in R_2} \frac{s}{\rho(s-\rho)} = O\left(M \sum_{\rho} \frac{1}{|\rho|^2}\right) = O(M)$$

and if finally $\rho \in R_3$ then for $s \in I$ we have $|s-\rho| > dM$ with some positive d and since the ρ's are bounded away from the real line we get by Theorem 5.2

$$S_3 = \sum_{\rho \in R_3} \frac{s}{\rho(s-\rho)} = O\left(\sum_{\rho \in R_3} \frac{1}{|\rho|}\right) = O(M^2).$$

Putting all this together we get

$$\frac{\zeta'}{\zeta}(\sigma + iT) = S_1 + S_2 + S_3 + O(\log T) = O(M^3) + O(\log T) = O(T^3)$$

as asserted. □

Proof of the Prime Number Theorem. Now let T be one of the numbers whose existence is asserted in the preceding lemma and consider the integral

$$-\frac{1}{2\pi i} \int_X \frac{\zeta'}{\zeta}(s) \Gamma(s) y^{-s} ds$$

over the rectangle X with vertices $A = 2 - Ti$, $B = 2 + Ti$, $C = -1 + Ti$ and $D = -1 - Ti$. Denote the interior of X by Ω. Computing the residues we see that our integral equals

$$\frac{1}{y} - \sum_{\rho \in \Omega} \frac{\Gamma(\rho)}{y^\rho} - \frac{\zeta'}{\zeta}(0).$$

The preceding lemma implies that the integrals over BC, CD and DA tend for $T \to \infty$ to zero because of $\Gamma(s) = O(\exp(-c|T|))$. This shows that for positive y one has (after multiplication by y)

$$\frac{y}{2\pi i} \int_{2-i\infty}^{2+i\infty} \frac{\zeta'}{\zeta}(s) \Gamma(s) y^{-s} ds = -1 + \sum_{\rho} \Gamma(\rho) y^{1-\rho} + O(y).$$

Due to (6.30) the integral on the left-hand side equals

$$-\sum_{n=1}^{\infty} \Lambda(n)e^{-ny},$$

hence[28]

$$y\sum_{n=1}^{\infty} \Lambda(n)e^{-ny} = 1 - \sum_{\rho} \Gamma(\rho)y^{1-\rho} + O(y).$$

Now note that $|y^{1-\rho}| < 1$ and as Theorem 5.1 and Stirling's formula imply that the series $\sum_\rho \Gamma(\rho)$ converges absolutely, we obtain the uniform convergence of $\sum_\rho y^{1-\rho}\Gamma(\rho)$. This implies that

$$\lim_{y\to 0+0} \sum_{\rho} y^{1-\rho}\Gamma(\rho) = 0$$

and the relation

$$\lim_{y\to 0+0} \sum_{n=1}^{\infty} \Lambda(n)e^{-ny} = 1$$

results. It suffices now to apply Theorem 6.13 to obtain

$$\lim_{N\to\infty} \frac{1}{N} \sum_{n=1}^{N} \Lambda(n) = 1$$

which implies the Prime Number Theorem. □

A variant of the Hardy-Littlewood tauberian theorem was proved by M.Baran (1991) who utilized it to give another proof of the Prime Number Theorem.

5. It came as a surprise when it turned out that one can formulate and prove tauberian results similar to Theorems 6.11 and 6.15 without any growth conditions. This was discovered by S.Ikehara (1931) who applied Wiener's[29] method in handling Fourier integrals to obtain the following result:

Theorem 6.17. *If*

$$f(s) = \sum_{n=1}^{\infty} a_n n^{-s}$$

is a Dirichlet series with nonnegative coefficients, convergent in the half-plane Re $s > 1$ *and, with a certain positive* A, *the function*

$$g(s) = f(s) - \frac{A}{s-1}$$

---

[28]This equality was used for the first time in Hardy, Littlewood (1918). For a more precise version of it see Wigert (1920).

[29]Wiener, Norbert (1894–1964), Professor at the Massachusetts Institute of Technology.

can be continued to a function regular in the closed half-plane Re $s \geq 1$ *then*

$$A(x) = \sum_{n \leq x} a_n = (A + o(1))x.$$

Actually Ikehara established a more general result, dealing with functions of the form

$$f(s) = \int_0^\infty e^{-st} d\alpha(t),$$

where $\alpha(t)$ is nondecreasing in $[0, \infty)$ and the integral converges in the half-plane Re $s > 1$. If with a suitable $A > 0$ the difference

$$f(s) - \frac{A}{s-1}$$

can be continued to a function regular in Re $s \geq 1$, then

$$\lim_{t \to \infty} \alpha(t) e^{-t} = A$$

holds.

The original proof (which earned a very critical review by Heilbronn in Zentralblatt f.Mathematik) and the proof given in Wiener's classical paper (Wiener 1932) were both rather complicated, but soon a much simpler proof was found by S.Bochner (1933). Further simplifications were introduced in Heilbronn, Landau (1933a), Landau (1932a,b) and it is this form that usually appears in textbooks (see e.g. Chandrasekharan (1968), Knopfmacher (1975) and Lang (1964)).

Other proofs of Ikehara's theorem and/or its generalizations may be found in Delange (1954), Diamond (1975), Ellison (1975), Ingham (1935), Raikov (1938), Tenenbaum (1990a) and Thurnheer (1980).

We present here the argument given by E.Landau (1932b), which seems to be the simplest available:

Proof. Let

$$H(y) = e^{-y} \sum_{n \leq e^y} a_n.$$

We have to establish

$$\lim_{y \to \infty} H(y) = A$$

and it is clear that we may assume $A = 1$ replacing, if necessary, the function $f(s)$ by $f(s)/A$.

Observe first that for Re $s > 1$ one has the identity

$$\int_0^\infty H(y) e^{-(s-1)y} dy = \frac{1}{s} f(s).$$

Indeed, the left-hand side equals

$$\int_0^\infty A(e^y)e^{-sy}dy = \int_1^\infty A(u)\frac{du}{u^{1+s}}$$

and it remains to apply Lemma 4.12. Using now

$$\int_0^\infty e^{-(s-1)y}dy = \frac{1}{s-1}$$

we get for Re $s > 1$ the equality

$$\int_0^\infty (H(y)-1)e^{-(s-1)y}dy = \frac{g(s)-1}{s}.$$

This gives for $\sigma > 1$ and every positive a

$$\frac{1}{2}\int_{-2a}^{2a} e^{yti}\left(1-\frac{|t|}{2a}\right)\frac{g(\sigma+it)-1}{\sigma+it}dt$$
$$= \frac{1}{2}\int_0^\infty (H(u)-1)e^{-(\sigma-1)u}du\int_{-2a}^{2a}\left(1-\frac{|t|}{2a}\right)e^{t(y-u)i}dt$$
$$= \int_0^\infty H(u)e^{-(\sigma-1)u}\frac{\sin^2(a(y-u))}{a(y-u)^2}du - \int_0^\infty e^{-(\sigma-1)u}\frac{\sin^2(a(y-u))}{a(y-u)^2}du.$$

Letting σ tend to 1 we get

$$\frac{1}{2}\int_{-2a}^{2a} e^{yti}\left(1-\frac{|t|}{2a}\right)\frac{g(1+it)-1}{1+it}dt$$
$$= \int_0^\infty H(u)\frac{\sin^2(a(y-u))}{a(y-u)^2}du - \int_0^\infty \frac{\sin^2(a(y-u))}{a(y-u)^2}du$$
$$= \int_{-\infty}^{ay} H(y-\frac{v}{a})\frac{\sin^2 v}{v^2}dv - \int_{-\infty}^{ay}\frac{\sin^2 v}{v^2}dv$$

For y tending to infinity the left hand-side of this equality tends to zero due to the theorem of Riemann-Lebesgue and the second integral in the last term tends to π, hence we arrive at

$$\lim_{y\to\infty}\int_{-\infty}^{ay} H(y-\frac{v}{a})\frac{\sin^2 v}{v^2}dv = \pi. \tag{6.31}$$

(For our purposes it suffices merely to know that this limit is positive.) Observing that $y_2 \geq y_1$ implies $H(y_2) \geq H(y_1)e^{y_1-y_2}$ we can now write

$$\pi \geq \limsup_{y\to\infty}\int_{-\sqrt{a}}^{\sqrt{a}} H(y-\frac{v}{a})\frac{\sin^2 v}{v^2}dv$$
$$\geq \limsup_{y\to\infty}\int_{-\sqrt{a}}^{\sqrt{a}} H(y-\frac{\sqrt{a}}{a})e^{-2/\sqrt{a}}\frac{\sin^2 v}{v^2}dv.$$

Thus we get

$$\limsup_{y\to\infty} H(y) \le \pi \frac{e^{2/\sqrt{a}}}{\int_{-\sqrt{a}}^{\sqrt{a}} \frac{\sin^2 v}{v^2} dv}$$

and letting a tend to infinity we arrive at

$$\limsup_{y\to\infty} H(y) \le 1. \tag{6.32}$$

This implies in particular the boundedness of $H(y)$, so we may write $H(y) \le C$ with a suitably chosen constant C. Choose again $a > 0$, put $b = \sqrt{a}$ and observe that we have

$$\begin{aligned} &\int_{-\infty}^{ay} H(y - \frac{v}{a}) \frac{\sin^2 v}{v^2} dv \\ \le &\int_{-\infty}^{-b} + \int_{-b}^{b} + \int_{b}^{\infty} H(y - \frac{v}{a}) \frac{\sin^2 v}{v^2} dv = I_1 + I_2 + I_3. \end{aligned}$$

Now

$$|I_1| + |I_3| \le 2C \int_b^\infty \frac{\sin^2 v}{v} dv \le \frac{2C}{b^2}$$

and thus using (6.31) we get

$$\pi \le \liminf_{y\to\infty} \int_{-b}^{b} H(y + \frac{b}{a}) e^{2b/a} \frac{\sin^2 v}{v^2} dv + \frac{2C}{b}$$

which implies

$$\liminf_{y\to\infty} H(y) \ge \frac{\pi - 2C/b}{\int_{-\sqrt{a}}^{\sqrt{a}} \frac{\sin^2 v}{v^2} dv} e^{2b/a}.$$

And since for a tending to infinity the right hand-side of the last equality tends to 1 we get finally

$$\liminf_{y\to\infty} H(y) \ge 1$$

which jointly with (6.32) gives the assertion. □

Proof of the Prime Number Theorem. Observe that the assumptions of Ikehara's theorem are satisfied with $A = 1$ by

$$f(s) = -\frac{\zeta'}{\zeta}(s) = \sum_{n=1}^{\infty} \frac{\Lambda(n)}{n^s}$$

hence

$$\psi(x) = \sum_{n\le x} \Lambda(n) = (1 + o(1))x$$

results and this is equivalent to the Prime Number Theorem. □

A modified version of the above proof appears in Errera (1956,1958). The proof of the Prime Number Theorem given in Čižek (1981) is a variant of Ikehara's approach and provides the evaluation $O(x\log^{-N} x)$ (with arbitrary N) of the error term (cf. Grant (1987a,b). See also Wiener, Geller (1950) where a variant of Ikehara's Theorem is used to deduce the non-vanishing of $\zeta(s)$ on Re $s=1$.

Sylvester's way of deducing the relation $\pi(x)\log x/x \to 1$ from the asymptotic equality of $\theta(x)$ and x can be used to obtain the following result, which is actually a particular case of Delange's generalization of Theorem 6.17 (Delange (1954), see also Narkiewicz (1983, Corollary to Theorem 3.10)), and from which Prime Number Theorem follows immediately if one recalls that equality (5.11) gives the regularity of the difference

$$\sum_p \frac{1}{p^s} - \log\left(\frac{1}{1-s}\right)$$

in the closed half-plane Re $s \ge 1$.

Corollary. *If for* Re $s>1$ *one has*

$$f(s) = \sum_{n=1}^{\infty} \frac{a_n}{n^s} = A\log\left(\frac{1}{s-1}\right) + g(s)$$

with nonnegative coefficients a_n *and* $g(s)$ *regular in the closed half-plane* Re $s\ge 1$ *then*

$$\sum_{n\le x} a_n = (1+o(1))\frac{x}{\log x}.$$

Proof. The assumptions imply for Re $s>1$ the equality

$$-f'(s) = \sum_{n=1}^{\infty} \frac{a_n\log n}{n^s} = \frac{A}{s-1} - g'(s)$$

and since the derivative of g is regular in Re $s\ge 1$, Ikehara's theorem gives

$$\sum_{n\le x} a_n\log n = (A+o(1))x.$$

Now on one hand we have

$$\liminf_{x\to\infty} \frac{\log x\sum_{n\le x} a_n}{x} \ge \lim_{x\to\infty} \frac{\sum_{n\le x} a_n\log n}{x} = A.$$

Onn the other hand for every $\delta<1$ we have

$$\sum_{n\le x^\delta} a_n \le \sum_{n\le x^\delta} a_n\log n = O(x^\delta)$$

and

$$\sum_{x^\delta<n\le x} a_n \le \frac{\sum_{n\le x} a_n \log n}{\log(x^\delta)} = (\frac{A}{\delta}+o(1))\frac{x}{\log x}$$

which leads to

$$\limsup_{x\to\infty} \frac{\log x \sum_{n\le x} a_n}{x} \le \frac{A}{\delta}.$$

The assertion follows by letting δ tend to 1. □

Later, other ways of deducing the Prime Number Theorem from various kind-s of tauberian results were found. One of them was given by N.Wiener (1932) and a special case of Wiener's tauberian theorem, which suffices for the proof of the Prime Number Theorem, appears in Levinson (1973). Another proof based on Wiener's tauberian theorem appears in Ingham (1945). Cf. Gordon (1958) and Segal (1969a,b). For application of Wiener's theorem in the case of primes in progressions see Glatfeld (1957,1969).

We mention here another tauberian theorem due to E.Landau (1908e,1912b) which in its simplest form states that if $a_n \ge 0$ and the summatory function $F(x) = \sum_{n\le x} a_n$ satisfies

$$\sum_{n\le x} F(x/n) = x\log x + Ax + O(\log x)$$

then

$$\lim_{x\to\infty} \frac{F(x)}{x} = 1.$$

This theorem permits us to deduce the Prime Number Theorem in the form $\psi(x) = (1+o(1))x$ from Lemma 3.3 (ii) and Stirling's formula. Later, A.Balog (1981) proved a version of Landau's result which provides a non-trivial error term.

6.5. Zeros of the Zeta-function

1. The explicit formula of Riemann-Mangoldt (Theorem 5.4) demonstrates that the distribution of primes depends heavily on the non-trivial zeros of the zeta-function. This observation spurted the study of these zeros[30] After von Mangoldt's first significant result in that respect, giving asymptotics for the number of zeros of $\zeta(s)$ in horizontal strips (Theorem 5.2), there were for a long time only scattered numerical results, which we have already mentioned: J.P.Gram (1895) found the first three and then (Gram 1903) the first ten zeros (cf. Vallée-Poussin 1899, Lindelöf 1903). This was extended by R.J.Backlund (1912) who computed the first 58 zeros and noted that they all lie on the critical line. Later (Backlund 1914,1916) he did this for all zeros ρ

[30]Note that much less effort was spent on calculation of non-real zeros of Dirichlet's L-functions. In particular there are only a few results (see e.g. Davies 1965, Davies, Haselgrove 1961, McCurley 1984c, Rumely 1993 and Spira 1969.) asserting that certain L-functions do not have zeros in some horizontal strips to the right of the line Re $s = 1/2$.

with Im $\rho \le 100$, resp. Im $\rho \le 200$. In 1912 H.Bohr (1912c) gave a new proof of the fact that $\zeta(s)$ has infinitely many zeros in the critical strip $0 \le \text{Re } s \le 1$ not utilizing Hadamard's theory of entire functions.

The first step in the direction of Riemann's Hypothesis was taken in 1914 in a joint paper of H.Bohr and E.Landau (Bohr, Landau 1914a) who established that the majority of non-trivial zeros of $\zeta(s)$ lie very close to the line Re $s = 1/2$. They proved the following result:

Theorem 6.18. *If for a positive δ we denote by $N_\delta(T)$ the number of zeros of $\zeta(s)$ in the rectangle $|\text{Im } s| \le T$, $1/2 + \delta < \text{Re } s < 1$ then*

$$N_\delta(T) = O(T),$$

the implied constant depending on δ.

In view of Theorem 5.2 this shows that almost all zeros of $\zeta(s)$ lie arbitrarily close to the critical line.

The proof is based on two results of W.Schnee concerning Dirichlet series, which are immediate in the case when x lies in the half-plane of absolute convergence of the series:

Lemma 6.19. (Schnee 1909a) *Assume that the series*

$$f(s) = \sum_{m=1}^{\infty} \frac{a_m}{m^s}$$

converges for Re $s > \alpha$.

(i)If $x > \alpha$ then for every integer $m > 0$ one has uniformly in m

$$\lim_{T\to\infty} \frac{1}{2T} \int_{-T}^{T} f(x+it) m^{it} dt = \frac{a_m}{m^x}.$$

(ii) If $x > a + 1/2$ then

$$\lim_{T\to\infty} \frac{1}{2T} \int_{-T}^{T} |f(x+iy)|^2 dy = \sum_{m=1}^{\infty} \frac{|a_m|^2}{m^{2x}}.$$

The case $m = 1$ of (i) appears earlier in Landau (1908b,§§14,15) and a proof in the general case can be found in Landau (1909a, Satz 35 and 41).

Corollary. (Bohr, Landau 1914a) *If the series*

$$f(s) = \sum_{m=1}^{\infty} \frac{a_m}{m^s}$$

converges in the half-plane Re $s > 0$ *and $\epsilon > 0$ and $A > 1$ are given, then*

$$\int_{1/2+\epsilon}^{A} dx \int_{-T}^{T} |f(x+iy)|^2 dy = O(T).$$

Proof. Obviously it suffices to obtain for $x \in [1/2+\epsilon, A]$ the inequality

$$\int_{-T}^{T} |f(x+iy)|^2 dy \leq BT \tag{6.33}$$

with some $B = B(\epsilon, A)$ not depending on x, so a uniform version of part (ii) of the last lemma would be helpful. This was obtained by Bohr and Landau in the following way:

Putting $g(s) = \sum_{m=1}^{\infty} \bar{a}_m m^{-s}$ one gets

$$\frac{1}{2T}\int_{-T}^{T} f(\epsilon+it)g(2x-\epsilon-it)dt - \sum_{m=1}^{\infty} \frac{|a_m|^2}{m^{2x}}$$

$$= \sum_{m=1}^{\infty} \frac{\overline{a_m}}{m^{2x-\epsilon}} \left(\frac{1}{2T}\int_{-T}^{T} m^{it} f(\epsilon+it)dt - \frac{a_m}{m^{\epsilon}} \right).$$

The justification of the interchange of summation and integration here is a consequence of the fact, proved by H.Bohr (1913b), that if a Dirichlet series has its abscissa of convergence equal to α, then its abscissa of uniform convergence does not exceed $\alpha + 1/2$. This shows that in our case the series $g(2x-\epsilon-it)$ converges uniformly.

Part (i) of the preceding lemma implies the uniform convergence to zero of the right hand-side of the last equality and thus it suffices to show that

$$\lim_{T\to\infty} \frac{1}{2T} \left(\int_{-T}^{T} f(\epsilon+it)g(2x-\epsilon-it)dt - \int_{-T}^{T} f(x+it)g(x-it)dt \right) = 0,$$

holds uniformly for $x \in [1/2+\epsilon, A]$. This is easily established by considering the integral of the product $f(s)g(2x-s)$ over the rectangle with vertices $\epsilon \pm iT$ and $x \pm iT$. □

Proof of Theorem 6.18. Consider the function

$$f(s) = (1-2^{1-s})\zeta(s) = \sum_{n=1}^{\infty} \frac{(-1)^{n+1}}{n^s},$$

the Dirichlet series being convergent in the half-plane Re $s > 0$. For any $x_0 > 1$ put

$$A = A(x_0) = 2 \inf_{\mathrm{Re} s = x_0} |f(s)|$$

and observe that A is positive. For any integer τ let K_τ be the circle with center $s_0 = x_0 + (1/2+\tau)i$ and radius $r = |x_0 - 1/2 - \delta + i/2|$. This circle passes through the points $1/2+\delta+\tau i$ and $1/2+\delta+(1+\tau)i$ and if we choose x_0 in a suitable manner then it will lie to the right of the line Re $s=(1+\delta)/2$. Now put $R=x_0-(1+\delta)/2$. Corollary 2 to Theorem 6.3 implies that the number of zeros of $f(s)$ in the rectangle $1/2+\delta \le \text{Re } s \le x_0$, $\tau < \text{Im } s \le \tau+1$ is

$$O\left(\int\int_{|x+iy-s_0|\le R} |f(x+iy)|^2 dxdy\right)$$
$$= O\left(\int_{(1+\delta)/2}^{x_0+R} dx \int_{\tau+1/2-R}^{\tau+1/2+R} |f(x+iy)|^2 dy\right).$$

Summing over integers $\tau \in [-T,T]$ we arrive finally at

$$n(T) = O\left(\int_{(1+\delta)/2}^{x_0+R} dx \int_{-T+1/2-R}^{T+1/2+R} |f(x+iy)|^2 dy\right)$$

and Corollary to Lemma 6.19 gives $n(T)=O(T)$. It suffices now to observe that every non-trivial zero of $\zeta(s)$ is also a zero of $f(s)$. □

The same argument is applicable also to Dirichlet's L-functions. Later it turned out (Bohr, Landau 1914b) that a modification of the above proof permits to replace the evaluation $O(T)$ in the assertion of Theorem 6.18 by $o(T)$, and the same holds also for L-functions. Much later N.Levinson (1975b) showed that the analogue of Theorem 6.18 holds also for solutions of the equation $\zeta(s)=a$ for arbitrary a. Earlier H.Bohr (1914b,1915) proved that the zeta-function attains every non-zero value infinitely often in every strip $1/2<\alpha<\text{Re } s<\beta<1$ and the same applies (Bohr 1911b,1912a) to every strip $1<\text{Re } s<1+\delta$ (for weaker earlier results see Bohr, Landau (1910), Bohr (1911a)). This underlines the special role played by the zeros of $\zeta(s)$. Note that the situation is completely different for the logarithmic derivative of $\zeta(s)$, which attains every complex value without exceptions infinitely often in the half-plane Re $s>1$ (Bohr 1912b).

2. The next step was done by G.H.Hardy (1914) who established the existence of infinitely many zeros of $\zeta(s)$ on the critical line Re $s=1/2$. Hardy's proof started with the formula (6.30) in the case $n=1$ which leads to

$$1+2\sum_{n=1}^{\infty} e^{-n^2y} = 1+\frac{1}{\pi i}\int_{a-i\infty}^{a+i\infty} \Gamma(s)y^{-s}\zeta(2s)ds$$

for $a>1$. Moving the integral to the line Re $s=1/4$ and using Cauchy's theorem as well as Riemann's formula (4.10) one gets a rather complicated expression from which, after using a lemma from the theory of elliptic functions, one gets for integral p the relation

$$\lim_{\alpha\to\pi/2}\int_0^{\infty}\frac{(e^{\alpha t}+e^{-\alpha t})t^{2p}\xi(1/2+2it)}{1/4+4t^2}dt=(-1)^p\frac{\pi}{4^{2p}}\cos(\frac{\pi}{8}).$$

If the zeta-function were to have only finitely many zeros on the critical line, then the function $\xi(1/2+2it)$ would be of constant sign for all large real values of t and this would give a contradiction, as the sign of the right hand-side of the last formula depends on the parity of p.

A variant of Hardy's argument appears in Bouliguine (1914). See Mellin (1917) for a simplification.

E.Landau (1915a) discussed and modified Hardy's proof so that it could be applied also to Dirichlet's L-functions (the original method was inapplicable to the case of L-functions corresponding to even characters (i.e. $\chi(-1)=1$) with respect to an even modulus) and to certain Epstein zeta-functions. A streamlined version of Hardy's proof, which in particular avoids any reference to elliptic functions, appears in Titchmarsh (1951, Chap.10), where also several other proofs due to M.Fekete (1926), G.Pólya (1927), E.Landau (1927a, Vol.II), and E.C.Titchmarsh (1934) may be found. Another proof was given by N.Levinson (1940). A related result was obtained by J.R.Wilton (1916) who showed that for any fixed $x\in(0,1)$ both Re $\zeta(s)$ and Im $\zeta(s)$ have infinitely many zeros on the line Re $s=x$ (cf. Berlowitz 1967/68, Berndt 1969 and Levinson 1971).

3. The number of zeros $1/2+it$ $(0<t\le T)$ of $\zeta(s)$ is usually denoted by $N_0(T)$. Thus Hardy's result shows that $N_0(T)$ tends to infinity and the truth of Riemann's Hypothesis jointly with Theorem 5.2 would imply that $N_0(T)$ is asymptotic to $T\log T/2\pi$. The first non-trivial lower bound for $N_0(T)$ was obtained by E.Landau (1915a) who established the inequality $N_0(T)\ge c_1\log\log T$ with a certain positive c_1. Later C.de la Vallée-Poussin (1916) utilized Hardy's idea to show that the ratio $N_0(T)/\sqrt{T}$ does not tend to zero and G.H.Hardy and J.E.Littlewood (1918,1921) proved for large T at first $N_0(T)\ge c_2T^a$ for every $a<3/4$ and then $N_0(T)\ge c_3T$ with certain positive $c_2=c_2(a)$ and c_3. For a proof of the last result see Titchmarsh (1951, Chap.10.9). The next improvement was made by A.Selberg (1942) who showed that a positive fraction of all roots of the zeta-function lies on the critical line, i.e. $N_0(T)\ge AN(T)$ holds with a certain positive A. (For other proofs see Titchmarsh (1947) and Moser (1987)). The first explicit value of A was, according to Titchmarsh (1951), calculated in the thesis of S.H.Min and was rather small. Later N.Levinson (1974,1975a,1975c) showed that at least for large T one can take $A=0.3474$, which was subsequently improved to 0.35 (Lou,Yao 1980) (cf.Lou 1981), 0.3658 (Conrey 1983) and 0.4 (Conrey 1989). An outline of the proof of Levinson's result appears in Bombieri (1976) and in the second edition of Titchmarsh (1951). Cf. Conrey,Ghosh (1985), where a simplification of the proof is given. (A similar result for zeros of L-functions was obtained by T.Hilano (1976,1978) who showed that at least 1/3 of zeros of an L-function lie on the critical line). Note also that D.R.Heath-Brown (1979a) showed that at least $0.3474N(T)$ roots $\rho = 1/2 + it$ ($0 < t \leq T$) are simple, R.J.Anderson (1983) improved this to $0.3532N(T)$ and H.L.Montgomery (1973) obtained $cN(T)$ for any $c < 2/3$. Under Riemann's Hypothesis one can even assert that more than two third roots of $\zeta(s)$ are simple (Montgomery 1974) and Cheer, Goldston 1993). Later it was shown in Conrey, Ghosh, Gonek (1989) that at least $0.955N(T)$ roots ρ with $0 < \text{Im}\,\rho \leq T$ are either simple or double. For a simple proof of the existence of infinitely many non-real simple zeros of $\zeta(s)$ see Conrey, Ghosh, Gonek (1988). It was shown in Farmer (1995) that there are at least $0.63952N(T)$ distinct roots of $\zeta(s)$ with $0 < \text{Im}\,\rho \leq T$.

For a long time the largest known zero-free region for Riemann's zeta-function had been that obtained by de la Vallée-Poussin in his proof of the Prime Number Theorem, namely

$$\{\sigma + it : \ \sigma > 1 - \frac{c}{\log|t|}, |t| \geq c_0\}$$

with certain positive constants c and c_0 (see Theorem 5.18).

The first essential improvement was obtained by J.E.Littlewood (1922) who showed that there are no zeros of $\zeta(s)$ in the region Ω defined by the inequalities

$$\sigma \geq 1 - c\frac{\log\log t}{\log t}, \quad t \geq t_0$$

with suitable positive c and t_0. A simplification of the proof was made by E.Landau (1924a). Both proofs are based on the bound

$$|\zeta(s)| \leq \log^B t \quad (t \geq 3, \ \sigma \geq 1 - \frac{\log\log^2 t}{\log t})$$

obtained by Hardy and Littlewood with the use of rather heavy tools. This bound was obtained in a simpler way by Landau (1924b) (reproduced in the second volume of Landau (1927a)). The main point of the argument is a clever use of Weyl's estimation of trigonometrical sums (Weyl 1916), which shortly before had led H.Weyl (1921) to the evaluation

$$\zeta(1+it) = O\left(\frac{\log t}{\log\log t}\right).$$

4. The largest manual calculation of the zeros ρ of $\zeta(s)$ was made by E.C.Titchmarsh (1935) who verified Riemann's Hypothesis for zeros ρ in the range $|\text{Im}\,\rho| < 1468$. (Earlier this had been done for $|\text{Im}\,\rho| < 300$ in Hutchinson (1925).) The advent of computers essentially speeded up this task. A.Turing[31] (1953) was the first to use them. He found more than 1100 zeros, all on the critical line. Then D.H.Lehmer (1956a,b) covered the range

[31] Turing, Alan Mathison (1912–1954).

$0 < \text{Im}\,\rho < 21\,943$, finding 25 000 zeros there and N.A.Meller (1958) found the first 35 537 zeros. A list of the first 1600 zeros can be found in Haselgrove, Miller (1960), where also several values of the zeta-functions are tabulated. The first quarter of a million of zeros was found by R.S.Lehman (1966) ($|\text{Im}\,\rho| < 170\,571.3$) and the next record was achieved by J.B.Rosser, J.M.Yohe, L.Schoenfeld (1968) ($3.5 \cdot 10^6$ zeros). A 28-digit table of the first 15000 zeros was published by H.J.J.te Riele (1979b). R.P.Brent (1979) showed that the first 75 000 001 non-trivial zeros (having their imaginary parts smaller than 32 585 736.4) are simple and lie on the critical line. This was later superseded by J.van der Lune, H.J.J.te Riele and D.T.Winter (1986) who found the first $1.5 \cdot 10^9$ zeros (none outside $\text{Re}\, s = 1/2$) and hold the current record.

A fast algorithm for finding the zeros can be obtained using the methods of evaluating $\zeta(s)$ proposed in Odlyzko, Schönhage (1988).

It was established by A.E.Ingham (1937) (for another proof see Turán (1954)) that Lindelöf's hypothesis implies for every $\epsilon > 0$ and $1/2 \le \sigma \le 1$ the bound

$$N(\sigma, T) = O(T^{(2+\epsilon)(1-\sigma)}) \tag{6.34}$$

for the number $N(\sigma, T)$ of zeros of $\zeta(s)$ lying in the rectangle $\text{Re}\, s \ge \sigma$, $0 < \text{Im}\, s \le T$. He obtained this by proving the inequality

$$\mu_1(\sigma) \le 2 + 4\mu(1/2),$$

where $\mu(\sigma)$ is Lindelöf's function and $\mu_1(\sigma)$ denotes the largest lower bound for exponents μ for which

$$N(\sigma, T) = O(T^{\mu}).$$

Ingham's result shows that the first bound for $\mu_1(\sigma)$ obtained by F.Carlson (1920), namely

$$\mu_1(\sigma) \le 4\sigma(1 - \sigma),$$

improved later by E.C.Titchmarsh (1930b) to

$$\mu_1(\sigma) \le 4\sigma(1 - \sigma)/(3 - 2\sigma),$$

is a consequence of Lindelöf's bound for $\zeta(s)$.

Ingham also proved $\mu_1(\sigma) \le 1 + 2\sigma$ and later (Ingham 1940) obtained $\mu_1(\sigma) \le 3/(2 - \sigma)$. A.Selberg (1946) got the bound $\mu_1(\sigma) \le 1 - (\sigma - 1/2)/4$ which improves Ingham's result for σ close to 1.

The inequality (6.34) is called *the density hypothesis* for Riemann's zeta-function. It can be used to prove results which are unattainable unconditionally. An example is furnished by the theorem of Lu Ming Gao (1985), who used the density hypothesis to deduce that for almost all n one has

$$p_{n+1} - p_n = O(n^{1/6} \log^c n)$$

with a certain c slightly larger than 21, p_n being the nth prime.

The truth of (6.34) for $\sigma \geq a$ with a certain $a < 1$ as well as its analogue for L-functions was proved by G.Halász and P.Turán (1969,1970). The value of a was later reduced to 0.9 (Montgomery 1969b), $5/6 = 0.833\ldots$ (Huxley 1972b), $21/26 = 0.8076\ldots$ (Ramachandra 1975a), 0.8079 (Forti,Viola 1973), 0.80118 (Huxley 1975a), 0.8 (Huxley 1975b), $43/54 = 0.79629\ldots$ (Jutila 1976) and $11/14 = 0.7857\ldots$ (Jutila 1977b). The current record is held by D.R.Heath-Brown (1979b) who proved the truth of the density hypothesis for $\sigma \in [3/4, 1]$.

The sharpest bound for $N(\sigma, T)$ of the form $O(T^{\alpha(1-\sigma)} \log^{\gamma} T)$ valid for $1/2 \leq \sigma \leq 1$ is due to S.A.Gritsenko (1994) who got $\alpha = 2.4$, $\beta = 18.2$ improving thus his previous result (Gritsenko 1992) ($\beta = 33.5$) and the result of M.N.Huxley (1972b) ($\beta = 44$). See also Forti, Viola (1973) and Karatsuba (1991). In particular one has

$$\mu_1(\sigma) \leq 2.4(1 - \sigma).$$

Previously H.L.Montgomery (1969b) had $\mu_1(\sigma) \leq 2.5(1-\sigma)$ (cf. Ivić 1979/80).

It was shown by G.Hálasz and P.Turán (1969) that Lindelöf's hypothesis implies for $\sigma > 3/4$ the equality $\mu_1(\sigma) = 0$. Another proof follows from an inequality relating $\mu(\sigma)$ and $\mu_1(\sigma)$ proved by E.Bombieri (1969).

The analogue of the density hypothesis for Dirichlet's L-functions is the relation

$$\sum_{\chi \bmod q} N(\sigma, T, \chi) = O\left((qT)^{2(1-\sigma)+\epsilon}\right),$$

$N(\sigma, T, \chi)$ denoting the number of zeros of $\prod_{\chi \bmod q} L(s, \chi)$ in the rectangle $\mathrm{Re}\, s \geq \sigma$, $|\mathrm{Im}\, s| \leq T$. It was shown to be true for $\sigma \geq 0.9$ (Montgomery 1971), $\sigma \geq 5/6 = 0.83333\ldots$ (Balasubramanian, Ramachandra 1974, Huxley 1973a, Jutila 1972), $\sigma \geq 21/26 = 0.8076\ldots$ (Jutila 1977b), $\sigma \geq 0.8$ (Huxley, Jutila 1977) and the best result, viz. $\sigma \geq 15/19 = 0.78947\ldots$ is due to D.R.Heath-Brown (1979b).

A short history of early bounds for $\sum_{\chi \bmod q} N(\sigma, T, \chi)$ is given in Montgomery (1971, Chap.12). These bounds are important in various questions concerning primes. In particular, if for every $\epsilon > 0$ one has

$$\sum_{\chi \bmod q} N(\sigma, T, \chi) = O\left((qT)^{\omega(1-\sigma)+\epsilon}\right)$$

uniformly in $1/2 \leq \sigma < 1$, $T \geq 1$, $q \geq 2$ then for the least prime $p(q, l)$ in the progression $l \bmod q$ with q being a power of a fixed prime p and $(l, p) = 1$ one has the evaluation

$$p(q, l) = O(q^{\omega+\epsilon})$$

with the implied constant depending only on p and ϵ. In Montgomery (1969b,1971) the value $\omega = 2.5$ was obtained and it was improved in Jutila (1972) to $\omega = 3(9 + \sqrt{17}/16) = 2.46058\ldots$.

There are similar results for

$$S(\sigma, T, Q) = \sum_{q \le Q} \sum_{\substack{\chi \bmod q \\ \chi \text{ primitive}}} N(\sigma, T, \chi),$$

the density hypothesis in this case having the form

$$S(\sigma, T, Q) = O\left((Q^2T)^{2(1-\sigma)+\epsilon}\right).$$

It is known to hold for $\sigma \ge 7/9 = 0.777\ldots$ (Jutila 1977b). (Cf. also Bombieri (1965), Heath-Brown (1979d,e), Huxley (1976), Jutila (1969/70), Montgomery (1969a,b), Gallagher (1970) and Pintz (1988).

The reader is referred to Ivić (1983,1985) or the second edition of Titchmarsh (1951) for a thorough exposition of questions concerning the growth of $\zeta(s)$ and the distribution of its zeros.

6.6. The sign of $\pi(x) - \mathrm{li}(x)$

1. One of the assertions made by Riemann in his memoir implies the inequality $\pi(x) > \mathrm{li}(x)$ for all sufficiently large integers. Although this inequality is satisfied in a very long initial interval, it is nevertheless wrong. This was deduced by E.Schmidt (1903) (see Corollary 2 to Theorem 5.22) from the negation of Riemann's Hypothesis and in the opposite case this is an immediate corollary of the following result of J.E.Littlewood (1914), a detailed proof of which appeared later in Hardy, Littlewood[32] (1918). We present a sketch of the original argument. Full details, with some simplifications, are provided in Prachar (1957, Chap. 7, Theorems 8.3, 8.4).

Theorem 6.20. *(i) One has*

$$\psi(x) - x = \Omega_\pm(\sqrt{x} \log\log\log x)$$

and

$$\pi(x) - \mathrm{li}(x) = \Omega_\pm(\sqrt{x}\frac{\log\log\log x}{\log x}).$$

(ii) The difference $\pi(x) - \mathrm{li}(x)$ changes its sign infinitely often.

Proof. (i) Assume first that the Riemann Hypothesis is false, hence there is a zero $\rho = \alpha + it$ of $\zeta(s)$ with $\alpha > 1/2$. Put $\beta = 1/4 + \alpha/2$ and assume that, with a certain positive C, we have

$$\psi(x) - x \le Cx^\beta$$

[32] At the end of that paper we find a curious correction to one of their previous papers, which we quote *verbatim*: "For $o\left\{\sqrt{\frac{1}{1-r}}\right\}$ read $o\left\{\sqrt{\frac{1}{1-r}}\right\}$."

for all sufficiently large x. Applying Theorem 5.16 to the integral

$$\int_1^\infty \frac{\psi(x) - x - Cx^\beta}{x^{s+1}}\,dx$$

one is easily led to a contradiction and the same result is obtained if we assume that

$$\psi(x) - x \geq Cx^\beta$$

holds with a certain constant C and all large x.

The same method is applicable to the difference $\pi(x) - \mathrm{li}(x)$, hence we may assume that all non-trivial roots of the zeta-function lie on the critical line Re $s = 1/2$. Let $\gamma_1 < \gamma_2 < \dots$ be the imaginary parts of the roots of $\zeta(s)$ lying in the upper half-plane. From the second explicit formula of von Mangoldt (Theorem 5.5) one gets

$$\frac{\psi(x) - x}{\sqrt{x}} = -2\sum_{n=1}^\infty \frac{\sin(\gamma_n \log x)}{\gamma_n} + O(1) \tag{6.35}$$

and Theorem 5.2 implies

$$\gamma_n = g(n) + O(1),$$

with $g(t)$ being the inverse function of

$$\frac{1}{2\pi} t \log t - \frac{1 + \log(2\pi)}{2\pi} t$$

hence

$$g(t) = (2\pi + o(1))\frac{t}{\log t}.$$

A fundamental role in the proof is played by the function $F(s)$ defined in the half-plane Re $s > 0$ by the formula

$$F(s) = \sum_{n=1}^\infty \frac{\exp(-\gamma_n s)}{\gamma_n}.$$

Its connection with our problem is due to the observation that for every fixed large x the formula (6.35) implies

$$\lim_{\sigma\to 0} \operatorname{Im} F(\sigma + i\log x) = -\sum_{n=1}^\infty \frac{\sin(\gamma_n \log x)}{\gamma_n} = \frac{1}{2}\frac{\psi(x) - x}{\sqrt{x}} + O(1).$$

2. The main properties of $F(s)$ are expressed in the following auxiliary result:

Lemma 6.21. *(i) There exist two sequences* $1 < t_1 < t_2 < \dots$ *and* $1 < u_1 < u_2 < \dots$*, both tending to infinity, two sequences* $\sigma_1, \sigma_2, \dots$ *and* $\tau_1, \tau_2, \dots$ *in* $(0,1)$ *and a positive constant c, such that*

$$-\mathrm{Im}\ F(\sigma_j + it_j) \geq c \log\log t_j \quad (j = 1, 2, \dots)$$

and

$$-\mathrm{Im}\ F(\tau_j + iu_j) \leq -c \log\log u_j \quad (j = 1, 2, \dots).$$

In particular one has

$$\mathrm{Im}\ F(\sigma + it) = \Omega_\pm(\log\log t)$$

for $\sigma \in (0, 1]$ *and* $t \longrightarrow \infty$.

(ii) If one defines

$$\mathrm{Im}\ F(it) = \lim_{\sigma \to 0} \mathrm{Im}\ F(\sigma + it)$$

then the assertions of (i) hold also for $\sigma_j = \tau_j = 0$.

Proof. (i) The first step of the proof consists in obtaining the asymptotic relation

$$-\mathrm{Im}\ \ F(\sigma(1+i)) = \left(\frac{1}{8} + o(1)\right) \log \sigma \tag{6.36}$$

for σ tending to zero. For this aim observe that if $n \leq u < n+1$, then the difference $g(u) - g(n)$ is bounded because the derivative of g equals

$$g'(u) = \frac{2\pi}{\log(g(u)/2\pi)} = O(\frac{1}{\log n}) = O(1).$$

Thus

$$\begin{aligned} \frac{\exp(-(1+i)\sigma g(u))}{g(u)} &= \frac{\exp(-(1+i)\sigma\gamma_n + O(\sigma))}{\gamma_n} \left(1 + O\left(\frac{\log n}{n}\right)\right) \\ &= \frac{\exp(-(1+i)\sigma\gamma_n)}{\gamma_n} \left(1 + O(\sigma) + O\left(\frac{\log n}{n}\right)\right), \end{aligned}$$

with the constants implied by the O-symbol being absolute, i.e. not depending on u, n and σ. This equality leads to

$$\begin{aligned} F((1+i)\sigma) &= \int_c^\infty \frac{\exp(-(1+i)\sigma g(u))}{g(u)} du \\ &+ O\left(\int_c^\infty \frac{\exp(-\sigma g(u))}{g(u)} \left(\sigma + \frac{\log u}{u}\right) du\right) + O(1) \end{aligned}$$

with any fixed $c > 0$ and it is now an exercise in integration to obtain

$$F((1+i)\sigma) = \frac{1}{2\pi}\int_{g(c)}^{\infty} \frac{\exp(-(1+i)\sigma g)}{g}(\log g - \log(2\pi))dg + O(1).$$

After taking imaginary parts of both sides of the last equality one obtains

$$\begin{aligned} \mathrm{Im}\ F(\sigma(1+i)) &= -\frac{1}{2\pi}\int_{g(c)}^{\infty} \frac{\exp(-\sigma x)\sin(\sigma x)}{x}\log x dx \\ &+ \frac{\log(2\pi)}{2\pi}\int_{g(c)}^{\infty} \frac{\exp(-\sigma x)\sin(\sigma x)}{x}dx + O(1) \\ &= -\frac{1}{2\pi}\int_{0}^{\infty} \frac{\exp(-x)\sin x}{x}(\log x - \log\sigma)dx + O(1) \\ &= -\frac{\log\sigma}{2\pi}\int_{0}^{\infty} \frac{\exp(-x)\sin x}{x}dx + O(1) = -\frac{1}{8}\log\sigma + O(1) \end{aligned}$$

and (6.36) follows readily.

Observe moreover that Lemma 6.9 implies that for positive a the sum

$$\sum_{\sigma\gamma_n > a} \frac{\exp(-\gamma_n\sigma)}{\gamma_n}$$

is

$$\begin{aligned} O\left(\sum_{n>[a\sigma^{-1}]} \frac{\exp(-n\sigma)\log n}{n}\right) &= O\left(\int_{a\sigma^{-1}}^{\infty} \frac{\exp(-\sigma u)\log u}{u}du\right) \\ = O\left(\int_{a}^{\infty} \frac{\exp(-u)\log u}{u}du\right) &+ O\left(|\log(\sigma)|\int_{a}^{\infty} \frac{\exp(-u)}{u}du\right) \end{aligned}$$

hence if $a = a(\epsilon)$ is sufficiently large then

$$\sum_{\sigma\gamma_n > a(\epsilon)} \frac{\exp(-\gamma_n\sigma)}{\gamma_n} < \epsilon|\log(\sigma)| \tag{6.37}$$

where $\epsilon > 0$ is arbitrarily given.

The rest of the argument is based on a result in diophantine approximations[33], stating that if $a_1, \dots, a_N$ are given real numbers then for any given positive q and x_0 one can find an integer x such that $x_0 < x < x_0 q^N$ and

$$\| a_j x \| < \frac{1}{q} \quad (j = 1, 2, \dots, N) \tag{6.38}$$

holds, $\| y \|$ denoting the distance of y to the nearest integer.

Choose a small $\sigma \in (0,1)$, put $N = a/\sigma$ with $a = a(1/32)$ and apply Kronecker's result with $a_j = \gamma_j/2\pi$ (for $j = 1, 2, \dots, N$), $q = 1/\sigma$ and a large x_0. This furnishes an integer $x \in (x_0, x_0\sigma^{-N})$ and an integer y_n such that

[33]See Dirichlet 1842b, Kronecker 1884.

$$\left|\frac{x\gamma_n}{2\pi} - y_n\right| < \sigma.$$

Put $\eta = x + \sigma$ and $\omega_n = x\gamma_n - 2\pi y_n$. Because of $|\omega_n| < 2\pi\sigma$ and

$$\sin(\gamma_n\eta) = \sin(\gamma_n\sigma + \omega_n)$$

we get

$$|\sin(\gamma_n\eta) - \sin(\gamma_n\sigma)| < 2\pi\sigma.$$

Now

$$|\text{Im } F(\sigma(1+i)) - \text{Im } F(\sigma + i\eta)| = \left|\sum_{n=1}^{\infty} \frac{\sin(\gamma_n\eta) - \sin(\gamma_n\sigma)}{\gamma_n} \exp(-\gamma_n\sigma)\right|$$

$$\leq \sum_{n<N} \left|\frac{\sin(\gamma_n\eta) - \sin(\gamma_n\sigma)}{\gamma_n}\right| + 2\sum_{n=N}^{\infty} \frac{\exp(-\gamma_n\sigma)}{\gamma_n}$$

$$\leq 2\pi\sigma \sum_{n<N} \frac{1}{\gamma_n} + 2\sum_{n=N}^{\infty} \frac{\exp(-\gamma_n\sigma)}{\gamma_n}. \tag{6.39}$$

As $\gamma_n = (2\pi + o(1))n/\log n$ we see that

$$\sum_{n\leq N} \frac{1}{\gamma_n} = O\left(\sum_{n\leq N} \frac{\log n}{n}\right) = O(\log^2 N)$$

hence the first summand in the last term in (6.39) is

$$O(\sigma \log^2(a/\sigma)) = O(1).$$

To evaluate the second summand observe that $n \geq N$ implies

$$\gamma_n\sigma \geq (1 + o(1))\frac{2\pi N\sigma}{\log N} \geq (1 + o(1))\frac{2\pi a}{\log(a/\sigma)}$$

and, since the last term here exceeds a for sufficiently small σ, (6.37) implies

$$2\sum_{n=N}^{\infty} \frac{\exp(-\gamma_n\sigma)}{\gamma_n} < \frac{|\log\sigma|}{16}$$

for small σ. Hence (6.39) gives (always for sufficiently small σ) the inequality

$$-\text{Im } F(\sigma + i\eta) \geq -\text{Im } F((1+i)\sigma) - \frac{1}{10}|\log\sigma|.$$

Now we may invoke (6.36) to get

$$-\text{Im } F(\sigma + i\eta) \geq \frac{1}{10}|\log\sigma| \tag{6.40}$$

for all sufficiently small σ and suitable η satisfying

$$x_0 + \sigma < \eta < \sigma + x_0 \sigma^{-a/\sigma}.$$

If one would have

$$-\mathrm{Im}\ F(\sigma + i\eta) < \epsilon \log\log \eta$$

then (6.40) would give

$$\frac{1}{10} |\log \sigma| \leq \epsilon \log\log(\sigma + x_0 \sigma^{-a/\sigma}) = O(\epsilon |\log \sigma|)$$

which is contradictory for sufficiently small ϵ.

This establishes the first part of (i) and the second part can be dealt with similarly.

The proof of (ii) uses a modification of the theorem of Phragmén–Lindelöf which in its original form cannot be applied since the function $F(s)$ is not continuous on the line $\mathrm{Re}\ s = 0$. Thus Hardy and Littlewood had to go around this difficulty and showed that if a function f is regular in the region

$$\Omega = \{s : \ 0 < \mathrm{Re}\ s < 1, \mathrm{Im}\ s \geq \theta > 0\},$$

bounded on the boundary of Ω, satisfies in the interior of Ω (with a suitable λ) the inequality

$$f(s) = O(\exp(s^\lambda)) \tag{6.41}$$

and moreover the limit

$$f(it) = \lim_{\sigma \to 0} f(\sigma + it)$$

exists and for every $Y \geq \theta$ and suitable $X = X(Y)$ the ratio

$$\left| \frac{f(\sigma + it)}{f(it)} \right|$$

is uniformly bounded in $\theta \leq \mathrm{Im}\ s \leq Y$, $0 < \mathrm{Re}\ s \leq X$ then f is bounded in Ω.

The proof is not much different from the proof of Theorem 6.5 but involves certain delicate arguments to take care of the behaviour of f near the boundary of Ω.

Fix now $\epsilon > 0$ and assume the bound $-\mathrm{Im}\ F(it) < \epsilon \log\log t$ for large t. The above quoted result should be now applied to the function

$$f(s) = \exp(iF(s))/(\log s)^c$$

with a fixed $c > \epsilon$ in the strip $0 < \mathrm{Re}\, s < 1$, $\mathrm{Im}\ s \geq 2$. To check the assumptions one has to use von Koch's Theorem 5.21, which implies the truth of (6.41) in our case (remember that we work under the assumption of Riemann's Hypothesis).

As a result one gets the boundedness of $f(s)$ in the considered strip and since this implies $\exp(iF(s)) = O(\log^k s)$ the inequality $-\mathrm{Im}\ F(s) < 2k \log\log s$ follows. As k can be made arbitrarily small by a suitable choice of ϵ this contradicts the already proved part (i) and hence the first assertion of (ii) follows. The proof of the second proceeds analogously. □

3. The first part of the assertion (i) of the theorem follows now immediately from part (ii) of the lemma.

To prove the second part of (i) observe that if one puts

$$f(x) = \pi(x) + \sum_{n=2}^{\infty} \frac{1}{n}\pi(x^{1/n})$$

that in view of

$$f(x) = \pi(x) + O(\sqrt{x})$$

and the already proved first part of (i) it suffices to establish

$$f(x) - \mathrm{li}(x) - \frac{\psi(x) - x}{\log x} = O\left(\frac{\sqrt{x}}{\log^2 x}\right). \tag{6.42}$$

To prove this write for integral x

$$\begin{aligned} f(x) &= \sum_{2\leq n\leq x} \frac{\psi(n) - \psi(n-1)}{\log n} \\ &= \sum_{2\leq n\leq x} \frac{1}{\log n} + \sum_{2\leq n\leq x} \frac{\psi(n) - \psi(n-1) - 1}{\log n} \\ &= \mathrm{li}(x) + O(1) + \sum_{2\leq n\leq x} \frac{(\psi(n) - n) - (\psi(n-1) - (n-1))}{\log n} \end{aligned}$$

and use partial summation to deduce

$$\begin{aligned} &f(x) - \mathrm{li}(x) - \frac{\psi(x) - x}{\log x} \\ &= \sum_{2\leq n\leq x} \frac{\psi(n) - n}{n\log^2 n} + O\left(\sum_{2\leq n\leq x} \frac{\psi(n) - n}{n^2\log^2 n}\right) + O(1) \\ &= \sum_{2\leq n\leq x} \frac{\psi(n) - n}{n\log^2 n} + O(1). \end{aligned}$$

At this point the evaluation

$$\sum_{2\leq n\leq x} (\psi(n) - n) = O(x^{3/2}) \tag{6.43}$$

(to which we shall return in a while) is utilized and using again partial summation one gets

$$f(x) - \mathrm{li}(x) - \frac{\psi(x) - x}{\log x} = O\Big(\sum_{2 \le n \le x} \frac{1}{\sqrt{n} \log^2 n}\Big) + O\left(\frac{\sqrt{x}}{\log^2 x}\right) = O\left(\frac{\sqrt{x}}{\log^2 x}\right),$$

thus establishing (6.42).

The proof of (6.43) given by Hardy and Littlewood is only outlined. They consider the Riesz-mean of the sequence $\{\Lambda(n)\}$, defined for positive δ by

$$R^\delta(x) = \sum_{n \le x} \left(1 - \frac{n}{x}\right)^\delta \Lambda(n).$$

Using a formula due to F.Riesz[34] (see Hardy, Riesz 1915, Theorem 40) they obtain, for $c > 1$,

$$R^\delta(x) = -\frac{1}{2\pi i} \int_{c-i\infty}^{c+i\infty} \frac{\Gamma(1+\delta)\Gamma(\delta)}{\Gamma(1+\delta+\rho)} \frac{\zeta'}{\zeta}(s) x^s ds$$

which can be transformed to the form

$$R^\delta(x) = \frac{x}{1+\delta} - \sum_\rho \frac{\Gamma(1+\delta)\Gamma(\delta)}{\Gamma(1+\delta+s)} x^\rho + \Phi\left(\frac{1}{x}\right),$$

where ρ runs over all non-real roots of the zeta-function and $\Phi(t)$ is either a power series convergent inside the unit circle (if δ is not an integer) or a finite sum with at most logarithmic singularities.

Subtracting from $R^\delta(x)$ the Riesz transform of the constant sequence $a_n = 1$ and using the fact that Riemann's Hypothesis implies

$$\sum_\rho \frac{\Gamma(1+\delta)\Gamma(\delta)}{\Gamma(1+\delta+\rho)} x^\rho = O(\sqrt{x})$$

one arrives at

$$\sum_{n \le x} (\Lambda(n) - 1) \left(1 - \frac{n}{x}\right)^\delta = O(\sqrt{x}).$$

The deduction of (6.43) from this relation is not given in the paper but this is very easy. Indeed, choosing $\delta = 1$ and assuming x to be an integer (which does not affect the generality) we see that the left-hand side of the last relation equals

$$\psi(x) - \frac{x}{2} - \frac{1}{x} \sum_{n \le x} n\Lambda(n) + \frac{1}{2}.$$

Since partial summation gives

$$\sum_{n \le x} n\Lambda(n) = x\psi(x) - \sum_{n \le x-1} \psi(n)$$

[34]Riesz, Frigyes (1880–1956), Professor in Cluj, Szeged and Budapest.

we get

$$\sum_{n\leq x}\psi(n)=\frac{x^2}{2}+O(x^{3/2})$$

and this implies immediately (6.43).

The assertion (ii) is now immediate. □

Note that it can be shown that certain mean values of $\pi(x)-\mathrm{li}(x)$ are negative. So it was proved by J.Pintz (1991) that for sufficiently large y one has, with a suitable positive c,

$$\int_1^\infty(\pi(x)-\mathrm{li}(x))\exp\left(-\frac{\log^2 x}{y}\right)dx<-\frac{c}{y}\exp(\frac{9}{16}y)<0.$$

Another deduction of (6.43) is given in Prachar's book (Prachar 1957). For simpler proofs of Theorem 6.20 see Ingham (1936) and Diamond (1975). In particular Ingham's proof does not use the modification of the Phragmén–Lindelöf theorem but utilizes the explicit formula for $\psi(x)$ whereas Diamond's argument also disposes of that tool and its main weapon is an extension of Wiener's tauberian theorem.

Theorem 6.20 (i) implies in particular that the ratio $(\psi(x)-x)/\sqrt{x}$ is unbounded, however it was established by H.Cramér (1920) that if Riemann's Hypothesis is true that this ratio is bounded in mean, i.e. one has

$$\frac{1}{x}\int_2^x\left(\frac{\psi(t)-t}{\sqrt{t}}\right)^2dt=O(1).$$

In a later paper (Cramér 1922) he proved (under the same assumption) the existence of

$$\lim_{x\to\infty}\frac{1}{\log x}\int_2^x\left(\frac{\psi(t)-t}{t}\right)^2dt.$$

A real-valued function f which is locally integrable on the real line is said to have *bounded mean oscillation* if there exists a constant C such that for every bounded interval I on the real line there exists a constant c_I such that

$$\int_I|f(x)-c_I|dx\leq C|I|,$$

$|I|$ denoting the length of I. Recently it was shown by J.Kaczorowski (1998) that if one puts

$$E(x)=\begin{cases}(\psi(x)-x)\sqrt{x} & \text{if } x\geq 2\\ 0 & \text{otherwise}\end{cases}$$

then $E(x)$ does not belong to this class of functions.

4. Numerical experiments show that for small values the difference $\pi(x)-\mathrm{li}(x)$ tends to be negative. S.Skewes (1933) sketched the proof that under the assumption of Riemann Hypothesis there exists an x not exceeding $\exp(\exp(\exp(79)))$ for which the converse inequality is valid, postponing the details to a later paper. A few years later J.E.Littlewood (1937) claimed[35] to have an explicit bound for such x without any hypotheses but did not present any details. After twenty years Skewes returned to this topic (Skewes 1955) and showed, using a method of A.E.Ingham (1936), that under Riemann Hypothesis the above bound could be reduced to $x < S = \exp(\exp(\exp(7.703)))$ and assuming the negation of this hypothesis obtained the rather large bound

$$x < 10^{10^{10^{1000}}}.$$

Later it was shown unconditionally that the difference $\pi(x) - \mathrm{li}(x)$ changes its sign below $1.65 \cdot 10^{1165}$ (Lehman 1966) and $6.69 \cdot 10^{360}$ (te Riele 1987). An effective proof of Theorem 6.20 was also given by G.Kreisel (1952).

Denote by $\nu(T)$ and $\omega(T)$ the number of sign changes of the difference $\pi(x) - \mathrm{li}(x)$ resp. $\psi(x) - x$ in the interval $[2, T]$. The relation

$$\lim_{T\to\infty} \omega(T) = \infty$$

already appears in Phragmén (1891). It was proved by G.Pólya (1930) that one has

$$c(\omega) = \liminf_{T\to\infty} \frac{\omega(T)}{\log T} > 0.$$

Pólya's argument contained a gap closed by J.Steinig (1969).

The explicit lower bound

$$c(\omega) \geq \frac{\gamma_0}{\pi}$$

with $\gamma_0 = 14.13\ldots$ being the imaginary part of the smallest zero of the zeta-function was obtained by J.Kaczorowski (1984), who later (Kaczorowski 1991b) improved this to

$$c(\omega) \geq \frac{\gamma_0}{\pi} + 10^{-250}$$

(cf. Kaczorowski,Pintz 1986,1987, Szydło 1989, 1991).

The function $\nu(T)$ was studied by A.E.Ingham (1936) who obtained the following result:

Let θ be the least upper bound of the set of real parts of zeros of the zeta-function. If there is a zero of $\zeta(s)$ on the line Re $s = \theta$ *(in particular if the Riemann Hypothesis is true), then there exists a constant C such that every interval (T, CT) contains a pair of integers at which the difference $\pi(x) - \mathrm{li}x$ attains different signs.*

---

[35] "*My object was to determine an upper bound to the value of x for which the well-known difference $\pi(x) - \mathrm{li}(x)$ first changes sign from negative to positive; this upper bound to be an explicit absolute constant, and to be obtained without the assumption of the Riemann hypothesis. ... presently [I] found a ... solution of the main problem.*" (Littlewood 1937, p.218).

This shows in particular that Riemann's Hypothesis implies the relation

$$\liminf_{T\to\infty} \frac{\nu(T)}{\log T} > 0.$$

For another proof see Kaczorowski (1991a) (who gave details only in the case $\theta = 1/2$). The constant C appearing in Ingham's result is not effectively given but a weaker effective result is contained in Dette, Meier, Pintz (1985), where it was proved under the Riemann Hypothesis that, for y exceeding the Skewes constant S, the interval $[y^{0.1}, y]$ contains a sign change of $\pi(x) - \mathrm{li}(x)$.

The first lower bound for $\nu(T)$ free from any unproved assumptions was obtained by S.Knapowski (1961, 1962b) who proved

$$\liminf_{T\to\infty} \frac{\nu(T)}{\log\log T} > 0$$

and got a weaker but effective lower bound

$$\nu(T) \ge e^{35} \log\log\log\log T$$

valid for $T > T_0$ with a very large explicit T_0. Later N.Levinson (1976) showed

$$\limsup_{T\to\infty} \frac{\nu(T)}{\log T} = \infty$$

and S.Knapowski and P.Turán (1976,I) obtained the bound

$$\nu(T) \ge c \frac{\log^{1/4} T}{\log\log^4 T}$$

for large T, the constant $c > 0$ being ineffective. Later the same authors showed (Knapowski,Turán 1976,II) that with an effective constant c_1 one has for large T the inequality

$$\nu(T) \ge c_1 \log\log T \log T.$$

A further improvement was made by J.Pintz (1976a,1977a,b,1978), who showed that

$$\nu(T) \ge 10^{-11} \frac{\log T}{(\log\log T)^3}$$

for $T \ge T_0$ with some ineffective T_0 and

$$\nu(T) \ge c_3 \frac{\log^{1/2} T}{\log\log T}$$

for $T \ge T_1$ with effective c_3 and T_1. One can take here c_3 equal to $\exp(-\exp\exp 3.55)$ and $T_1 = \exp\exp\exp\exp 3.57$ (Dette, Meier 1984).

Finally J.Kaczorowski (1985) was able to prove Ingham's assertion without using any unproved hypotheses. He obtained also a similar result for the number of sign changes of $\theta(x) - x$.

Note that certain similar questions are very hard to settle. So e.g. G.Robin (1984) proved that Riemann Hypothesis is equivalent to the positivity of the difference $\mathrm{li}(\theta(x)) - \pi(x)$ for all sufficiently large x.

For analogues of Theorem 6.20 for primes in progressions see Kátai (1968).

6.7. The Conjectures of Hardy and Littlewood

1. In 1922 Hardy and Littlewood published a path-breaking paper[36] (Hardy, Littlewood 1923a) in which they attacked Goldbach's Conjecture and presented a list of fifteen other conjectures concerning prime numbers. We conclude our book with a short survey of the present status of them.

Goldbach's Conjecture has its origin in a letter of Christian Goldbach to Euler dated June, 7th 1742 (Fuss 1843,I,letter 43, 125–129; Euler,Goldbach 1965) where we read[37]:

"*Es*[39] *scheinet ... dass eine jede Zahl, die grösser als 2, ein aggregatum trium numerorum primorum*[40] *sei.*"

Euler replied in a letter of June 30th (Fuss 1843,I,letter 44, 130–136; Euler,Goldbach 1965):

"*[Dass] ... ein*[41] *jeder numerus par eine summa duorum primorum sey, halte ich für ein ganz gewisses theorema, ungeachtet ich dasselbe nicht demonstriren kann.*"

They never returned to this subject, except for an off-hand remark of Goldbach in his letter of March 12th 1753 (Fuss 1843,I,letter 152, 601–603; Euler,Goldbach 1965) pointing out that one did not find yet an even integer which could not be written as a sum of two primes.

Nowadays one actually speaks about two Goldbach's conjectures, binary and ternary: the binary conjecture asserts that every even integer exceeding 4 is a sum of two odd primes and the ternary conjecture states that every odd integer exceeding 7 is a sum of three odd primes.

For a very long time no progress had been made towards the solution of any of these questions, only some numerical evidence was collected. It seems that the first published computation concerning Goldbach's Conjecture

[36]It appeared in February 1922 in the 44th volume of *Acta Mathematica*, the last issue of which was published in November 1923.

[37]Actually the same assertion was later found in an unpublished manuscript of Descartes[38] (see Descartes, *Oeuvres*, vol.10. p.298, Paris 1908).

[39]"*It seems ... that every number larger than 2 is a sum of three prime numbers.*"

[40]Note that Goldbach regarded 1 as a prime number.

[41]"*... every even integer a is a sum of two primes. I regard this as a completely certain theorem, although I cannot prove it*".

was made by A.Desboves (1855) who checked that every even integer n not exceeding 10^4 is a sum of two primes. This result has not become widely known since later calculations in that or smaller range were published (see e.g. Cantor[42] 1896 and Haussner 1899).

A larger range ($n \leq 50816$) was covered in Ripert (1903) and for several large numbers n the checking was done by A.Cunningham (1906/07) and N.M.Shah and B.M.Wilson (1919). Later some numerical experiments were done in Shen (1964) ($n \leq 33 \cdot 10^6$), Stein, Stein (1965) ($n \leq 10^8$), Light, Forest, Hammond, Roe (1980) ($n \leq 10^8$), Granville, van der Lune, te Riele (1989) ($n \leq 2 \cdot 10^{10}$) and Sinisalo (1993) ($n \leq 4 \cdot 10^{11}$).

2. J.J.Sylvester (1871a) gave some heuristical arguments concerning Goldbach's Conjecture and one of his statements may be interpreted as a forecast of the method used later by Hardy and Littlewood (see also Sylvester 1896). Sylvester wrote:

"*The exact number of the equation $x + y = n$ in prime numbers may be expressed algebraically by means of the method of generating functions in terms of the inferior primes to n. The expression will be found to consist of two parts — one a constant multiple of n, the other a function of the roots of unity corresponding to the several inferior primes and their combinations.*"

He did not pursue this matter further, so we do not know exactly what he had in mind. He stated also a conjectural asymptotical formula for the number $P_2(n)$ of representations of an even integer n as a sum of two primes:

$$P_2(n) = (2 + o(1))\pi(n) \prod_{\substack{3 \leq p \leq \sqrt{n} \\ (p,n)=1}} \frac{p-2}{p-1}.$$

Using Corollary to Theorem 3.11 it is not difficult to show that the right hand-side of this formula equals

$$\left(4Ae^{-\gamma} + o(1)\right) \frac{n}{\log^2 n} \prod_{p|n} \frac{p-1}{p-2}$$

with

$$A = \prod_{p \neq 2} \left(1 - \frac{1}{(p-1)^2}\right).$$

It was shown later, by G.H.Hardy and J.E.Littlewood (1919b, 1923a), that Sylvester's formula cannot be correct.

Based on numerical investigations P.Stäckel (1896) conjectured that the number of representations of an even integer as a sum of two primes is asymptotically equal to

[42] Cantor, Georg (1845–1918), Professor in Halle, inventor of set theory.

$$\frac{\pi^2(n)}{\varphi(n)}$$

but this was shown to be false in Landau (1900b) (see Exercise 13). Three other formulas appearing in Brun (1915), Stäckel (1916) and Shah, Wilson (1919) were refuted by Hardy and Littlewood (1919b, 1923a).

3. The first attempt to apply the Eratosthenian sieve to the problem of Goldbach was made by J.Merlin (1911), who also considered the problem of twin primes, i.e. prime pairs differing by 2. His paper contained only the statements of his assertions and arguments leading to them were published by J.Hadamard (1915), since Merlin was killed at the beginning of the First World War, but the reasoning presented here is only heuristical. (This paper contains the assertion that there exist infinitely many prime pairs with a given even difference.) A few years later the paper of Viggo Brun (1919b) (see Brun 1919a) appeared, in which Merlin's idea obtained the correct form. Brun gave there the first non-trivial upper bound for the number $P_2(n)$, namely

$$P_2(n) \leq c\frac{n(\log\log n)^2}{\log^2 n} \tag{6.44}$$

with some effective constant c.

4. Hardy and Littlewood in the third part of their series '*Partitio Numerorum*' (Hardy, Littlewood 1923a) applied to the ternary Goldbach's Conjecture a method of obtaining asymptotic formulas in additive problems introduced by G.H.Hardy and S.Ramanujan (1918) to solve the question of asymptotic behaviour of the partition function. This method was also succesfully applied (Hardy, Littlewood 1919a,1920) to give a new solution of Waring's problem. They did not succeed in proving Goldbach's Conjecture but they showed that if no Dirichlet's L-function has a zero in the half-plane Re $s > c$ for some $c < 3/4$ then every sufficiently large odd integer n is a sum of three odd primes and that the number of such representations is asymptotically equal to

$$C_1\frac{n^2}{\log^3 n}\prod_{\substack{p|n\\ p>2}}\frac{(p-1)(p-2)}{p^2-3p-3}$$

where

$$C_1 = \prod_{p>2}\left(1+\frac{1}{(p-1)^3}\right).$$

Important simplifications of the proof of the qualitative part of this theorem were made by E.Landau (1922) and later an unconditional proof was found by I.M.Vinogradov (1937). Other proofs based on the same main idea were given in Estermann (1937), Mardzhanišvili (1941), Linnik (1946), Čudakov (1947a), Čudakov, Rodosskij (1949), Pan Chen Biao (1977) (see Pan Chen Dong, Pan Chen Biao 1992) and Heath-Brown (1985). See Hua

(1947). A completely different proof, based on Linnik's dispersion method was sketched by B.M.Bredihin (1975).

It was stated by K.G.Borozdkin[43] (1956) that Vinogradov's result applies for all odd integers exceeding exp(exp(16.038)), however it seems that no proof of that assertion ever appeared. Borozdkin's enormous bound was reduced in Chen, Wang (1989a,b,1991,1996) to exp(exp 9.715)) and a weaker result, with the constant equal to exp(exp 13.465), appears in Kasimov (1992). In Wang,Chen (1993) it was shown that the Extended Riemann Hypothesis permits us to reduce this bound to exp(114). Quite recently D.Zinoviev (1997) showed that if the Extended Riemann Hypothesis is true, then the above bound can be reduced to 10^{20}. Since it was shown recently (Deshouillers, Effinger, te Riele, Zinoviev (1997), Saouter (1998)) that every odd integer from the interval $[9, 10^{20}]$ is a sum of three primes it follows that the ternary Goldbach conjecture follows from the Extended Riemann Hypothesis.

The result of Wang and Chen was utilized by L.Kaniecki (1995) to show that under the assumptions of Riemann Hypothesis every odd integer is a sum of at most 5 primes and thus every integer is a sum of at most six primes.

A different way to approach the problem of representation of integers as sums of primes was found by L.G.Schnirelman[44] (1930,1933), who used a bound for the number of representations of an even integer as a sum of two primes, resulting from the application of Brun's sieve, to deduce the existence of a constant $S \leq 8 \cdot 10^4$ with the property that every integer exceeding 1 is a sum of at most S primes. A simplification of Schnirelman's proof was presented by Landau (1930), whose argument however did not lead to any explicit value for S.

Later work diminished Schnirelman's bound for S:

Šanin (1964): $S < 2 \cdot 10^{10}$,
Klimov (1969): $6\dot{1}0^9$,
Deshouillers (1972/73): $S \leq 159$,
Klimov, Pil'tjai , Šeptickaja (1972): $S \leq 115$,
Klimov (1978): $S \leq 61$,
Klimov (1975): $S \leq 55$,
Vaughan (1977): $S \leq 27$
Deshouillers (1975/76): $S \leq 26$,
Zhang, Ding (1983): $S \leq 24$,
Riesel,Vaughan (1983): $S \leq 19$,
Ramaré (1995): $S \leq 7$.

Denote by S_1 the smallest number such that every sufficiently large integer is a sum of at most S_1 primes. It follows from Vinogradov's solution

[43]According to Čudakov (1947b) Borozdkin proved this already in 1939.
[44]Schnirelman (Šnirelman),Lev Genrikovič (1905–1938), worked in Moscow.

of Goldbach's ternary problem that $S_1 \leq 4$ and the truth of both Goldbach's conjectures would imply $S_1 = 3$. It is of interest to determine upper bounds for this constant which can be achieved with Schnirelman's elementary method. The first reasonable bound in this case ($S_1 \leq 71$) is contained in the paper of H.Heilbronn, E.Landau and P.Scherk (1936) and it has been subsequently improved in Ricci (1936,1937) ($S_1 \leq 67$), Shapiro,Warga (1950) ($S_1 \leq 20$), Yin (1956) ($S_1 \leq 18$), Siebert (1968), Kozjašev, Čečuro (1969) ($S_1 \leq 10$) and Vaughan (1976) ($S_1 \leq 6$).

5. We shall now present the conjectures stated by Hardy and Littlewood. The first concerns the binary case of Goldbach's Conjecture.

Conjecture A. *Every large even integer n is a sum of two primes and for the number $P_2(n)$ of such representations one has*

$$P_2(n) = (2C_2 + o(1))\frac{n}{\log^2 n} \prod_{2<p|n} \frac{p-1}{p-2}, \tag{6.45}$$

where

$$C_2 = \prod_{p>2} \left(1 - \frac{1}{(p-1)^2}\right).$$

The first result concerning this conjecture was obtained in Hardy, Littlewood (1924) who deduced from the Extended Riemann Hypothesis that the number $N_2(x)$ of even integers $n \leq x$ which are not represented as a sum of two primes is $O(x^{1/2+\epsilon})$ for every $\epsilon > 0$. Thus the binary Goldbach's Conjecture is true for almost all even integers. The fact that the set of integers which are sums of two primes has positive density was established by Schnirelman (1930,1933) and the first unconditional non-trivial upper bound for $N_2(x)$, viz.

$$N_2(x) = O\left(\frac{x}{\log^a x}\right),$$

valid for every positive a was established independently by J.G.van der Corput (1937), T.Estermann (1938), and N.G.Čudakov (1938b). Other proofs were given in Linnik (1946) and Čudakov (1947a). R.C.Vaughan (1972) improved this to

$$N_2(x) = O(N \exp(-c \log^{1/2} x))$$

(with a certain $c > 0$) and later (Montgomery, Vaughan 1975) the bound

$$N_2(x) = O(x^{1-\delta})$$

with some unspecified positive δ was achieved (see Fouvry 1975/76, Pintz 1988b). In Chen, Pan (1980) the effective bound $\delta \geq 0.01$ was found and it was later improved to $\delta \geq 0.04$ (Chen 1983) and $\delta \geq 0.05$ (Chen 1989).

G.Dufner (1995) showed that almost all even integers n can be represented as a sum of two primes, the smaller being $O(\log^a n)$ for every $a > 44$ (earlier K.Prachar (1965) indicated how to establish this assertion with an unspecified value of a). An elementary proof of $N_2(x) = o(x)$ appeared in Bredihin, Jakovleva (1975).

Brun's first upper bound (6.44) for $P_2(n)$ was later improved to

$$P_2(n) = O\left(\frac{n \log\log n}{\log^2 n}\right)$$

(Schnirelman 1930) and it follows from a result of K.Prachar (1954, 1957, Theorem 3.2) that this bound is best possible. The obvious bound is

$$P_2(n) \leq \pi(n) - \pi(n/2)$$

and it was shown in Deshouillers, Granville, Narkiewicz, Pomerance (1993) that the maximal integer n for which one has equality here equals 210.

The best known upper bound for $P_2(n)$ is due to Wu Dong Hua (1987): if we denote by $P_2^*(n)$ the right-hand side of (6.45) then

$$P_2(n) \leq K P_2^*(n)$$

with some constant $K \leq 7.8165$. Earlier J.R.Chen (1978b) obtained this with $K \leq 7.8342$ and the bound $K \leq 8$ follows from an earlier result of E.Bombieri and H.Davenport (1966) (see Halberstam, Richert 1974).

Another asymptotic formula for $P_2(n)$ was proposed by E.S.Selmer (1943a) and a numerical check of it appeared in Bohman, Fröberg (1975).

V.Brun (1920) proved that every sufficiently large even integer is a sum of two numbers, each of which has at most nine prime factors. If we denote, as is usually done, by P_k any number having at most k prime factors (counted with multiplicities), then this result can be expressed in the form: every large even integer can be written as $P_r + P_s$ with $r = s = 9$. Later the values of r and s were diminished: the first improvement was obtained by H.Rademacher (1924) who gave $r = s = 7$ and later several authors worked on this problem (Estermann 1932, Ricci 1933b, 1936, 1937, Buhštab[45] 1940, 1965, 1967, Rényi 1947, 1948, A.I.Vinogradov 1957, 1965, Wang Yuan 1958, Pan Chen Dong 1962, 1963, 1964, Barban 1963, Levin 1963, 1964/65, Halberstam, Jurkat, Richert 1967, Uchiyama 1967) and the best known result ($r = 1, s = 2$) is due to Chen Jing Run (1966, 1973, 1978a). An important simplification of Chen's proof was given by P.M.Ross (1975) and it found its place in the book of Halberstam and Richert (1974) where further information on this topic may be found. Other proofs of Chen's theorem appeared in Fouvry, Grupp (1989) and Pan Cheng Dong, Ding Xia Xi, Wang Yuan (1975).

[45]Buhštab, Aleksandr Adolfovič (1905–1990), Professor in Moscow.

6. The second conjecture of Hardy and Littlewood deals with pairs of primes with a given difference:

Conjecture B. *For every even k there are infinitely many prime pairs $p, p+k$ and the number $\pi_k(x)$ of such pairs less than x is*

$$(C_3(k) + o(1))\frac{x}{\log^2 x}$$

with

$$C_3(k) = 2C_2 \prod_{2<p|k} \frac{p-1}{p-2},$$

the constant C_2 having the same meaning as in Conjecture A.

This includes in particular the question of *twin primes*, i.e. primes differing by two, like 3 and 5, 5 and 7,

$$6797727 \cdot 2^{15328} \pm 1 \quad \text{(Forbes 1997)}$$

or

$$697053813 \cdot 2^{16352} \pm 1 \quad \text{(Indlekofer,Járai 1996)}.$$

For other examples of large twin pairs see Parady, Smith, Zarantonello (1990).

The asymptotic formula proposed in Conjecture **B** is in certain sense true for almost all integers, as established by A.F.Lavrik (1961a,b) who proved that if $c \geq 3$ and $M > 0$ are given, then for all even integers $k \leq x \log^{-c} x$ with at most $O(x \log^{M+c} x)$ exceptions the conjectured formula holds (see Grosswald 1968/69). Note that this does not imply its truth for any fixed value of k, since the exceptional set depends on x.

It is not known, whether the existence of infinitely many twin primes is a consequence of the Riemann Hypothesis, but D.R.Heath-Brown (1983) proved that it is implied by the existence of Siegel zeros[46] of L-functions (i.e. real solutions ρ_n of the equation $L(x, \chi_n) = 0$ satisfying $\rho_n > 1 - c/\log k_n$ with a certain positive constant c and χ_n being a primitive character mod k_n).

In 1919 V.Brun (1919b) obtained with the use of his sieve method the first non-trivial result concerning twin primes. He showed that one has

$$\pi_2(x) = O\left(\frac{x(\log\log x)^2}{\log^2 x}\right).$$

This implies in particular the convergence of the series whose terms are the reciprocals of all twin primes. For the value of the sum of that series see Wrench (1961) and Shanks,Wrench (1974).

---

[46]The assumption of the existence of these zeros implies in an easy way the Prime Number Theorem, as shown by J.Pintz (1976/79). These two results may be used as an argument supporting the conjecture that there are no Siegel zeros.

Note that in Brun (1919a) a stronger bound

$$\pi_2(x) = O\left(\frac{x}{\log^2 x}\right)$$

for $x \geq 2$ was announced, but the first proof of it appeared only later (Brun 1920). The constant B is usually written as a product $cC_3(2)$ and the smallest known value of c is 3.418, found by J.Wu (1990). For previous bounds see Selberg (1947) ($c = 8$), Pan (1964) ($c = 6$), Halberstam, Richert (1974) ($c = 4$), Chen (1978) ($c = 3.9171$), Fouvry, Iwaniec (1983) ($c = 34/9 = 3.77\ldots$), Fouvry (1984) ($c = 64/17 = 3.76\ldots$), Bombieri, Friedlander, Iwaniec (1986) ($c = 3.5$) and Fouvry, Grupp (1986) ($c = 3.454\ldots$).

No non-trivial lower bounds for $\pi_2(x)$ are known but it was shown that there are infinitely many primes p such that the number $p + 2$ is either prime or has two prime factors. This was first established by Chen Jing Run (1966,1972). Previously it was known that for infinitely many primes p the number $p+2$ has at most k factors (Rényi (1947,1948) with unspecified k and Buhštab (1967) with $k = 3$). For the number $\pi_2^{(2)}(x)$ of primes $p \leq x$ with $p+2$ having at most two prime factors the best known lower bound is due again to J.Wu (1990), who proved that for large values of x one has

$$\pi_2^{(2)}(x) \geq C_3(2)ax\log^{-2} x$$

with $a = 1.05$. Previously Chen Jing Run (1973,1978a) had $a = 0.67$, resp. $a = 0.81$.

Small values of the differences $d_n = p_{n+1} - p_n$ are still a mystery. One does not even know whether for every positive ϵ one can find infinitely many primes p_n with $d_n \leq \epsilon \log p_n$. Several authors were concerned with the ratio $d_n/\log p_n$. If we put

$$E = \liminf_{n\to\infty} \frac{d_n}{\log p_n}$$

then the inequality $E \leq 1$ is an immediate consequence of the Prime Number Theorem and it is widely believed that one has $E = 0$. The first step in that direction was taken by G.H.Hardy and J.E.Littlewood who in an unpublished manuscript of around 1926 (which should have been the seventh part of the series *Partitio Numerorum*) deduced the inequality $E \leq 2/3$ from the Extended Riemann Hypothesis. They established in fact that if Θ is such that there is no zero of any L-function in the open half-plane Re $s > \Theta$, then $E \leq (1+2\Theta)/3$. This was improved in Rankin (1940) to $E \leq (1+4\Theta)/5$ and in Rankin (1950) to

$$E \leq \frac{42}{43}\left(\frac{1+4\Theta}{5}\right)$$

which under the assumption of the Extended Riemann Hypothesis gives $E \leq 126/215 = 0.5860\ldots$.

The first non-trivial unconditional bound for E was given by P.Erdős (1940) who obtained $E < 1$. Later the upper bound for E was first slowly reduced to $57/59 = 0.966\ldots$ (Rankin 1947), $15/16 = 0.9375$ (Ricci 1954), $29/32 = 0.906\ldots$ (Wang Yuan, Xie Sheng-gang, Yu Kun-rui 1965) and then a big step was made by E.Bombieri and H.Davenport (1966) who utilized the large sieve to get $E \leq (2+\sqrt{3})/8 = 0.466\ldots$ (for another proof see Goldston (1992)). This bound was subsequently improved to $(2\sqrt{2}-1)/4 = 0.4571\ldots$ (Pilt'jai 1972), $0.4463\ldots$ (Huxley 1973b), 0.4426 (Huxley 1977), 0.4394 (Huxley 1984), 0.4342 (Fouvry, Grupp 1986) (they showed also that the inequality $d_n \leq 0.4342 \log p_n$ holds for a positive proportion of primes) and the best result at this time is $E \leq 0.248$, due to H.Maier (1985).

Heuristic arguments in favour of conjecture **B** were given in Rubinstein (1993).

7. The next two conjectures extend the two preceding:

Let a be a positive integer. For positive integral b denote by $N_{a,b}(n)$ the number of representations of an integer n in the form

$$n = ap_1 + bp_2$$

with prime numbers p_1, p_2. For b negative let $N_{a,b}$ be the number of such representations with the restriction $p_2 < n$.

Conjecture C. *Let b be positive. If $(a,b) = (n,a) = (n,b) = 1$ and exactly one of the numbers a, b, n is even then*

$$N_{a,b}(n) = \left(\frac{2C_2}{ab} + o(1)\right)\frac{n}{\log^2 n}\prod_{2<p|abn}\frac{p-1}{p-2}.$$

If these conditions are not satisfied then

$$N_{a,b}(n) = o(n/\log^2 n).$$

Conjecture D. *Let b be negative. If $(a,b) = (n,a) = (n,b) = 1$ and exactly one of the numbers a, b, n is even then*

$$N_{a,b}(n) = \frac{2C_2}{a}\frac{n}{\log^2 n}\prod_{2<p|abn}\frac{p-1}{p-2}.$$

Otherwise $N_{a,b}(n) = o(n/\log^2 n)$

The truth of these two conjectures for almost all n was established by J.G.van der Corput (1939). Later it was shown by M.D.Coleman (1990) that under the assumptions of Conjecture **D** the equation

$$ap - by = n$$

is solvable with prime $p < (ab)^C n$ (with a fixed computable C) and y being either a prime or a product of two primes.

The analogous equation

$$n = \sum_{j=1}^{k} a_j p_j$$

with pairwise coprime coefficients a_j was shown to be solvable in the case $k \geq 3$ with odd primes p_j for sufficiently large n, satisfying the necessary congruence conditions, by R.D.James, H.Weyl (1942) and H.E.Richert (1953) who also obtained an asymptotic formula for the number of solutions. The question of getting an upper bound for the solutions p_j of this equation was, in the case $k = 3$, considered in Baker (1967), Choi (1997), Choi, Liu, Tsang (1992), Liu (1985,1987), Liu, Tsang (1989,1993) and Liu, Wang (1998)

8. Four conjectures deal with primes represented by quadratic polynomials. The first makes precise the *Landau conjecture*, which asserts that the polynomial x^2+1 represents infinitely many primes (Landau 1912c), the second generalizes it to arbitrary quadratic polynomials, the third is essentially a variant of the second and the third deals with twin primes of the form x^2+1, x^2+3.

Conjecture E. *The number of primes of the form $x^2 + 1 \leq n$ is asymptotically equal to*

$$C_3 \frac{\sqrt{n}}{\log n}$$

where

$$C_3 = \prod_{p>2} \left(1 - \frac{1}{p-1}\left(\frac{-1}{p}\right)\right).$$

Conjecture F. *Let $a > 0$, $(a, b, c) = 1$. Assume that $(a + b, c)$ is odd and $D = b^2 - 4ac$ is not a square. Then the number of primes of the form $am^2 + bm + c \leq n$ is asymptotically equal to*

$$\epsilon \frac{C_4}{\sqrt{a}} \frac{\sqrt{n}}{\log n} \prod_{2<p|(a,b)} \frac{p}{p-1}$$

where

$$\epsilon = \begin{cases} 1 & \text{if } a+b \text{ is odd} \\ 2 & \text{otherwise} \end{cases}$$

and

$$C_4 = \prod_{2<p\nmid a} \left(1 - \frac{1}{p-1}\left(\frac{D}{p}\right)\right).$$

Conjecture G. *Let $a > 0$ and let $N(n)$ denote the number of representations of n in the form $am^2 + bm + p$ with prime p. If $(a, b, n) = 1$, $(n, a+b)$ is odd and $D = b^2 + 4an$ is not a square then*

$$N(n) = \frac{\epsilon}{\sqrt{a + o(1)})} \frac{\sqrt{n}}{\log n} \prod_{2<p|(a,b)} \frac{p}{p-1} \prod_{\substack{p>2 \\ p\nmid a}} \left(1 - \frac{1}{p-1}\left(\frac{D}{p}\right)\right),$$

where ϵ has the same meaning as in Conjecture **F**.

Otherwise

$$N(n) = o\left(\sqrt{n}\log n\right).$$

The last in this series of conjectures deals with simultaneous representation of primes by two particular quadratic polynomials:

Conjecture P. *There are infinitely many prime pairs m^2+1, m^2+3 and the number of such pairs less than n is given asymptotically by*

$$\frac{6\sqrt{n}}{\log^2 n} \prod_{p \geq 5} \frac{p(p-\nu(p))}{(p-1)^2},$$

with $\nu(p)$ denoting the number of quadratic residues mod p *contained in the set $\{-1,-3\}$.*

An early numerical study of Conjecture **E** was made by A.E.Western (1922). It is not difficult to see (see Exercise 12) that the existence of a quadratic polynomial representing infinitely many primes would follow from the existence of a constant c such that for infinitely many primes p the fractional part $\{\sqrt{p}\}$ of $\sqrt{p}$ does not exceed $c/\sqrt{p}$. Evaluations of the form

$$\{\sqrt{p}\} \leq \frac{c}{p^{\alpha}}$$

with positive a are known, the first (with any $a < 1/15$) being due to I.M.Vinogradov (1940). Later it was established that one can take here any $\alpha < 1/10$ (Vinogradov 1976, Chap.4) and R.M.Kaufman (1979) improved this to $\alpha \leq \sqrt{15}/(2(8+\sqrt{15})) = 0.163\ldots$. The best known result (with any $\alpha < 1/4$) is due to A.Balog (1983) and G.Harman (1983). See also Harman (1991).

There is also a conjecture, due to A.Schinzel (Schinzel, Sierpiński 1958), which generalizes the qualitative part of Conjecture **P** and states that if $f_1, \ldots, f_k$ are irreducible polynomials in $\mathbf{Z}[X]$ and their product does not have a fixed factor, then for infinitely many integers n all values $f_i(n)$ are prime.

P.T.Bateman and R.A.Horn (1962,1965) proposed a quantitative version of Schinzel's conjecture considering in their first paper the case of one polynomial:

If $f_1, f_2, \ldots, f_r$ are distinct irreducible polynomials of respective degrees equal to $n_1, n_2, \ldots, n_r$ with integral coefficients and positive leading coefficients and moreover their product does not have a fixed factor then the number $Q(f_1, \ldots, f_r; x)$ of integers $n \leq x$ for which the values $f_j(n)$ for $j = 1, 2, \ldots, r$ are primes equals asymptotically to

$$\frac{1}{n_1 n_2 \cdots n_r} \prod_p \left(\left(1-\frac{1}{p}\right)^{-r} \left(1-\frac{\omega(p)}{p}\right)\right) \int_2^x \frac{dt}{\log^r t},$$

where $\omega(p)$ denotes the number of solutions of the congruence

$$f_1(x)f_2(x)\cdots f_r(x) \equiv 0 \pmod p.$$

It was shown in Bateman,Stemmler (1962) that Selberg's sieve implies an upper bound for $Q(f_1,\dots,f_r;x)$ which equals the product of the above expression by $2^r r! h_1 \cdots h_r$, with h_i being the degree of f_i. For the case of one polynomial see Wang Yuan (1957).

An early bound for the number of primes represented by a polynomial $f(x)$ for $x = 1,2,\dots N$ was obtained by T.Nagell[47](1922) who showed that this number is $o(N)$. This bound was later improved to $O(N/\log N)$ by H.Heilbronn (1931). See also Bantle (1986), Davenport, Schinzel (1966), Grosswald (1987a), Ricci (1933c,1936,1937), Schinzel (1963), Shanks (1960, 1961,1963) and Shanks, Lal (1972).

There is a large literature concerning the representation of numbers with few prime factors by a polynomial $f(X) \in \mathbf{Z}[X]$ with a positive leading coefficient and which is assumed to be irreducible over the rationals and without a fixed divisor (i.e. if d divides all values of f at integers, then $d = \pm 1$). Let $p(f)$ be the minimal number such that f represents infinitely many integers having at most $p(f)$ prime factors and denote by N the degree of f. The first result concerning $p(f)$ is due to H.Rademacher (1924) who proved $p(f) \le 4N-1$. This bound was later reduced to $3N-1$ (Ricci 1936), $N + c\log N$ with a certain constant c (Kuhn 1953,1954, Wang Yuan 1959), $1.8N$ (Miech 1964/65), $1.0007N+3$ (Levin 1964/65), $N+2+\log(2N+1)$ (Miech 1966) and the best known result is $p(f) \le N+1$, stated in Buhštab (1967). The first published proof of Buhštab's result is due to H.-E.Richert (1969) (see also Vakhitova 1992). The case of small N was treated in Kuhn (1953, 1954), Wang Yuan (1959) and Levin (1961). If $N=2$ and $f(0)$ is odd then $p(f) \le 2$, as proved by H.Iwaniec (1978). See also Halberstam, Richert (1972,1974) for strong lower bounds for the number of integers $n \le x$ such that $f(n)$ has few prime factors.

Denote by $P(n)$ the maximal prime divisor of n. Several authors were concerned with the question of how large may $P(f(n))$ be for a polynomial f. Note that the bound $P(n^2+1) \ge cn^2$ (with some positive c) for infinitely many n would give a rather close approximation to Conjecture **E**. Already in 1895 A.Markov[48](1895) had published (see also Størmer[49]1898) a proof (due essentially to Čebyšev) of the relation

[47]Nagell, Trygve (1895–1988), Professor in Uppsala.
[48]Markov,Andrei Andreevič (1856–1922), Professor in Sankt Petersburg.
[49]Størmer, Carl (1874–1957), Professor in Oslo.

$$\lim_{x\to\infty} \frac{1}{x} P\Bigl(\prod_{n\le x} (n^2+1)\Bigr) = \infty.$$

G.Pólya (1918) extended this to arbitrary irreducible quadratic polynomials and T.Nagell (1921) showed that if f has at least one irrational root, then

$$P\Bigl(\prod_{n\le x} f(n)\Bigr) \ge x \log^{1-\epsilon} x$$

holds for every positive ϵ and P.Erdős (1952) replaced the right-hand side of the last inequality by $x \exp(c \log\log x \log\log\log x)$. Better bounds were later obtained, from which we quote only the results of P.Erdős, A.Schinzel (1990) implying for any irreducible non-linear f the inequality

$$P\Bigl(\prod_{n\le x} f(n)\Bigr) > x \exp\exp(c(\log\log x)^{1/3})$$

with some $c > 0$ and G.Tenenbaum (1990b) who showed that for every $a < 2 - \log 4 = 0.61\ldots$ and sufficiently large x one has

$$P\Bigl(\prod_{n\le x} f(n)\Bigr) > x \exp(\log^a x).$$

In the particular case $f(x) = x^2 + 1$, J.M.Deshouillers, H.Iwaniec (1982) established

$$P(n^2+1) \ge n^{\vartheta}$$

with $\vartheta = 1.202468\ldots$ for infinitely many n, improving the result of C.Hooley (1967) who had $\vartheta = 11/10$.

It was proved by C.L.Siegel (1921, Satz 7) that for every polynomial f with integral coefficients and at least two distinct zeros one has

$$\lim_{n\to\infty} P(f(n)) = \infty.$$

K.Mahler[50] (1933) obtained a similar result for binary forms.

For the polynomial $f(x) = x^2 + 1$ Mahler (1935a,b) and S.Chowla (1935) proved $P(f(n)) \ge c \log\log n$ (with some $c > 0$), and this was extended to arbitrary irreducible nonlinear polynomials by S.V.Kotov (1973) (see Nagell 1937,1955 and Keates 1968/69).

9. Addition of primes and squares is the subject of three next conjectures:

Conjecture H. *Every large integer n is either*[51] *a square or a sum of a prime and a square, the number of representations being asymptotic to*

[50] Mahler, Kurt (1904–1988), Professor in Manchester and Canberra.

[51] It happens sometimes that the words 'either a square' in this conjecture is omitted and this leads to trivial 'counterexamples', as evidently there are infinitely many squares which cannot be written as a sum of a square and a prime number. See e.g. the review 89i:11111 in Mathematical Reviews.

$$\frac{\sqrt{n}}{\log n}\prod_{p>2}\left(1-\frac{1}{p-1}\left(\frac{n}{p}\right)\right).$$

In a later paper (Hardy,Littlewood 1924) stated that their methods can be used to show that the truth of the conjectured formula for the number of representations of almost all integers as a sum of a prime and a square is deducible from the Extended Riemann Hypothesis (ERH) and few years later this was confirmed by G.K.Stanley (1928). She proved that under ERH the number $N_2(x)$ of integers $n \le x$ which are not sums of a prime and a square is $O(x^a)$ for every $a > 1/2$. Recently this evaluation was made more precise by H.Mikawa (1993) who showed that ERH implies $N_2(x) = O(x^{1/2}\log^5 x)$. N.P. Romanov (1934) proved that the set of integers which can be written in the form $p+a^2$ with prime p has a positive lower density and the same applies to integers of the form $p+a^k$ with fixed k. H.Davenport and H.Heilbronn (1937) obtained unconditionally the bound $N_2(x) = O(x\log^{-c} x)$ for a certain positive c. Later R.J.Miech (1968b) proved that this holds for every positive c and I.V.Polyakov (1981) got

$$N_2(x) = O(x\exp(-c\sqrt{\log N}))$$

with a certain positive c. The best known evaluation for this quantity was obtained by A.I.Vinogradov (1985) who got $N_2(x) = O(x^a)$ with a certain a smaller than 1. Other proofs of this assertion were given in Brünner, Perelli, Pintz (1989) and Polyakov (1990). Recently Wang Tianze (1995) provided the explicit value 0.99 for a. For early numerical results see Selmer (1943b). It is not difficult to check that below 10^5 there are 36 non-squares which are not sums of a prime and a square, the largest being 21679.

Conjecture I. *Every large odd integer n is the sum of a prime and the double of a square, the number of representations being asymptotic to*

$$\frac{\sqrt{2n}}{\log n}\prod_{p>2}\left(1-\frac{1}{p-1}\left(\frac{2n}{p}\right)\right).$$

Note that the integers 3, 17, 137, 227, 977, 1187, 1493, 5777 and 5993 are not of the asserted form. They are the only exceptions below 10^5 and no larger exceptions seem to be known.

Conjecture J. *If $N_1(n)$, $N_2(n)$ denote the numbers of solutions of the equations*

$$n = x^2 + y^2 + p$$

resp.

$$n = x^2 + y^2 + z^2 + t^2 + p$$

with prime p and $x, y, z, t \in \mathbf{Z}$ then

$$N_1(n) = (C_1 + o(1))n \prod_{\substack{p\mid n\\ p\equiv 1 \pmod 4}} \left(\frac{(p-1)^2}{p^2-p+1}\right) \prod_{\substack{p\mid n\\ p\equiv 3 \pmod 4}} \left(\frac{p^2-1}{p^2-p-1}\right)$$

with

$$C_1 = \pi \sum_{p>2} \left(1 + \frac{1}{p(p-1)}\left(\frac{-1}{p}\right)\right)$$

and

$$N_2(n) = (C_2 + o(1))n^2 \prod_{2<p\mid n} \left(\frac{(p-1)^2(p+1)}{p^3-p^2+1}\right)$$

with

$$C_2 = \frac{\pi^2}{2} \prod_{p>2} \left(1 + \frac{1}{p^2(p-1)}\right).$$

G.K.Stanley (1928) deduced the second part of this conjecture from the Extended Riemann Hypothesis. She obtained, more generally, asymptotic formulas for the number of representations of an integer as a sum of r squares and s primes in the following cases: $r = 1, s = 2$; $r \geq 3, s = 2$;$r \geq 4, s = 1$; $r \geq 1, s \geq 3$. The second of these conjectures ($r = 4, s = 1$) was proved unconditionally by S.Chowla (1936) and the asymptotic formulas obtained by Stanley were, in the case $r \geq 5, s \geq 1$, proved by A.Walfisz (1936a). The case $r = 1, s = 2$ was settled by T.Estermann (1936) and H.Halberstam (1951) obtained the asymptotic formula in all cases with $r + 2s \geq 5$. In the case $r = s = 2$ it was shown by G.Greaves (1976) that the number of representations of a positive integer $N \not\equiv 0, 1, 5 \pmod 8$ exceeds $cN \log^{-5/2} N$ with a certain positive c.

The first part of the conjecture turned out to be much more difficult. C.Hooley (1957) showed that the conjectured formula for $N_1(n)$ can be deduced from the Extended Riemann's Hypothesis and subsequently J.V.Linnik (1960b) used his dispersion method to give an unconditional proof of that formula. Earlier he proved (Linnik 1959, 1960a) that for all sufficiently large integers n the number $N_1(n)$ is positive. A proof of the second part of the conjecture was given by P.D.T.A.Elliott and H.Halberstam (1966). They observed that this can be achieved by Hooley's method, the use of Extended Riemann's Hypothesis being replaced by a theorem of Bombieri concerning the mean value of the remainder term in Theorem 5.14.

For a generalization, in which the sum of two squares is replaced by a rather general quadratic polynomial in two variables see Iwaniec (1975).

10. The next four conjectures are devoted to primes and cubes.

Conjecture K. *If k is not a cube then the number of primes not exceeding n which are of the form $x^3 + k$ is asymptotically equal to*

$$\frac{n^{1/3}}{\log n} \prod_{\substack{p\equiv 1 \pmod 3 \\ p\nmid k}} \left(1 - \frac{2}{p-1}\eta(k,p)\right)$$

where

$$\eta(k,p) = \begin{cases} 1 & \text{if } k \text{ is a cubic residue mod } p \\ -\frac{1}{2} & \text{otherwise.} \end{cases}$$

Conjecture L. *Every large integer n is either a cube or the sum of a prime and a positive cube, the number of representations being asymptotic to*

$$\frac{n^{1/3}}{\log n} \prod_{\substack{p\equiv 1 \pmod 3 \\ p\nmid k}} \left(1 - \frac{2}{p-1}\eta(-n,p)\right).$$

A.I.Vinogradov (1985) proved that the number of integers $\leq x$ which do not have such a representation is $O(x^a)$ with a certain $a < 1$. A more general result is contained in a paper of A.Zaccagnini (1992) who showed that for every fixed $k \geq 2$ the number $N(k,x)$ of integers not exceeding x which are not sums of a prime and a kth power is $O(x^{a_k})$ with a suitable $a_k < 1$. See Perelli, Zaccagnini (1995). A direct check shows that in the interval $[1, 48\,000]$ there are 2173 integers which are not cubes and cannot be represented as a sum of a prime and a cube.

To formulate the next conjecture some definitions are needed. Let p be a prime congruent to 1 mod 3 and define the integers $a(p)$, $b(p)$ by

$$p = a^2 - ab + b^2, \qquad a \equiv 2 \pmod 3, \quad b \equiv 0 \pmod 3.$$

It follows from the theory of quadratic forms that these conditions determine a and b uniquely. Put moreover

$$A(p) = 2a(p) - b(p), \quad B(p) = b(p)/3$$

and let $\omega_p = a(p) + b(p)\zeta_3$. The number ω_p is a prime factor of p in the ring of integers in the field generated by ζ_3, so the cubic residue symbol $\left(\frac{-n}{\omega_p}\right)_3$ is well-defined and attains values $1, \zeta_3$ and ζ_3^2. If now n is a non-zero integer then put

$$\epsilon(n,p) = \begin{cases} 1 & \text{if } \left(\frac{-n}{\omega_p}\right)_3 = \zeta_3 \\ -1 & \text{otherwise.} \end{cases}$$

Finally let

$$C(p) = -\frac{A(p)-2}{p(p-1)} \quad \text{if} \quad \left(\frac{-n}{\omega_p}\right)_3 = 1$$

and

$$C(p) = \frac{A(p) + 9\epsilon(n,p)B(p) - 4}{2p(p-1)}$$

otherwise.

Conjecture M. *If $k \neq 0$ then there are infinitely many primes of the form $x^3 + y^3 + k$ with positive x, y. The number of such primes not exceeding n each counted according to their number of representations is asymptotically equal to*

$$C_3 \frac{n^{2/3}}{\log n} \prod_{\substack{p\equiv 1 \pmod 3 \\ p\mid k}} \left(1 - \frac{2}{p}\right) \prod_{\substack{p\equiv 1 \pmod 3 \\ p\nmid k}} (1 + C(p))$$

where

$$C_3 = \frac{\Gamma^2(4/3)}{\Gamma(5/3)}.$$

Conjecture N. *There are infinitely many primes of the form $x^3 + y^3 + z^3$ with positive x, y, z. The number $P(n)$ of such primes not exceeding n, counted according to the number of their representations is asymptotically equal to*

$$\Gamma^3\left(\frac{4}{3}\right)\frac{n}{\log n} \prod_{p\equiv 1 \pmod 3} \left(1 - \frac{A(p)}{p^2}\right),$$

$A(p)$ being as in Conjecture **M.**

The only known result concerning this conjecture was obtained recently by C.Hooley (1997) who proved that for sufficiently large n the ratio of $P(n)$ and the conjectured value does not exceed $4 + \epsilon$ for every positive ϵ and if one assumes the truth of the analogue of Riemann's conjecture for a class of Dedekind's zeta-functions then one can replace the constant 4 by 3.

11. Hardy and Littlewood also considered the problem of the existence of infinitely many sequences $x, x+a_1, \dots, x+a_m$ consisting exclusively of primes. Here one must obviously assume that for every prime p the congruence

$$x \prod_{j=1}^{m} (x - a_j) \equiv 0 \pmod p$$

has less than p solutions. It was conjectured by L.E.Dickson (1904) that this condition is also sufficient. Hardy and Littlewood considered the function

$$f(x) = \sum_n \Lambda(n)\Lambda(n + a_1)\cdots\Lambda(n + a_m)x^n,$$

defined in the interval $(-1,1)$ and stated the conjecture (**Conjecture X**) that for x tending to 1 the product $(x-1)f(x)$ tends to an explicitly given limit $S(a_1,a_2,\dots,a_m)$. From this conjecture they deduced that the number of sequences $x, x+a_1,\dots,x+a_m$ consisting of primes and satisfying $x+a_j \le T$ is asymptotically equal to

$$S(a_1,a_2,\dots,a_m)\frac{T}{\log^m T}.$$

Exercises

1. (Landau 1913) Let σ be the convergence abscissa of the Dirichlet series $f(s) = \sum_{n=1}^{\infty} a_n n^{-s}$ with at least one non-zero coefficient and let $\delta > 0$. Use Jensen's formula to show that the number of zeros of $f(s)$ satisfying $\mathrm{Re}\ s \ge \sigma + \delta$ and $T < \mathrm{Im}\ s < T+1$ is $O(\log T)$.

2. (Lindelöf 1908) Prove that the Lindelöf function defined for real σ by

$$\mu(\sigma) = \inf\{\mu : \ \zeta(\sigma+it) = O(|t|^{\mu})\}$$

is continuous, convex and satisfies

$$\mu(\sigma) = \mu(1-\sigma) + \frac{1}{2} - \sigma.$$

3. (Backlund 1918/19) Prove that if $1/2 < \sigma \le 1$ then Lindelöf's conjecture implies

$$\lim_{T\to\infty} \frac{N(\sigma,T+1) - N(\sigma,T)}{\log T} = 0,$$

$N(\sigma,T)$ being the number of zeros of $\zeta(s)$ in the rectangle $\mathrm{Re}\ s \ge \sigma$, $0 < \mathrm{Im}\ s \le T$.

4. (Littlewood 1912) Let $\lambda(n)$ be the function of Liouville. Prove that Riemann's conjecture implies the convergence of the series $\sum_{n=1}^{\infty} \lambda(n)n^{-s}$ in the half-plane $\mathrm{Re}\ s > 1/2$.

5. Prove the following extension of part (i) of the Lemma 6.11:

If $f(x)$ is positive, nondecreasing and

$$\lim_{x\to\infty} \frac{1}{x}\int_1^x \frac{f(t)}{t}dt = 1$$

then

$$\lim_{x\to\infty} \frac{f(x)}{x} = 1.$$

6. (Littlewood 1912) (i) Use the three circle theorem to deduce from Riemann's Hypothesis the evaluation

$$\log \zeta(\sigma + it) = O\left(\log^{2(1-\sigma)+\epsilon}(|t| + 2)\right),$$

valid for every fixed positive ϵ and δ uniformly for $\sigma \in [1/2 + \delta, 1]$.

(ii) Prove that under Riemann Hypothesis one has for every $\delta > 0$ the bounds

$$\zeta(s) = O(t^{\epsilon})$$

and

$$\zeta(s)^{-1} = O(t^{\epsilon})$$

in the half-plane Re $s \geq 1/2 + \delta$.

(iii) Show that the Riemann Hypothesis implies the convergence of the series

$$\sum_{n=1}^{\infty} \frac{\mu(n)}{n^s}$$

in the open half-plane Re $s > 1/2$.

7. (Heilbronn,Landau 1933a,Landau 1932b) Assume that the series

$$f(s) = \sum_{n=1}^{\infty} \frac{a_n}{n^s}$$

has nonnegative coefficients and there exists a positive constant λ such that for $|t| \leq \lambda$ the limit

$$\lim_{\epsilon \to 0+0} \left(f(1 + \epsilon + it) - \frac{A}{\epsilon + it}\right)$$

exists uniformly in t. Moreover let $A(x) = \sum_{n \leq x} a_n$. Then one has

$$P_1(\lambda) \leq \liminf_{x \to \infty} \frac{A(x)}{x} \leq \limsup_{x \to \infty} \frac{A(x)}{x} \leq P_2(\lambda),$$

where $P_1(\lambda)$, $P_2(\lambda)$ are functions defined for $\lambda > 0$ and not depending on f which both tend to 1 when λ tends to infinity.

8. (Landau 1906c) Deduce the relations

$$\sum_{n \leq x} \frac{\mu(n)}{n} = o\left(\frac{1}{\log x}\right)$$

and

$$\sum_{n \leq x} \mu(n) = o\left(\frac{x}{\log x}\right)$$

from

$$\sum_{n=1}^{\infty} \frac{\mu(n) \log n}{n} = -1$$

without the use of complex analysis.

9. Deduce the asymptotic formula

$$\pi(x;k,l) = \frac{1+o(1)}{\phi(k)}\frac{x}{\log x}$$

from Ikehara's theorem.

10. (Cahen 1894, Mellin 1910) Prove that for $k > 0$ and Re $s > 0$ one has

$$e^{-s} = \frac{1}{2\pi i}\int_{k-i\infty}^{k+i\infty} \Gamma(t)s^{-t}dt.$$

11. (Hardy,Littlewood 1915) Use the Theorems 6.13 and 6.14 to deduce the equality

$$\sum_{n\leq T} \mu(n) = o(T).$$

12. (i) Prove that if there is a constant c such that for infinitely many primes p one has

$$\{\sqrt{p}\} < \frac{c}{\sqrt{p}}$$

then there is a quadratic polynomial with integral coefficients representing infinitely many primes.

(ii) Show[52] that the assertion of (i) remains valid if there is a constant c and a real ϑ such that for infinitely many primes p one has

$$\{\sqrt{p}-\vartheta\} < \frac{c}{\sqrt{p}}\,.$$

13. (Landau 1900b) (i) Let $\nu(N)$ be the number of representations of an even integer N as a sum of two primes. Use Exercise 3 of Chap.5. to prove that

$$S(n) = \nu(2)+\nu(4)+\cdots+\nu(n) = (1+o(1))\frac{n^2}{2\log^2 n}.$$

(ii) Prove the identity

$$\frac{1}{\varphi(N)} = \frac{1}{N}\sum_{d|N}\frac{\mu^2(d)}{\varphi(d)}.$$

(iii) Show that one has

$$\sum_{n\leq x}\frac{1}{\varphi(n)} = C\log x + O(1).$$

with

---

[52]It follows from the results obtained in Nair, Perelli (1990) that for every $\epsilon > 0$ and algebraic ϑ there exist infinitely many primes p satisfying $\{\sqrt{p}-\vartheta\} < \frac{\log^{\epsilon} p}{\sqrt{p}}$.

$$C = \frac{315\zeta(3)}{2\pi^4}.$$

(iv) Prove

$$\sum_{\substack{n\le x\\ 2|n}} \frac{1}{\varphi(n)} = \frac{C}{3}\log x + O(1).$$

(iv) Let

$$G(n) = \frac{n^2}{\log^2 n\varphi(n)}.$$

Show that

$$\sum_{n\le x} G(n) = \left(\frac{105\zeta(3)}{2\pi^4} + o(1)\right) x^2 \log^2 x$$

and deduce that Stäckel's hypothetical formula

$$\nu(n) = (1+o(1))G(n)$$

is incorrect.

14. (Scherk 1833, Pillai 1927–28, Sierpiński 1952b, Teuffel 1955, Brown 1967). Show that if p_k denotes the kth prime then for $n \ge 2$ one can write

$$p_n = 1 + \sum_{k=1}^{n-2} \epsilon_k p_k + p_{n-1}$$

if n is even and

$$p_n = 1 + \sum_{k=1}^{n-2} \epsilon_k p_k + 2p_{n-1}$$

if n is odd, with $\epsilon_k \in \{-1, 0, 1\}$.

References

Abel,N.H. (1823): Solution de quelques problèmes à l'aide d'intégrales définies. Magazin for Naturvidenskaberne. **1₂**. [Oeuvres complétes, **1**, 11–27. Christiania 1881.]

Abel,N.H. (1825): L'intégrale definie $\sum_n \varphi x$ exprimée par une intégrale définie simple. Magazin for Naturvidenskaberne. **3₂**. [Oeuvres complétes, **1**, 34–39. Christiania 1881.]

Abel,N.H. (1826): Recherches sur la série $1+\frac{m}{1}x+\frac{m(m-1)}{1\cdot 2}x^2+\frac{m(m-1)(m-2)}{1\cdot 2\cdot 3}x^3+\cdots$. J. Reine Angew. Math., **1**, 311–339. [Oeuvres complétes, **1**, 219–250. Christiania 1881.]

Abel,U.,Siebert,H. (1993): Sequences with large number of prime values. Amer. Math. Monthly, **100**, 167–169. MR 94b:11092.

Adams, J.C. (1878): On the value of Euler's constant. Proc. Roy. Soc. London, **27**, 88–94.

Ageev,A.A. (1994): Sierpinski's theorem is deducible from Euler and Dirichlet. Amer. Math. Monthly, **101**, 659–660. MR 95h:11096.

Ahlfors,L.V. (1953): Complex analysis. McGraw-Hill, New York. MR 14 p.857. [2nd ed. 1966. MR 32#5844; 3rd ed. 1978. MR 80c:30001.]

Ajello,C. (1896): Sul numero dei numeri primi inferiori ad un dato limite. Giornale di Mat., **34**, 14–20.

Allison,D. (1970): On obtaining zero-free regions for the zeta-function from estimates of $M(x)$. Proc. Cambridge Philos. Soc., **67**, 333–337. MR 40#4211.

Almansa,J.,Prieto,L. (1994): New formulas for the nth prime. Lecturas Mat. **15**, 227–231. (Spanish) MR 96a:11009.

Amitsur,S.A. (1956/57): On arithmetic functions. J. Analyse Math., **5**, 273–314. MR 21#4138.

Amitsur,S.A. (1962): On a lemma in elementary proofs of the prime number theorem. Bull. Res. Council Israel, **10F**, 101–108. MR 26#1290.

Ananda-Rau,K. (1924): The infinite product for $(s-1)\zeta(s)$. Math. Zeitschr., **20**, 156–164.

Anderson,R.J. (1983): Simple zeros of the Riemann zeta-function. J. Number Th., **17**, 176–182. MR 85c:11075.

Anderson,R.,Stark,H.M. (1981): Oscillation theorems. Lecture Notes in Math., **899**, 79–106. Springer, Berlin–Heidelberg. MR 83h:10082.

Andrica,D.,Tóth,L. (1991): Some remarks on Stieltjes constants of the zeta function. Stud. Cerc. Mat., **43**, 3–9. MR 93c:11066.

Ansari,A.R. (1951): On prime representing function. Ganita, **2**, 81–82. MR 15 p.11.

Apéry,R. (1979): Irrationalité de $\zeta 2$ et $\zeta 3$. Astérisque, **61**, 11–13.

Apostol,T.M. (1951a): Remark on the Hurwitz zeta function. Proc. Amer. Math. Soc., **2**, 690–693. MR 13 p.328.

Apostol,T.M. (1951b): On the Lerch zeta function. Pacific J. Math., **1**, 161–167. MR 13 p.328; Addendum: *ibidem*, **2**, 1952, p.10. MR 13 p.725.

Apostol,T.M. (1954): Some series involving the Riemann zeta function. Proc. Amer. Math. Soc., **5**, 239–243. MR 15 p.685.

Apostol,T.M. (1965): A characteristic property of the Möbius function. Amer. Math. Monthly, **72**, 279–282. MR 31#2193.

Apostol,T.M. (1970): Dirichlet L-functions and character power sums. J. Number Th., **2**, 223–234. MR 41#3412.

Apostol,T.M. (1972): Dirichlet L-functions and primitive characters. Proc. Amer. Math. Soc., **31**, 384–386. MR 44#2717.

Apostol,T.M. (1973a): Another elementary proof of Euler's formula for $\zeta(2n)$. Amer. Math. Monthly, **80**, 425–431. MR 47#3330.

Apostol,T.M. (1973b): Identities for series of the type $\sum f(n)\mu(n)n^{-s}$. Proc. Amer. Math. Soc., **40**, 341–345. MR 47#8409.

Apostol,T.M. (1975): A note on periodic completely multiplicative arithmetical functions. Amer. Math. Monthly, **83**, 39–40. MR 52#5538.

Apostol,T.M. (1976): Introduction to Analytic Number Theory. Springer, Berlin–Heidelberg. MR 55#7892.

Apostol,T.M. (1985a): Note on the trivial zeros of Dirichlet L-functions. Proc. Amer. Math. Soc., **94**, 29–30. MR 86f:11063.

Apostol,T.M. (1985b): Formulas for higher derivatives of the Riemann zeta function. Math. Comp., **44**, 223–232. MR 86c:11063.

Apostol,T.M.,Chowla,S. (1959): On the non-existence of certain Euler products. Norske Vid. Selsk. Forh. Trondheim, **32**, 65–67. MR 23#A3726.

Arndt,C.F. (1846): Disquisitiones de residuis cujusvis ordinis. J. Reine Angew. Math., **31**, 333–342.

Artin,E. (1955): The orders of the linear groups. Comm. Pure Appl. Math., **8**, 355–365. MR 17 p.12. [Collected Papers, 387–397. Addison-Wesley 1965. Reprint: Springer, Berlin–Heidelberg, 1982.]

Arwin,A. (1923): A functional equation from the theory of the Riemann $\zeta(s)$-function. Ann. of Math., (2), **24**, 359–366.

Auric,A. (1915): Suite illimitée des nombres premiers. L'Intermédiaire des Math., **22**, 252–253.

Axer,A. (1910): Beitrag zur Kenntnis der zahlentheoretischen Funktionen $\mu(n)$ und $\lambda(n)$. Prace Mat.-Fiz., **21**, 65–95.

Ayoub,R. (1963): An Introduction to the Analytic Theory of Numbers. American Math.Soc., Providence. MR 28#3954.

Ayoub,R. (1967): On L-functions. Monatsh. Math., **71**, 193–202. MR 35#5401.

Ayoub,R. (1974): Euler and the zeta function. Amer. Math. Monthly, **81**, 1067–1086. MR 50#12566.

Ayoub,R.,Chowla,S. (1981): On Euler's polynomial. J. Number Th., **13**, 443–445. MR 83h:10013.

Bach,E.,Sorenson,J. (1994): Explicit bounds for primes in residue classes. Proc. Symp. Appl. Math., **48**, 535–539. MR 96e:11152.

Bach,E.,Sorenson,J. (1996): Explicit bounds for primes in residue classes. Math. Comp., **65**, 1717–1735. MR 97a:11143.

Bachmann,P. (1894): Zahlentheorie, **2**, Die analytische Zahlentheorie. Teubner, Leipzig.

Backlund,R.J. (1912): Einige numerische Rechnungen, die Nullpunkte der Riemannschen ζ-Funktion betreffend. Öfversigt af Finska Vetenskaps. Soc. Förh. **54 A**, No.3, 1–7.

Backlund,R.J. (1914): Sur les zéros de la fonction $\zeta(s)$ de Riemann. Comptes Rendus Acad. Sci. Paris, **158**, 1979–1981.

Backlund,R.J. (1916): Über die Nullstellen der Riemannscher ζ-Funktion. Dissertation, Helsingfors.

Backlund,R.J. (1918): Über die Nullstellen der Riemannscher ζ-Funktion. Acta Math., **41**, 345–375.
Backlund,R.J. (1918/19): Über die Beziehung zwischen Anwachsen und Nullstellen der Zetafunktion. Öfversigt Finska Vetensk. Soc. Förh., **61A**, No.9, 1–8.
Backlund,R.J. (1929): Über die Differenzen zwischen den Zahlen, die zu den ersten n Primzahlen teilerfremd sind. Commentationes in honorem E.L. Lindelöf, Annales Acad. Sci. Fenn., **32**, Nr.2, 1–9.
Badea,C. (1986): On a stronger Bertrand's postulate. An. Stiint. Univ."Al.I.Cuza". Iaşi, sect I1. Mat., (N.S.), **32**, no.3, 3–5. MR 88j:11005.
Baillaud,B.,Bourget,H. (editors) (1905): Correspondance d'Hermite et de Stieltjes. Gauthier-Villars, Paris.
Baker,A. (1966): Linear forms in the logarithms of algebraic numbers, I. Mathematika, **13**, 204–216. MR 36#3732.
Baker,A. (1967): On some diophantine inequalities involving primes. J. Reine Angew. Math., **228**, 166–181. MR 36#111.
Baker,A.,Schinzel,A. (1971): On the least integers represented by genera of binary quadratic forms. Acta Arith., **18**, 137–144. MR 47#8437.
Baker,H.F. (1889/90): On Euler's φ-function. Proc. London Math. Soc., **21**, 30–32.
Baker,R.C. (1986): The greatest prime factor of the integers in an interval. Acta Arith., **47**, 193–231. MR 88a:11086.
Baker,R.C. (1996): The Brun-Titchmarsh theorem. J. Number Th., **56**, 343–365. MR 97a:11150.
Baker,R.C.,Harman,G. (1996a): The Brun-Titchmarsh theorem on average. Progress in Math., **85**, 39–103. MR 97h:11096.
Baker,R.C.,Harman,G. (1996b): The difference between consecutive primes. Proc. London Math. Soc., (3), **72**, 261–280. MR 96k:11111.
Balasubramanian,R.,Ramachandra,K. (1974): Two remarks on a result of Ramachandra. J.Indian Math.Soc., **38**, 395–397. MR 52#13674.
Balasubramanian,R.,Ramachandra,K. (1977): On the frequency of Titchmarsh phenomenon for $\zeta(s)$. Proc. Indian Acad.Sci., **A86**, 341–351. MR 58#21968.
Balasubramanian,R.,Ramachandra,K. (1987): On an analytic continuation of $\zeta(s)$. Indian J.Pure Appl.Math., **18**, 790–793. MR 88m:11069.
Balog,A. (1980): Numbers with a large prime factor. Studia Sci. Math. Hungar., **15**, 139–146. MR 84e:10049.
Balog,A. (1984): Numbers with a large prime factor,II. Topics in Classical Number Theory, Proc. J.Bólyai Colloq., **34**, 49–67. North-Holland, Amsterdam. MR 87e:11106a.
Balog,A. (1981): An elementary Tauberian theorem and the prime number theorem. Acta Math. Acad. Sci. Hung., **37**, 285–299. MR 82k:10052.
Balog,A. (1983): On the fractional part of p^{ϑ}. Archiv Math., **40**, 434–440. MR 85e:11063.
Balog,A. (1990): The prime k-tuplets conjecture on average. Progress in Math., **85**, 47–75. MR 92e:11105.
Balog,A.,Harman,G.,Pintz,J. (1983): Numbers with a large prime factor,IV. J. London Math. Soc., (2), **28**, 218–226. MR 87e:11106b.
Bang,A.S. (1886): Taltheoretiske undersøgelser. Tidsskr. f.Math. (5), **4**, 70–80, 130–137.
Bang,A.S. (1937): Elementaere Beviser for specielle Tilfaelde af Dirichlets Saetning om Differensraekker. Thesis, København.
Bang,Th. (1952): A function representing all prime numbers. Norsk Mat. Tidsskr., **34**, 117–118. MR 14 p.621.

Bang,Th. (1964): An inequality for real functions of a real variable and its application to the prime number theorem. Proc. Conf. Oberwolfach 1963, 155–160. Birkhäuser, Basel. MR 32#98.

Bantle,G. (1986): On the A-dependence of the linear sieve with application to polynomial sequences. J. Number Th., **24**, 134–153. MR 88a:11090.

Baran,M. (1991): A Karamata method.I. Elementary properties and applications. Canad. Math. Bull., **34**, 147–157. MR 92e:11092.

Baranowski,A. (1895): O wzorach służących do obliczenia liczby liczb pierwszych, nie przekraczających danej granicy. Rozprawy Akad. Umiejętn. w Krakowie, (2), **8**, 192–219.

Barban,M.B. (1963): The density of zeros of Dirichlet L-series and the problem of the sum of primes and near primes. Mat. Sb., **61**, 418–425. (Russian)MR 30#1992.

Barban,M.B.,Linnik.Yu.V.,Čudakov, N.G. (1964): On prime numbers in an arithmetic progression with a prime-power difference. Acta Arith., **9**, 375–390. MR 30#1993. [Yu.V.Linnik, Izbrannye Trudy, **2**, 151–165. Nauka, Leningrad 1980.]

Barbeau,E.J.,Leah,P.J. (1976): Euler's 1760 paper on divergent series. Historia Math., **3**, 141–160; Errata: *ibidem*, **5**, 1978, p.332. MR 58#21162ab.

Barnes,C.W. (1976): The infinitude of primes; a proof using continued fractions. Enseign. Math., (2), **22**, 313–316. MR 54#12615.

Barnes,E.S.,Swinnerton-Dyer,H.P.F. (1952): The inhomogeneous minima of binary quadratic forms. Acta Math., **87**, 259–323. MR 14 p.730.

Barnes,E.W. (1905): The Maclaurin sum formula, Proc. London Math. Soc., (2), **3**, 253–272.

Bartz,K. (1990): On an effective order estimate for the Riemann zeta-function in the critical strip. Monatsh. Math., **109**, 267–270. MR 91g:11093.

Bašmakova,I.G. (1948): Arithmetical books of Euclid's "Elementa". Istor.-Mat. Issled., **1**, 296–328. (Russian)

Bateman,P.T. (1959): Theorems implying the non-vanishing of $\sum \chi(m)m^{-1}$ for real residue-characters. J.Indian Math. Soc. (N.S.), **23**, 101–115. MR 23#A1613.

Bateman,P.T. (1966): Lower bounds for $\sum h(m)/m$ for arithmetical functions h similar to real residue characters. J. Math. Anal. Appl., **15**, 1966, 2–20. MR 33#7313.

Bateman,P.T. (1997): A theorem of Ingham implying that Dirichlet's L-functions have no zeros with real part one. Enseign. Math., (2), **43**, 281–284. MR 98h:11107.

Bateman,P.T.,Horn,R.A. (1962): A heuristic asymptotic formula concerning the distribution of prime numbers. Math. Comp., **16**, 363–367. MR 26#6139.

Bateman,P.T.,Horn,R.A. (1965): Primes represented by irreducible polynomials in one variable. Proc. Symp. Pure Math., **8**, 119–132. MR 31#1234.

Bateman,P.T.,Low,M.E. (1965): Prime numbers in arithmetic progressions with difference 24. Amer. Math. Monthly, **72**, 139–143. MR 30#3859.

Bateman,P.T.,Stemmler,R.M. (1962): Waring's problem in algebraic number fields and primes of the form $(p^r-1)/(p^d-1)$. Illinois J. Math., **6**, 142–156. MR 25#2059.

Bauer,J.-P. (1971): Méthodes de la théorie des ensembles en arithmétique élémentaire. Comptes Rendus Acad. Sci. Paris, **273**, 756–759. MR 45#3299.

Bauer,M. (1903): Zur Theorie der arithmetischen Progression. Archiv Math. Phys., (3), **5**, 274–277.

Bauer,M. (1906): Über die arithmetische Reihe. J. Reine Angew. Math., **131**, 265–267.

Baxa,C. (1992): Über Gandhis Primzahlformel. Elem. Math., **47**, 82–84. MR 93h: 11007.

Bays,C.,Hudson,R.H. (1977): The segmented sieve of Eratosthenes, and primes in arithmetic progressions to 10^{12}. BIT, **17**, 121–127. MR 56#5405.
Bays,C.,Hudson,R.H. (1978a): The appearance of tens of billions of integers x with $\pi_{24,13}(x) < \pi_{24,1}(x)$ in the vicinity of 10^{12}. J. Reine Angew. Math., **299/300**, 234–237. MR 57#12418.
Bays,C.,Hudson,R.H. (1978b): On the fluctuations of Littlewood for primes of the form $4n \pm 1$. Math. Comp., **32**, 281–286. MR 57#16174.
Bays,C.,Hudson,R.H. (1978c): Details of the first region of integers x with $\pi_{3,2}(x) < \pi_{3,1}(x)$. Math. Comp., **32**, 571–576. MR 57#16175.
Bays,C.,Hudson,R.H. (1979): Numerical and graphical description of all axis crossing regions for the moduli 4 and 8 which occur before 10^{12}. Internat. J. Math. Math. Sci., **2**, 111–119. MR 80h:10003.
Bays,C.,Hudson,R.H. (1983): The cyclic behavior of primes in the arithmetic progressions modulo 11. J. Reine Angew. Math., **339**, 215–220. MR 84d:10048.
Behrend,F.A. (1940): On obtaining an elementary estimate of the frequency of the primes by means of the elementary properties of the integers. J. London Math. Soc., **15**, 257–259. MR 2 p.249.
Bellman,R. (1943): A note on the divergence of a series. Amer. Math. Monthly, **50**, 318–319.
Bellman,R. (1947): A note on relatively prime sequences. Bull. Amer. Math. Soc., **53**, 778–779. MR 9 p.11.
Benedetto,J.J. (1979/80): Fourier analysis of Riemann distributions and explicit formulas. Math. Ann., **252**, 141–164. MR 82c:10048.
Bentz,H.J. (1982): Discrepancies in the distribution of prime numbers. J. Number Th., **15**, 252–274. MR 83k:10078.
Bentz,H.J.,Pintz,J. (1980a): Über eine Verallgemeinerung des Tschebyschef–Problems, Math. Zeitschr., **174**, 35–41. MR 81m:10087.
Bentz,H.J.,Pintz,J.(1980b): Quadratic residues and the distribution of prime numbers. Monatsh. Math., **90**, 91–100. MR 82e:10072.
Bentz,H.J.,Pintz,J. (1982): Über das Tschebyschef–Problem. Resultate Math. **5**, 1–5. MR 83h:10079.
Berlowitz,B. (1967/68): Extension of a theorem of Hardy. Acta Arith., **14**, 203–207. MR 37#1324.
Bernays,P. (1912): Über die Darstellung von positiven, ganzen Zahlen durch die primitiven, binären quadratischen Formen einer nicht-quadratischen Diskriminante. Dissertation, Göttingen.
Berndt,B.C. (1969): On the zeros of the Riemann zeta-function. Proc. Amer. Math. Soc., **22**, 183–188. MR 39#4104.
Berndt,B.C. (1971): The functional equation of some Dirichlet series. Proc. Amer. Math. Soc., **29**, 457–460. MR 43#1929; II. *ibidem*, **31**, 1972, 24–26. MR 44#2718.
Berndt,B.C. (1972a): On the Hurwitz zeta-function. Rocky Mount. J. Math., **2**, 151–157. MR 44#6622.
Berndt,B.C. (1972b): Two new proofs of Lerch's functional equation. Proc. Amer. Math. Soc., **32**, 403–408. MR 45#6773.
Berndt,B.C. (1973): A new proof of the functional equation of Dirichlet L-functions. Proc. Amer. Math. Soc., **37**, 355–357. MR 47#155.
Berndt,B.C. (1975): Elementary evaluation of $\zeta(2n)$. Math. Mag., **48**, 148–154. MR 51#3078.
Berndt,B.C.,Evans,R.J. (1981): The determination of Gauss sums. Bull. Amer. Math. Soc., **5**, 107–129. MR 82h:10051.
Bertrand,J. (1845): Mémoire sur le nombre de valeurs que peut prendre une fonction quand on y permute les lettres qu'elle enferme. J.École Roy. Polytechnique, **18**, 123–140.

Besenfelder,H.J. (1977): Zur Nullstellenfreiheit der Riemannscher Zeta-funktion auf der Geraden $\sigma = 1$. J. Reine Angew. Math., **295**, 116–119. MR 56#11924.

Besenfelder,H.J. (1979): Über eine Vermutung von Tschebyscheff, I,II. J. Reine Angew. Math., **307/308**, 411–417; **313**, 1980, 52–58. MR 81e:10037.

Besenfelder,H.J.,Palm,G. (1977): Einige Äquivalenzen zur Riemannscher Vermutung. J. Reine Angew. Math., **293/294**, 109–115. Addendum: *ibidem*, **297**, 1978, 91. MR 57#246a,b.

Bessel,F.W. (1811): Königsberger Archiv Naturwiss. Math., Januar 1811.

Beukers,F. (1979): A note on the irrationality of $\zeta(2)$ and $\zeta(3)$. Bull. London Math. Soc., **11**, 268–272. MR 81j:10045.

Beurling,A. (1937): Analyse de la loi asymptotique de la distribution des nombres premiers généralisés,I. Acta Math., **68**, 255–291. [Collected Works, **2**, 1–37. Birkhäuser 1989.]

Bhattacharjee,N.R.,Bhattacharjee,T. (1990): Extension of zeta-function $\zeta(2m)$. Indian J.Pure Appl. Math., **21**, 977–980. MR 92d:11091.

Birch,B.,Singmaster,D. (1984): An elementary number theory result. Irish Math. Soc. Newsletter, 10–13. MR 87g:11012a.

Birkhoff,G.D.,Vandiver,H.S. (1904): On the integral divisors of $a^n - b^n$. Ann. of Math., (2), **5**, 173–180. [G.D.Birkhoff, Collected Mathematical Papers, **III**, 145–155. Dover, New York 1968.]

Blanchard,A. (1969): Initiation à la théorie analytique des nombres premiers. Dunod, Paris. MR 39#5487.

Bochner,S. (1933): Ein Satz von Landau und Ikehara. Math. Zeitschr., **37**, 1–9. [Selected Mathematical Papers, 71–79. Benjamin, New York 1969; Collected Papers, **2**, 639–647. American Math. Soc., Providence 1991.]

Bochner,S.,Chandrasekharan,K. (1956): On Riemann's functional equation. Ann. of Math., (2), **63**, 336–360. [S.Bochner, Selected Mathematical Papers, 165–189. Benjamin, New York 1969; S.Bochner, Collected Papers, **2**, 715–739. American Math. Soc., Providence 1991.]

Bohman,J. (1972): On the number of primes less than a given limit. BIT, **12**, 576–577. MR 52#10644.

Bohman,J.,Fröberg,C.-E. (1975): Numerical results on the Goldbach conjecture. BIT, **15**, 239–243. MR 52#10644.

Bohnenblust,H.F.,Hille,E. (1931a): Sur la convergence absolue des séries de Dirichlet. Comptes Rendus Acad. Sci. Paris, **192**, 30–32.

Bohnenblust,H.F.,Hille,E. (1931b): On the absolute convergence of Dirichlet series. Ann. of Math., 600–622.

Bohr,H. (1909): Sur la série de Dirichlet. Comptes Rendus Acad. Sci. Paris, **148**, 75–80. [Collected Mathematical Works, **1**, A1. Dansk Mat. Forening, København 1952.]

Bohr,H. (1910a): Sur la convergence des séries de Dirichlet. Comptes Rendus Acad. Sci. Paris, **151**, 375–377. [Collected Mathematical Works, **1**, A4. Dansk Mat. Forening, København 1952.]

Bohr,H. (1910b): Bidrag til de Dirichlet'ske Rækkers theori. G.E.C.Gad, København. [Collected Mathematical Works, **1**, A3. Dansk Mat. Forening, København 1952.]

Bohr,H. (1911a): Sur l'existence de valeurs arbitrairement petites de la fonction $\zeta(s) = \zeta(\sigma + it)$ de Riemann pour $\sigma > 1$. Oversigt Kgl. Danske Videnskabernes Selskabs Forhandlinger, Nr.3. [Collected Mathematical Works, **1**, B3. Dansk Mat.Forening, København 1952.]

Bohr,H. (1911b): Über das Verhalten von $\zeta(s)$ in der Halbebene $\sigma > 1$. Nachr. Ges. Wiss. Göttingen, 409–428. [Collected Mathematical Works, **1**, B6. Dansk Mat.Forening, København 1952.]

Bohr,H. (1912a): Sur la fonction $\zeta(s)$ dans la demi-plan $\sigma > 1$. Comptes Rendus Acad. Sci. Paris, **154**, 1078–1081. [Collected Mathematical Works, **1**, B7. Dansk Mat.Forening, København 1952.]

Bohr,H. (1912b): Über die Funktion $\frac{\zeta'(s)}{\zeta(s)}$. J. Reine Angew. Math., **141**, 217–234. [Collected Mathematical Works, **1**, B8. Dansk Mat.Forening, København 1952.]

Bohr,H. (1912c): Et nyt Bewis for, at den Riemann'ske Zetafunktion $\zeta(s) = \zeta(\sigma + it)$ har uendelig mange Nulpunkter indenfor Parallelstrimlen $0 \le \sigma \le 1$. Nyt Tidsskr.Mat. B, **23**, 81–85. [Collected Mathematical Works, **1**, B9. Dansk Mat.Forening, København 1952.]

Bohr,H. (1913a): Über die gleichmässige Konvergenz Dirichletscher Reihen. J. Reine Angew. Math., **143**, 203–213. [Collected Mathematical Works, **1**, A8. Dansk Mat.Forening, København 1952.]

Bohr,H. (1913b): Über die Bedeutung der Potenzreihen unendlich vieler Variabeln in der Theorie der Dirichletschen Reihen. Nachr. Ges. Wiss. Göttingen, 441–488. [Collected Mathematical Works, **1**, A9. Dansk Mat.Forening, København 1952.]

Bohr,H. (1913c): Darstellung der gleichmäßigen Konvergenzabszisse einer Dirichletschen Reihe $\sum_{n=1}^{\infty} \frac{a_n}{n^s}$ als Funktion der Koeffizienten der Reihe. Archiv Math., (3), **21**, 326–330. [Collected Mathematical Works, **1**, A10. Dansk Mat.Forening, København 1952.]

Bohr,H. (1914b): Sur la fonction $\zeta(s)$ de Riemann. Comptes Rendus Acad. Sci. Paris, **158**, 1986–1988. [Collected Mathematical Works, **1**, B17. Dansk Mat. Forening, København 1952.]

Bohr,H. (1915): Zur Theorie der Riemannschen Zetafunktion im kritischen Streifen. Acta Math., **40**, 67–100. [Collected Mathematical Works, **1**, B19. Dansk Mat. Forening, København 1952.]

Bohr,H. (1942): Über das Koeffizientendarstellungsproblem Dirichletscher Reihen. Det Kgl.Danske Videnskab. Selskab, Mat.-Fys. Medd. **20**, No.2,1–12. MR 7 p.438. [Collected Mathematical Works, **1**, A16. Dansk Mat.Forening, København 1952.]

Bohr,H.,Cramér,H. (1923): Die neuere Entwicklung der analytischen Zahlentheorie. Enzyklopädie der Mathematischen Wissenschaften, **2**, C8, 722–849. [H.Bohr, Collected Mathematical Works, **III**, H. Dansk Mat.Forening, København 1952; H.Cramér, Collected Works, **1**, 289–416, Springer, Berlin–Heidelberg 1994.]

Bohr,H.,Landau,E. (1910): Über das Verhalten von $\zeta(s)$ und $\zeta_\kappa(s)$ in der Nähe der Geraden $\sigma = 1$. Nachr. Ges. Wiss. Göttingen, 303–330. [H.Bohr, Collected Mathematical Works, **1**, B1. Dansk Mat.Forening, København 1952; E.Landau, Collected Works, **4**, 221–248. Thales Verlag.]

Bohr,H.,Landau,E. (1913): Beiträge zur Theorie der Riemannscher Zetafunktion. Math. Ann., **74**, 3–30. [H.Bohr, Collected Mathematical Works, **1**, B11. Dansk Mat.Forening, København 1952; E.Landau, Collected Works, **5**, 454–481. Thales Verlag.]

Bohr,H.,Landau,E. (1914a): Ein Satz über Dirichletsche Reihen mit Anwendung auf die ζ-Funktion und die L-Funktionen. Rend. Circ. Mat. Palermo, **37**, 269–272. [H.Bohr, Collected Mathematical Works, **1**, B13. Dansk Mat.Forening, København 1952; E.Landau, Collected Works, **6**, 45–48. Thales Verlag.]

Bohr,H.,Landau,E. (1914b): Sur les zéros de la fonction $\zeta(s)$ de Riemann. Comptes Rendus Acad. Sci. Paris, **158**, 158–162. [H.Bohr, Collected Mathematical Works, **1**, B14. Dansk Mat.Forening, København 1952; E.Landau, Collected Works, **6**, 56–60. Thales Verlag.]

Bohr,H.,Landau,E. (1923): Über das Verhalten von $\frac{1}{\zeta(s)}$ auf der Geraden $\sigma = 1$. Nachr. Ges. Wiss. Göttingen, 71–80. [H.Bohr, Collected Mathematical Works, **1**, B21. Dansk Mat.Forening, København 1952; E.Landau, Collected Works, **8**, 49–58. Thales Verlag.]

Bohr,H.,Landau,E. (1924): Nachtrag zu unseren Abhandlungen aus den Jahrgängen 1910 and 1923. Nachr. Ges. Wiss. Göttingen, 168–172. [H.Bohr, Collected Mathematical Works, **1**, B21. Dansk Mat.Forening, København 1952; E.Landau, Collected Works, **8**, 159–163. Thales Verlag.]

Bohr,H.,Landau,E.,Littlewood,J.E. (1913): Sur la fonction $\zeta(s)$ dans le voisinage de la droite $\sigma = 1/2$. Acad. Roy. Belgique, Bulletin de la classe des sciences, 1144–1175. [H.Bohr, Collected Mathematical Works, **1**, B12. Dansk Mat.Forening, København 1952; E.Landau, Collected Works, **6**, 61–93. Thales Verlag; J.E.Littlewood, Collected Papers, **2**, 797–828. Clarendon Press, Oxford 1982.]

Bois-Reymond,P.du (1870): Neue Lehrsätze über die Summen unendlichen Reihen. Antrittsprogramm, Freiburg in Breisgau.

Bombieri,E. (1960): Limitazioni riguardanti somme di caratteri reali e somme di funzioni completamente moltiplicative. Ist. Lombardo Ac. Sci. Lett. Rend., **A 94**, 642–649. MR 23#A3725.

Bombieri,E. (1962a): Maggiorazione del resto nel "Primzahlsatz" col metodo di Erdős-Selberg. Ist. Lombardo Ac. Sci. Lett. Rend., **A 96**, 343–350. MR 27#120.

Bombieri,E. (1962b): Sulle formule di A.Selberg generalizzate per classi di funzioni aritmetiche e le applicazioni al problema del resto nel "Primzahlsatz". Riv. Mat. Univ. Parma, (2), **3**, 393-440. MR 27#4804.

Bombieri,E. (1965): On the large sieve. Mathematika, **12**, 201–225. MR 33#5590.

Bombieri,E. (1969): Density theorems for the zeta-function. Proc. Symposia Pure Math., **20**, 352–358. American Math. Soc., Providence 1971. MR 47#4945.

Bombieri,E. (1974): Le grand crible dans la théorie analytique des nombres. Astérisque, **18**. MR 51#8057. 2nd edition, 1987. MR 88g:11064.

Bombieri,E. (1976): A lower bound for the zeros of Riemann's zeta function on the critical line (following N.Levinson). Séminaire Bourbaki, 1974/75, exp.465, Lecture Notes in Math., **514**, 176–182. Springer, Berlin–Heidelberg. MR 56#2935.

Bombieri,E.,Davenport,H. (1966): Small differences between consecutive primes. Proc. Roy. Soc. London, **A293**, 1–18. MR 33#7314.

Bombieri,E.,Davenport,H. (1968): On the large sieve method. Abhandlungen aus Zahlentheorie und Analysis zur Erinnerung an Edmund Landau, 11–21. Deutscher Verlag der Wissenschaften, Berlin 1968. [Number Theory and Analysis (Papers in Honor of Edmund Landau), 9–22. Plenum, New York 1969.] MR 41#5327.

Bombieri,E.,Friedlander,J.,Iwaniec,H. (1986): Primes in arithmetic progressions to large moduli. Acta Math., **156**, 203–252. MR 88b:11058.

Bombieri,E.,Iwaniec,H. (1986): On the order of $\zeta(1/2+it)$. Ann. Scuola Norm. Sup. Pisa, Cl. Sci., (4), **13**, 449–472. MR 88i:1105a.

Bonse,H. (1907): Über die bekannte Eigenschaft der Zahl 30 und ihre Verallgemeinerung. Archiv Math. Phys., (3), **12**, 292–295.

Borevič,Z.I.,Šafarevič,I.R. (1964): Theory of Numbers. Moskva. (Russian)MR 30 #1080. [2nd ed. Moskva 1972. MR 50#4526; 3rd ed. Moskva 1985; English translation: Academic Press, New York 1966. MR 33#4401; French translation: Gauthier-Villars, Paris 1967. MR 34#5734; German translation: Birkhäuser, Basel-Stuttgart 1966. MR 33#4000.]

Borning,A. (1972): Some results for $k! \pm 1$ and $2 \cdot 3 \cdot 5 \cdot \ldots \cdot p + 1$. Math. Comp., **26**, 567–570. MR 46#7133.

Borozdkin,K.G. (1956): On the problem of Vinogradov's constant. Trudy III Vsesojuzn. S'ezda, **1**, p.3. (Russian)

Boston,N.,Greenwood,M.L. (1995): Quadratics representing primes. Amer. Math. Monthly, **102**, 595–599. MR 96g:11154.

Bougaïeff,N.V.(1876): The theory of numerical derivatives. Mamontov, Moskva. (Russian)[French summary written by the author: Bull. Sci. Math., **10**, 1876, 13–32.]

Bougaïeff,N.V. (1888a): The properties of a numerical integral according to divisors and its various applications. The logarithmic number function. Mat. Sb., **13**, 757–777. (Russian)

Bougaïeff,N.V. (1888b): Sur une intégrale numérique suivant les diviseurs. Comptes Rendus Acad. Sci. Paris, **106**, 652–653.

Bouliguine,B. (1914): Sur une propriéte de la fonction $\xi(t)$ de Riemann. Comptes Rendus Acad. Sci. Paris, **158**, 1666–1667.

Bouniakowski,V. (1857): Sur les diviseurs numeriques invariables des fonctions rationnelles entieres. Mem. Acad. Sci. St.Pétersbourg, **6**, 305–329.

Bouwkamp,C.J. (1933): Über die Riemannsche Zetafunktion für positive, gerade Werte des Argumentes. Nieuw Arch. Wisk., **19**, 50–58.

Brauer,A. (1946): On the exact number of primes below a given limit. Amer. Math. Monthly, **53**, 521–523. MR 8 p.136.

Brauer,A.,Zeitz,H. (1930): Über eine zahlentheoretische Behauptung von Legendre. SBer. Berliner Math. Ges., **29**, 116–125.

Braun,J. (1899): Das Fortschreitungsgesetz der Primzahlen durch eine transzendente Gleichung exakt dargestellt. JBer. Gymnasium Trier, Wiss.Beilage. 1–96.

Bredihin,B.M. (1962): Binary additive problems with prime numbers. Doklady Akad. Nauk SSSR, **142**, 766–768. (Russian)MR 24#A2572.

Bredihin,B.M. (1963a): Binary additive problems of indeterminate type,II. Analogue of the problem of Hardy and Littlewood. Izv. Akad. Nauk SSSR, **27**, 577–612. (Russian)MR 27#121.

Bredihin,B.M. (1963b): Applications of the dispersion method in binary additive problems. Doklady Akad. Nauk SSSR, **149**, 9–11. (Russian)MR 26#2419.

Bredihin,B.M. (1964): Binary additive problems of indeterminate type,IV. Analogue of the generalized Hardy–Littlewood problem. Izv. Akad. Nauk SSSR, **28**, 1409–1440. (Russian)MR 30#1099.

Bredihin,B.M. (1975): On the ternary Goldbach problem. Issled. po teorii čisel, Saratov, **6**, 5–18. (Russian)

Bredihin,B.M.,Jakovleva,N.A. (1975): Applications of the dispersion method in the Goldbach problem. Acta Arith., **27**, 253–263. (Russian)MR 51#3094.

Brent,R.P. (1973): The first occurence of large gaps between successive primes. Math. Comp., **27**, 959–963. MR 48#8360.

Brent,R.P. (1975): Irregularities in the distribution of primes and twin primes. Math. Comp., **9**, 43–56. MR 51#5522. Corr.: *ibidem*, **30**, 1976, p.198. MR 53#302.

Brent,R.P. (1979): On the zeros of the Riemann zeta function in the critical strip. Math. Comp.. **33**, 1361–1372. MR 80g:10033.

Brent,R.P. (1980): The first occurence of certain large prime gaps. Math. Comp., **35**, 1435–1436. MR 81g:10002.

Brent,R.P. (1989): Factorization of the eleventh Fermat number. Abstracts Amer. Math. Soc., **10**, 89T–11–73.

Brent,R.P. (1990): Parallel algorithms for integer factorization. Number Theory and Cryptography, 26–37. Cambridge University Press. MR 91h:11148.

Brent,R.P. (1999): Factorization of the tenth Fermat number. Math. Comp., **68**, 429–451. MR 99e:11154.

Brent,R.,Pollard,J.M. (1981): Factorizations of the eighth Fermat number. Math. Comp., **36**, 627–630. MR 83h:10014.

Breusch,R. (1932): Zur Verallgemeinerung des Bertrandschen Postulates, dass zwischen x und $2x$ stets Primzahlen liegen, Math. Zeitschr., **34**, 505–526.

Breusch,R. (1954): Another proof of the prime number theorem. Duke Math. J., **21**, 49–53. MR 16 p.904.

Breusch,R. (1960) An elementary proof of the prime number theorem with remainder term. Pacific J. Math., **10**, 487–497. MR 22#4685.

Briggs,W.E. (1954): An elementary proof of a theorem about the representation of primes by quadratic forms. Canad. J. Math., **6**, 353–363. MR 16 p.117.

Briggs,W.E.,Chowla,S. (1955): The power series coefficients of $\zeta(s)$. Amer. Math. Monthly, **62**, 323–325. MR 16 p.999.

Brika,M. (1933): Über eine Gestalt der Riemannscher Reihe $\zeta(s)$ für s = gerade ganze Zahl. Bull. Soc. Math. Grèce, **14**, 36–38.

Brocard,H. (1880): Sur la fréquence et la totalité des nombres premiers. Nouv. Corresp. Math., **6**, 255–266, 481–488, 529–542.

Brocard,H. (1915): L'Intermédiaire des Math., **22**, p.253.

Broderick,T.S. (1939): On obtaining an estimate of the frequency of the primes by means of the elementary properties of the integers. J. London Math. Soc., **14**, 303–310. MR 1 p.41.

Brown,J.L.Jr. (1967): Proof of Scherk's conjecture on the representation of primes. Amer. Math. Monthly, **74**, 31–33. MR 34#7434.

Brown,M. (1953): A countable connected Hausdorff space. Bull. Amer. Math. Soc., **59**, p.367.

Brown,T.,Shiue,P.J.-S.,Yu,X.Y. (1998): Sequences with translates containing many primes. Canad. Math. Bull., **41**, 15–19. MR 99b:11105.

Bruijn,N.G. de (1971): A survey of generalisations of Pólya's enumeration theorem. Nieuw Arch. Wisk., (3), **19**, 89–112. MR 47#1624.

Brun,V. (1915): Über das Goldbachsche Gesetz und die Anzahl der Primzahlpaare. Archiv for Mathem. og Naturw., **34**, Nr.8, 1–19.

Brun,V. (1916): On the distribution of primes in certain classes of numbers. Nyt Tidsskr. for Math., **27**, B, 45–58.

Brun,V. (1917): Sur les nombres premiers de la forme $ap + b$. Archiv for Mathem. og Naturw., **34**, Nr.14, 1–9.

Brun,V. (1919a): Le crible d'Eratosthéne et le théorème de Goldbach, Comptes Rendus Acad. Sci. Paris, **168**, 544–546.

Brun,V. (1919b): La série $\frac{1}{5}+\frac{1}{7}+\frac{1}{11}+\frac{1}{13}+\dots$ où les dénominateurs sont "nombres premiers jumeaux" est convergente ou finie. Bull. Sci. Math., (2), **43**, 100–104, 124–128.

Brun,V. (1920): Le crible d'Eratosthéne et le théorème de Goldbach. Videnselsk. Skr. **1**, Nr.3.

Brun,V. (1925): Untersuchungen über das Siebverfahren des Eratosthenes. Jahresber. Deutsche Math. Ver., **33**, 81–96.

Brun,V. (1928): Étude directe de la fonction $f(x)$ de Riemann. Comptes Rendus Acad. Sci. Paris, **177**, 810–813.

Brun,V. (1931): Algorithme pour calculer le n-iéme nombre premier. Forhandl. Norske Vid. Selsk., **4**, 66–69.

Brun,V. (1935): Umkehrprobleme. Avh.Norges Tekn. Høiskole, Jubil.Bd. 1–14.

Brünner,R.,Perelli,A.,Pintz,J. (1989): The exceptional set for the sum of a prime and a square. Acta Math. Acad. Sci. Hung., **53**, 347–365. MR 91b:11104.

Buck,R.C. (1946): Prime-representing functions. Amer. Math. Monthly, **53**, p.265. MR 8 p.11.

Buhler,J.P.,Crandall,R.E.,Penk,M.A. (1982): Primes of the form $n! \pm 1$ and $2 \cdot 3 \cdot 5 \cdots p \pm 1$. Math. Comp., **38**, 639–643. MR 83c:10006. Corr.: *ibidem*, **40**, 1983, p.727. MR 85b:11119.

Buhštab,A.A. (1940): Sur la décomposition des nombres pairs en somme de deux composantes dont chacune est formée d'un nombre borné de facteurs premiers. Doklady Akad. Nauk SSSR, **29**, 544–548. MR 2 p.348.

Buhštab,A.A. (1965): New results in the investigation of Goldbach-Euler's problem and the problem of twin prime numbers. Doklady Akad. Nauk SSSR, **162**, 735–738. (Russian)MR 31#3226.

Buhštab,A.A. (1967): A combinatorial strengthening of the Eratosthenian sieve method. Usp. Mat. Nauk, **22**, no.3, 199–226. (Russian) MR 36#1413.

Burlačenko,V.P. (1968): A method for the analytic continuation of the Riemann eta-function., Ukrain. Mat. Zh., **20**, 238–243. (Russian) MR 37#388.

Buschman,R.G.,Briggs,W.E. (1961): The power series coefficients of functions defined by Dirichlet series. Illinois J. Math., **5**, 43–44. MR 22#10956.

Byrnes,J.S.,Giroux,A.,Shisha,O. (1984): Riemann sums and improper integrals of step functions related to the Prime Number Theorem. J.Approx.Th., **40**, 180–192. MR 85f:40002.

Cahen,E. (1893): Sur la somme des logarithmes des nombres premiers qui ne depassent pas x. Comptes Rendus Acad. Sci. Paris, **116**, 85–88.

Cahen,E. (1894): Sur la fonction $\zeta(s)$ de Riemann et sur des fonctions analogues. Ann. Sci. Ec. Norm. Sup., (3), **11**, 75–164.

Cahen,E. (1900): Eléments de la théorie des nombres. Gauthier-Villars, Paris.

Cahen,E. (1911): Sur une question proposée par M.Fontené. Nouv. Ann. Math., (4), **11**, 70–72.

Caldwell,C.K. (1995): On the primality of $n! \pm 1$ and $2 \cdot 3 \cdot 5 \cdots p \pm 1$. Math. Comp., **64**, 889–890. MR 95g:11003.

Cantor,G. (1880): Zur Theorie der zahlentheoretischen Funktionen. Math. Ann., **16**, 583–588.

Cantor,G. (1896): Vérification jusqu'á 1000 du théorème empirique de Goldbach. Rep. Assoc. Franc. pour l'Avancement des Sciences, 18 pp.

Carlitz,L. (1961): A recurrence formula for $\zeta(2n)$. Proc. Amer. Math. Soc., **12**, 991–992. MR 21#A3140.

Carlson,F. (1920): Über die Nullstellen der Dirichletschen Reihen und der Riemannscher ζ-Funktion. Arkiv f. Math. Astr. Fys., **15**, No.20.

Carmichael,R.D. (1905): Six propositions on prime numbers. Amer. Math. Monthly, **12**, 106–108.

Carmichael,R.D. (1909): On the numerical factors of certain arithmetic forms. Amer. Math. Monthly, **16**, 153–159.

Carmichael,R.D. (1913): On the numerical factors of the arithmetic forms $\alpha^n \pm \beta^n$. Ann. of Math., **15**, 30–70.

Cassels.J.W.S., Fröhlich,A. (editors) (1967): Algebraic Number Theory. Academic Press. MR 35#6500.

Catalan,E. (1839): Note sur la théorie des nombres. J. math. pures appl., **4**, 7–8.

Catalan,E. (1888): Sur le postulatum de Bertrand. Mém. Soc. Roy. Sci. Liége, (2), **15**, 108–111.

Čebyšev,P.L. (1848): Sur la fonction qui détermine la totalité des nombres premiers inférieurs à une limite donnée. Mémoires des savants étrangers de l'Acad. Sci. St.Pétersbourg, **6**, 1–19. [Also in J. math. pures appl., **17**, 1852, 341–365; Oeuvres, **1**, 27–48. S.Pétersbourg 1899, reprinted by Chelsea. Russian translation: Polnoe sobranie sočinenii, **1**, 173–190. Moskva–Leningrad 1946.]

Čebyšev,P.L. (1850): Mémoire sur nombres premiers. Mémoires des savants étrangers de l'Acad. Sci. St.Pétersbourg, **7**, 17–33. [Also in J. math. pures appl., **17**, 1852, 366–390; Oeuvres, **1**, 49–70. S.Pétersbourg 1899, reprinted by Chelsea. Russian translation: Polnoe sobranie sočinenii, **1**, 191–207. Moskva–Leningrad 1946.]

Čebyšev,P.L. (1851): Note sur différentes séries. J. math. pures appl., **16**, 337–346. [Oeuvres, **1**, 97–108. S.Pétersbourg 1899, reprinted by Chelsea. Russian translation: Polnoe sobranie sočinenii, **1**, 229–236. Moskva–Leningrad 1946;
Čebyšev,P.L. (1853): Lettre de M. le professeur Tchébychev à M.Fuss, sur un noveau théorème relatif aux nombres premiers contenus dans les formes $4n+1$ et $4n+3$. Bull. Cl. phys.-math. Acad. Imp. Sci. St.Pétersbourg, **11**, p.208. [Oeuvres, **1**, 697–698. S.Pétersbourg 1899, reprinted by Chelsea. Russian translation: Polnoe sobranie sočinenii, **1**, p.276. Moskva–Leningrad 1946.]
Čebyšev,P.L. (1857): Sur une formule d'analyse. J. Reine Angew. Math., **53**, p.286. [Oeuvres, **1**, 701–702. S.Pétersbourg 1899, reprinted by Chelsea. Russian translation: Polnoe sobranie sočinenii **2**, 99–100. Moskva–Leningrad 1947.]
Čebyšev,P.L. (1879): Sur une transformation de séries numeriques. Nouv. Corresp. Math., **4**, 305–308. [Oeuvres, **2**, 705–707. S.Pétersbourg 1907, reprinted by Chelsea. Russian translation: Polnoe sobranie sočinenii, **1**, p.280–282. Moskva–Leningrad 1946.]
Césaro,E. (1883): Sur diverses questions d'arithmétique. Mém. soc. roy. sci. Liége, (2), **10**, 1–355.
Césaro,E. (1885a): Sull'inversione delle identità aritmetiche. Giornale di mat., **23**, 168–174.
Césaro,E. (1885b): Généralisation de l'identité de MM Tchebychew et de Polignac. Nouv. Ann. Math., (3), **4**, 418–422.
Césaro,E. (1888): Sur une fonction arithmétique. Comptes Rendus Acad. Sci. Paris, **106**, 1340–1343.
Césaro,E. (1893): Nuova contribuzione ai principii fondamentali dell'aritmetica assintotica. Atti R.Accad. sci.fis.mat. Napoli, (2), **6**.
Césaro,E. (1896): Sulla distribuzione dei numeri primi. Rendiconti Accad. sci.fis. mat. Napoli, (3), **2**, pp. 297 seq.
Chakravarty,I.C. (1974): On the functional equation of the Riemann zeta-function. Bull. Math. Soc. Sci. Math.R.S.Roumanie, (N.S.), **18**, 17–23. MR 53#3279.
Chandrasekharan,K. (1941): On Hadamard's factorization theorem. J.Indian Math. Soc., **5**, 128–132. MR 3 p.201.
Chandrasekharan,K. (1968): Introduction to Analytic Number Theory. Springer, Berlin-Heidelberg. MR 40#2593.
Chandrasekharan,K. (1970): Arithmetical Functions. Springer, Berlin–Heidelberg. MR 43#3223.
Chandrasekharan,K.,Mandelbrojt,S. (1957): On Riemann's functional equation. Ann. of Math., (2), **66**, 285–296. MR 19 p.635.
Chandrasekharan,K.,Mandelbrojt,S. (1959): On solutions of Riemann's functional equation. Bull. Amer. Math. Soc., **65**, 358–362. MR 22#2589.
Chang,G.J.,Lih,K.W. (1977): Polynomial representation of primes. Tamkang J. Math., **8**, 197–198. MR 58#27718.
Chao, Chung Yun (1982): Generalizations of theorems of Wilson, Fermat and Euler. J. Number Th., **15**, 95–114. MR 83k:10010.
Cheer,A.Y.,Goldston,D.A. (1993): Simple zeros of the Riemann zeta-function. Proc. Amer. Math. Soc., **118**, 365–372. MR 93g:11092.
Chen Jing Run (1965): On the least prime in an arithmetic progression. Sci. Sinica, **14**, 1868–1871. MR 32#5611.
Chen Jing Run (1966): On the representation of a large even integer as the sum of a prime and the product of at most two primes. Kexue Tongbao, **17**, 385–386. MR 34#7483.
Chen Jing Run (1973): On the representation of a large even integer as the sum of a prime and the product of at most two primes. Sci. Sinica, **16**, 157–176. MR 55#7959.

Chen Jing Run (1975): On the distribution of almost primes in an interval. Sci. Sinica, **18**, 611–627. MR 56#15584.
Chen Jing Run (1977): On the least prime in an arithmetic progression and two theorems concerning the zeros of Dirichlet's L-functions,I. Sci. Sinica, **20**, 529–562. MR 57#16227.
Chen Jing Run (1978a): On the representation of a large even integer as the sum of a prime and the product of at most two primes,II. Sci. Sinica, **21**, 421–430. MR 80e:10037.
Chen Jing Run (1978b): On the Goldbach's problem and the sieve methods. Sci. Sinica, **21**, 701–739. MR 80b:10069.
Chen Jing Run (1979a): On the distribution of almost primes in an interval, II. Sci. Sinica, **22**, 253–275. MR 82d:10065.
Chen Jing Run (1979b): On the least prime in an arithmetic progression and theorems concerning the zeros of Dirichlet's L-functions,II. Sci. Sinica, **22**, 859–889. MR 80k:10042.
Chen Jing Run (1983): The exceptional set of Goldbach numbers,II. Sci. Sinica, **26**, 714–731. MR 85m:11059.
Chen Jing Run (1989): The exceptional set of Goldbach numbers,III. Chinese Quart. J. Math., **4**, 1–15. MR 90k:11129.
Chen Jing Run,Liu Jan Min (1989): On the least prime in an arithmetic progression, III,IV. Sci. Sinica, **32**, 654–673, 792–807. MR 91h:11090ab.
Chen Jing Run,Liu Jan Min (1991): On the least prime in an arithmetic progression, V. International Symposium in Memory of Hua Loo Keng, **1**, 19–42, Springer, Berlin–Heidelberg 1991. MR 92m:11093.
Chen Jing Run,Pan Cheng Dong (1980): The exceptional set of Goldbach numbers. Sci. Sinica, **23**, 416–430. MR 82f:10060.
Chen Jing Run,Wang,T. (1989a): On the Goldbach problem. Acta Math. Sinica, **32**, 702–718. (Chinese) MR 91e:11108.
Chen Jing Run,Wang,T. (1989b): A theorem on exceptional zeros of L-functions. Acta Math. Sinica, **32**, 841–858. MR 91j:11069.
Chen Jing Run,Wang,T. (1991): A note on the Goldbach problem. Acta Math. Sinica, **34**, 143–144. (Chinese) MR 92g:11101.
Chen Jing Run,Wang,T. (1996): The Goldbach problem for odd numbers. Acta Math. Sinica, **39**, 169–174. MR 97k:11138.
Chen Ming Po (1975): An elementary evaluation of $\zeta(2m)$. Chinese J.Math., **3**, 11–15. MR 56#280.
Choi,K.K. (1997): A numerical bound for Baker's constant — some explicit estimates for small solutions of linear equations. Bull. Hong Kong Math.Soc., **1**, 1–19. MR 98k:11137.
Choi,K.K.,Liu,M.C.,Tsang,K.M. (1992): Conditional bounds for small prime solutions of linear equations. Manuscr. Math., **74**, 321–340. MR 93a:11084.
Chowla,S. (1934a): On the least prime in the arithmetical progression. J.Indian Math. Soc., (2), **1**, 1–3.
Chowla,S. (1934b): Primes in an arithmetic progression. Indian Phys.-Math.J., **5**, 35–43.
Chowla,S. (1935): The greatest prime factor of x^2+1. J. London Math. Soc., **10**, 117–120.
Chowla,S. (1936a): Note on Dirichlet's L-functions. Acta Arith., **1**, 113–114.
Chowla,S. (1936b): The representation of a number as a sum of four squares and a prime. Acta Arith., **1**, 115–122.
Chowla,S. (1950): A new proof of a theorem of Siegel. Ann. of Math., (2), **51**, 120–122. MR 11 p.420.

Chowla,S.,Mordell,L.J. (1961): Note on the nonvanishing of $L(1)$. Proc. Amer. Math. Soc., **12**, 283–284. MR 22#10972.

Chowla,S.,Selberg.A. (1949): On Epstein's zeta function,I. Proc. Nat. Acad. Sci. U.S.A. , **25**, 371–374. MR 11 p.84. [A.Selberg, Collected Papers, **I**, 367–370. Springer, Berlin–Heidelberg 1989.]

Cipolla,M. (1902): La determinazione assintotica dell' n^{imo} numero primo. Rend. Accad. Sci. fis. mat. Napoli, (3), **8**, 132–166.

Cipolla,M. (1905): Estensione delle formole di Meissel–Rogel e di Torelli sulla totalità dei numeri primi che non superano un numero assegnato. Annali di Mat. Pura Appl., (3), **11**, 253–267.

Čižek,J. (1981): On the proof of the prime number theorem. Časopis Pěst. Mat., **106**, 395–401. MR 82m:10061.

Clarkson,J.A. (1966): On the series of prime reciprocals. Proc. Amer. Math. Soc., **17**, 541. MR 32#5573.

Clausen,T. (1858): Beweis des von Schlömilch Archiv Bd.XII, No. XXXV aufgestellten Lehrssatzes; — über die Ableitung des Differentials von $\log \Gamma x$; und — über eine allgemeine Aufgabe über die Funktionen von Abel. Archiv Math. Phys., **30**, 166–170.

Clement,P.A. (1949): Congruence for sets of primes. Amer. Math. Monthly, **56**, 23–25. MR 10 p.353.

Cohen,E. (1969): Legendre's identity. Amer. Math. Monthly, **76**, 611–616. MR 39#4094.

Coleman,M.D. (1990): On the equation $b_1p - b_2P_2 = b_3$. J. Reine Angew. Math., **403**, 1–66. MR 90j:11088.

Coleman,M.D. (1993): The Rosser-Iwaniec sieve in number fields, with an application. Acta Arith., **65**, 53–83. MR 94h:11086.

Comessatti,A. (1906): Una dimostrazione della formola di Meissel. Periodico di Matematica insegn. sec., (3), **3**, 183–186.

Conrey,J.B. (1983): Zeros of derivatives of Riemann's ξ-function on the critical line. J. Number Th., **16**, 49–74. MR 84g:10070.

Conrey,J.B. (1989): More than two fifth of the zeros of the Riemann zeta function are on the critical line. J. Reine Angew. Math., **399**, 1-26. MR 90g:11120.

Conrey,J.B.,Ghosh,A. (1985): A simpler proof of Levinson's theorem. Math. Proc. Cambridge Philos. Soc., **97**, 385–395. MR 86g:11049.

Conrey,J.B.,Ghosh,A.,Gonek,S.M. (1988): Simple zeros of the zeta functions. Colloque de Théorie Analytique des Nombres "Jean Coquet", (Marseille 1985), 77–83. Publ. Math. Orsay, **88-02**. MR 89h:11056.

Conrey,J.B.,Ghosh,A.,Gonek,S.M. (1989): Mean values of the Riemann zeta-function with application to the distribution of zeros. Number Theory, Trace Formulas and Discrete Groups (Oslo 1987), 185–199. Academic Press, Boston. MR 90h:11077.

Cook,R.J. (1991): Chebyshev functions and inclusion-exclusion. J.Indian Math. Soc., (N.S.), **57**, 109–116. MR 93c:11004.

Corput,J.G.van der (1937): Sur l'hypothése de Goldbach pour presque tous les nombres pairs. Acta Arith., **2**, 266–290.

Corput,J.G.van der (1939): Über Summen von Primzahlen und Primzahlquadraten. Math. Ann., **116**, 1–50.

Corput,J.G.van der (1956): Sur la reste dans la démonstration élementaire du théorème des nombres premiers. Colloque sur la Théorie des Nombres, Bruxelles 1955, 163–182. G.Thon, Liège, Masson et Cie, Paris. MR 18 p.112.

Corput,J.G.van der,Koksma,J.F. (1930): Sur l'ordre de grandeur de la fonction $\zeta(s)$ de Riemann dans le bande critique. Annales de Toulouse, (3), **22**, 1–39.

Corput, J.G.van der,Schaake,G. (1932): Über den Viggo Brunschen Algorithmus zur Berechnung der n-ten Primzahl. Forhandl. Norske Vid. Selsk., **5**, 36–37.
Costa Pereira,N. (1981): An elementary method for estimating Chebyshev functions θ and ψ in prime number theory. Ph.D.dissertation, Columbia University.
Costa Pereira,N. (1985a): Estimates for the Chebyshev function $\psi(x) - \theta(x)$. Math. Comp., **44**, 211–221. MR 86k:11005. Corr.: *ibidem*, **48**,1987,p.447. MR 87k:11006.
Costa Pereira,N. (1985b): A short proof of Chebyshev's theorem. Amer. Math. Monthly, **92**, 494–495. MR 87a:11084.
Costa Pereira,N. (1985/86): Sharp elementary estimates for the sequence of primes. Portugalia Math., **43**, 399–400. MR 88k:11058.
Costa Pereira,N. (1989): Elementary estimates for the Chebyshev function $\psi(x)$ and for the Möbius function $M(x)$. Acta Arith., **52**, 307–337. MR 91e:11099.
Cox,C.D.,Poorten,A.J. van der (1968): On a sequence of prime numbers. J. Austral. Math. Soc., **8**, 571–574. MR 37#3998.
Craig,C.F. (1923): On the Riemann zeta-function. Bull. Amer. Math. Soc., **29**, 337–340.
Cramér,H. (1918a): Über die Herleitung der Riemannscher Primzahlformel. Arkiv f. Math. Astr. Fys., **13**, No.24, 1–7. [Collected Works, **1**, 85–91, Springer, Berlin–Heidelberg 1994.]
Cramér,H. (1918b): Über die Nullstellen der Zetafunktion. Math. Zeitschr., **2**, 237–241. [Collected Works, **1**, 92–96. Springer, Berlin–Heidelberg 1994.]
Cramér,H. (1919a): Sur les zéros de la fonction $\zeta(s)$. Comptes Rendus Acad. Sci. Paris, **168**, 539–541. [Does not appear in Collected Works.]
Cramér,H. (1919b): Studien über die Nullstellen der Riemannscher Zetafunktion. Math. Zeitschr., **4**, 104–130. [Collected Works, **1**, 97–123. Springer, Berlin–Heidelberg 1994.]
Cramér,H. (1920): Some theorems concerning prime numbers, Arkiv f. Math. Astr. Fys., **15**, No.5. 1–33. [Collected Works, **1**, 138–170. Springer, Berlin–Heidelberg 1994.]
Cramér,H. (1922): Ein Mittelwertsatz der Primzahltheorie. Math. Zeitschr., **12**, 147–153. [Collected Works, **1**, 229–235. Springer, Berlin–Heidelberg 1994.]
Cramér,H. (1936): On the order of magnitude of the difference between consecutive primes. Acta Arith., **2**, 23–46. [Collected Works, **2**, 871–894. Springer, Berlin–Heidelberg 1994.]
Cramér,H.,Landau,E. (1921): Über die Zetafunktion auf der Mittellinie des kritischen Streifens. Arkiv f. Math. Astr. Fys., **15**, nr.28,1–4. [H.Cramér, Collected Works, **1**, 180–183. Springer, Berlin–Heidelberg 1994; E.Landau,Collected Works, **7**, 336–339. Thales Verlag.]
Crandall,R.E.,Doenias,J.,Norrie,C.,Young,J. (1995): The twenty-second Fermat number is composite. Math. Comp., **64**, 863–868. MR 95f:11104.
Crelle,A.L. (1845): Encyclopädische und elementare Darstellung der Theorie der Zahlen. (III Teil). J. Reine Angew. Math., **29**, 58–95.
Čudakov,N.G. (1936a): On the zeros of Dirichlet's L-functions. Mat. Sb., **1**, 591–602.
Čudakov,N.G. (1936b): On the difference between two neighbouring prime numbers. Mat. Sb., **1**, 799–814.
Čudakov,N.G. (1938): On the density of the set of even numbers which are not representable as a sum of two primes. Izv. Akad. Nauk SSSR, **2**, 25–40.
Čudakov,N.G. (1942): On Siegel's theorem. Izv. Akad. Nauk SSSR, **6**, 135–142. (Russian)MR 4 p.240.
Čudakov,N.G. (1947a): On Goldbach-Vinogradoff's theorem. Ann. of Math., (2), **48**, 515–545. MR 9 p.11.

Čudakov,N.G. (1947b): Introduction to the theory of Dirichlet's L-functions. OGIZ, Moskva,Leningrad. (Russian)MR 11 p.234.

Čudakov,N.G. (1953): On a class of completely multiplicative functions. Usp. Mat. Nauk, **8**, no.3, 149–150. (Russian)MR 15 p.289.

Čudakov,N.G.,Linnik,J.V. (1950): On a class of completely multiplicative functions. Doklady Akad. Nauk SSSR, **74**, 193–196. (Russian)MR 12 p.393.

Čudakov,N.G.,Rodosskii,K.A. (1949): New methods in the theory of Dirichlet's L-functions. Usp. Mat. Nauk, **4**, no.2, 22–56. (Russian)MR 13 p.824.

Čudakov,N.G.,Rodosskii,K.A. (1950): On generalized characters. Doklady Akad. Nauk SSSR, **73**, 1137–1139. (Russian)MR 12 p.393.

Cunningham,A. (1906/07): Evidence of Goldbach's theorem. Messenger of Math., **36**, 17—30.

Cunningham,A. (1911): Number of primes of given linear form. Proc. London Math. Soc., (2), **10**, 249–253.

Cunningham,A. (1913): On the number of primes of same residuacity. Proc. London Math. Soc., (2), **13**, 258–272.

Daboussi,H. (1984): Sur le théorème des nombres premiers. Comptes Rendus Acad. Sci. Paris, **198**, 161–164. MR 85f:11065.

Daboussi,H. (1989): On the Prime Number Theorem for arithmetic progressions. J. Number Th., **31**, 243–254. MR 90h:11085.

Davenport,H. (1966): Eine Bemerkung über Dirichlet's L-Funktionen. Nachr. Ges. Wiss. Göttingen, 203–212. MR 36#120.

Davenport,H. (1967): Multiplicative Number Theory. Markham, Chicago. MR 36 #117. [2nd ed. Springer, Berlin–Heidelberg 1980. MR 82m:10001.]

Davenport,H.,Halberstam,H. (1966): Primes in arithmetic progressions. Michigan J. Math., **13**, 485–489. MR 34#156. Corr.: *ibidem*, **15**, 1968, p.505. MR 38#2099.

Davenport,H.,Heilbronn,H. (1936): On the zeros of certain Dirichlet series,I,II. J. London Math. Soc., (2), **11**, 181–185, 307–313. [The Collected Papers of Hans Arnold Heilbronn, 272–282. J.Wiley 1988.]

Davenport,H.,Heilbronn,H. (1937): Note on a result in the additive theory of numbers. Proc. London Math. Soc., **43**, 142–151.

Davenport,H.,Schinzel,A. (1966): A note on certain arithmetical constants. Illinois J. Math., **10**, 181–185. MR 32#5632.

Davies,D. (1965): An approximate functional equation for Dirichlet L-functions. Proc. Roy. Soc. London, **A 284**, 224–236. MR 30#3565.

Davies,D,Haselgrove,C.B. (1961): The evaluation of Dirichlet L-functions. Proc. Roy. Soc. London, **A 264**, 122–132. MR 24#B2091.

Davis,M.,Putnam,H.,Robinson,J. (1961): The decision problem for exponential Diophantine equations. Ann. of Math., **74**, 425–436. MR 24#A3061.

Day,J.W.R. (1952): A theorem on Bernoulli numbers and the zeta function. J. London Math. Soc., **27**, 502–504. MR 14#248.

Dedekind,R. (1857): Abriss einer Theorie der höheren Congruenzen in Bezug auf einen rellen Primzahl-Modulus. J. Reine Angew. Math., **54**, 1–26. [Gesammelte mathematische Werke, **I**, 40–67. Braunschweig 1930; reprint Chelsea 1969.]

Delange,H. (1954): Généralisation du théorème de Ikehara. Ann. Sci. Ec. Norm. Sup., (3), **71**, 213–242. MR 16 p.921.

Deléglise,M.,Rivat,J. (1996): Computing $\pi(x)$: the Meissel, Lehmer, Lagarias, Miller, Odlyzko method. Math. Comp., **65**, 235–245. MR 96d:11139.

Delsarte,S. (1948): Fonctions de Möbius sur les groupes abéliens. Ann. of Math., (2), **49**, 600–609. MR 10 p.9.

Denjoy,A. (1954): Une démonstration de l'identité fondamentale de la fonction $\zeta(s)$ de Riemann. J. Analyse Math., **3**, 197–206. MR 15 p.948.

Desboves,A. (1855): Sur un théorème de Legendre et son application à la recherche de limites qui comprennent entre elles des nombres premiers. Nouv. Ann. Math., **14**, 281–295.
Deshouillers,J.-M. (1972/73): Amélioration de la constante de Šnirelman dans le probléme de Goldbach. Sém. Delange, Pisot, Poitou, **14**, exp. 14. MR 54#5149.
Deshouillers,J.-M. (1975/76): Sur la constante de Šnirelman. Sém. Delange, Pisot, Poitou, **17**, exp. 16. MR 57#5933.
Deshouillers,J.-M. (1977): Retour sur la methode de Čebyšev dans la théorie des nombres premiers. Bull. Soc. Math. France, Mémoire **49–50**, 41–45. MR 58#548.
Deshouillers,J.-M., Effinger,G., te Riele,H., Zinoviev,D. (1997): A complete Vinogradov 3–primes theorem under the Riemann hypothesis. Electron. Res. Announc. Amer.Math.Soc., **3**, 99–104. MR 98g:11112.
Deshouillers,J.-M.,Granville,A.,Narkiewicz,W.,Pomerance,C. (1993): An upper bound in Goldbach's problem, Math. Comp., **61**, 209–213. MR 94b:11101.
Deshouillers,J.-M.,Iwaniec,H. (1982): On the greatest prime factor of $n^2 + 1$. Ann. Inst. Fourier, **32**, 1–11. MR 84m:10033.
Deshouillers,J.-M.,Iwaniec,H. (1984): On the Brun-Titchmarsh theorem on average. Topics in Classical Number Theory, Proc. J.Bólyai Colloq., **34**, 319–333. North-Holland, Amsterdam. MR 86e:11085.
Dette,W.,Meier,J. (1984): Comments on: "On the remainder term of the prime number formula. III" by J.Pintz. Studia Sci. Math. Hungar., **19**, 125–130. MR 87h:11084.
Dette,W.,Meier,J.,Pintz,J. (1985): Bemerkungen zu einer Arbeit von Ingham über die Verteilung der Primzahlen. Acta Math. Acad. Sci. Hung., **45**, 121–132. MR 86h:11070.
Deuring,M. (1968): Imaginäre quadratische Zahlkörper mit der Klassenzahl Eins. Invent. math., **5**, 169–179. MR 37#4044.
Diamond,H.G. (1970): A set of generalized numbers showing Beurling's theorem to be sharp. Illinois J. Math., **14**, 29–34. MR 40#5556.
Diamond,H.G. (1973): Chebyshev estimates for Beurling generalized primes. Proc. Amer. Math. Soc., **39**, 503–508. MR 47#3332.
Diamond,H.G. (1975): Changes of sign of li $x - \pi(x)$. Enseign. Math., (2), **21**, 1–14. MR 51#12741.
Diamond,H.G. (1982): Elementary methods in the study of the distribution of prime numbers. Bull. Amer. Math. Soc., (N.S.), **7**, 553–589. MR 83m:10002.
Diamond,H.G.,Erdős,P. (1980): On sharp elementary prime number estimates. Enseign. Math., (2), **26**, 313–321. MR 83i:10055.
Diamond,H.G.,McCurley,K. (1981): Constructive elementary estimates for $M(x)$. Analytic Number Theory, Lecture Notes in Math., **899**, 239–253, Springer, Berlin–Heidelberg. MR 84b:10061.
Diamond,H.G.,Steinig,J. (1970): An elementary proof of the prime number theorem with a remainder term. Invent. math., **11**, 199–258. MR 43#6169.
Dickson,L.E. (1904): A new extension of Dirichlet's theorem on prime numbers. Messenger of Math., (2), **33**, 155–161.
Dickson,L.E. (1905): On the cyclotomic function. Amer. Math. Monthly, **12**, 86–89.
Dickson,L.E. (1919): History of the Theory of Numbers, I. Washington. [Reprint: Chelsea, New York 1952, 1966.]
Dilcher,K. (1992): Generalized Euler constants for arithmetic progression. Math. Comp., **59**, 259–282. MR 92k:11145.
Dirichlet,P.G.L. (1829): Sur la convergence des séries trigonométriques, qui servent à représenter une fonction arbitraire entre des limites données. J. Reine Angew. Math., **4**, 157–169. [Werke, **I**, 117–132. G.Reimer, Berlin 1889.]

Dirichlet,P.G.L. (1835): Über eine neue Anwendung bestimmter Integrale auf die Summation endlicher oder unendlicher Reihen. Abhandl. Kgl. Preuß. Akad.Wiss., 391–407. [Werke, **I**, 237–256. G.Reimer, Berlin 1889.]

Dirichlet,P.G.L. (1837a): Sur l'usage des intégrales définies dans la sommation des séries finies ou infinies. J. Reine Angew. Math., **17**, 57–67. [Werke, **I**, 257–270. G.Reimer, Berlin 1889.]

Dirichlet,P.G.L. (1837b): Beweis eines Satzes ueber die arithmetische Progression. Ber. Verhandl. Kgl. Preuß. Akad. Wiss., 108–110. [Werke, **I**, 307–312. G.Reimer, Berlin 1889.]

Dirichlet,P.G.L. (1837c): Beweis des Satzes, dass jede unbegrenzte arithmetische Progression, deren erstes Glied und Differenz ganze Zahlen ohne gemeinschaftlichen Factor sind, unendlich viele Primzahlen enthält. Abhandl. Kgl. Preuß. Akad. Wiss., 45–81. [Werke, **I**, 313—-342. G.Reimer, Berlin 1889; French translation: J. math. pures appl., **4**, 1839,393–422.]

Dirichlet,P.G.L. (1838a): Über die Bestimmung asymptotischer Gesetze in der Zahlentheorie. Ber. Verhandl. Kgl. Preuß. Akad. Wiss., 13–15. [Werke, **I**, 351–356. G.Reimer, Berlin 1889.]

Dirichlet,P.G.L. (1838b): Sur l'usage des séries infinies dans la théorie des nombres. J. Reine Angew. Math., **18**, 259–274. [Werke, **I**, 357–374. G.Reimer, Berlin 1889.]

Dirichlet,P.G.L. (1839): Recherches sur diverses applications de l'analyse infinitésimale a la théorie des nombres. J. Reine Angew. Math., **19**, 324–369; *ibidem*, **21**, 1840, 1–12, 134–155. [Werke, **I**, 411–496. G.Reimer, Berlin 1889.]

Dirichlet,P.G.L. (1840a): Über eine Eigenschaft der quadratischen Formen. Ber. Verhandl. Kgl. Preuß. Akad. Wiss., 49–52. [Werke, **I**, 497–502. G.Reimer, Berlin 1889.]

Dirichlet,P.G.L. (1840b): Sur la théorie des nombres. Comptes Rendus Acad. Sci. Paris, **10**, 285–288. [Werke, **I**, 619–623. G.Reimer, Berlin 1889.]

Dirichlet,P.G.L. (1841): Untersuchungen über die Theorie der complexen Zahlen. Abhandl. Kgl. Preuß. Akad. Wiss., 141–161. [Werke, **I**, 509–532. G.Reimer, Berlin 1889.]

Dirichlet,P.G.L. (1842a): Recherches sur les formes quadratiques a coefficients et a indéterminées complexes. J. Reine Angew. Math., **24**, 291–371. [Werke, **I**, 533–618. G.Reimer, Berlin 1889.]

Dirichlet,P.G.L. (1842b): Verallgemeinerung eines Satzes aus der Lehre von Kettenbrüchen nebst einigen Anwendungen auf die Theorie der Zahlen. Ber. Verhandl. Kgl. Preuß. Akad. Wiss., 93–95. [Werke, **I**, 633–638. G.Reimer, Berlin 1889.]

Dirichlet,P.G.L. (1849): Über die Bestimmung der mittleren Werte in der Zahlentheorie. Abhandl. Kgl. Preuß. Akad. Wiss., 69–83. [Werke, **II**, 49-66. G.Reimer, Berlin 1897.]

Dirichlet,P.G.L. (1856): Sur un théorème relatif aux séries. J. math. pures appl., (2), **1**, 80–81. [J. Reine Angew. Math., **53**, 130–132; Werke, **II**, 195–200. G.Reimer, Berlin 1897.]

Dirichlet,P.G.L. (1879): Vorlesungen über Zahlentheorie. 3rd edition, published by R.Dedekind, Vieweg, Braunschweig. [I ed. 1863, II ed. 1871, IV ed. 1893; reprint: Chelsea 1968. MR 38#5573.]

Drach (1844): On the empirical law in the enumeration of prime numbers. Philosoph. Magazine. (3), **24**, pp. 192 seq.

Dress,F. (1984): Théorèmes d'oscillations et fonction de Möbius. Sém. Th. Nombres Bordeaux, 1983–1984, exposé 33. MR 86f:11070.

Dress,F. (1993): Fonction sommatoire de la fonction de Möbius,I. Majorations expérimentales. Experiment. Math., **2**, 89–98. MR 95d:11177.

Dress,F.,El Marraki,M. (1993): Fonction sommatoire de la fonction de Möbius, II. Majorations asymptotiques élémentaires. Experiment. Math., **2**, 99–112. MR 95d:11178.
Dressler,R.E. (1972): A stronger Bertrand's postulate with an application to partitions. Proc. Amer. Math. Soc., **33**, 226–228. MR 45#1828. Addendum *ibidem*, **38**, 1973, p.667. MR 46#8947.
Dressler,R.E. (1973): Sums of distinct primes. Nordisk Mat. Tidskr., **21**, 31–32. MR 48#3851.
Dressler,R.E.,van der Lune,J. (1975): A note on a formula of Chebychef. Amer. Math. Monthly, **82**, 506–507. MR 51#3083.
Dudley,U. (1969): History of a formula for primes. Amer. Math. Monthly, **76**, 23–28. MR 38#4270.
Dudley,U. (1983): Formulas for primes. Math.Mag., **56**, 17–22. MR 84g:01084.
Dufner,G. (1995): Binäres Goldbachproblem mit Einschränkung an die Summanden. Period. Math. Hungar., **30**, 105–134. MR 96g:11122.
Dupré,A. (1859): Examen d'une proposition de Legendre relative à la théorie des nombres. Mallet-Bachelier, Paris.
Durand,L. (1856): Solution de la question 316. Nouv. Ann. Math., **15**, 296–297.
Dusart,P. (1999): The kth prime is greater that $k(LectureNotesinMath.k + LectureNotesinMath.LectureNotesinMath.k - 1)$ for $k \geq 2$. Math. Comp., **68**, 411–415, MR 99d:11133.
Dusumbetov,A. (1962): An elementary proof of the asymptotic law for the distribution of prime numbers. Izv. Akad. Nauk Uzb. SSR., ser. Fiz.-Mat. Nauk, **2**, 24–31. (Russian)MR 27#117.
Dusumbetov,A. (1963): An elementary proof of the asymptotic law for the distribution of primes in an arithmetic progression. Izv. Akad. Nauk Uzb. SSR., ser. Fiz.-Mat. Nauk, **3**, 5–12. (Russian)MR 27#2491.
Dux,E. (1956): Ein kurzer Beweis der Divergenz der unendlichen Reihe $\sum_{r=1}^{\infty} \frac{1}{p_r}$. Elem. Math., **11**, 50–51. MR 17 p.960.
Ecklund Jr.,E.F.,Eggleton,R.B. (1972): Prime factors of consecutive integers. Amer. Math. Monthly, **79**, 1082–1089. MR 47#6599.
Eda,Y. (1953): On the prime number theorem. Sci. Rep. Kanazawa Univ., **2**, 23–33. MR 17 p.587.
Edgorov,Zh. (1972): Applications of evaluations of trigonometric sums to the divisor problem and the mean-value of Riemann's zeta-function. Mat. Zametki, **12**, 539–548. (Russian)MR 47#6628.
Edwards,A.W.F. (1964): Infinite prime sequences. Math. Gaz., **48**, 416–422. MR 32#1180.
Edwards,H.M. (1974): Riemann's Zeta Function. Academic Press, New York. MR 57 #5922.
Ehlich,H. (1959): Ein elementarer Beweis des Primzahlsatzes für binäre quadratische Formen. J. Reine Angew. Math., **201**, 1–36. MR 21#4135.
Elliot,D.D. (1983): The Two-Year College Math.J., **14**, Nr.2, p.57.
Elliott,P.D.T.A., Halberstam,H. (1966): Some applications of Bombieri's theorem. Mathematika, **13**, 193–203. MR 34#5788.
Elliott,P.D.T.A., Halberstam,H. (1971): The least prime in an arithmetic progression. Studies in Pure Mathematics (Presented to R.Rado), 59–61. Academic Press, London. MR 42#7609.
Ellis,J.R. (1981): The computation of prime numbers by Gandhi's formula for the n-th prime. Math.Gaz., **65**, 212–214. MR 82j:10010.
Ellison,W.J. (in collaboration with M.Mendes-France) (1975): Les nombres premiers. Hermann, Paris. MR 54#5138. [English translation: Ellison,F., Ellison,W., Prime Numbers. J.Wiley 1985. MR 87d:11082.]

El Marraki,M. (1995): Fonction sommatoire de la fonction de Möbius. III. Majorations asymptotiques effectives fortes. J. Théor. Nombres Bordeaux, **7**, 407–433. MR 97e:11113.

Eneström,G. (1890): Note historique sur le somme des valeurs inverses des nombres carrés. Bibliotheca Math., **4**, 22–24.

Epstein,P. (1903): Zur Theorie allgemeiner Zetafunctionen. Math. Ann., **56**, 615–644.

Erdős.P. (1932): Beweis eines Satzes von Tschebyscheff. Acta Sci. Math., **5**, 194–198.

Erdős,P. (1934): A theorem of Sylvester and Schur. J. London Math. Soc., **9**, 282–288.

Erdős,P. (1935a): On the difference of consecutive primes. Quart. J.Math. Oxford ser., **6**, 124–128.

Erdős,P. (1935b): Über die Primzahlen gewisser arithmetischer Reihen. Math. Zeitschr., **39**, 473–491.

Erdős,P. (1938): Über die Reihe $\sum \frac{1}{p}$, Mathematica (Zutphen), B**7**, 1–2.

Erdős,P. (1940): The difference of consecutive primes. Duke Math. J., **6**, 438–441. MR 1 p.292.

Erdős,P. (1946): On the distribution of additive functions. Ann. of Math., (2), **47**, 1–20. MR 7 p.416.

Erdős,P. (1949a): On a new method in elementary number theory which leads to an elementary proof of the prime number theorem. Proc. Nat. Acad. Sci. U.S.A. , **35**, 374–384. MR 10 p.595.

Erdős,P. (1949b): On a Tauberian theorem connected with the new proof of the prime number theorem. J.Indian Math. Soc., (N.S.), **13**, 131–144. Suppl. 145–147. MR 11 p.420.

Erdős,P. (1949c): On some applications of Brun's method. Acta Sci. Math., **13**, 57–63. MR 10 p.684.

Erdős,P. (1952): On the greatest prime factor of $\prod_{k=1}^{x} f(k)$. J. London Math. Soc., **27**, 379–384. MR 13 p.914.

Erdős,P. (1976): Problems in number theory and combinatorics. Proc.6th Conf. Numer. Math., Manitoba. Congr. Numer., **18**, 35–58. MR 80e:10005.

Erdős,P.,Schinzel,A. (1990): On the greatest prime factor of $\prod_{k=1}^{x} f(k)$. Acta Arith., **55**, 191–200. MR 91h:11100.

Erdős,P.,Stewart,C.L. (1976): On the greatest and least prime factors of $n! + 1$. J. London Math. Soc., (2), **13**, 513–519. MR 53#13093.

Erman,A. (1852): Briefwechsel zwischen W.Olbers und F.W.Bessel. Avenarius u. Mendelssohn, Leipzig.

Ernvall,R. (1975): A formula for the least prime greater than a given integer. Elem. Math., **30**, 13–14. MR 54 #12616.

Errera,A. (1956): Sur le théorème fondamental des nombres premiers. Colloque Th. des Nombres, Bruxelles 1955, 111–118, Liége 1956. MR 18 p.112.

Errera,A. (1958): Une modification de la démonstration de Landau du théorème des nombres premiers. Mathesis, **67**, 321–337. MR 21#1285.

Estermann,T. (1932): Eine neue Darstellung und neue Anwendungen der Viggo Brunschen Methode. J. Reine Angew. Math., **168**, 106–116.

Estermann,T. (1936): Proof that every large integer is a sum of two primes and a square. Proc. London Math. Soc., (2), **42**, 501–526.

Estermann,T. (1937): A new result in the additive prime-number theory. Quart. J.Math. Oxford ser., **8**, 32–38.

Estermann,T. (1938): On Goldbach's problem: Proof that almost all even positive integers are sums of two primes. Proc. London Math. Soc., (2), **44**, 307–314.

Estermann,T. (1947): Elementary evaluation of $\zeta(2k)$. J. London Math. Soc., **22**, 10–13. MR 9 p.234.
Estermann,T. (1948): On Dirichlet's L functions. J. London Math. Soc., **23**, 275–279. MR 10 p.356.
Estermann,T. (1952): Introduction to Modern Prime Number Theory. Cambridge University Press. MR 13 p.915.
Estermann,T. (1962/63): Note on a paper of A.Rotkiewicz. Acta Arith., **8**, 465–467. MR 28#2076.
Eswarathasan,A.,Levine,E. (1991): p-integral harmonic sums. Discrete Math. **91**, 249–257. MR 93b:11039.
Euclid (300 B.C.): Elementa, I–XIII. [Euclidis Opera Omnia, edited by J.L.Heiberg. Teubner, Leipzig 1883–1885; English translation: Heath, T.L., The Thirteen books of Euclid's Elements, Cambridge 1926, reprint: Dover 1956. MR 17 p.814; French translation: Les Éléments, CNRS, Paris 1978, MR 80f:10125ab; German translation: Thaer,C.,Die Elemente, Akademische Verlagsgesellschaft, Leipzig 1933–1937; reprint: Wiss. Buchgemeinschaft Darmstadt 1973. MR 50#6769; Russian translation of books VII–IX: Mordukhai-Boltovskii,D.D., GITT, Moskva 1949. MR 14 p.524.]
Euler,L. (1730): De progressionis transcendentibus seu quarum termini generales algebraica dari nequeunt. Comment. Acad. Sci. Petropol., **5**, 36–57. [Opera omnia, **I.14**, 1–24, Teubner, Lipsiae et Berolini 1924.]
Euler,L. (1732/33a): Methodus generalis summandi progressiones. [Opera omnia, **I.14**, 42–72,Comment. Acad. Sci. Petropol., **6**, 42–72 (1738)[1]. Teubner, Lipsiae et Berolini 1924.]
Euler,L. (1732/33b): Observationes de theoremate quodam Fermatiano aliisque ad numeros primos spectantibus. Comment. Acad. Sci. Petropol., **6**, 103–107 (1738). [Opera omnia, **I.2**, 1–5, Teubner, Lipsiae et Berolini 1915.]
Euler,L. (1734/35a): De summis serierum reciprocarum. Comment. Acad. Sci. Petropol., **7**, 123–134 (1740). [Opera omnia, **I.14**, 73–86, Teubner, Lipsiae et Berolini 1924.]
Euler,L. (1734/35b): De progressionibus harmonicis observationes. Comment. Acad. Sci. Petropol., **7**, 150–161 (1740). [Opera omnia, **I.14**, 87–100, Teubner, Lipsiae et Berolini 1924.]
Euler,L. (1736): Inventio summae cuiusque seriei ex dato termino generale. Comment. Acad. Sci. Petropol., **8**, 9–22 (1741). [Opera omnia, **I.14**, 108–123, Teubner, Lipsiae et Berolini 1924.]
Euler,L. (1737): Variae observationes circa series infinitas. Comment. Acad. Sci. Petropol., **9**, 160–188 (1744). [Opera omnia, **I.14**, 216–244, Teubner, Lipsiae et Berolini 1924.]
Euler,L. (1740): De seriebus quibusdam considerationes. Comment. Acad. Sci. Petropol., **12**, 53–96 (1750). [Opera omnia, **I.14**, 407–462, Teubner, Lipsiae et Berolini 1924]
Euler,L. (1743): Demonstration de la somme de cette suite $1+\frac{1}{4}+\frac{1}{9}+\frac{1}{16}+\frac{1}{25}+\frac{1}{36}+$etc. J. littèraire d'Allemagne, de Suisse et du Nord, **2**, 115–127. [Bibl.mat., **8**, 1907/08, 54–60; Opera omnia, **I.14**, 177–186, Teubner, Lipsiae et Berolini 1924]
Euler,L. (1748): Introductio in analysin infinitorum, I. Lausanne. [Opera omnia, **I.8**, Teubner, Lipsiae et Berolini 1922.]
Euler,L. (1754): De seriebus divergentibus. Novi Comment. Acad. Sci. Petropol., **5**, 205–237 (1760). [Opera omnia, **I.14**, 585–617, Teubner, Lipsiae et Berolini 1924.]
Euler,L. (1755): Institutiones calculi differentiali cum eius usu in analysi finitorum ac doctrina serierum. [Opera omnia, **I.10**, Teubner, Lipsiae et Berolini 1913.]

[1] The second date indicates the actual date of publication.

Euler,L. (1760/61): Theoremata arithmetica nova methodo demonstrata. Novi Comment. Acad. Sci. Petropol., **8**, 74–104 (1763). [Opera omnia, **I.2**, 531–555. Teubner, Lipsiae et Berolini 1915.]

Euler,L. (1761): Remarques sur un beau rapport entre les series des puissances tant directes que reciproques. Mémoires Acad. Sci., **17**, 83–106, (1768). [Opera omnia, **I.15**, 70–90, Teubner, Lipsiae et Berolini 1927.]

Euler,L. (1762/63): De numeris primis valde magni. Novi Acta Acad. Sci. Petropol., **9**, 99–153 (1764). [Opera omnia, **I.3**, 1–45, Teubner, Lipsiae et Berolini 1917.]

Euler,L. (1769): De summis serierum numeros Bernouillianos involventiam. Novi Comment. Acad. Sci. Petropol., **14**, 129–167 (1770). [Opera omnia, **I.15**, 91–130, Teubner, Lipsiae et Berolini 1927.]

Euler,L. (1772): Extrait d'une lettre de M.Euler le pére à M.Bernoulli concernant la mémoire imprimé parmi ceux de 1771 p.318. Nouvelles Mémoires Acad. Sci. Berlin, 35–36 (1774). [Opera omnia, **I.3**, 335–337, Teubner, Lipsiae et Berolini 1917.]

Euler,L. (1773): Demonstrationes circa residua ex divisione potestatum per numeros primos resultantia. Novi Comment. Acad. Sci. Petropol., **18**, 85–135 (1774). [Opera Omnia **I.3**, 240–281. Teubner, Lipsiae et Berolini 1917.]

Euler,L. (1774): Nova methodus, quantitates integrali determinandi. Novi Comment. Acad. Sci. Petropol., **19**, 66–102. [Opera omnia, **I.17**, 421–457, Teubner, Lipsiae et Berolini 1915.]

Euler,L. (1775): De tabula numerorum primorum usque ad millionem ed ultra continuanda, in qua simul omnium numerorum non primorum minimi divisores exprimantur. Novi Comment. Acad. Sci. Petropol., **19**, 132–180. [Opera omnia, **I.3**, 359–404, Teubner, Lipsiae et Berolini 1917.]

Euler,L. (1780): Speculationes circa quasdam insignes proprietates numerorum. Acta Acad. Sci. Petropol., **4**, 18–30 (1784). [Opera omnia, **I.4**, 105–115, Genevae 1941.]

Euler L. (1785): De summa seriei ex numeris primis formatae $\frac{1}{3} - \frac{1}{5} + \frac{1}{7} + \frac{1}{11} - \frac{1}{13} - \frac{1}{17} + \frac{1}{19} + \frac{1}{23} - \frac{1}{29} + \frac{1}{31} -$ etc. ubi numeri primi formae $4n-1$ habent signum positivum, formae autem $4n+1$ signum negativum. Opuscula analytica, **2**, p.240. [Opera omnia, **I.4**, 146–162, Genevae 1941.]

Euler,L. (1849): Tractatus de numerorum doctrina. Commentationes Arithmeticae Collectae, **II**, 504–575, Petropoli. [Opera omnia, **I.5**, 182–283, Genevae 1944.]

Euler,L.,Goldbach,C. (1965): Briefwechsel 1729–1764. Akademie Verlag, Berlin. MR 34#1144.

Fanta,E. (1901a): Beweis, daß jede lineare Funktion, deren Koeffizienten dem kubischen Kreisteilungskörper entnommene ganze teilerfremde Zahlen sind, unendlich viele Primzahlen dieses Körpers darstellt. Monatsh. Math. Phys., **12**, 1–44.

Fanta,E. (1901b): Über die Verteilung der Primzahlen. Monatsh. Math. Phys., **12**, 299–301.

Farmer,D.W. (1995): Counting distinct zeros of the Riemann zeta-function. The Electronic J. of Combinatorics, **2**, #R1, 1–5. MR 96b:11120.

Faulkner,M. (1966): On a theorem of Sylvester and Schur. J. London Math. Soc., **41**, 107–110. MR 32#7487.

Fekete,M. (1910): Sur les séries de Dirichlet. Comptes Rendus Acad. Sci. Paris, **150**, 1033–1035.

Fekete,M. (1926): The zeros of Riemann's zeta-function on the critical line. J. London Math. Soc., **1**, 15–19.

Felgner,U. (1991): Estimates for the sequence of primes. Elem. Math., **46**, 17–25. MR 91m:11003.

Ferguson,R.P. (1963): An application of Stieltjes integration to the power series coefficients of the Riemann zeta function. Amer. Math. Monthly, **70**, 60–61. MR 26#2408.

Fermat,P. (1640a): Letter to Frenicle (August 1640). [Oeuvres, **II**, 205–206, Gauthier-Villars, Paris 1894.]

Fermat,P. (1640b): Letter to Frenicle (October 18th, 1640). [Oeuvres, **II**, 206–212, Gauthier-Villars, Paris 1894.]

Fermat,P. (1654): Letter to Pascal (August 29th, 1654). [Oeuvres, **II**, 307–310, Gauthier-Villars, Paris 1894.]

Fermat,P. (1658): Letter to Digby (June 1658). [Oeuvres, **II**, 402–408, Gauthier-Villars, Paris 1894.]

Fermat,P. (1659): Letter to Carcavi (August 1659). [Oeuvres, **II**, 431–436, Gauthier-Villars, Paris 1894.]

Fermat,P. (1679): De solutione problematum geometricorum per curvas simplicissimas et unicuique problematum generi proprie convenientes, dissertatio tripartita. Varia Opera Mathematica, 111–115. [Reprint: Friedlaender, Berlin 1861; Oeuvres, **I**, 118–131, Gauthier-Villars, Paris 1891; French translation: Oeuvres, **III**, 109–120, Gauthier-Villars, Paris 1896.]

Filaseta,M. (1988): Prime values of irreducible polynomials. Acta Arith., **50**, 133–145. MR 89g:11079.

Fine,N.J. (1951): Note on the Hurwitz zeta-function. Proc. Amer. Math. Soc., **2**, 361–364. MR 13 p.220.

Finsler,P. (1945): Über die Primzahlen zwischen n und $2n$. Festschrift zum 60 Geburtstag von Prof.Dr. Andreas Speiser, Füssli, Zürich. 118–122. MR 7 p.243.

Flett,T.M. (1950): On the function $\sum_{n=1}^{\infty} \frac{1}{n} \sin \frac{t}{n}$. J. London Math. Soc., **25**, MR 11 p.420.

Fogels,E. (1950): On an elementary proof of the prime number theorem. Latvijas PSR Zin. Akad. Fiz. Mat. Inst. Rakšti, **2**, 14–45. (Russian)MR 13#824.

Forbes,T. (1997): A large pair of twin primes. Math. Comp., **66**, 451–455. MR 97c:11111.

Forman,R. (1992): Sequences with many primes. Amer. Math. Monthly, **99**, 548–557. MR 93e:11104.

Forti,M.,Viola,C. (1973): Density estimates for the zeros of L-functions. Acta Arith., **23**, 379–391. MR 51#3081.

Fousserau,G. (1892): Sur la frèquence des nombres premiers. Ann. Sci. Ec. Norm. Sup., (3), **9**, 31–34.

Fouvry,É. (1975/76): Un résultat nouveau en théorie additive des nombres premiers. Sém. Delange–Pisot–Poitou, exp.7, 1–11. MR 55#12672.

Fouvry,É. (1984): Sur le théorème de Brun-Titchmarsh. Acta Arith., **43**, 417–424. MR 86c:11065.

Fouvry,É. (1984): Autour du thérème de Bombieri-Vinogradov. Acta Math., **152**, 219–244. MR 85m:11052.

Fouvry,É. (1990): Nombres presque premiers dans les petits intervalles. Analytic Number Theory (Tokyo 1988), Lecture Notes in Math., **1434**, 65–85. Springer, Berlin–Heidelberg. MR 91g:11104.

Fouvry,É.,Grupp,F. (1986): On the switching principle in sieve theory. J. Reine Angew. Math., **370**, 101–126. MR 87j:11092.

Fouvry,É.,Grupp,F. (1989): Weighted sieves and twin prime type equations. Duke Math. J., **58**, 731–748. MR 91e:11100.

Fouvry,E.,Iwaniec,H. (1983): Primes in arithmetic progressions. Acta Arith., **42**, 197–218. MR 84k:10035.

Fouvry,E.,Iwaniec,H. (1997): Gaussian primes. Acta Arith., **79**, 249–287. MR 98a:11120.

Franel,J.(1896): Sur la fonction $\xi(s)$ de Riemann et son application à l'arithmétique. Vierteljahrschr. Naturforsch. Gesellsch. Zürich, **41**, part 2, 7–19.
Franel,J.(1899): Sur une formule utile dans la détermination de certaines valeurs asymptotiques. Math. Ann., **51**, 369–387.
French,S.H. (1966): Trigonometric polynomials in prime number theory. Illinois J. Math., **10**, 240–248. MR 35#5404.
Friedlander,J.B. (1985): Notes on Chebyshev's method. Rocky Mount. J. Math., **15**, 377–383. MR 88f:11086.
Friedlander,J.B.,Granville,A. (1989): Limitations to the equi-distribution of primes, I. Ann. of Math., **129**, 363–382. MR 90e:11125.
Friedlander,J.B.,Granville,A. (1992a): Relevance of the residue class to the abundance of primes. Proc. Amalfi Conference on Analytic Number Theory (September 25–29, 1989), 95–103. Universitá di Salerno. MR 94i:11067.
Friedlander,J.B.,Granville,A. (1992b): Limitations to the equi-distribution of primes, III. Compos. Math., **81**, 19–32. MR 93g:11097.
Friedlander,J.B.,Iwaniec,H. (1997): Using a parity-sensitive sieve to count prime values of a polynomial. Proc. Nat. Acad. Sci. U.S.A. , **94**, 1054–1058. MR 98b:11097.
Frobenius,G. (1912): Über quadratische Formen, die viele Primzahlen darstellen. SBer. Kgl. Preuß. Akad. Wiss. Berlin, 966–980. [Gesammelte Abhandlungen, **III**, 573–587. Springer, Berlin–Heidelberg 1968.]
Fujii,A. (1988): Some generalization of Chebyshev's conjecture. Proc. Japan Acad. Sci., **64**, 260–263. MR 90h:11081.
Fujiwara,M. (1914): On the convergence-abscissa of general Dirichlet's series. Tôhoku Math. J., **6**, 140–142.
Fujiwara,M. (1920): Über Konvergenzabszisse der Dirichletschen Reihe. Tôhoku Math. J., **17**, 344-350.
Funakura,T. (1996): On characterization of Dirichlet L-functions. Acta Arith., **76**, 305–315. MR 97h:11093.
Fung,G.W.,Williams,H.C. (1990): Quadratic polynomials which have a high density of prime values. Math. Comp., **55**, 343–353. MR 90j:11090.
Furstenberg,H. (1955): On the infinitude of primes. Amer. Math. Monthly, **62**, 353. MR 16 p.904.
Furtwängler,P. (1907): Allgemeiner Existenzbeweis für den Klassenkörper eines beliebigen algebraischen Zahlkörpers. Math. Ann., **63**, 1–37.
Fuss,P.-H. (1843): Correspondance mathématique et physique de quelques célèbres géomètres du XVIIIéme siécle. St.Pétersbourg. [Reprint: Johnson Reprint Co. 1968.]
Gallagher,P.X (1970): A large sieve density estimate near $\sigma = 1$. Invent. math., **11**, 329–339. MR 43#4775.
Gallagher,P.X (1980): Some consequences of the Riemann hypothesis. Acta Arith., **37**, 339–343. MR 82j:10071.
Gandhi,J.M. (1971): Formulae for the nth prime. Proc. Washington State Univ. Conf. on Number Theory, 96–106. Pullman,Wash. MR 48#218.
Garrison,B. (1981): Consecutive integers for which $n^2 + 1$ is composite. Pacific J. Math., **97**, 93–96. MR 83h:10023.
Garrison,B. (1990): Polynomials with large number of prime values. Amer. Math. Monthly, **97**, 316–317. MR 91i:11124.
Gatteschi,L. (1947): Un perfezionamento di un teorema di I.Schur sulla frequenza dei numeri primi. Boll. Un. Mat. Ital., (3), **2**, 123–125. MR 9 p.332.
Gauger (1882): Ueber die Verallgemeinerung des Jakobi'schen Ausdruckes der Wurzeln einer Gleichung durch bestimmte Integrale. Dissertation, Greifswald.

Gauss,C.F. (1791): Asymptotische Gesetze der Zahlentheorie. Werke, **10.1**, 11–16, Teubner, Leipzig 1917.
Gauss,C.F. (1801) Disquisitiones Arithmeticae. G.Fleischer, Lipsiae. [Werke, **1**, 1–474, Kgl. Ges. Wiss. Göttingen 1863; French translation: Courcier, Paris 1807; English translation: Yale Univ. Press, New Haven 1966. MR 33#5545; German translation: Springer, Berlin 1889 (Reprint: Chelsea 1965. MR 32#5488); Russian translation: Trudy po teorii chisel, 7–583, Izdat. AN SSSR, Moskva 1959. MR 23#A2352.]
Gauss,C.F. (1811): Summatio quarumdam serierum singularium. Comm. Soc. Reg. Sc. Gotting., **1**. [Werke, **2**, 9–45, Kgl. Ges. Wiss., Göttingen 1863.]
Gauss,C.F. (1846): Disquisitiones generales de congruentiis. Werke, **2**, 212–242, Kgl. Ges. Wiss., Göttingen, 1863.
Gauss,C.F. (1849): Gauss an Encke. Werke, **2**, 444–447, Kgl. Ges. Wiss., Göttingen, 1863. [English translation in Goldstein (1973), pp.612–614.]
Gegenbauer,L. (1888): Note über die Anzahl der Primzahlen. SBer. Kais. Akad. Wissensch. Wien, **97**, Abt.2a, 374–377.
Gegenbauer,L. (1892a): Über die Tchebychef–de Polygnac'sche Identität. Monatsh. Math. Phys., **3**, 319–335.
Gegenbauer,L. (1892b): Über eine arithmetische Formel. Monatsh. Math. Phys., **3**, p.336.
Gegenbauer,L. (1896): Bemerkung über reelle Primzahlen. Monatsh. Math. Phys., **7**, 73–76.
Gegenbauer,L. (1899): Bemerkung über einen die Anzahl der Primzahlen eines bestimmten Intervals betreffended Satz des Herrn J.J.Sylvester. Monatsh. Math. Phys., **10**, 370–373.
Gelfond,A.O. (1953): On an elementary approach to certain problems in the theory of distribution of prime numbers. Vestnik Moskovs. Univ., ser. Fiz. Mat. Eststv. N., **8**, no.2, 21–26. (Russian) MR 15 p.402. [Izbrannye Trudy, 279–283, Moskva 1973.]
Gelfond,A.O. (1956): On the arithmetical equivalent of the analyticity of Dirichlet's L-series on the line Re $s = 1$. Izv. Akad. Nauk SSSR, **20**, 145–166. MR 22#4681. (Russian)[Izbrannye Trudy, 310–328, Moskva 1973.]
Gelfond,A.O.,Linnik,Yu.V. (1962): Elementary methods in analytic number theory. Moskva. MR 32#5575a. (Russian)[English translations: Rand McNally, Chicago 1965. MR 32#5575b; Pergamon Press 1966. MR 34#1252.]
Genocchi,A. (1860) Formule par determinare quanti siano i numeri primi fino ad un dato limite. Annali di Mat. Pura Appl., **3**, 52–59.
Genocchi,A. (1868/1869): Intorno ad alcune forme di numeri primi. Annali di Mat. Pura Appl., (2), **2**, 256–267.
Gentile,G. (1950): Numeri primi in un intervallo particolare. Period. Mat. (4), **28**, 130. MR 12 p.243.
Gerig,S. (1976): A simple proof of the prime number theorem. J. Number Th., **8**, 131–136. MR 53#10743.
Giordano,G. (1987): The generalization and proof of Bertrand's postulate. Intern. J. Math. Sci., **10**, 821–823. MR 88k:11059.
Giordano,G. (1989): The existence of primes in small intervals. Indian J.Math., **31**, 105–110. MR 91d:11111.
Gjeddebaek,N.F. (1966): On the difference between consecutive primes. Skand. Aktuarietidskr., 122–131. MR 36#6345.
Glaisher,J. (1870): On the calculation of Euler's constant. Proc. Roy. Soc. London, **19**, 514–524.
Glaisher,J. (1872): History of Euler's constant. Messenger of Math., **1**, 25–30.
Glaisher,J. (1876/77): Proc. London Math. Soc., **8**, p.261.

Glaisher,J. (1879–1883): Factor tables for the fourth, fifth, sixth million. London 1879, 1880, 1883.
Glatfeld,M. (1957): Anwendungen der Wienerschen Methode auf den Primzahlsatz für die arithmetische Progression. J. London Math. Soc., **32**, 67–73. MR 18 p.874.
Glatfeld,M. (1969): Primzahlverteilung und Taubersche Asymptotik. Überblicke Math., **3**, 71–95. MR 42#4504.
Goetgheluck,P. (1989): On cubic polynomials giving many primes. Elem. Math., **44**, 70–73. MR 90j:11014.
Goldfeld,D.M. (1974): A simple proof of Siegel's theorem. Proc. Nat. Acad. Sci. U.S.A. , **71**, p.1055. MR 49#8962.
Goldfeld,D.M. (1975): A further improvement of the Brun Titchmarsh theorem. J. London Math. Soc., (2), **11**, 434–444. MR 52#3083.
Goldfeld,D.M.,Schinzel,A. (1975): On Siegel's zero. Ann. Scuola Norm. Sup. Pisa, Cl. Sci., **2**, 571–583. MR 53#8016.
Goldsmith,D.L. (1965): On a result of E.Bombieri concerning Dirichlet's L-functions. Ist. Lombardo Ac. Sci. Lett. Rend., **A99**, 662–668. MR 32#7518.
Goldstein,L.J. (1973): A history of the prime number theorem. Amer. Math. Monthly, **80**, 599–615. MR 47#1726. Corr. *ibidem*, p.1115. MR 48#8355.
Goldston,D.A. (1982): On a result of Littlewood concerning prime numbers. Acta Arith., **40**, 263–271. MR 83m:10076.
Goldston,D.A. (1983): On a result of Littlewood concerning prime numbers,II. Acta Arith., **43**, 49–51. MR 85k:11042.
Goldston,D.A. (1992): On Bombieri and Davenport's theorem concerning small gaps between primes. Mathematika, **39**, 10–17. MR 93h:11102.
Golomb,S.W. (1955): Sets of primes with intermediate density. Math. Scand., **3**, 264–274. MR 17 p.828.
Golomb,S.W. (1959): A connected topology for integers. Amer. Math. Monthly, **66**, 663–665. MR 21#6347.
Golomb,S. (1962): On the ratio of N to $\pi(N)$. Amer. Math. Monthly, **69**, 36–37.
Golomb,S.W. (1974): A direct interpretation of Gandhi's formula. Amer. Math. Monthly, **81**, 752–754. MR 50#7003.
Golomb,S.W. (1976): Formulas for the next prime. Pacific J. Math., **63**, 401–404. MR 53#13094.
Gomes-Teixeira,F. (1904): Note sur deux travaux d'Abel relatifs a l'intégration des différences finies. Acta Math., **28**, 235–242.
Gonek,S.M. (1993): An explicit formula of Landau and its application to the theory of the zeta-function. Contemp. Math., **143**, 395–413. MR 94a:11134.
Goodstein,R.L. (1967): Formulae for primes. Math. Gaz., **51**, 35–36.
Gordon,B. (1958): On a tauberian theorem of Landau. Proc. Amer. Math. Soc., **9**, 693–696. MR 21#2846.
Graham,S.W. (1981): The greatest prime factor of the integers in an interval. J. London Math. Soc., **24**, 427–440. MR 82k:10054.
Gram,J.P. (1884): Undersøgesler angaaende Maengden af Primtal under en given Graense. K.Danske Vidensk. Selskabs Skrifter, Nat.Math.Afd., **6**, 183–308.
Gram,J.P. (1893): Rapport sur quelques calculs entrepris par M.Bertelsen et concernant les nombres premiers. Acta Math., **17**, 301–314.
Gram,J.P. (1895): Note sur le calcul de la fonction $\zeta(s)$ de Riemann. Oversigt Kong. Dansk.Vid.Selsk.Forh., 303–308.
Gram,J.P. (1897/98): Note sur la problème des nombres premiers. Oversigt Kong. Danske Vid. Selsk. Forh., 235–251.
Gram,J.P. (1903): Note sur les zéros de la fonction $\zeta(s)$. Acta Math., **27**, 289–304. [Oversigt Kong. Dansk.Vid.Selsk.Forh., 3–15, 1902.]

Grandjot,K.,Landau,E. (1924): Bemerkungen zu einer Arbeit des Herrn Onicescu. Bull. Sect. Scient. Acad. Roumaine, **8**, 166–168. [E.Landau, Collected Works, **8**, 67–69. Thales Verlag.]

Grant,A. (1987a): Fourier transform and the P.N.T. error term. Časopis Pěst. Math., **112**, 337–347. MR 89c:11139.

Grant,A. (1987b): A further note on the P.N.T. error term. Časopis Pěst. Math., **112**, 348–350. MR 89c:11140.

Granville,A. (1989): Least primes in arithmetic progressions. Théorie des nombres; Number Theory (Laval 1987), 306–321. de Gruyter, Berlin. MR 91c:11052.

Granville,A. (1992): On elementary proofs of the Prime Number Theorem for arithmetic progressions, without characters. Proc. Amalfi Conference on Analytic Number Theory (September 25–29, 1989), 157–194. Universitá di Salerno. MR 94k:11103.

Granville,A.,van de Lune,J.,te Riele, H.J.J. (1989): Checking the Goldbach conjecture on a vector computer. Number Theory and Applications, 423–433. Kluwer. MR 93c:11085.

Granville,A.,Pomerance,C. (1990): On the least prime in certain arithmetic progressions. J. London Math. Soc., (2), **41**, 193–200. MR 91i: 11119.

Greaves,G. (1976): On the representation of a number in the form $x^2+y^2+p^2+q^2$ where p and q are odd primes. Acta Arith., **29**, 257–274. MR 53#7985.

Grimson,W.E.L.,Hanson,D. (1978): Estimates for the product of the primes not exceeding x. Proc. VII Manitoba Conf. on Numerical Math. and Comput., Winnipeg. Congr. Numer. **20**, 407–416. MR 80i:10056.

Gritsenko,S.A. (1992): On the density theorem. Mat. Zametki, **51**, 32-40. (Russian)MR 93m:11088.

Gritsenko,S.A. (1994): Sharpening a constant in a density theorem. Mat. Zametki, **55**, no.2, 59–61. (Russian)MR 95a:11077.

Gronwall,T.H. (1913a): Über das Verhalten der Riemannscher Zeta-funktion auf der Geraden $\sigma=1$. Archiv Math. Phys., (3), **21**, 231–238.

Gronwall,T.H. (1913b): Sur la fonction $\zeta(s)$ de Riemann au voisinage de $\sigma=1$. Rend. Circ. Mat. Palermo, **35**, 95–102.

Grossmann,J. (1913): Über die Nullstellen der Riemannscher ζ-Funktion und der Dirichletschen L-Funktionen. Dissertation, Göttingen.

Grosswald,E. (1964): A proof of the prime number theorem. Amer. Math. Monthly, **71**, 736–743. MR 29#4742.

Grosswald,E. (1965): Sur l'ordre de grandeur des differences $\psi(x)-x$ et $\pi(x)-\mathrm{li}x$. Comptes Rendus Acad. Sci. Paris, **260**, 3813–3816. MR 31#3397.

Grosswald,E. (1967): Oscillation theorems of arithmetical functions. Trans. Amer. Math. Soc., **126**, 1–28. MR 34#2545.

Grosswald,E. (1968/69): On some conjectures of Hardy and Littlewood. Publ. Ramanujan Inst., **1**, 75–89. MR 42#4505.

Grosswald,E. (1972): Oscillation theorems. Lecture Notes in Math., **251**, 141–168. Springer, Berlin–Heidelberg. MR 48#11101.

Grosswald,E. (1987a): On prime representing polynomials. Proc. Indian Acad. Sci., Math. Sci., **97**, 75–84. MR 90b:11090.

Grosswald,E. (1987b): Some number theoretical products. Rev.Colombiana Mat. **21**, 231–242. MR 90e:11129.

Grosswald,E.,Schnitzler,F.J. (1978): A class of modified ζ and L-functions. Pacific J. Math., **74**, 357–364. MR 58#21970.

Guerin,E.E.,Buschman,R.G. (1975): Elementary proofs of the prime number theorem. Math. Japon., **20**, 273–282. MR 53#7981.

Guerin,E.E.,Buschman,R.G. (1976): Elementary proofs of statements equivalent to Prime Number Theorem. Portugalia Math., **35**,41–59. MR 55#2724.

Gupta,H. (1940): On a table of values of $L(n)$. Proc. Indian Acad. Sci., **A12**, 407–409. MR 2 p.248.

Gupta,H. (1943): A formula for $L(n)$. J.Indian Math. Soc., (N.S.), **7**, 68–71. MR 5 p. 199.

Gupta,H.N. (1985): A theorem in combinatorics and Wilson's theorem. Amer. Math. Monthly, **92**, 575–576. MR 87a:05012.

Guy,R.K. (1994): Unsolved problems in number theory, 2nd ed. Springer, Berlin–Heidelberg. MR 96e:11002. [1st ed. 1981. MR 83k:10002.]

Guy,R.,Nowakowski,R. (1975): Discovering primes with Euclid. Delta (Waukesha), **5**, 49–63. MR 52#5548.

Hacks,J. (1893): Über einige für Primzahlen charakteristische Beziehungen. Acta Math., **17**, 205–208.

Hadamard,J. (1893): Étude sur le propriétés des fonctions entières et en particulier d'une fonction considérée par Riemann. J. math. pures appl., (4), **9**, 171–215. [Selecta, 52–93, Gauthier-Villars, Paris 1935.]

Hadamard,J. (1896a): Sur les zéros de la fonction $\zeta(s)$ de Riemann. Comptes Rendus Acad. Sci. Paris, **122**, 1470–1473.

Hadamard,J. (1896b): Sur la fonction $\zeta(s)$. Comptes Rendus Acad. Sci. Paris, **123**, p.93.

Hadamard,J. (1896c): Sur la distribution des zéros de la fonction $\zeta(s)$ et ses conséquences arithmétiques. Bull. Soc. Math. France, **24**, 199–220. [Selecta, 111–132, Gauthier-Villars, Paris 1935.]

Hadamard,J. (1896d): Sur les fonctions entières. Bull. Soc. Math. France, **24**, 186–187.

Hadamard,J. (1897): Sur les séries de Dirichlet. Procés-verbaux Séanc. Soc. Sci. Phys. Nat. Bordeaux. 41–45.

Hadamard,J. (1899): Théorème sur les séries entières. Acta Math., **22**, 55–63.

Hadamard,J. (1904): Sur les séries de la forme $\sum a_n e^{-\lambda_n z}$. Nouv. Ann. Math., (4), **4**, 529–533.

Hadamard,J. (1908): Sur les séries de Dirichlet. Rend. Circ. Mat. Palermo, **25**, 326–330. Corr.: *ibidem*, 395–396.

Hadamard,J. (1909): Sue une propriété fonctionnelle de la fonction $\zeta(s)$. Bull. Soc. Math. France, **37**, 59–60.

Hadamard,J. (1915): Un travail de Jean Merlin sur les nombres premiers. Bull. Sci. Math., **50**, 121–136.

Halász,G.,Turán,P. (1969): On the distribution of roots of Riemann zeta and allied functions,I. J. Number Th., **1**, 122–137. MR 38#4422.

Halász,G.,Turán,P. (1970): On the distribution of roots of Riemann zeta and allied functions,II. Acta Math. Acad. Sci. Hung., **21**, 403–419. MR 42#3035.

Halberstam,H. (1951): On the representation of large numbers as sums of squares, higher powers and primes. Proc. London Math. Soc., (2), **53**, 363–380. MR 13 p.112.

Halberstam,H.,Heath-Brown,D.R.,Richert,H.-E. (1979): Almost-primes in short intervals. Recent Progress in Analytic Number Theory, (Durham 1979), **I**, 69–101, Academic Press, London. MR 83a:10075.

Halberstam,H.,Jurkat,W.B.,Richert,H.-E. (1967): Un nouveau résultat de la méthode du crible. Comptes Rendus Acad. Sci. Paris, **264**, A920–A923. MR 36#6374.

Halberstam,H.,Richert,H.-E. (1972): The distribution of polynomial sequences. Mathematika, **19**, 25–50. MR 58#21980.

Halberstam,H.,Richert,H.-E. (1974): Sieve methods. Academic Press, London, New York, San Francisco. MR 54#12689.

Halberstam,H.,Roth,K.F. (1966): Sequences, I. Oxford University Press. MR 35 p.1565.

Hall,R.S. (1973): Beurling generalized prime number systems in which the Chebyshev inequality fails. Proc. Amer. Math. Soc., **40**, 79–82. MR 47#6634.

Hall,T. (1972): Some relations in connection with the Möbius function. Nordisk Mat. Tidskr., **20**, 34–36. MR 49#2504.

Halphen,G.H. (1883): Sur l'approximation des sommes de fonctions numériques. Comptes Rendus Acad. Sci. Paris, **96**, 634–637. [Oeuvres, **IV**, 96–98, Gauthier-Villars, Paris 1924.]

Halphen,G.H. (1885): Notice sur les travaux mathématiques. Gauthier-Villars, Paris. [Oeuvres, **I**, 1–47, Gauthier-Villars, Paris 1916.]

Halter-Koch,F. (1991): Prime-producing quadratic polynomials and class numbers of quadratic orders. Computational Number Theory (Debrecen 1989), 73–82, de Gruyter, Berlin. MR 93c:11101.

Hamburger,H. (1921a): Über die Riemannsche Funktionalgleichung der ζ-Funktion. SBer. Berlin.Math.Ges., **20**, 4–9.

Hamburger,H. (1921b): Über die Funktionalgleichung der L-Reihen. SBer. Berlin. Math. Ges., **20**, 10–13.

Hamburger,H. (1921c): Über die Riemannsche Funktionalgleichung der ζ-Funktion. Math. Zeitschr., **10**, 240-254.

Hamburger,H. (1922a): Über die Riemannsche Funktionalgleichung der ζ-Funktion,II. Math. Zeitschr., **11**, 224–245; III. *ibidem*, **13**, 283–311.

Hamburger,H. (1922b): Über einige Beziehungen, die mit der Funktionalgleichung der Riemannschen ζ-Funktion äquivalent sind. Math. Ann., **85**, 129–140.

Haneke,W. (1962/63): Verschärfung der Abschätzung von $\zeta(1/2+it)$. Acta Arith., **8**, 357-430. MR 28#1179.

Haneke,W. (1970): Ein neuer Beweis des Linnikschen Nullstellensatzes. Publ. Math., **17**, 249–257. MR 48#6028.

Haneke,W. (1973): Über die reellen Nullstellen der Dirichletschen L-Reihen. Acta Arith., **22**, 391–421. MR 47#6623. Corr.: *ibidem*, **31**, 1976, 99–100.

Hanson,D. (1972): On the product of primes. Canad. Math. Bull., **15**, 33–37. MR 47#1734.

Hanson,D. (1973): On a theorem of Sylvester and Schur. Canad. Math. Bull., **16**, 195–199. MR 49#4918.

Harborth,H.,Kanold,H.-J.,Kemnitz,A. (1981): Abschätzung der Primzahlfunktion mit elementaren Methoden. Elem. Math., **36**, 167–170. MR 83a:10079.

Harborth,H.,Kemnitz,A. (1981): Calculations for Bertrand's postulate. Math. Mag., **54**, 33–34. MR 82c:10001.

Hardy,G.H. (1906): A formula for the prime factors of any numbers. Messenger of Math., **35**, 145–146. [Collected Papers, **II**, 4–5. Clarendon Press, Oxford 1967.]

Hardy,G.H. (1910): Theorems relating to the summability and convergence of slowly oscillating series. Proc. London Math. Soc., (2), **8**, 301–320. [Collected Papers, **VI**, 291–312. Clarendon Press, Oxford 1967.]

Hardy,G.H. (1914): Sur les zéros de la fonction $\zeta(s)$ de Riemann. Comptes Rendus Acad. Sci. Paris, **158**, 1012–1014. [Collected Papers, **II**, 6–8, Clarendon Press, Oxford 1967.]

Hardy,G.H. (1916): On the average order of the arithmetical functions $p(n)$ and $\Delta(n)$. Proc. London Math. Soc., (2), **15**, 1–25, 192–213. [Collected Papers, **II**, 294–315. Clarendon Press, Oxford 1967.]

Hardy,G.H. (1922a): A new proof of the functional equation for the Zeta-function. Mat.Tidsskrift, B, 71–73. [Collected Papers, **II**, 161–163. Clarendon Press, Oxford 1967.]

Hardy,G.H. (1922b): Notes on some points in the integral calculus. LV. On the integration of Fourier series. Messenger of Math., **51**, 186–192. [Collected Papers, **III**, 506–512. Clarendon Press, Oxford 1967.]

Hardy,G.H. (1927): Note on a theorem of Mertens. J. London Math. Soc., **2**, 70–72. [Collected Papers, **II**, 210–212. Clarendon Press, Oxford 1967.]

Hardy,G.H. (1929): An introduction to the theory of numbers. Bull. Amer. Math. Soc., **35**, 778–818.

Hardy,G.H. (1935): Second note on a theorem of Mertens. J. London Math. Soc., **10**, 91–94. [Collected Papers, **II**, 230–233. Clarendon Press, Oxford 1967.]

Hardy,G.H. (1949): Divergent Series. Clarendon Press, Oxford. MR 11 p.25.

Hardy,G.H.,Littlewood,J.E. (1913a): Tauberian theorems concerning series of positive terms. Messenger of Math., **42**, 191–192. [G.H.Hardy, Collected Papers, **VI**, 508–509. Clarendon Press, Oxford 1974.]

Hardy,G.H.,Littlewood,J.E. (1913b): Tauberian theorems concerning power series and Dirichlet's series whose coefficients are positive. Proc. London Math. Soc., (2), **13**, 174–191. [G.H.Hardy, Collected Papers, **VI**, 510–529. Clarendon Press, Oxford 1974.]

Hardy,G.H.,Littlewood,J.E. (1914): Some theorems concerning Dirichlet's series. Messenger of Math., **43**, 542–555. [G.H.Hardy, Collected Papers, **VI**, 542–555. Clarendon Press, Oxford 1974.]

Hardy,G.H.,Littlewood,J.E. (1915): New proofs of the prime number theorem and similar theorems. Quart. J.Math. Oxford ser., **46**, 215–219. [Hardy,G.H., Collected Papers, **II**, 9–13, Oxford 1967.]

Hardy,G.H.,Littlewood,J.E. (1918): Contributions to the theory of Riemann zeta-function and the theory of the distribution of primes. Acta Math., **41**, 119–196. [Hardy,G.H., Collected Papers, **II**, 20–99, Oxford 1967.]

Hardy,G.H.,Littlewood,J.E. (1919a): A new solution of Waring's problem. Quart. J.Math. Oxford ser., **48**, 272–293. [G.H.Hardy, Collected Papers, **I**, 382–404. Clarendon Press, Oxford 1966.]

Hardy,G.H.,Littlewood,J.E. (1919b): Note on Messrs Shah and Wilson's paper entitled: 'On an empirical formula connected with Goldbach's Theorem.' Proc. Cambridge Philos. Soc., **19**, 245–254. [G.H.Hardy, Collected Papers, **I**, 535–544. Clarendon Press, Oxford 1966.]

Hardy,G.H.,Littlewood,J.E. (1920): Some problems of 'Partitio Numerorum', I. A new solution of Waring's problem. Nachr. Ges. Wiss. Göttingen, 33–54. [G.H.Hardy, Collected Papers, **I**, 405–426. Clarendon Press, Oxford 1966.]

Hardy,G.H.,Littlewood,J.E. (1921): The zeros of Riemann's zeta-function on the critical line. Math. Zeitschr., **10**, 283–317. [G.H.Hardy, Collected Papers, **II**, 115–150. Clarendon Press, Oxford 1967.]

Hardy,G.H.,Littlewood,J.E. (1922): The approximate functional equation in the theory of the zeta-function, with applications to the divisor problem of Dirichlet and Piltz. Proc. London Math. Soc., (2), **21**, 39–74. [G.H.Hardy, Collected Papers, **II**, ???–???. Clarendon Press, Oxford 1966.]

Hardy,G.H.,Littlewood,J.E. (1923a): Some problems of 'Partitio Numerorum',III: On the expression of a number as a sum of primes. Acta Math., **44**, 1–70. [G.H.Hardy, Collected Papers, **I**, 561–630. Clarendon Press, Oxford 1966.]

Hardy.G.H.,Littlewood,J.E. (1923b): On Lindelöf's hypothesis concerning the Riemann zeta-function. Proc. Royal Society, **A103**, 403–412. [G.H.Hardy, Collected Papers, **II**, 164–173. Clarendon Press, Oxford 1967.]

Hardy,G.H.,Littlewood,J.E. (1924): Some problems of 'Partitio Numerorum',V: A further contribution to the study of Goldbach's problem. Proc. London Math. Soc., (2), **22**, 46–56. [G.H.Hardy, Collected Papers, **I**, 632–642. Clarendon Press, Oxford 1966.]

Hardy.G.H.,Littlewood,J.E. (1929): The approximate functional equation for $\zeta(s)$ and $\zeta^2(s)$. Proc. London Math. Soc., (2), **29**, 81–97. [G.H.Hardy, Collected Papers, **II**, 213–229. Clarendon Press, Oxford 1967.]

Hardy.G.H.,Littlewood,J.E.,Pólya,G. (1934): Inequalities. University Press, Cambridge. [2nd ed. 1952]
Hardy,G.H.,Ramanujan,S. (1918): Asymptotic formulae in combinatory analysis. Proc. London Math. Soc., (2), **17**, 75–115. [G.H.Hardy, Collected Papers, **II**, 306–339. Clarendon Press, Oxford 1966.]
Hardy,G.H.,Riesz,F. (1915): The General Theory of Dirichlet Series. Cambridge University Press.
Hardy,G.H.,Wright,E.M. (1960): The Introduction to the Theory of Numbers. 4th ed. Clarendon Press, Oxford. [1st ed. 1938, 2nd ed. 1945, 3rd ed. 1954. MR 16 p.673; German translation: Oldenburg, München 1958. MR 20#828.]
Hargreave,C.J. (1849): Analytical researches concerning numbers. Philosophical Magazine, (3), **35**, 36–53.
Harman,G. (1981): Almost-primes in short intervals. Math. Ann., **258**, 107–112. MR 83d:10048.
Harman,G. (1982): Primes in short intervals. Math. Zeitschr., **180**, 335–348. MR 84f:10055.
Harman,G. (1983): On the distribution of $\sqrt{p}$ modulo one. Mathematika, **30**, 104–116. MR 85e:11051.
Harman,G. (1991): Fractional and integral parts of p^λ. Acta Arith., **58**, 141–152. MR 92g:11089.
Harris,V.C. (1956): Another proof of the infinitude of primes. Amer. Math. Monthly, **63**, 711. MR 18 p.379.
Hartmann,F. (1931): Miszellen zur Primzahltheorie,I. Jahresber. Deutsche Math. Ver., **40**, 228–232.
Härtter,E. (1961): Verallgemeinerung eines Satzes von Sierpiński. Elem. Math., **16**, 123–126. MR 24#A1869.
Haselgrove,C.B. (1949): A connection between the zeros and the mean values of $\zeta(s)$. J. London Math. Soc., **24**, 215–222. MR 11 p.162.
Haselgrove,C.B. (1958): A disproof of a conjecture of Pólya. Mathematika, **5**, 141–145. MR 21#3391.
Haselgrove,C.B.,Miller,J.C.P. (1960): Tables of the Riemann Zeta Function. Royal Society Mathematical Tables, **6**. Cambridge University Press. MR 22#8679.
Hasse,H. (1950): Vorlesungen über Zahlentheorie. Springer, Berlin–Heidelberg. MR 14 p.534. [2nd ed. 1964. MR 32#5569.]
Hasse,H. (1966): Über die Dichte der Primzahlen p, für die eine vorgegebene rationale Zahl $a \neq 0$ von gerader bzw. ungerader Ordnung mod p ist. Math. Ann., **168**, 19–23. MR 32#4111.
Haussner,R. (1899): Tafeln für den Goldbachschen Gesetz. Abhandlungen der Kais. Leopold. Carol. Deutsch. Akad. der Naturforscher, **1**, Halle, 214 pp.
Hawkins,T. (1970/71): The origins of the theory of group characters. Archive Hist. Exact Sci., **7**, 142–170.
Hayashi,T. (1901a): An expression of the number of primes lying between two given integers. Archiv Math. Phys., (3), **1**, 246–247.
Hayashi,T. (1901b): On some theorems concerning with prime numbers. Archiv Math. Phys., (3), **1**, 248–251.
Heath-Brown,D.R. (1978): Almost primes in arithmetic progressions and short intervals. Proc. Cambridge Philos. Soc., **83**, 357–375. MR 58#10789.
Heath-Brown,D.R. (1979a): Simple zeros of the Riemann zeta function on the critical line. Bull. London Math. Soc., **11**, 17–18. MR 81f:10051.
Heath-Brown,D.R. (1979b): Zero density estimates for the Riemann zeta-function and Dirichlet L-functions. J. London Math. Soc., (2), **19**, 221–232. MR 80i:10055.
Heath-Brown,D.R. (1979c): The fourth power moment of the Riemann Zeta-function. Proc. London Math. Soc., (3), **38**, 385–422. MR 81f:10052.

Heath-Brown,D.R. (1979d): A large value estimate for Dirichlet polynomials. J. London Math. Soc., (2), **20**, 8–18. MR 81a:10052.
Heath-Brown,D.R. (1979e): The density of zeros of Dirichlet's L-functions. Canad. J. Math., **31**, 231–240. MR 80g:10037.
Heath-Brown,D.R. (1982): Prime numbers in short intervals and a generalized Vaughan identity. Canad. J. Math., **34**, 1365–1377. MR 84g:10075.
Heath-Brown,D.R. (1983): Prime twins and Siegel zeros. Proc. London Math. Soc., (3), **47**, 193–224. MR 84m:10029.
Heath-Brown,D.R. (1985): The ternary Goldbach problem. Rev. Mat. Iberoamericana, **1**, 45–59. MR 87f:11078.
Heath-Brown,D.R. (1988a): The number of primes in a short interval. J. Reine Angew. Math., **389**, 22–63. MR 89i:11099.
Heath-Brown,D.R. (1988b): Differences between consecutive primes. Jahresber. Deutsche Math. Ver., **90**, 71–89. MR 89i:11100.
Heath-Brown,D.R. (1990): Siegel zeros and the least prime in an arithmetic progression. Quart. J.Math. Oxford ser., **41**, 405–418. MR 91m:11073.
Heath-Brown,D.R. (1992): Zero-free regions for Dirichlet L-functions and the least prime in an arithmetic progression. Proc. London Math. Soc., (3), **64**, 265–338. MR 93a:11075.
Heath-Brown,D.R. (1993): Exponential sums and the Riemann zeta-function, IV. Proc. London Math. Soc., (3), **66**, 1–40.
Heath-Brown,D.R. (1996): The largest prime factor of the integers in an interval. Sci.China, ser.A, **39**, 449–476. MR 97j:11044.
Heath-Brown,D.R.,Iwaniec,H. (1979): On the difference between consecutive prime numbers. Invent. math., **55**, 49–69. MR 81h:10064.
Heath-Brown,D.R.,Jia Chaohua (1998): The largest prime factor of the integers in an interval,II. J. Reine Angew. Math., **498**, 35–59. MR 99d:11103.
Hecke,E. (1917a): Über die Zetafunktion beliebiger algebraischer Zahlkörper. Nachr. Ges. Wiss. Göttingen, 77–89. [Math. Werke, 159–171, Vandenhoeck & Ruprecht, Göttingen 1959.]
Hecke,E. (1917b): Über die L-Funktionen und den Dirichletschen Primzahlsatz für einen beliebigen Zahlkörper. Nachr. Ges. Wiss. Göttingen, 1917, 299–318. [Math. Werke, 178–197, Vandenhoeck & Ruprecht, Göttingen 1959.]
Hecke,E. (1936): Über die Bestimmung Dirichletscher Reihen durch ihre Funktionalgleichung. Math. Ann., **112**, 1936, 664–699. [Math. Werke, 591–626, Vandenhoeck & Ruprecht, Göttingen 1959.]
Hecke,E. (1944): Herleitung des Euler-Produktes der Zetafunktion und einiger L-Reihen aus ihrer Funktionalgleichung. Math. Ann., **119**, 266–287. MR 6 p.173. [Math. Werke, 919–940, Vandenhoeck & Ruprecht, Göttingen 1959.]
Heegner,K. (1952): Diophantische Analysis und Modulfunktionen. Math.Zeitschr., **56**, 227–253. MR 14 p.725.
Heilbronn,H. (1931): Über die Verteilung der Primzahlen in Polynomen. Math. Ann., **104**, 794–799. [The Collected Papers of Hans Arnold Heilbronn, 57–62, J.Wiley 1988.]
Heilbronn,H. (1933): Über den Primzahlsatz von Herrn Hoheisel. Math. Zeitschr., **36**, 394–423. [The Collected Papers of Hans Arnold Heilbronn, 70–99, J.Wiley 1988.]
Heilbronn,H. (1938): On Dirichlet series which satisfy a certain functional equation. Quart. J.Math. Oxford ser., **9**, 194–195. [The Collected Papers of Hans Arnold Heilbronn, 343–344, J.Wiley 1988.]

Heilbronn,H.,Landau,E. (1933a): Bemerkungen zur vorstehenden Arbeit von Herrn Bochner. Math. Zeitschr., **37**, 10–16. [E.Landau, Collected Works, **9**, 215–221, Thales-Verlag, Duisburg; The Collected Papers of Hans Arnold Heilbronn, 143–149, J.Wiley 1988.]

Heilbronn,H.,Landau,E.,Scherk,P. (1936): Alle großen ganzen Zahlen lassen sich als Summe von höchstens 71 Primzahlen darstellen. Časopis pěst. Mat.Fys., **65**, 117–141. [E.Landau, Collected Works, **9**, 351–375, Thales Verlag; The Collected Papers of Hans Arnold Heilbronn, 197–211, J.Wiley 1988.]

Heine,E. (1846): Summation der Reihe $\frac{1}{(b+a)^{1+\rho}} + \frac{1}{(b+2a)^{1+\rho}} + \frac{1}{(b+3a)^{1+\rho}} + \dots$ für $\rho = 0$. J. Reine Angew. Math., **31**, 133–135.

Helson,H. (1973): Convergence of Dirichlet series. L'analyse harmonique dans le domaine complexe, Lecture Notes in Math., **336**, 153–160. Springer, Berlin–Heidelberg. MR 52#8059.

Hendy,M.D. (1974): Prime quadratics associated with complex quadratic fields of class number two. Proc. Amer. Math. Soc., **43**, 253–260. MR 49#2648.

Hendy,M.D. (1975): Euclid and the fundamental theorem of arithmetic. Historia Math., **2**, 189–191. MR 58#26701.

Hensley,D.,Richards,I. (1973): On the incompatibility of two conjectures concerning primes. Proc. Symp. Pure Math. **24**, 123–127. MR 49#4950.

Hensley,D.,Richards,I. (1973/74): Primes in intervals. Acta Arith., **25**, 375–391. MR 53#305.

Hering,Ch. (1974): Transitive linear groups and linear groups which contain irreducible subgroups of prime order. Geom. Dedicata, **2**, 425–460. MR 49#439.

Hermite,C. (1885): Note au sujet de la communication de M.Stieltjes sur une fonction uniforme. Comptes Rendus Acad. Sci. Paris, **101**, 112– 115. [Oeuvres, **4**, 184–187, Gauthier-Villars, Paris 1917.]

Hilano,T. (1976): On the distribution of zeros of Dirichlet's L-functions,I. Proc. Japan Acad. Sci., **52**, 537–540. MR 56#8507.

Hilano,T. (1978): On the distribution of zeros of Dirichlet's L-functions,II. Tokyo J.Math., **1**, 285–304. MR 80a:10057.

Hildebrand,A. (1986): The Prime Number Theorem via the large sieve. Mathematika, **33**, 23–30. MR 88a:11085.

Hille,E. (1937): The inversion problem of Möbius. Duke Math. J., **3**, 549–569. [Classical Analysis and Functional Analysis, Selected Papers, 218–237, MIT Press 1975.]

Hille,E.,Szász,O. (1936): On the completeness of Lambert functions. Bull. Amer. Math. Soc., **42**, 411–418; II, Ann. of Math., (2), **37**, 801–815.

Hiroshi,I. (1984): On Bonse's theorem. Rep. Toyama Univ., **7**, 115–117. MR 86b:11004.

Hobby,D.,Silberberger,D.M. (1993): Quotients of primes. Amer. Math. Monthly, **100**, 50–52. MR 94a:11007.

Hoffmann,K.E. (1879): Ueber die Anzahl der unter einer gegebener Grenze liegenden Primzahlen. Archiv Math. Phys., **64**, 333–336.

Hoheisel,G. (1930): Primzahlprobleme in der Analysis. SBer. Preuss. Akad. Wiss. Berlin, 580-588.

Hölder,O. (1882): Grenzwerte von Reihen an der Convergenzgrenze. Math. Ann., **20**, 535–549.

Holme,F. (1970): En enkel beregning av $\sum_{k=1}^{\infty} 1/k^2$. Nordisk Mat. Tidskr., **18**, 91–92. MR 42#8123.

Holmgren,E. (1902/03): On Primtalens Fördelning. Öfversigt Kgl. Vetensk. Acad. Stockholm, **59**, 221–225.

Hooley,C. (1957): On the representation of a number as the sum of two squares and a prime. Acta Math., **97**, 189–210. MR 19 p.532.

Hooley,C. (1967): On the greatest prime factor of a quadratic polynomial. Acta Math., **117**, 281–299. MR 34#4225.
Hooley,C. (1972): On the Brun-Titchmarsh theorem. J. Reine Angew. Math., **255**, 60–82. MR 46#3463.
Hooley,C. (1975a): On the Barban-Davenport-Halberstam theorem,I. J. Reine Angew. Math., **274/275**, 206–223; II. J. London Math. Soc., (2), **9**, 625–636. MR 52#3090.
Hooley,C. (1975b): On the Brun-Titchmarsh theorem,II. Proc. London Math. Soc., (3), **30**, 114–128. MR 51#5531.
Hooley,C. (1997): On a problem of Hardy and Littlewood. Acta Arith., **79**, 289–311. MR 98b:11100.
Hoppe,E. (1911): Mathematik und Astronomie im klassischen Altertum. Heidelberg.
Hua,Loo Keng (1947): Additive prime number theory. Trudy Mat. Inst. Steklov.**22**, 1–179. (Russian)MR 10 p.597. [German translation: Teubner, Leipzig 1959. MR 23#A1620; English translation: Amer. Math. Soc. Providence 1965. MR 33#2614.]
Hua,Loo Keng (1959): Die Abschätzung von Exponentialsummen und ihre Anwendung in der Zahlentheorie. Enzyklopädie der mathematischen Wissenschaften, $\mathbf{I_2}$, Heft 13_I, 1–123, B.G.Teubner,Leipzig. MR 24#A94. [Russian translation: Mir, Moscow 1964. MR 31#4768.]
Hudson,R.H. (1977): A formula for the exact number of primes below a given bound in any arithmetic progression. Bull. Austral. Math. Soc., **16**, 67–73. MR 55#12663.
Hudson,R.H. (1980): A common combinatorial principle underlies Riemann's formula, the Chebyshev phenomenon, and other subtle effects in comparative prime number theory,I. J. Reine Angew. Math., **313**, 133–150. MR 83b:10053.
Hudson,R.H. (1985): Averaging effects on irregularities in the distribution of primes in arithmetic progressions. Math. Comp., **44**, 561–571. MR 86h:11074.
Hudson,R.H.,Bays,C. (1977): The mean behavior of primes in arithmetic progressions. J. Reine Angew. Math., **296**, 80–99. MR 57#255.
Hudson,R.H.,Brauer, A. (1977): On the exact number of primes in the arithmetic progressions $4n \pm 1$ and $6n \pm 1$. J. Reine Angew. Math., **291**, 23–29. MR 56#283.
Hurwitz,A. (1882): Einige Eigenschaften der Dirichlet'schen Functionen

$$F(s) = \sum \left(\frac{D}{n}\right) \cdot \frac{1}{n^s},$$

die bei der Bestimmung der Classenanzahlen binärer quadratischer Formen auftreten. Zeitschr. Math. Phys., **27**, 86–101. [Mathematische Werke, **I**, 72—88, Birkhäuser, Basel, 1932.]
Hutchinson,J.I. (1925): On the roots of the Riemann zeta-function. Trans. Amer. Math. Soc., **27**, 49–60.
Huxley,M.N. (1972a): The Distribution of Prime Numbers. Clarendon Press, Oxford. MR 56#2943.
Huxley,M.N. (1972b): On the difference between consecutive primes. Invent. math., **15**, 164–170. MR 45#1856.
Huxley,M.N. (1973a): Large values of Dirichlet polynomials. Acta Arith., **24**, 329–346. MR 48#8406.
Huxley,M.N. (1973b): Small differences between consecutive primes. Mathematika, **20**, 229—232, MR 50#4509.
Huxley,M.N. (1975a): Large values of Dirichlet polynomials,II. Acta Arith., **27**, 159–169. MR 50#12934.

Huxley,M.N. (1975b): Large values of Dirichlet polynomials,III. Acta Arith., **26**, 435—444. MR 52#300.
Huxley,M.N. (1976): An imperfect hybrid zero-density theorem. J. London Math. Soc., (2), **13**, 53–56. MR 53#2863.
Huxley,M.N. (1977): Small differences between consecutive primes, II. Mathematika, **24**, 1977, 142–152. MR 57#5925.
Huxley,M.N. (1984); An application of the Fouvry-Iwaniec theorem. Acta Arith., **43**, 441–443. MR 85k:11043.
Huxley,M.N. (1993a): Exponential sums and the Riemann zeta function,IV. Proc. London Math. Soc., (3), **66**, 1-40. MR 93j:11056.
Huxley,M.N. (1993b): Exponential sums and lattice points,II. Proc. London Math. Soc., (3), **66**, 279–301. MR 94b:11100. Corr.: *ibidem*, 68 (1994), p.264. MR 95d:11134.
Huxley,M.N.,Jutila,M. (1977): Large values of Dirichlet polynomials, V. Acta Arith., **32**, 297–312. MR 58#5550.
Huxley,M.N.,Kolesnik,G. (1991): Exponential sums and the Riemann zeta function,III. Proc. London Math. Soc., (3), **62**, 449–468. MR 92d:11089. Corr.: *ibidem*, **66**, (1993), p.302. MR 93k:11073.
Huxley,M.N.,Watt,N. (1988): Exponential sums and the Riemann zeta function. Proc. London Math. Soc., (3), **57**, 1–24. MR 90g:11110.
Ikehara,S. (1931): An extension of Landau's theorem in the analytical theory of numbers. J. of Math. and Physics, **10**, 1–12.
Indlekofer,K.H.,Járai,A. (1996): Largest known twin primes. Math. Comp., **65**, 427–428. MR 96d:11009.
Ingham,A.E. (1927): Mean-value theorems in the theory of the Riemann zeta-function. Proc. London Math. Soc., (2), **27**, 273–300.
Ingham,A.E. (1930): Note on Riemann's ζ-function and Dirichlet's L-functions. J. London Math. Soc., **5**, 107–112.
Ingham,A.E. (1932): The Distribution of Prime Numbers. Cambridge University Press. [Reprint: Stechert-Hafner, New York 1964; Hafner, New York 1971. MR 32#2391; Cambridge University Press 1990. MR 91f:11064.]
Ingham,A.E. (1935): On Wiener's method in tauberian theorems. Proc. London Math. Soc., (2), **38**, 458–480.
Ingham,A.E. (1936): Note on the distribution of primes. Acta Arith., **1**, 201–211.
Ingham,A.E. (1937): On the difference between consecutive primes. Quart. J.Math. Oxford ser., **8**, 255–266.
Ingham,A.E. (1940): On the estimation of $N(\sigma, T)$. Quart. J.Math. Oxford ser., **11**, 291–292. MR 2 p.249.
Ingham,A.E. (1942): On two conjectures in the theory of numbers. Amer. J. Math, **64**, 313–319. MR 3 p.271.
Ingham,A.E. (1945): Some Tauberian theorems connected with the prime number theorem. J. London Math. Soc., **20**, 171–180. MR 8 p.147.
Inojatova,K.G.,Sobirov,A.Š. (1972): On the question of providing an elementary proof of the asymptotic law of the distribution of primes. Izv. Akad. Nauk Uzb. SSR., ser. Fiz.-Mat. Nauk, **16**, no.5, 13–19. (Russian) MR 27#4804; **17**, 1973, no.2, 10–16. (Russian) MR 47#8458.
Isenkrahe,C. (1900): Über eine Lösung der Aufgabe, jede Primzahl als Funktion der vorhergehenden Primzahlen durch einen geschlossenen Ausdruck darzustellen. Math. Ann., **53**, 42–44.
Israilov,M.I. (1979): Coefficients of the Laurent expansion of Riemann zeta functions. Dokl. Akad. Nauk UzSSR., no.12, 9–10. (Russian) MR 81m:10075.
Ivić,A. (1979/80): A note on the zero-density estimate for the zeta-function. Archiv Math., **33**, 155–164. MR 81f:10053.

Ivić,A. (1980): Exponent pairs and the zeta function of Riemann. Studia Sci. Math. Hungar., **15**, 157–181. MR 84c:10035.

Ivić,A. (1983): Topics in Recent Zeta Function Theory. Publicat. Math. d'Orsay, **83–6**. MR 86b:11055.

Ivić,A. (1985): The Riemann Zeta-function. Wiley, New York. MR 87d:11062.

Ivić,A.,Motohashi,Y. (1994): The mean square of the error term for the fourth moment of the zeta-function. Proc. London Math. Soc., (3), **69**, 309–329. MR 95g:11081.

Ivić,A.,Motohashi,Y. (1995): On the fourth moment of the Riemann zeta-function. J. Number Th., **51**, 16–45. MR 96c:11099.

Iwaniec,H. (1972a): Primes of the type $\varphi(x,y)+A$ where φ is a quadratic form. Acta Arith., **21**, 203–234. MR 46#3466.

Iwaniec,H. (1972b): Primes represented by quadratic polynomials in two variables. Bull. Acad. Polon. Sci., **20**, 195–202. MR 47#132.

Iwaniec,H. (1973/74): Primes represented by quadratic polynomials in two variables. Acta Arith., **24**, 435–459. MR 49#7210.

Iwaniec,H. (1974): On zeros of Dirichlet L-series. Invent. math., **23**, 97–104. MR 49#8947.

Iwaniec,H. (1975): The generalized Hardy-Littlewood's problem involving a quadratic polynomial with coprime discriminants. Acta Arith., **27**, 421–446. MR 51#3093.

Iwaniec,H. (1978): Almost primes represented by quadratic polynomials. Invent. math., **47**, 171–188. MR 58#5533.

Iwaniec,H. (1982): On the Brun-Titchmarsh theorem. J. Math. Soc. Japan, **34**, 95–123. MR 83a:10082.

Iwaniec,H.,Jutila,M. (1979): Primes in short intervals. Arkiv Mat., **17**, 167–176. MR 80j:10047.

Iwaniec,H.,Laborde,M. (1981): P_2 in short intervals. Ann. Inst. Fourier, **31**, 4, 37–56. MR 83e:10061.

Iwaniec,H.,Mozzochi,C.J. (1988): On the divisor and circle problems. J. Number Th., **29**, 60–93. MR 89g:11091.

Iwaniec,H.,Pintz,J. (1984): Primes in short intervals. Monatsh. Math., **98**, 115–143. MR 86e:11078.

Izumi,M.,Izumi,S. (1976): On the Riemann zeta-function,I. Bull. Austral. Math. Soc., **15**, 161–211. MR 55#2798.

Jacobi,C.G.J. (1827): Ueber den Ausdruck der verschiedenen Wurzeln einer Gleichung durch bestimmte Integrale. J. Reine Angew. Math., **2**, 1–8. [Gesammelte Werke, **6**, 12–20, G.Reimer,Berlin 1891.]

Jacobi,C.G.J. (1828): Suite des notices sur les fonctions elliptiques. J. Reine Angew. Math., **3**, 303–310. [Gesammelte Werke, **1**, 255–263, G.Reimer, Berlin 1881.]

Jacobi,C.G.J. (1834): De usu legitimo formulae summmatoriae Maclaurinianae. J. Reine Angew. Math., **12**, 263–272. [Gesammelte Werke, **6**, 64–75, G.Reimer, Berlin 1891.]

Jacobi,C.G.J. (1837): Über die Kreistheilung und ihre Anwendung auf die Zahlentheorie. SBer. Kgl. Preuß. Akad. Wiss. Berlin, 127–136. [J. Reine Angew. Math., **30**, 166–182; Gesammelte Werke, **6**, 254–274, G.Reimer, Berlin 1891.]

Jacobi,C.G.J. (1839): Canon Arithmeticus. Regiomonti. [Extended version: Akademie Verlag, Berlin 1956. MR 18 p.421.]

Jager,H. (1973): On the number of Dirichlet characters with modulus not exceeding x. Indag. Math., **35**, 452–455. MR 48#6035.

James,R.D.,Weyl,H. (1942): Elementary note on prime number problems of Vinogradoff's type. Amer. J. Math, **64**, 539–552. MR 4 p.35.

Jansson,T. (1921): Über die Grössenordnung Dirichletschen Reihen. Arkiv f. Math. Astr. Fys., **15**, No.6, 1–11.
Jensen,J.L.W.V. (1884): Om Rækkers Konvergens. Tidsskrift f. Math., (5), **2**, 63–72.
Jensen,J.L.W.V. (1887): Sur la fonction $\zeta(s)$ de Riemann. Comptes Rendus Acad. Sci. Paris, **104**, 1156–1159.
Jensen,J.L.W.V. (1888): Sur une généralisation d'un théorème de Cauchy. Comptes Rendus Acad. Sci. Paris, **106**, 833–836.
Jensen,J.L.W.V. (1899): Sur un nouvel et important théorème de la théorie des fonctions. Acta Math., **22**, 359–364.
Jia,Chaohua (1986): The greatest prime factor of the product of the integers in a short interval,I. Acta Math. Sinica, **29**, 815–825. (Chinese) MR 88f:11083.
Jia,Chaohua (1995): Difference between consecutive primes. Sci. China, Ser.A., **38**, 1163–1186. MR 96m:11081.
Jia,Chaohua (1996a): Almost all short intervals containing prime numbers. Acta Arith., **76**, 21–84. MR 97e:11110.
Jia,Chaohua (1996b): The greatest prime factor of the product of the integers in a short interval,IV. Acta Math. Sinica, (N.S.) **12**, 433–445. MR 98f:11098.
Jones,J.P. (1975): Formula for the nth prime number. Canad. Math. Bull., **18**, 433–434. MR 57#9641.
Jones,J.P. (1979): Diophantine representation of Mersenne and Fermat primes. Acta Arith., **35**, 209–221. MR 81a:10020.
Jones,J.P.,Sato,D.,Wada,H.,Wiens,D. (1976): Diophantine representation of the set of prime numbers. Amer. Math. Monthly, **83**, 449–464. MR 54#2615.
Jonquiéres,E.de (1882a): Formule pour déterminer combien il y a de nombres premiers n'excédant pas un nombre donnée. Comptes Rendus Acad. Sci. Paris, **95**, 1144–1146.
Jonquiéres,E.de (1882b): Sur le formule recemment communiquée à l'Académie au sujet des nombres premiers. Comptes Rendus Acad. Sci. Paris, **95**, 1343–1344.
Jonquiéres,E.de (1883): Addition à une note sur les nombres premiers. Comptes Rendus Acad. Sci. Paris, **96**, 231–232.
Jordan,C. (1893–1896): Cours d'analyse de l'École Polytechnique. 2nd ed. I, 1893; II, 1894; III, 1896. Gauthier-Villars, Paris. [3rd ed. 1909.]
Jordan,C. (1922): On a new demonstration of Maclaurin's or Euler's summation formula. Tôhoku Math. J., **21**, 244–246.
Jurkat,W. (1972): On the Mertens conjecture and related general Ω-theorem. Proc. Symposia Pure Math., **24**, 147–158. MR 50#4514.
Jurkat,W.,Peyerimhoff,A. (1976): A constructive approach to Kronecker approximations and its application to the Mertens conjecture. J. Reine Angew. Math., **286/287**, 322–340. MR 55#2799.
Jutila,M. (1969): On two theorems of Linnik concerning the zeros of Dirichlet's L-functions. Ann. Acad. Sci. Fenn. Ser. A.1, **458**, 1–32. MR 43#1930.
Jutila,M. (1969/70): A statistical density theorem for L-functions with applications. Acta Arith., **16**, 207–216. MR 40#5557.
Jutila,M. (1970): A new estimate for Linnik's constant. Ann. Acad. Sci. Fenn. Ser. A.1, **471**. MR 42#5939.
Jutila,M. (1972): On a density theorem of H.L.Montgomery for L-functions. Ann. Acad. Sci. Fenn. Ser. A.1, **520**, 1–13. MR 48#6023.
Jutila,M. (1973): On numbers with a large prime factor,I. J.Indian Math.Soc., (N.S.), **37**, 43–53. MR 50#12936.
Jutila,M. (1976): On large values of Dirichlet polynomials. Topics in Number Theory, Debrecen 1974, Proc. J.Bólyai Colloq., **13**, 129–140. North-Holland, Amsterdam. MR 55#7951.

Jutila,M. (1977a): On Linnik's constant. Math.Scand., **41**, 45–62. MR 57#16230.
Jutila,M. (1977b): Zero-density estimates for L-functions. Acta Arith., **32**, 55–62. MR 55#2800.
Kaczorowski,J. (1984): On sign-changes in the remainder term of the prime-number formula. I. Acta Arith., **44**, 365–377. MR 87c:11081a.
Kaczorowski,J. (1985): On sign-changes in the remainder term of the prime-number formula. II. Acta Arith., **45**, 65–74. MR 87c:11081b.
Kaczorowski,J. (1991a): The k-functions in multiplicative number theory, IV. On a method of A.E.Ingham. Acta Arith., **57**, 231–244. MR 92f:11117b.
Kaczorowski,J. (1991b): The k-functions in multiplicative number theory, V. Changes of sign of some arithmetical error terms. Acta Arith., **59**, 37–58. MR 92j:11096.
Kaczorowski,J. (1992): The boundary values of generalized Dirichlet series and a problem of Chebyshev. Astérisque, **209**, 227–235. MR 94a:11144.
Kaczorowski,J. (1993): A contribution to the Shanks-Rényi race problem. Quart. J.Math. Oxford ser., **44**, 451–458. MR 94m:11105.
Kaczorowski,J. (1994): Results on the distribution of primes. J. Reine Angew. Math., **446**, 89–113. MR 95f:11070.
Kaczorowski,J. (1995a): On the Shanks-Rényi race problem mod 5. J. Number Th., **50**, 106–118. MR 95j:11089.
Kaczorowski,J. (1995b): On the distribution of primes mod 4. Analysis, **15**, 159–171. MR 96h:11095.
Kaczorowski,J. (1996): On the Shanks-Rényi race problem. Acta Arith., **74**, 31–46. MR 96k:11113.
Kaczorowski,J. (1998): Bounded mean oscillation and the distribution of primes. Math. Proc. Cambridge Philos. Soc., to appear.
Kaczorowski,J.,Perelli,A. (1999): On the structure of the Selberg class,I: $0 \le d \le 1$. Acta Math., **182**, 207–241.
Kaczorowski,J.,Pintz,J. (1986): Oscillatory properties of arithmetical functions,I. Acta Math. Acad. Sci. Hung., **48**, 173–185. MR 88c:11053.
Kaczorowski,J.,Pintz,J. (1987): Oscillatory properties of arithmetical functions,II. Acta Math. Acad. Sci. Hung., **49**, 441–453. MR 88i:11062.
Kahane,J.P. (1966): Sur trois notes de Stieltjes relatives aux séries de Dirichlet. Ann.Fac.Sci.Toulouse, Math., (6), special issue, 33–56. MR 99f:11106.
Kakeya,S. (1917): On the convergence abscissa of general Dirichlet series. Tôhoku Math. J., **11**, 232–238.
Kalecki,M. (1966/67): Uogólnienie postulatu Bertranda. Wiadom. Mat., (2), **9**, 219–220. MR 35#1539.
Kalecki,M. (1967): A simple elementary proof of $M(x) = \sum_{n \le x} \mu(n) = o(x)$. Acta Arith., **13**, 1–7. MR 36#2574.
Kalecki,M. (1969): A short elementary proof of the prime number theorem. Prace Mat., **13**, 51–55. MR 41#6772.
Kaniecki,L. (1995): On Šnirelman's constant under the Riemann hypothesis. Acta Arith., **72**, 361–374. MR 96i:11112.
Kanold,H.J. (1950): Sätze über Kreisteilungspolynome und ihre Anwendung auf einige zahlentheoretische Probleme,I. J. Reine Angew. Math., **187**, 169–182. MR 12 p.592.
Kanold,H.J. (1984): Über einige Ergebnisse aus der kombinatorischen Zahlentheorie. Abh. Braunschweig. Wiss. Ges., **36**, 203–229. MR 86j: 11012.
Karamata,J. (1930): Über die Hardy-Littlewoodsche Umkehrung des Abelschen Stetigkeitssatzes. Math. Zeitschr., **32**, 319–320.

Karamata,J. (1931): Neuer Beweis und Verallgemeinerung der Tauberschen Sätze, welche die Laplacesche und Stieltjessche Transformation betreffen. J. Reine Angew. Math., **164**, 27–40.
Karamata,J. (1954): Sur un lemme de Mertens relatif aux nombres premiers. C.R. Congr. Soc. Sav. Paris, 277–284. MR 23#A1615.
Karanikolov,Ch. (1971): On some properties of function $\pi(x)$. Univ. Beograd, Publ. Elektroteh. Fak. Ser. Mat.Fiz., No.357–380, 29–30. MR 45#6771.
Karatsuba,A.A. (1991): On a density theorem. Trudy Mat. Inst. Steklov., **200**, 185–196. (Russian) MR 92m:11087.
Karatsuba,A.A.,Voronin,S.M. (1992): The Riemann Zeta-function. Walter de Gruyter, Berlin. MR 93h:11096.
Karst,E. (1973): New quadratic forms with high density of primes. Elem. Math., **28**, 116–118. MR 49#2539.
Kasimov,A.M. (1992): On the I.M.Vinogradov constant in the Goldbach ternary problem. Uzbek. Mat. Zh, no.3–4, 55–65. MR 95f:11082.
Kátai,I. (1964): Eine Bemerkung zur "Comparative prime-number theory, I–VIII" von S.Knapowski und P.Turán. Ann. Univ. Sci. Budapest. **7**, 33–40. MR 31#1235.
Kátai.I. (1967): On the comparative theory of prime numbers. Acta Math. Acad. Sci. Hung., **18**, 133–149. (Russian) MR 34#7480.
Kátai,I. (1968): On oscillation of the number of primes in an arithmetical progression. Acta Sci. Math., **29**, 271–282. MR 38#2103.
Kaufman,R.M. (1979): The distribution of $\{\sqrt{p}\}$. Mat. Zametki, **26**, 497–504. (Russian) MR 81a:10055.
Keates,M. (1968/69): On the greatest prime factor of a polynomial. Proc. Edinburgh Math. Soc., **16**, 301–303. MR 41#1689.
Keiper,J.B. (1992): Power series expansions of Riemann's ξ function. Math. Comp., **58**, 765–773. MR 92f:11116.
Kienast,A. (1924): Extension to other series of Abel's and Tauber's theorems on power series. Proc. London Math. Soc., (2), **25**, 45–52.
Kienast,A. (1926): Über die Äquivalenz zweier Ergebnisse der analytischen Zahlentheorie. Math. Ann., **95**, 426–445.
Kinkelin,H. (1861/62): Allgemeine Theorie der harmonischen Reihen, mit Anwendungen auf die Zahlentheorie. Programm der Gewerbeschule Basel, 1–32.
Kirch,A.M. (1969): A countable, connected, locally connected Hausdorff space. Amer. Math. Monthly, **76**, 169–171. MR 39#920.
Klimov,N.I. (1969): On the computation of Šnirelman's constant. Volžskij Mat. Sbornik, **7**, 32–40. (Russian) MR 44#6633.
Klimov,N.I. (1975): A refinement of the estimate of the absolute constant in the Goldbach-Šnirelman problem. Number theory: Collection of Studies in Additive Number Theory, Naučn. Trudy Kuibyšev. Gos. Ped. Inst. **158**, 14–30. (Russian) MR 58#27865.
Klimov,N.I. (1978): A new estimate of the absolute constant in the Goldbach-Šnirelman problem. Izv.VUZ, no.1, 25–35. (Russian) MR 80a:10067.
Klimov,N.I. (1980): The small sieve. Mat. Zametki, **27**, 161–174. (Russian) MR 81g:10059.
Klimov,N.I.,Pil'tjai, G.Z.,Šeptickaja,T.A. (1972): An estimate for the absolute constant in the Goldbach-Šnirelman problem. Issled. po teorii čisel, Saratov, No.4, 35–51. (Russian) MR 54#2607.
Kline,M. (1983): Euler and infinite series. Math.Mag., **56**, 307–314. MR 86a:01020.
Kluyver,J.C. (1896): Sur les valeurs que prend la fonction $\zeta(s)$ de Riemann, pour s entier positif et impair. Bull. Sci. Math., (2), **20**, 116–119.
Kluyver,J.C. (1904): Series derived from the series $\sum \mu(m)/m$. Koninkl. Akad. Wetensch. Amsterdam, **6**, 305–312.

Knapowski,S. (1958): On prime numbers in an arithmetic progression. Acta Arith., **4**, 57–70. MR 20#3015.

Knapowski,S. (1959): On an explicit lower estimate in prime number theory, J. London Math. Soc., **34**, 437–441. MR 22#2585.

Knapowski,S. (1960/61): Contributions to the theory of the distribution of prime numbers in arithmetical progressions,I. Acta Arith., **6**, 415–434. MR 23#A3119.

Knapowski,S. (1961): On sign-changes in the remainder-term in the prime-number formula. J. London Math. Soc., **36**, 451–460. MR 24#A3143.

Knapowski,S. (1961/62): Contributions to the theory of the distribution of prime numbers in arithmetical progressions,II. Acta Arith., **7**, 325–335. MR 26#89.

Knapowski,S. (1962a): On Linnik's theorem concerning exceptional L-zeros. Publ. Math., **9**, 168–178. MR 26#6131.

Knapowski,S. (1962b): On sign changes of the difference $\pi(x) - \mathrm{li}(x)$. Acta Arith., **7**, 107–120. MR 24#A3142.

Knapowski,S. (1962c): Contributions to the theory of the distribution of prime numbers in arithmetical progressions,III. Acta Arith., **8**, 97–105. MR 26#90.

Knapowski,S. (1968): On Siegel's theorem. Acta Arith., **14**, 417–424. MR 37#4034.

Knapowski,S.,Turán,P (1962): Comparative prime-number theory, I, II, III. Acta Math. Acad. Sci. Hung., **13**, 299–314, 315–342, 343–364. MR 26#3682. [P.Turán, Collected Papers, **II**, 1329–1343, 1344–1371, 1372–1393. Akadémiai Kiado, Budapest 1990.]

Knapowski,S.,Turán,P (1963): Comparative prime-number theory, IV, V, VI, VII, VIII. Acta Math. Acad. Sci. Hung., **14**, 31–42, 43–63, 65–78, 241–250, 251–268. MR 26#3683a-c, MR 28#70a,b. [P.Turán, Collected Papers, **II**, 1408–1419, 1420–1440, 1441–1454, 1465–1474, 1475–1492. Akadémiai Kiado, Budapest 1990.]

Knapowski,S.,Turán,P (1964): Further developments in the comparative prime-number theory, I. Acta Arith., **9**, 23–40. MR 29#75; II. *ibidem*, **10**, 293–313. MR 30#4739. [P.Turán, Collected Papers, **II**, 1534–1551, 1572–1592, Akadémiai Kiado, Budapest 1990.]

Knapowski,S.,Turán,P (1965a): Further developments in the comparative prime-number theory, III, IV, V. Acta Arith., **11**, 115–127, 147–161, 193–202. MR 31#4773, MR 32#99. [P.Turán, Collected Papers, **II**, 1619–1631, 1656–1670, 1671–1680, Akadémiai Kiado, Budapest 1990.]

Knapowski,S.,Turán,P (1965b): On an asssertion of Chebyshev, J. Analyse Math., **14**, 267–274. MR 31#2221. [P.Turán, Collected Papers, **II**, 1605–1612, Akadémiai Kiado, Budapest 1990.]

Knapowski,S.,Turán,P. (1966): Further developments in the comparative prime-number theory,VI. Acta Arith., **12**, 85–96. MR 34#149. [P.Turán, Collected Papers, **II**, 1738–1749, Akadémiai Kiado, Budapest 1990.]

Knapowski,S.,Turán,P. (1972): Further developments in the comparative prime-number theory,VII. Acta Arith., **21**, 193–201. MR 46#1729. [P.Turán, Collected Papers, **II**, 2230–2238, Akadémiai Kiado, Budapest 1990.]

Knapowski,S.,Turán,P. (1976): On the sign changes of $(\pi(x) - \mathrm{li}\ x)$, I: Topics in Number Theory, Debrecen 1974, Proc. J.Bólyai Colloq. **13**, 153–169, North Holland, Amsterdam; II: Monatsh. Math., **82**, 163–175. MR 55#12657ab. [P.Turán, Collected Papers, **II**, 2378–2394, 2404–2416, Akadémiai Kiado, Budapest 1990.]

Knapowski,S.,Turán,P. (1977): On prime numbers $= 1$ resp. $= 3 \bmod 4$. Number Theory and Algebra, 157–165. Academic Press. MR 57#5926. [P.Turán, Collected Papers, **III**, 2534–2542, Akadémiai Kiado, Budapest 1990.]

Knopfmacher,J. (1975): Abstract Analytic Number Theory. North-Holland, Amsterdam. MR 54#7404.

Knopfmacher,J. (1977/78): Recursive formulae for prime numbers. Archiv Math., **33**, 144–149. MR 81j:10008.

Knopp,K. (1910): Über die Abszisse der Grenzgeraden einer Dirichletschen Reihe. Sber. Berliner Math. Ges., **10**, 1–7.
Knopp,M. (1994): On Dirichlet series satisfying Riemann's functional equation. Invent. math., **117**, 361–372. MR 95h:11087.
Knorr,W. (1976): Problems in the interpretation of Greek number theory: Euclid and the "fundamental theorem of arithmetic". Studies in Historical and Philosophical Sciences, **7**, 353–368. MR 58#15806.
Knuth,D.E. (1962): Euler's constant to 1271 places. Math. Comp., **16**, 275–281. MR 26#5763.
Knuth,D.E. (1969): The Art of Computer Programming. Vol.2: Seminumerical algorithms. Addison-Wesley. MR 44#3531.
Koch,H.von (1894): Sur la distribution des nombres premiers. Comptes Rendus Acad. Sci. Paris, **118**, 850–853.
Koch,H.von (1900): Sur la distribution des nombres premiers. Comptes Rendus Acad. Sci. Paris, **130**, 1243–1246.
Koch,H.von (1901): Sur la distribution des nombres premiers. Acta Math., **24**, 159–182.
Koch,H.von (1902): Ueber die Riemannsche Primzahlfunction. Math. Ann., **55**, 441–464.
Koch,H.von (1903/04): Sur un théorème concernant les nombres premiers. Arkiv f. Math. Astr. Fys., **1**, 481–488.
Koch,H.von (1910): Contribution à la théorie des nombres premiers. Acta Math., **33**, 293–320.
Koessler,M. (1917): On a recurrential formula for prime numbers. Rozpravy Česke Akademie Věd, **26**, No.48, 1–6. (Czech)
Kojima,T. (1914): On the convergence-abscissa of general Dirichlet series. Tôhoku Math. J., **6**, 134–139.
Kojima,T. (1916): Note on the convergence-abscissa of Dirichlet series. Tôhoku Math. J., **9**, 28–37.
Kolesnik,G. (1973): On the estimation of some exponential sums. Acta Arith., **25**, 7–30. MR 48#11002.
Kolesnik,G. (1982): On the order of $\zeta(1/2+it)$ and $\Delta(R)$. Pacific J. Math., **98**, 107–122. MR 83d:10052.
Kolesnik,G. (1985): On the method of exponent pairs. Acta Arith., **45**, 115–143. MR 86k:11047.
Kondrat'ev,V.P. (1977): Some extremal properties of positive trigonometric polynomials. Mat. Zametki, **22**, 371–374. (Russian) MR 58#5547.
Kopetzky,H.G.,Schwarz,W. (1979): Two conjectures of B.R.Santos concerning totatives. Math. Comp., **33**, 841–844. MR 80i:10009.
Korevaar,J. (1982): On Newman's quick way to the prime number theorem. Math. Intel., nr.4, 108–115. MR 84b:10063.
Korfhage,R.R. (1964): On a sequence of prime numbers. Bull. Amer. Math. Soc., **70**, 341–342. Errata: *ibidem* p.747.
Korobov,N.M. (1958a): On the zeros of the function $\zeta(s)$. Doklady Akad. Nauk SSSR, **118**, 231–232. (Russian) MR 20#6394.
Korobov,N.M. (1958b): Weyl's estimates of sums and the distribution of primes. Doklady Akad. Nauk SSSR, **123**, 28–31. (Russian) MR 21#2625.
Korobov,N.M. (1958c): Evaluations of trigonometrical sums and their applications. Usp. Mat. Nauk, **13**, no.4, 185–192. (Russian) MR 21#4939.
Koshiba,Z.,Uchiyama,S. (1966): On the existence of prime numbers in arithmetic progressions. Proc. Japan Acad. Sci., **42**, 696–701. MR 35#1561.
Koshiba,Z.,Uchiyama,S. (1967): On the existence of prime numbers in arithmetic progressions. J. Math. Anal. Appl., **19**, 431–443. MR 35#4177.

Kotov,S.V. (1973): The greatest prime factor of polynomial. Mat. Zametki, **13**, 515–522. (Russian) MR 48#6012.

Kozjašev,A.A.,Čečuro,E.F. (1969): The representation of large integers by sums of primes. Issled. po teorii čisel, Saratov, **3**, 46–50. (Russian) MR 42#4517.

Kraus,L. (1886): The proof of the theorem that if p is a prime number then there exist infinitely many primes $kp+1$. Časopis Mat.Fys., **15**, 61.

Kreisel,G. (1952): On the interpretation of non-finitist proof. II. Interpretation of number theory. J.Symbolic Logic, **17**, 43–58. MR 14 p.440.

Kronecker,L. (1878): Notiz über Potenzreihen. Monatsber. Preuss. Akad. Wiss. Berlin, 53–58. [Werke, **V**, 197–201, Teubner, Berlin, Leipzig 1930.]

Kronecker,L. (1884): Die Periodensysteme von Functionen reeller Variabeln. S-Ber. Kgl. Preuß. Akad. Wiss. Berlin, 1071–1080. [Werke, **III.1**, 31–46. Teubner, Leipzig 1899.]

Kronecker,L. (1885): Zur Theorie der elliptischen Functionen, III. SBer. Kgl. Preuß. Akad. Wiss. Berlin, 761–784. [Werke, **IV**, 347–495. Teubner, Leipzig 1929.]

Kronecker,L. (1889): Bemerkungen über die Darstellung von Reihen durch Integrale. J. Reine Angew. Math., **105**, 157–159, 345–354. [Werke, **V**, 327–342. Teubner, Berlin, Leipzig 1930.]

Kronecker,L. (1891): Eine analytisch-arithmetische Formel. J. Reine Angew. Math., **108**, p.348. [Werke, **III.2**, 199–202. Teubner, Leipzig–Berlin 1931.]

Kronecker,L. (1901): Vorlesungen über Zahlentheorie, I. Teubner, Leipzig.

Kueh,K.L. (1990): A local prime number theorem. Chinese J.Math., **18**, 251–255. MR 91i:11117.

Kuhn,P. (1953): Neue Abschätzungen auf Grund der Viggo Brunschen Siebmethode. 12 Skand. Mat. Kongr., 160–168. MR 16 p.676.

Kuhn,P. (1954): Über die Primteiler eines Polynoms, Proc. ICM Amsterdam, **2**, 35–37.

Kuhn,P. (1955): Eine Verbesserung des Restgliedes beim elementaren Beweis des Primzahlsatzes. Math. Scand., **3**, 75–89. MR 17 p.587.

Kuipers,L. (1950): Prime-representing functions. Indag. Math., **12**, 57–58. MR 11 p.644.

Kummer,E.E. (1856): Theorie der idealen Primfaktoren der complexen Zahlen, welche aus der Wurzeln der Gleichung $\omega^n=1$ gebildet sind, wenn n eine zusammengesetzte Zahl ist. Abhandl. Kgl. Preuß. Akad. Wiss., 1–47. [Collected Papers, **I**, 583–629, Springer, Berlin–Heidelberg 1975.]

Kummer,E.E. (1878): Neuer elementarer Beweis des Satzes, daß die Anzahl aller Primzahlen eine unendliche ist. Monatsber. Preuß. Akad. Wiss. Berlin, 777–778. [Collected Papers, **II**, 669–670, Springer, Berlin–Heidelberg 1975.]

Kuniyeda,M. (1916): Uniform convergence-abscissa of general Dirichlet's series. Tôhoku Math. J., **9**, 7–27.

Kuo,H.-T. (1949): A recurrence formula for $\zeta(2n)$. Bull. Amer. Math. Soc., **55**, 573–574. MR 10 p.683.

Lacroix,S.F. (1819): Traité des différences et séries. 2nd ed. Paris.

Laeng,F. (1900): Nombres entiers N jouissant de cette propriété, que les $\varphi(N)$ nombres premiers à N et non supérieurs à N soient tous des nombres premiers. L'Intermédiaire des Math., **7**, 253–254.

Lagarias,J.C. (1985): The set of primes dividing the Lucas numbers has density 2/3. Pacific J. Math., **118**, 449–461. MR 86i:11007. Errata: *ibidem*, **162**, 1994, 393–396. MR 94m:11015.

Lagarias,J.C.,Miller,V.S.,Odlyzko,A.M. (1985): Computing $\pi(x)$: the Meissel-Lehmer method. Math. Comp., **44**, 537–560. MR 86h:11111.

Lagarias,J.C.,Odlyzko,A.M. (1984): New algorithms for computing $\pi(x)$. Number Theory: New York 1982, 176–193, Lecture Notes in Math., **1052**, Springer, Berlin–Heidelberg. MR 85j:11182.

Lagarias,J.C.,Odlyzko,A.M. (1987): Computing $\pi(x)$: an analytic method. J. of Algorithms, **8**, 173–191. MR 88k:11095.

Lagrange,J.L. (1768): Nouvelle méthode pour résoudre les problèmes indétérminés en nombres entiers. Mémoires de l'Acad. Roy. des sciences de Berlin, **24**. [Oeuvres **II**, p.655–726, Paris, Gauthier-Villars 1868].

Lagrange,J.L. (1773): Recherches d'arithmétique I,II. Nouveaux Mémoires de l'Acad. Roy. des sciences de Berlin 1773, 1775. [Oeuvres, **III**, 695–795, Paris, Gauthier-Villars 1869.]

Laguerre,E.N. (1872/73): Sur quelques théorèmes d'arithmétique. Bull. Soc. Math. France, **1**, 77–81. [Oeuvres, 212–217. Paris 1898; reprint: Chelsea 1972.]

Lambek,J.,Moser,L. (1957): Math. Gaz., **41**, 287–288.

Lambert,J.H. (1769): Nova Acta Eruditorum, p.127.

Lammel,E. (1966): Ein Beweis, dass die Riemannsche Zetafunktion $\zeta(s)$ in $|s-1| \leq 1$ keine Nullstelle besitzt. Rev. Univ. Nac. Tucumán, **A16**, 209–217. MR 36#5090.

Landau E. (1899a): Neuer Beweis der Gleichung $\sum_1^\infty \mu(k)/k = 0$. Dissertation, Berlin. [Collected Works, **1**, 69–83, Thales Verlag.]

Landau E. (1899b): Contribution à la théorie de la fonction $\zeta(s)$ de Riemann. Comptes Rendus Acad. Sci. Paris, **129**, 812–815. [Collected Works, **1**, 84–88, Thales Verlag.]

Landau E. (1900a): Sur quelques problémes rélatifs à la distribution des nombres premiers. Bull. Soc. Math. France, **28**, 25–38. [Collected Works, **1**, 92–105, Thales Verlag.]

Landau E. (1900b): Ueber die zahlentheoretische Funktion $\varphi(n)$ und ihre Beziehung zum Goldbachschen Satz. Nachr. Ges. Wiss. Göttingen, 177–186. [Collected Works, **1**, 106–115, Thales Verlag.]

Landau,E. (1901a): Über einen zahlentheoretischen Satz. Archiv Math. Phys., (3), **1**, 138–142. [Collected Works, **1**, 136–140, Thales Verlag.]

Landau,E. (1901b): Ueber die asymptotische Werte einiger zahlentheoretischer Functionen. Math. Ann., **54**, 570–591. [Collected Works, **1**, 141–162, Thales Verlag.]

Landau,E. (1901c): Solutions de questions proposées,1075. Nouv. Ann. Math., (4), **1**, p.139. [Collected Works, **1**, 181–182, Thales Verlag.]

Landau,E. (1902): Über die zu einem algebraischen Zahlkörper gehörige Zetafunction und die Ausdehnung der Tschebyschefschen Primzahltheorie auf das Problem der Vertheilung der Primideale. J. Reine Angew. Math., **125**, 64–188. [Collected Works, **1**, 201–325, Thales Verlag.]

Landau E. (1903a): Neuer Beweis des Primzahlsatzes und Beweis des Primidealsatzes. Math. Ann., **56**, 645–670. [Collected Works, **1**, 327–352, Thales Verlag.]

Landau E. (1903b): Über die Primzahlen einer arithmetischer Progression. SBer. Kais. Akad. Wissensch. Wien, **112**, Abt.2a, 493–535. [Collected Works, **2**, 17–43, Thales Verlag.]

Landau E. (1903c): Über die zahlentheoretische Funktion $\mu(k)$. SBer. Kais. Akad. Wissensch. Wien, **112**, Abt.2a, 537–550. [Collected Works, **2**, 60–93, Thales Verlag.]

Landau,E. (1904a): Remarks on the paper of Mr.Kluyver on page 305 of Vol.VI. "*Series derived from the series* $\sum \frac{\mu(m)}{m}$". Koninkl. Akad. Wetensch., Amsterdam **13**, 71-83. [Collected Works, **2**, 116–128, Thales Verlag.]

Landau,E. (1904b): Über eine Verallgemeinerung des Picardschen Satzes. SBer. Kgl. Preuß. Akad. Wiss. Berlin, **38**, 1118-1133. [Collected Works, **2**, 129–144, Thales Verlag.]

Landau,E. (1905a): Sur quelques inégalités dans la théorie de la fonction $\zeta(s)$ de Riemann. Bull. Soc. Math. France, **33**, 229–241. [Collected Works, **2**, 167–179, Thales Verlag.]

Landau,E. (1905b): Über einen Satz von Tschebyschef. Math. Ann., **61**, 527–550. [Collected Works, **2**, 206–229, Thales Verlag.]

Landau,E. (1906a): Über das Nichtverschwinden einer Dirichletschen Reihe. SBer. Kgl. Preuß. Akad. Wiss. Berlin, **11**, 1906, 314–320. [Collected Works, **2**, 230–236, Thales Verlag.]

Landau,E. (1906b): Euler und die Funktionalgleichung der Riemannscher Zetafunktion. Bibl. Mat., (3), **7**, 69–79. [Collected Works, **2**, 335–345, Thales Verlag.]

Landau,E. (1906c): Über den Zusammenhang einiger neueren Sätze der analytischen Zahlentheorie. SBer. Kais. Akad. Wissensch. Wien, **115**, Abt.2a, 589–632. [Collected Works, **2**, 346–388, Thales Verlag.]

Landau,E. (1906d): Sur une inégalité de M.Hadamard. Nouv. Ann. Math., (4), **6**, 135–140. [Collected Works, **2**, 323–328, Thales Verlag.]

Landau,E. (1907a): Über die Verteilung der Primideale in den Idealklassen eines algebraischen Zahlkörpers. Math. Ann., **63**, 145–204. [Collected Works, **3**, 181–240, Thales Verlag.]

Landau,E. (1907b): Über die Multiplikation Dirichletscher Reihen. Rend. Circ. Mat. Palermo, **24**, 81–160. [Collected Works, **3**, 323–401, Thales Verlag.]

Landau,E. (1908a): Über die Approximation einer stetigen Funktion durch eine ganze rationale Funktion. Rend. Circ. Mat. Palermo, **25**, 337–346. [Collected Works, **3**, 402–410, Thales Verlag.]

Landau,E. (1908b): Beiträge zur analytischen Zahlentheorie. Rend. Circ. Mat. Palermo, **26**, 169–302. [Collected Works, **3**, 411–544, Thales Verlag.]

Landau,E. (1908c): Neuer Beweis der Riemannschen Primzahlformel. SBer. Kgl. Preuß. Akad. Wiss. Berlin, 737–745. [Collected Works, **4**, 11–19, Thales Verlag.]

Landau,E. (1908d): Zwei neue Herleitungen für die asymptotische Anzahl der Primzahlen unter einer gegebener Grenze. SBer. Kgl. Preuß. Akad. Wiss. Berlin, 746–764. [Collected Works, **4**, 21–39, Thales Verlag.]

Landau,E. (1908e): Über einen Grenzwertsatz. SBer. Kais. Akad. Wissensch. Wien, **117**, Abt.2a, 1089–1094. [Collected Works, **4**, 67–72, Thales Verlag.]

Landau,E. (1908f): Über die Primzahlen einer arithmetischer Progression und die Primideale in einer Idealklasse. SBer. Kais. Akad. Wissensch. Wien, **117**, Abt.2a, 1095–1107. [Collected Works, **4**, 73–85, Thales Verlag.]

Landau,E. (1908g): Nouvelle démonstration pour la formule de Riemann sur le nombre des nombres premiers inférieurs à une limite donnée, et démonstration d'une formule plus générale pour le cas des nombres premiers d'une progression arithmétique. Ann. Sci. Ec. Norm. Sup., (3), **25**, 399–408. [Collected Works, **4**, 87–130, Thales Verlag.]

Landau,E. (1908h): Über die Verteilung der Nullstellen der Riemannscher Zetafunktion und einer Klasse verwandter Funktiuonen. Math. Ann., **66**, 419–445. [Collected Works, **4**, 149–175, Thales Verlag.]

Landau,E. (1909a): Handbuch der Lehre von der Verteilung der Primzahlen. Teubner, Leipzig. [Reprint: Chelsea 1953.]

Landau,E. (1909b): Neue Beiträge zur analytischen Zahlentheorie. Rend. Circ. Mat. Palermo, **27**, 46–58. [Collected Works, **4**, 41–53, Thales Verlag.]

Landau,E. (1910): Über die Bedeutung einiger neuerer Grenzwertsätze der Herren Hardy und Axer. Prace Mat.-Fiz., **21**, 97–177. [Collected Works, **4**, 267–347, Thales Verlag.]

Landau,E. (1911a): Zur Theorie der Riemannscher Zetafunktion. Vierteljahrsschr. Naturf. Ges. Zürich, **56**, 125–148. [Collected Works, **5**, 22–45, Thales Verlag.]

Landau,E. (1911b):Über die Äquivalenz zweier Hauptsätze der analytischer Zahlentheorie. SBer. Kais. Akad. Wissensch. Wien, **120**, Abt.2a, 973–988. [Collected Works, **5**, 46–61, Thales Verlag.]

Landau,E. (1911c): Über das Nichtverschwinden der Dirichletschen Reihen, welche komplexen Charakteren entsprechen. Math. Ann., **70**, 69–78. [Collected Works, **4**, 249–258, Thales Verlag.]

Landau,E. (1911d): Über den Gebrauch bedingt konvergenter Integrale in der Primzahltheorie. Math. Ann., **71**, 368–379. [Collected Works, **5**, 86–97, Thales Verlag.]

Landau,E. (1912a): Über einige Summen, die von den Nullstellen der Riemannschen Zetafunktion abhängen. Acta Math., **35**, 271–294. [Collected Works, **5**, 62–85, Thales Verlag.]

Landau,E. (1912b): Über einige neuere Grenzwertsätze. Rend. Circ. Mat. Palermo, **34**, 121–131. [Collected Works, **5**, 145–155, Thales Verlag.]

Landau,E. (1912c): Gelöste und ungelöste Probleme aus der Theorie der Primzahlverteilung und der Riemannschen Zetafunktion. Jahresber. Deutsche Math. Ver., **21**, 208–228. [Proc. 5th Internat. Congress of Math., **I**, 93–108, Cambridge 1913; Collected Works, **5**, 240–255, Thales Verlag.]

Landau,E. (1913): Über die Nullstellen Dirichletscher Reihen., SBer. Kgl. Preuß. Akad. Wiss. Berlin, **41**, 897–907. [Collected Works, **6**, 33–43, Thales Verlag.]

Landau,E. (1914a): Über die Primzahlen in definiten quadratischen Formen und die Zetafunktion reiner kubischer Körper. H.A.Schwarz Festschrift, 244–273, Berlin. [Collected Works, **6**, 105–134. Thales Verlag.]

Landau,E. (1914b): Über eine Aufgabe der Funktionentheorie. Tôhoku Math. J., **5**, 97–116. [Collected Works, **6**, 167–186, Thales Verlag.]

Landau,E. (1915a): Über die Hardysche Entdeckung unendlich vieler Nullstellen der Zetafunktion mit reellem Teil 1/2. Math. Ann., **76**, 212–243. [Collected Works, **6**, 135–166, Thales Verlag.]

Landau,E. (1915b): Neuer Beweis eines analytischen Satzes des Herrn de la Vallée Poussin. Jahresber. Deutsche Math. Ver., **24**, 250–278. [Collected Works, **6**, 241–269, Thales Verlag.]

Landau,E. (1918a): Über einige ältere Vermutungen und Behauptungen in der Primzahltheorie. Math. Zeitschr., **1**, 1–24. [Collected Works, **6**, 469–492, Thales Verlag.]

Landau,E. (1918b): Über die Klassenzahl imaginär-quadratischer Zahlkörper.Nachr. Ges. Wiss. Göttingen, 285–295. [Collected Works, **7**, 150–160. Thales Verlag.]

Landau,E. (1920): Über die Nullstellen der Zetafunktion. Math. Zeitschr., **6**, 151–154. [Collected Works, **7**, 226–229, Thales Verlag.]

Landau,E. (1922): Zur additiven Primzahltheorie. Rend. Circ. Mat. Palermo, **46**, 344–356. [Collected Works, **7**, 436–443, Thales Verlag.]

Landau,E. (1924a): Über die Wurzeln der Zetafunktion. Math. Zeitschr., **20**, 98–104. [Collected Works, **8**, 70–76, Thales Verlag.]

Landau,E. (1924b): Über die ζ-Funktion und die L-Funktionen. Math. Zeitschr., **20**, 105–125. [Collected Works, **8**, 77–97, Thales Verlag.]

Landau,E. (1924c): Über die Möbiussche Funktion. Rend. Circ. Mat. Palermo. **48**, 277–280. [Collected Works, **8**, 171–174, Thales Verlag.]

Landau,E. (1927a): Vorlesungen über Zahlentheorie. I–III. S.Hirzel, Leipzig. [English translation of Vol.I: Chelsea, New York 1958.]

Landau,E. (1927b):Über die Zetafunktion und die Hadamardsche Theorie der ganzen Funktionen. Math. Zeitschr., **26**, 170–175. [Collected Works, **9**, 11-16, Thales Verlag.]

Landau,E. (1930): Die Golbachsche Vermutung und der Schnirelmannsche Satz. Nachr. Ges. Wiss. Göttingen, 255–276. [Collected Works, **9**, 167–188, Thales Verlag.]

Landau,E. (1932a): Über den Wienerschen neuen Weg zum Primzahlsatz. SBer. Preuss. Akad. Wiss. Berlin, 514–521. [Collected Works, **9**, 207–214, Thales Verlag.]

Landau,E. (1932b): Über Dirichletsche Reihen. Nachr. Ges. Wiss. Göttingen, 525–527. [Collected Works, **9**, 203–205, Thales Verlag.]

Landau,E. (1933): Eine Frage über trigonometrische Polynome. Ann. Scuola Norm. Sup. Pisa, Cl. Sci., (2), **2**, 209–210. Corr.: *ibidem*, (2), **5**, p.141. [Collected Works, **9**, 255–256, 264, Thales Verlag.]

Landau,E.,Walfisz,A. (1920): Über die Nichtfortsetzbarkeit einiger durch Dirichletsche Reihen definierte Funktionen. Rend. Circ. Mat. Palermo, **44**, 82–86. [E.Landau, Collected Works, **7**, 252–256, Thales Verlag.]

Lander,J.L., Parkin,T.R. (1967): On first appearance of prime difference. Math. Comp., **21**, 483–488. MR 37#6237.

Landry,F. (1853): Deuxième mémoire sur la théorie des nombres. Paris.

Landry,F. (1880a): Note sur la décomposition du nombre $2^{64}+1$. Comptes Rendus Acad. Sci. Paris, **91**, p.138.

Landry,F. (1880b): Décomposition de $2^{64}+1$. Nouv. Corresp. Math., **6**, p.417.

Landsberg,G. (1893): Zur Theorie der Gaussschen Summen und der linearen Transformation der Thetafunctionen. J. Reine Angew. Math., **111**, 234-253.

Lang,S. (1964): Algebraic Numbers. Addison-Wesley. MR 28#3974.

Lang,S. (1970): Algebraic Number Theory. Addison-Wesley. MR 44#181; 2nd ed.: Springer, Berlin–Heidelberg 1994. MR 95f:11085.

Langmann,K. (1974): Eine Formel für die Anzahl der Primzahlen. Archiv Math., **25**, p.40. MR 49#4951.

Laurent,H. (1898): Sur la théorie des nombres premiers. Comptes Rendus Acad. Sci. Paris, **126**, 809–810.

Laurent,H. (1899): Sur les nombres premiers. Nouv. Ann. Math., (3), **18**, 234–241.

Lavrik,A.F. (1961a): The number of k-twin primes lying in an interval of a given length. Doklady Akad. Nauk SSSR, **136**, 281–283. (Russian) MR 24#A1092.

Lavrik,A.F. (1961b): Binary problems of additive prime number theory connected with the method of I.M.Vinogradov. Vestnik LGU, **16**, no.13, 11–27. (Russian) MR 25#2050.

Lavrik,A.F. (1984): Methods of studying the law of distribution of primes. Trudy Mat. Inst. Steklov., **163**, 118–142. (Russian) MR 86d:11070.

Lavrik,A.F. (1990a): The arithmetic equivalent of Riemann's functional equation. Dokl. Akad. Nauk Uz. SSR, **1990**, No.6, 9-11. (Russian) MR 92d:11095.

Lavrik,A.F. (1990b): Functional equations with a parameter for zeta-functions. Izv. Akad. Nauk SSSR, **54**, 501–521. (Russian) MR 92d:11096.

Lavrik,A.F. (1991): Arithmetic equivalents of functional equations of Riemann type. Trudy Mat. Inst. Steklov., **200**, 213–221. (Russian) MR 93b:11108.

Lavrik,A.F.,Sobirov,A.Š. (1973): The remainder term in the elementary proof of the prime number theorem. Doklady Akad. Nauk SSSR, **211**, 534–536. (Russian) MR 48#2092.

Lebesgue,V.A. (1843): Théorèmes nouveaux sur l'équation indéterminée $x^5+y^5=az^5$. J. math. pures appl., **8**, 49–70.

Lebesgue,V.A. (1856): Remarques diverses sur les nombres premiers. Nouv. Ann. Math., **15**, 130–134, 236–239.
Lebesgue,V.A. (1867): Théorème sur les racines primitives. Comptes Rendus Acad. Sci. Paris, **64**, 1268–1269.
Lebon,E. (1904): Sur le nombre des nombres premiers de 1 à N. Rend. Circ. Mat. Palermo, **18**, 260–268.
Leech,J. (1957): Note on the distribution of prime numbers. J. London Math. Soc., **32**, 56–58. MR 18 p.642.
Lefébure,A. (1884): Mémoire sur la composition de polynômes entiers qui n'admettent que des diviseurs premiers d'une forme déterminée. Ann. Sci. Ec. Norm. Sup., (3), **1**, 389–404.
Lefébure,A. (1885): Mémoire sur la composition de polynômes entiers qui n'admettent que des diviseurs premiers d'une forme déterminée. Ann. Sci. Ec. Norm. Sup., (3), **2**, 113–122.
Legendre,A.M. (1798): Essai sur la théorie des nombres. Paris. [2nd ed. Courcier, Paris 1808.]
Legendre,A.M. (1830): Théorie des Nombres. Didot, Paris. [3rd edition of Legendre (1798); Reprints: Hermann, Paris 1900; Blanchard, Paris 1955. German translation: Zahlentheorie. Teubner, Leipzig 1886, 2nd ed. 1893.]
Lehman,R.S. (1960): On Liouville's function. Math. Comp., **14**, 311–320. MR 22#10955.
Lehman,R.S. (1966): On the difference $\pi(x) - \mathrm{li}(x)$. Acta Arith., **11**, 397-410. MR 34#2546.
Lehmer,D.H. (1936): On the function $x^2 + x + A$. Sphinx, **6**, 212–214.
Lehmer,D.H. (1956a): Extended computations of the Riemann zeta-function. Mathematika, **3**, 102–108. MR 19 p.121. [Selected Papers, **II**, 580–586. Winnipeg 1981.]
Lehmer,D.H. (1956b): On the roots of the Riemann zeta-function. Acta Math., **95**, 291–298. MR 19 p.121. [Selected Papers, **II**, 572–579. Winnipeg 1981.]
Lehmer,D.H. (1959): On the exact number of primes less than a given limit. Illinois J. Math., **3**, 381–388. MR 21#5613. [Selected Papers, **III**, 1104–1111. Winnipeg 1981.]
Lenstra,A.K.,Lenstra,H.W.Jr.,Manasse,M.S.,Pollard,J.M. (1993): The factorization of the ninth Fermat number. Math. Comp., **61**, 319—349. MR 93k:11116. Add.: *ibidem*, **64**, 1995, p.1357.
Lenstra,H.W.Jr.,Stevenhagen,P. (1991): Primes of degree one and algebraic cases of Čebotarev's theorem. Enseign. Math., **37**, 17–30. MR 92f:11162.
Leopoldt,H.W. (1966): Lösung einer Aufgabe von Kostrikhin. J. Reine Angew. Math., **221**, 160–161. MR 33#5550.
Lepistö,T. (1974): A zero-free region for certain L-functions. Ann. Acad. Sci. Fenn. Ser. A.1, **576**, 1–13. MR 50#251.
Lerch,M. (1887/88): Sur la fonction $\mathfrak{K}(w,x,s) = \sum_{k=0}^{\infty} \frac{\exp(2k\pi i x)}{(w+k)^s}$. Acta Math., **11**, 19–24.
Lerch,M. (1892a): Základove theorie Malmsténovych řad. Rozpravy České Akad. 2.Kl., **1**, No.27, 1–70.
Lerch,M. (1892b): O hlavni větě theorie funkci vytvořujicich. Rozpravy České Akad. 2.Kl., **1**, No.33, 1–7.
Lerch,M. (1893): Studie v oboru Malmsténovych řad a invariantů forem kvadratických. Rozpravy České Akad. 2.Kl., **2**, No.4, 1–12.
Lerch,M. (1894): Dalši studie v oboru Malmsténovych řad. Rozpravy České Akad. 2.Kl., **4**, No.28, 1–63.
Lerch,M. (1903): Sur un point de la théorie des fonctions géneratrices d'Abel. Acta Math., **27**, 339–351.

Lerch,M. (1914): Ëtablissement des coefficients dans le développement en série de puissances de la fonction $\zeta(s)$. Časopis pro pěstovani Mat. Fys., **43**, 513–522.
Levi-Civita,B. (1895): Di una espressione analitica atta a rappresentare il numero di numeri primi compresi in un determinato intervallo. Rend. Accad, dei Lincei, (5), **4**, 303–309. [Opere Matematiche, **I**, 153–158, N.Zanichelli, Bologna.]
Levin,B.V. (1961a): On the distribution of primes of an arithmetic progression. Izv. AN Uz.SSR, nr.5, 15–28. (Russian) MR 25#2048.
Levin,B.V. (1961b): The weak Landau problem and its generalization. Usp. Mat. Nauk, **16**, no.2, 123–125. (Russian) MR 26#3677.
Levin,B.V. (1963): Distribution of near primes in polynomial sequences. Mat. Sb., **61**, 389–407. (Russian) MR 30#1991.
Levin,B.V. (1964/65): A one-dimensional sieve. Acta Arith., **10**, 387–397. (Russian) MR 31#4774.
Levinson,N. (1964): Prime number theorem from $\log n!$. Proc. Amer. Math. Soc., **15**, 480–485. MR 29#81.
Levinson,N. (1966/67): On the elementary proof of the prime number theorem. Proc. Edinburgh Math. Soc., (2), **15**, 141–146. MR 36#1398.
Levinson,N. (1969): A motivated account of an elementary proof of the prime number theorem. Amer. Math. Monthly, **76**, 225–245. MR 39#2712.
Levinson,N. (1971): On theorems of Berlowitz and Berndt. J. Number Th., **3**, 502–504. MR 45#3338.
Levinson,N. (1972): Ω-theorems for the Riemann zeta-function. Acta Arith., **20**, 317–330. MR 46#5262.
Levinson,N. (1973): On the elementary character of Wiener's general Tauberian theorem. J. Math. Anal. Appl., **42**, 381–396. MR 48#4751.
Levinson,N. (1974): More than one third of zeros of Riemann's zeta-function are on $\sigma = 1/2$. Adv. Math., **13**, 383-436. MR 58#27837.
Levinson,N. (1975a): A simplification in the proof that $N_0(T) > \frac{1}{3}N(T)$ for Riemann zeta function. Adv. Math., **18**, 239–242. MR 53#7966.
Levinson,N. (1975b): Almost all roots of $\zeta(s) = a$ are arbitrarily close to $\sigma = 1/2$. Proc. Nat. Acad. Sci. U.S.A. , **72**, 1322—1324. MR 53#10737.
Levinson,N. (1975c): Deduction of semi-optimal mollifier for obtaining lower bounds for $N_0(T)$ for Riemann's zeta-function. Proc. Nat. Acad. Sci. U.S.A. , **72**, 294–297. MR 51#5516.
Levinson,N. (1976): On the number of sign changes of $\pi(x) - \text{li}\, x$. Topics in Number Theory, Debrecen 1974, Proc. J.Bólyai Colloq. **13**, 171–177, North Holland, Amsterdam. MR 55#12658.
Li Hong Ze (1994): Almost primes in short intervals. Sci. China, **A37**, 1428–1441. MR 96e:11116.
Li Hong Ze (1997): Primes in short intervals. Math. Proc. Cambridge Philos. Soc., **122**, 193–205. MR 98e:11103.
Liang,J.J.Y.,Todd,J. (1972): The Stieltjes constants. J. Res. Nat. Bur. Standards, **76B**, 161–178. MR 48#5316.
Lichtenbaum,P. (1931): Über die Funktion $\chi(s) = \sum_{k=0}^{\infty}(-1)^k(2k+1)^{-s}$. Math. Zeitschr., **33**, 641–647.
Light,W.A.,Forest,J.,Hammond,N.,Roe,S. (1980): A note on Goldbach's conjecture. BIT, **79**, p.525. MR 82h:10003.
Lindelöf,E. (1903): Sur une formule sommatoire générale. Acta Math., **27**, 305–312.
Lindelöf,E. (1908): Quelques remarques sur la croissance de la fonction $\zeta(s)$. Bull. sci. math., (2), **32**, 341–356.
Lindgren,H. (1963): The calculation of $\pi(N)$. J. Austral. Math. Soc., **3**, 257–266. MR 27#3612.

Linnik,Yu.V. (1944): On the least prime in an arithmetic progression, I,II. Mat. Sb., (N.S.), **15**, 139–178, 347–368. (Russian) MR 6 p.260. [Izbrannye Trudy, **I**, 336–399, Nauka, Leningrad 1979.]

Linnik,Yu.V. (1946): A new proof of the Goldbach-Vinogradov theorem. Mat. Sb., **19**, 3–8. (Russian) MR 8 p.317. [Izbrannye Trudy, **II**, 23–27, Nauka, Leningrad 1980.]

Linnik,Yu.V. (1950): An elementary proof of Siegel's theorem based on the method of I.M.Vinogradov. Izv. Akad. Nauk SSSR, **14**, 327–342. (Russian) MR 12 p.482. [Izbrannye Trudy, **II**, 44–59, Nauka, Leningrad 1980.]

Linnik,Yu.V. (1959): Hardy-Littlewood problem on representation as the sum of a prime and two squares. Doklady Akad. Nauk SSSR, **124**, 29–30. (Russian) MR 20#6401. [Izbrannye Trudy, **II**, 181–184, Nauka, Leningrad 1980.]

Linnik,Yu.V. (1960a): All large numbers are sums of a prime and two squares (A problem of Hardy and Littlewood),I. Mat. Sb., **52**, 661–700. (Russian) MR 22#10963; II. *ibidem*, **53**, (1961), 3–38 (Russian) #10964. [Izbrannye Trudy, **II**, 217–284, Nauka, Leningrad 1980.]

Linnik,Yu.V. (1960b): An asymptotic formula in an additive problem of Hardy-Littlewood. Izv. Akad. Nauk SSSR, **24**, 629–706. (Russian)MR 23#A130.

Lint,J.H.van,Richert,H.-E. (1965): On primes in arithmetic progressions. Acta Arith., **11**, 209–216. MR 32#5613.

Lionnet,F.J.E. (1872): Question 1075. Nouv. Ann. Math., (2), **11**, p.190.

Liouville,J. (1837): Solution d'un problème d'analyse. J. math. pures appl., **2**, 1–2.

Liouville,J. (1856): Ueber einen Satz der Zahlentheorie. Zeitschr. Math. Phys., **1**, 371–372.

Liouville,J. (1857): Sur l'expression $\varphi(n)$ qui marque combien la suite $1, 2, 3, \dots, n$ contient de nombres premiers a n. J. math. pures appl., (2), **2**, 110–112.

Lipschitz,R. (1857): Untersuchung einer aus vier Elementen gebildeten Reihe. J. Reine Angew. Math., **54**, 19–24.

Lipschitz,R. (1879): Sur des séries relatives à la théorie des nombres. Comptes Rendus Acad. Sci. Paris, **89**, 948–950.

Lipschitz,R. (1882): Sur une communication de M.du Jonquiéres relative aux nombres premiers. Comptes Rendus Acad. Sci. Paris, **95**, 1344–1346; *ibidem*, **96**, 1883, 58–61.

Lipschitz,R. (1883a): Addition à une note sur les nombres premiers. Comptes Rendus Acad. Sci. Paris, **96**, 114–115.

Lipschitz,R. (1883b): Application d'une méthode donnée par Legendre. Comptes Rendus Acad. Sci. Paris, **96**, 327–329.

Lipschitz,R. (1889): Untersuchungen der Eigenschaften einer Gattung von unendlichen Reihen. J. Reine Angew. Math., **105**, 127–156.

Lipschitz,R. (1890): Bemerkung zu dem Aufsatze: Untersuchungen der Eigenschaften einer Gattung von unendlichen Reihen. J. Reine Angew. Math., **106**, 27–29.

Littlewood,J.E. (1912): Quelques conséquences de l'hypothése que la fonction $\zeta(s)$ de Riemann n'a pas de zéros dans le demi-plan $R(s) > \frac{1}{2}$. Comptes Rendus Acad. Sci. Paris, **154**, 263–266. [Collected Papers, **II**, 793–796, Clarendon Press, Oxford 1982.]

Littlewood,J.E. (1914): Sur la distribution des nombres premiers. Comptes Rendus Acad. Sci. Paris, **158**, 1869–1872. [Collected Papers, **II**, 829–832, Clarendon Press, Oxford 1982.]

Littlewood,J.E. (1922): Researches in the theory of Riemann's ζ-function. Proc. London Math. Soc., (2), **20**, xxii–xxviii. [Collected Papers, **II**, 844–850, Clarendon Press, Oxford 1982.]

Littlewood,J.E. (1924a): Two notes on the Riemann zeta-function. Proc. Cambridge Philos. Soc., **22**, 239–248. [Collected Papers, **II**, 851–859, Clarendon Press, Oxford 1982.]

Littlewood,J.E. (1924b): On the zeros of the Riemann zeta-function. Proc. Cambridge Philos. Soc., **22**, 295–318. [Collected Papers, **II**, 860–883, Clarendon Press, Oxford 1982.]

Littlewood,J.E. (1925): On the Riemann zeta-function. Proc. London Math. Soc., (2), **24**, 175–201. [Collected Papers, **II**, 884–910, Clarendon Press, Oxford 1982.]

Littlewood,J.E. (1927): Mathematical notes: 3; On a theorem concerning the distribution of prime numbers. J. London Math. Soc., **2**, 41–45. [Collected Papers, **II**, 833–837, Clarendon Press, Oxford 1982.]

Littlewood,J.E. (1928): On the function $1/\zeta(1+ti)$. Proc. London Math. Soc., (2), **27**, 349–357. [Collected Papers, **II**, 911–919, Clarendon Press, Oxford 1982.]

Littlewood,J.E. (1937): Mathematical notes (12): An inequality for a sum of cosines. J. London Math. Soc., **12**, 217–221. [Collected Papers, **II**, 838–842, Clarendon Press, Oxford 1982.]

Littlewood,J.E. (1971): The quickest proof of the prime number theorem. Acta Arith., **18**, 83–86. MR 45#3343. [Collected Papers, **II**, 951–954, Clarendon Press, Oxford 1982.]

Liu Hong Quang (1993): The greatest prime factor of the integers in an interval. Acta Arith., **65**, 301–328. MR 95d:11117.

Liu Hong Quang (1996): Almost primes in short intervals. J. Number Th., **57**, 303–322. MR 97c:11092.

Liu Ming Chit (1985): A bound for prime solutions of some ternary equations. Math. Zeitschr., **188**, 313–323. MR 86g:11057.

Liu Ming Chit (1987): An improved bound for prime solutions of some ternary equations. Math. Zeitschr., **194**, 575–583. MR 88e:11097.

Liu Ming Chit,Tsang Kai Man (1989): Small prime solutions of some linear equations. Théorie des Nombres, 595–624, de Gruyter, Berlin. MR 90i:11112.

Liu Ming Chit,Tsang Kai Man (1993): Recent progress on a problem of A.Baker. Séminaire de Théorie des Nombres, Paris 1991/92, Progress in Math., **116**, 121–133. MR 95h:11107.

Liu Ming Chit,Wang Tianze (1989): A numerical bound for small prime solutions of some ternary linear equations. Acta Arith., **86**, 343–383.

Lou,S.T. (1981): A lower bound for the number of zeros of Riemann's zeta-function on $\sigma = 1/2$. Recent Progress in Analytic Number Theory (Durham 1979), **I**, 319–324, Academic Press, London. MR 83e:10059.

Lou,S.T.,Yao,Q. (1980): A lower bound for the number of zeros of Riemann's zeta-function on $\sigma = 1/2$. Kexue Tongbao, **25**, 374–377. MR 82a:10043.

Lou,S.T.,Yao,Q. (1992): A Chebychev's type of prime number theorem in a short interval, II. Hardy-Ramanujan J., **15**, 1–33. MR 95d:11115.

Lou,S.T.,Yao,Q. (1993): The number of primes in a short interval. Hardy-Ramanujan J., **16**, 21–43. MR 94b:11089.

Low,M.E. (1967/68): Real zeros of the Dedekind zeta function of an imaginary quadratic field. Acta Arith., **14**, 117–140. MR 38#4425.

Loxton,J.H.,Sanders,J.W. (1980/81): On an inversion theorem of Möbius. J. Austral. Math. Soc., A, **30**, 15–32. MR 81k:10006.

Lu, Ming Gao (1985): The difference between consecutive primes. Acta Math. Sinica, (N.S.), **1**, 109–118. MR 87j:11089.

Lucas,E. (1878): Théorie des fonctions numériques simplement périodiques. Ann. of Math., **1**, 1878, 184–240, 289–321.

Lugli,A. (1888): Sul numero dei numeri primi de 1 ad n. Giornale di Math., **26**, 86–95.

Lukes,R.F.,Patterson,C.D.,Williams,H.C. (1995): Numerical sieving devices: their history and some applications. Nieuw Arch. Wiskunde, (4), **13**, 113–139. MR 96m:11082.

Lune,J.van der (1979): A note on a formula of van der Pol. Acta Arith., **35**, 361–366. MR 81i:10050.

Lune,J.van der,Riele,H.J.J.te,Winter,D.T. (1986): On the zeros of the Riemann zeta function in the critical strip, IV. Math. Comp., **46**, 667–681. MR 87e:11102.

Lüneburg,H. (1981): Ein einfacher Beweis für den Satz von Zsigmondy über primitive Primteiler von A^N-1. Lecture Notes in Math., **893**, 219–222. MR 84d:10006.

Lüroth,J. (1905): Eine historische Bemerkung zur Funktionentheorie. Math. Ann., **60**, 398–401.

Maclaurin,C. (1742): Treatise of Fluxions, Edinburgh.

Mac Leod,R.A. (1967): A new estimate for the function $M(x) = \sum_{n\leq x}\mu(n)$. Acta Arith., **13**, 49–59. MR 36#2575. Errata: *ibidem*, **16**, 1969/70, 99–100. MR 39#6847.

Mahler,K. (1933): Zur Approximation algebraischen Zahlen, I. (Über den grössten Primteiler binärer Formen.) Math. Ann., **107**, 691–730.

Mahler,K. (1935a): Über den grössten Primteiler der Polynome $x^2 \mp 1$. Arch. Math. Naturvid., **41**, Nr.1, 1–8.

Mahler,K. (1935b): Über den grössten Primteiler spezieller Polynome zweiten Grades. Arch. Math. Naturvid., **41**, Nr.6, 1–26.

Maier,H. (1985): Primes in short intervals. Michigan J. Math., **32**, 221–225. MR 86i:11049.

Maier,H. (1985): Small differences between prime numbers. Michigan J. Math., **35**, 323–344. MR 90e:11126.

Maier,H.,Pomerance,C. (1990): Unusually large gaps between consecutive primes. Trans. Amer. Math. Soc., **322**, 201–237. MR 91b:11093.

Maillet,E. (1900): Nombres entiers N jouissant de cette propriété, que les $\varphi(N)$ nombres premiers à N et non supérieurs à N soient tous des nombres premiers. L'Intermédiaire des Math., **7**, p.254.

Mąkowski,A. (1960): Partitions into unequal primes. Bull. Acad. Pol. Sci., **8**, 125–126. MR 22#7991.

Mallik,A. (1981): New proofs of asymptotic formulae for Siegel's zero. Ann. Univ. Sci. Budapest, **24**, 21–28. MR 87m:11082.

Malmrot,B. (1920): Sur une formule de M.Fujiwara. Arkiv f. Math. Astr. Fys., **14**, No.4, 1–10.

Malmstén,C.J. (1849): De integralibus quibusdam definitis, seriebusque infinitis. J. Reine Angew. Math., **38**, 1–39.

Mamangakis,S.E. (1962): Remark on $\pi(x) = o(x)$. Proc. Amer. Math. Soc., **13**, 664–665. MR 26#1291.

Mandelbrojt.S. (1967): Théorie des fonctions et théorie des nombres dans l'oeuvre de Jacques Hadamard. Enseign. Math., (2), **13**, 25–34. MR 36#36226.

Mangoldt,H.von (1894): Auszug aus einer Arbeit unter dem Titel: "Zu Riemann's Abhandlung "Ueber die Anzahl der Primzahlen unter einer gegebener Größe" ". SBer. Kgl. Preuß. Akad. Wiss. Berlin, 883-896. [French translation: Ann. Sci. Ec. Norm. Sup., (3), **13**, 1896, 61–78.]

Mangoldt,H.von (1895): Zu Riemann's Abhandlung "Ueber die Anzahl der Primzahlen unter einer gegebener Größe". J. Reine Angew. Math., **114**, 255–305.

Mangoldt,H.von (1897): Beweis der Gleichung $\sum_{k=1}^{\infty}\frac{\mu(k)}{k} = 0$. SBer. Kgl. Preuß. Akad. Wiss. Berlin, 835-852. [French translation: Ann. Sci. Ec. Norm. Sup., (3), **15**, 1898, 431–454.]

Mangoldt,H.von (1898): Über eine Anwendung der Riemannscher Formeln für die Anzahl der Primzahlen unter einer gegebener Grösse. J. Reine Angew. Math., **119**, 65–71.

Mangoldt,H.von (1905): Zur Verteilung der Nullstellen der Riemannscher Funktion $\xi(t)$. Math. Ann., **60**, 1–19.

Mapes,D.C. (1963): Fast method for computing the number of primes less than a given limit. Math. Comp., **17**, 179–185. MR 28#1731.

Mardzhanišvili,K.K. (Mardjanishvili,C.) (1941): Sur la démonstration du théorème de Goldbach-Vinogradov. Doklady Akad. Nauk SSSR, **30**, 687–689. MR 3 p.165.

Markov,A. (1895): Dèmonstration d'un théorème de Tchébycheff. Comptes Rendus Acad. Sci. Paris, **120**, 1032–1034.

Martić,B., Vasić,P.M. (1973): A supplement to Karanikolov's paper. Publ. Elektrotehn. Fak., Univ. Beograd, ser. Mat.Fiz., No.412–460, 165–166. MR 48#5975.

Massias,J.-P.,Robin,G. (1996): Bornes effectives pour certaines fonctions concernant les nombres premiers. J. Théor. Nombres Bordeaux, **8**, 215–242. MR 97g:11099.

Mathews,G.B. (1892): Theory of Numbers, part 1, Deighton Bell, Cambridge. [Reprint Chelsea 1961].

Matijasevič,Yu.V. (1970): The diophantineness of enumerable sets. Doklady Akad. Nauk SSSR, **191**, 279–282. (Russian) MR 41#3390.

Matijasevič,Yu.V. (1971a): Diophantine representation of enumerable predicates. Izv. Akad. Nauk SSSR, **35**, 3–30. (Russian) MR 43#54.

Matijasevič,Yu.V. (1971b): A Diophantine representation of the set of prime numbers. Doklady Akad. Nauk SSSR, **196**, 770–773. (Russian) MR 43#1921.

Matijasevič,Yu.V. (1977): Primes are enumerated by a polynomial in 10 variables. Zap. Naučn. Sem. LOMI, **68**, 62–82. (Russian) MR 58#21534.

Matijasevič,Yu.V., Robinson,J. (1975): Reduction of an arbitrary Diophantine equation to one in 13 unknowns. Acta Arith., **27**, 521–553. MR 52#8033.

Matvievskaja,G.N. (1961): Bertrand's postulate in Euler's notes. Istor.-Mat. Issled., **14**, 285–288. (Russian) MR 32#1176.

Matvievskaja,G.N.,Ožigova,H.P. (1983): Leonhard Eulers handschriftlicher Nachlass zur Zahlentheorie. Leonhard Euler 1707–1783, 151–160, Birkhäuser, Basel.

Maxfield,J.,Maxfield,M. (1959/60): The existence of integers less than p belonging to ep^{r-1} (mod p^r). Math.Mag., **33**, 219–220. MR 24#A2551.

McCurley,K.S. (1984a): Explicit estimates for the error term in the Prime Number Theorem for arithmetic progressions. Math. Comp., **42**, 265–285. MR 85e:11065.

McCurley,K.S. (1984b): Prime values of polynomials and irreducibility testing. Bull. Amer. Math. Soc., (N.S.), **11**, 155–158. MR 85e:11066.

McCurley,K.S. (1984c): Explicit zero-free regions for Dirichlet L-functions. J. Number Th., **19**, 7–32. MR 85k:11041.

McCurley,K.S. (1984d): Explicit estimates for $\theta(X,3,l)$ and $\psi(X,3,l)$. Math.Comp. **42**, 287–296. MR 85g:11085.

McCurley,K.S. (1986a): The smallest prime value of $x^n + a$. Canad. J. Math., **38**, 925–936. MR 87i:1117.

McCurley,K.S. (1986b): Polynomials with no small prime values. Proc. Amer. Math. Soc., **97**, 393–395. MR 87i:11118.

McIntosh,R.J. (1995): On the converse of Wolstenholme's theorem. Acta Arith., **71**, 381–389. MR 96h:11002.

McKinzie,M.,Tuckey,C. (1997): Hidden lemmas in Euler's summation of the reciprocals of the squares. Arch. Hist. Ex. Sci., **51**, 29–57. MR 99c:11003.

Meissel,E. (1854): Observationes quaedam in theoria numerorum. J. Reine Angew. Math., **48**, 301–316.

Meissel,E. (1870): Ueber die Bestimmung der Primzahlmenge innerhalb gegebener Grenzen. Math. Ann., **2**, 636–642.

Meissel,E. (1871): Berechnung der Menge der Primzahlen, welche innerhalb der ersten Hundert Millionen natürlichen Zahlen vorkommen. Math. Ann., **3**, 523–525.

Meissel,E. (1883): Ueber Primzahlmengen. Math. Ann., **21**, 304.

Meissel,E. (1885): Berechnung der Menge der Primzahlen, welche innerhalb der ersten Milliarde natürlichen Zahlen vorkommen. Math. Ann., **25**, 251–257.

Meldercreutz,J. (1755): De summatione seriei reciprocae e quadratis numerorum naturalium. Holmiae.

Meller,N.A. (1958): Computations connected with the check of Riemann's hypothesis. Doklady Akad. Nauk SSSR, **123**, 246–248. (Russian) MR 20#6396.

Mellin,H. (1899): Über eine Verallgemeinerung der Riemannscher Funktion $\zeta(s)$. Acta soc. sci. Fenn., **24**, No.10, 1–50.

Mellin,H. (1900): Eine Formel für den Logarithmus transcendenter Funktionen von endlichen Geschlecht. Acta Soc. Sci. Fenn., **29**, Nr. 4. 1–50.

Mellin,H. (1902): Eine Formel für den Logarithmus transcendenter Funktionen von endlichen Geschlecht. Acta Math., **25**, 165–183.

Mellin,H. (1904): Die Dirichletschen Reihen, die zahlentheoretischen Funktionen und die unendlichen Produkte von endlichem Geschlecht. Acta Math., **28**, 37–64.

Mellin,H. (1910): Abriß einer einheitlichen Theorie der Gamma- und hypergeometrischen Funktionen. Math. Ann., **68**, 305–337.

Mellin,H. (1917): Bemerkungen im Anschluss an den Beweis eines Satzes von Hardy über die Zetafunktion. Acta Soc. Sci. Fenn., A, **11**, No.3.

Merlin,J. (1911): Sur quelques théorèmes d'arithmétique et un énoncé qui les contient. Comptes Rendus Acad. Sci. Paris, **153**, 516–518.

Mertens,F. (1874a): Ueber einige asymptotische Gesetze der Zahlentheorie. J. Reine Angew. Math., **77**, 289–338.

Mertens,F. (1874b): Ein Beitrag zur analytischen Zahlentheorie. J. Reine Angew. Math., **78**, 46–62.

Mertens,F. (1895a): Über Dirichletsche Reihen. SBer. Kais. Akad. Wissensch. Wien, **104**, Abt.2a, 1093–1153.

Mertens,F. (1895b): Über das Nichtverschwinden Dirichletscher Reihen mit reellen Gliedern. SBer. Kais. Akad. Wissensch. Wien, **104**, Abt.2a, 1158–1166.

Mertens,F. (1897a): Über Dirichlet's Beweis des Satzes, daß jede unbegrenzte ganzzahlige arithmetische Progression, deren Differenz zu ihren Gliedern teilerfremd ist, unendlich viele Primzahlen darstellt. SBer. Kais. Akad. Wissensch. Wien, **106**, Abt. 2a, 254–286.

Mertens,F. (1897b): Über eine zahlentheoretische Funktion. SBer. Kais. Akad. Wissensch. Wien, **106**, Abt. 2a, 761–830.

Mertens,F. (1897c): Über Multiplikation und Nichtverschwinden Dirichlet'scher Reihen. J. Reine Angew. Math., **117**, 169–184.

Mertens,F. (1898): Über eine Eigenschaft der Riemannscher ζ- Funktion. SBer. Kais. Akad. Wissensch. Wien, **107**, Abt.2a, 1429–1434.

Mertens,F. (1899a): Eine asymptotische Aufgabe. SBer. Kais. Akad. Wissensch. Wien, **108**, Abt.2a, 32–37.

Mertens,F. (1899b): Beweis, dass jede lineare Function mit ganzen complexen teilerfremden Coefficienten unendlich viele complexe Primzahlen darstellt. SBer. Kais. Akad. Wissensch. Wien, **108**, Abt.2a, 517–556.

Mertens,F. (1900): Über einen Satz von Dirichlet. SBer. Kais. Akad. Wissensch. Wien, **109**, Abt.2a, 415–480.

Metsänkyla,T. (1970): Zero-free regions of Dirichlet's L-functions near the point 1. Ann. Univ. Turku, AI, **139**, 3–10. MR 42#3036.

Meyer,A. (1888): Ueber einen Satz von Dirichlet. J. Reine Angew. Math., **103**, 98–117.

Miech,R.J. (1964/65): Almost primes generated by a polynomial. Acta Arith., **10**, 9–30. MR 29#1174.

Miech,R.J. (1966): A uniform result on almost primes. Acta Arith., **11**, 371–391. MR 34#1287.

Miech,R.J. (1968a): A number-theoretic constant. Acta Arith., **15**, 119–137. MR 39#142.

Miech,R.J. (1968b): On the equation $n = p + x^2$. Trans. Amer. Math. Soc., **130**, 494–512. MR 42#1775.

Mientka,W. (1961): An application of the Selberg sieve method. J. Indian Math. Soc., **25**, 129–138. MR 27#2490.

Mikawa,H. (1989): Almost-primes in arithmetic progressions and short intervals. Tsukuba J.Math., (2), **13**, 387–401. MR 90m:11136.

Mikawa,H. (1993): On the sum of a prime and a square. Tsukuba J.Math., (2), **17**, 299–310. MR 94k:11117.

Mikolás,M. (1957): A simple proof of the functional equation for the Riemann zeta-function and a formula of Hurwitz. Acta Sci. Math., **18**, 261—263. MR 20#24.

Mikolás,M. (1971): New proof and extension of the functional equation of Lerch's zeta-function. Ann. Univ. Sci. Budapest, **14**, 111–116. MR 47#157.

Miller,A.G. (1905): Some relations between number theory and group theory. Amer. J. Math, **27**, 315–322.

Mills, W.H. (1947): A prime representing function. Bull. Amer. Math. Soc., **53**, 604. MR 8 p.567.

Min,S.H. (1949): On the order of $\zeta(1/2+it)$. Trans. Amer. Math. Soc., **65**, 448–472. MR 11 p.84.

Mitchell,H.H. (1925): On classes of ideals in a quadratic field. Ann. of Math., (2), **27**, 297–314.

Mitrović,D. (1957): Sur la fonction $\zeta(s)$ de Riemann. Comptes Rendus Acad. Sci. Paris, **244**, 1602–1604. MR 19 p.393; **245**, 885–886. MR 19 p.839.

Möbius,A.F. (1832): Über eine besondere Art von Umkehrung der Reihen. J. Reine Angew. Math., **9**, 1832, 105–123. [Gesammelte Werke, **IV**, 589–612, Hirzel, Leipzig 1887. Reprint: Sändig, Wiesbaden, 1967.]

Mohanty,S.P. (1978): The number of primes is infinite. Fibonacci Quart., **16**, 381–384. MR 80c:10009.

Möller,H. (1976): Verallgemeinerung eines Satzes von Rabinowitsch über imaginär-quadratische Zahlkörper. J. Reine Angew. Math., **285**, 100–113. MR 54#7422.

Mollin,R.A. (1995): Quadratics. C.R.C.Press, Florida. MR 97e:11135.

Mollin,R.A. (1996): Quadratic polynomials producing consecutive distinct primes and class groups of complex quadratic fields. Acta Arith., **74**, 17–30. MR 96k:11123.

Mollin,R.A.,Williams, H.C. (1989): Quadratic non-residues and prime-producing polynomials. Canad. Math. Bull., **32**, 474–478. MR 91a: 11009.

Molsen,K. (1941): Zur Verallgemeinerung des Bertrandschen Postulates. Deutsche Math., **6**, 248–256. MR 8 p.197.

Monsky,P. (1993): Simplifying the proof of Dirichlet's theorem. Amer. Math. Monthly, **100**, 861–862. MR 95b:11085.

Montgomery,H.L. (1969a): Mean and large values of Dirichlet polynomials. Invent. math., **8**, 334–345. MR 42#3029.

Montgomery,H.L. (1969b): Zeros of L-functions. Invent. math., **8**, 346–354. MR 40#2620.

Montgomery,H.L. (1970): Primes in arithmetic progressions. Michigan J. Math., **17**, 33–39. MR 41#1660.

Montgomery,H.L. (1971): Topics in multiplicative number theory. Lecture Notes in Math., **227**, Springer, Berlin–Heidelberg. MR 49#2616
Montgomery,H.L. (1972): The pair correlation of zeros of the zeta function. Proc. Symposia Pure Math., **24**, 181–193. MR 49#2590.
Montgomery,H.L. (1977): Extreme values of the Riemann zeta function. Comment. Math. Helv., **52**, 511–518. MR 57#249.
Montgomery,H.L. (1974): Distribution of zeros of the Riemann zeta function. Proc. ICM Vancouver 1974, 379–381.
Montgomery,H.L.,Vaughan,R.C. (1973): The large sieve. Mathematika, **20**, 119–134. MR 51#10260.
Montgomery,H.L.,Vaughan,R.C. (1975): The exceptional set in Goldbach's problem. Acta Arith., **27**, 353–370. MR 51#10263.
Mordell,L.J. (1928): Some applications of Fourier series in the analytic theory of numbers. Proc. Cambridge Philos. Soc., **34**, 585–590.
Mordell,L.J. (1929): Poisson's summation formula and the Riemann zeta-function. J. London Math. Soc., **4**, 285–291.
Moreau,C. (1873): Extrait d'une lettre. Nouv. Ann. Math., (2), **12**, 323–324.
Moree,P. (1993): Bertrand's postulate for primes in arithmetical progressions. Comput. Math. Appl., **26**, 35–43. MR 94f:11089.
Moreno,L.J. (1973): The Hoheisel phenomenon for generalized Dirichlet series. Proc. Amer. Math. Soc., **40**, 47–51. MR 48#6024.
Morrison,M.A.,Brillhart,J. (1971): The factorization of F_7. Bull. Amer. Math. Soc., **77**, p.264. MR 42#3012.
Morrison,M.A.,Brillhart,J. (1975): A method of factoring and the factorization of F_7. Math. Comp., **29**, 183–205. MR 51#8017.
Morton,H.R. (1995): Fibonacci-like sequences and greatest common divisors. Amer. Math. Monthly, **102**, 731–734. MR 96g:11014.
Moser,J. (1987): Remarks on A.Selberg's theorems on the zeros of the function $\zeta(1/2+it)$. Acta Math. Univ. Comen., **50/51**, 111–132. (Russian) MR 90g:11122.
Moser,L. (1949): A theorem on the distribution of primes. Amer. Math. Monthly, **56**, 624–625. MR 11 p.332.
Moser,L. (1950): A prime representing function. Math. Mag. **23**, 163–164.
Moser,L. (1951): On the equation $\varphi(n) = \pi(n)$. PiMuEpsilon J., 177–180.
Moser,L. (1958): On the series $\sum 1/p$. Amer. Math. Monthly, **65**, 104–105. MR 21#232.
Moser,L. (1959): Notes on number theory. I. On the product of the primes not exceeding x. Canad. Math. Bull., **2**, 119–121. MR 21#4127.
Motohashi,Y. (1970a): On the distribution of prime numbers which are of the form $x^2 + y^2 + 1$. Acta Arith., **16**, 351–363. MR 44#5284.
Motohashi,Y. (1970b): A note on the least prime in an arithmetic progression with a prime difference. Acta Arith., **17**, 283–285. MR 42#3030.
Motohashi,Y. (1971): On the distribution of prime numbers which are of the form $x^2 + y^2 + 1$, II. Acta Math. Acad. Sci. Hung., **22**, 207–210. MR 44#5285.
Motohashi,Y. (1974): On some improvements of the Brun-Titchmarsh theorem. J. Math. Soc. Japan, **26**, 306–323. MR 49#2617.
Motohashi,Y. (1975): On some improvements of the Brun-Titchmarsh theorem,III. J. Math. Soc. Japan, **27**, 444–453. MR 58#10788b.
Motohashi,Y. (1978a): On Vinogradov's zero-free region for the Riemann zeta-function. Proc. Japan Acad. Sci., **54**, 300–302. MR 80a:10056.
Motohashi,Y. (1978b): On the zero-free region of Dirichlet's L-functions. Proc. Japan Acad. Sci.. **54**, 332–334. MR 80a:10058.
Motohashi,Y. (1979a): A note on Siegel's zeros. Proc. Japan Acad. Sci., **55**, 190–192. MR 80d:10061.

Motohashi,Y. (1979b): A note on almost-primes in short intervals. Proc. Japan Acad. Sci., **55**, 225–226. MR 81a:10056.
Motohashi,Y. (1981): An elementary proof of Vinogradov's zero-free region for the Riemann zeta-function. Recent Progress in Analytic Number Theory, (Durham 1979), **I**, 257–267. Academic Press, London. MR 83b:10047.
Motohashi,Y. (1983): Lectures on Sieve Methods and Prime Numbers. Springer, Berlin–Heidelberg. MR 85k:11045.
Mozzochi,C.J. (1986): On the difference between consecutive primes. J. Number Th., **24**, 181–187. MR 88b:11057.
Mullin,A.A. (1963): Recursive function theory. Bull. Amer. Math. Soc., **69**, p.737.
Mumford,D. (1983): Tata Lectures on Theta,I. Birkhäuser. MR 85h:14026.
Murty,M.Ram, (1988): Primes in certain arithmetic progressions. J.Madras Univ., sect.B, **51**, 161–169.
Nagell,T. (Nagel,T.) (1921): Généralisation d'un théorème de Tchebycheff. J. math. pures appl., (8), **4**, 343–356.
Nagell,T. (Nagel,T.) (1922): Zur Arithmetik der Polynome. Abh. Math. Sem. Hamburg, **1**, 180–194.
Nagell,T. (1937): Über den grössten Primteiler gewisser Polynome dritten Grades. Math. Ann., **114**, 284–292.
Nagell,T. (1955): Contribution to the theory of a category of Diophantine equations of the second degree with two unknowns. Nova Acta Soc. Sci. Upsal., **16**, no.2, 1–38. MR 17 p.13.
Nagura,J. (1952): On the interval containing at least one prime number. Proc. Japan Acad. Sci., **28**, 177–181. MR 14 p.355.
Nair,M. (1982): On Chebyshev-type inequalities for primes. Amer. Math. Monthly, **89**, 126–129. MR 83f:10043.
Nair,M.,Perelli,A. (1990): On the distribution of $p^{1/2}$ modulo one. Colloq. Math. J.Bólyai, **51**, 393–435. North-Holland, Amsterdam. MR 91g:11101.
Nakajima,M. (1987): The Taylor coefficients of $\zeta(s)$, $(s-1)\zeta(s)$ and $(z/(1-z))\zeta(1/(1-z))$. Math. J. Okayama Univ., **29**, 207–219. MR 89f:11120.
Nakajima,M. (1988): The Taylor coefficients of some Dirichlet series. Prospects of Mathematical Science (Tokyo 1986), 181–188. World Scientific Publ., Singapore. MR 89h:11053.
Namboodiripad,K.S.S. (1971): A note on the formulae for the n-th prime. Monatsh. Math., **75**, 252–262. MR 46#126.
Narasimhan,R. (1968): Une remarque sur $\zeta(1+it)$. Enseign. Math., **14**, 189–191. MR 40#2618.
Narkiewicz,W. (1983): Number Theory. World Scientific Publ., Singapore. MR 85j:11107.
Narkiewicz,W. (1986): Classical Problems in Number Theory. PWN, Warszawa. MR 90e:11002.
Narkiewicz,W. (1990): Elementary and Analytic Theory of Algebraic Numbers. 2nd ed., Springer, Berlin–Heidelberg & PWN, Warszawa. MR 91h:11107. [1st ed. PWN 1974. MR 50#268.]
Naur,T. (1984): Mullin's sequence of primes is not monotonic. Proc. Amer. Math. Soc., **90**, 43–44. MR 85i:11011.
Neder,L. (1922): Über die Lage der Konvergenzabszissen einer Dirichletschen Reihe zur Beschränkheitsabszisse ihrer Summe. Arkiv f. Math. Astr. Fys., **16**, No.20.
Neill,T.B.M., Singer,M. (1965): Letter to the editor. Math. Gaz., **49**, p.303.
Neubauer,G. (1963): Eine empirische Untersuchung zur Mertensscher Funktion. Numer. Math., **5**, 1–13. MR 27#5721.
Neukirch,J. (1992): Algebraische Zahlentheorie. Springer, Berlin–Heidelberg.

Nevanlinna,F.,Nevanlinna,R. (1923): Über die Nullstellen der Riemannscher Zetafunktion. Math. Zeitschr., **20**, 253–263.

Nevanlinna,V. (1962) Über den elementaren Beweis des Primzahlsatzes. Soc. Sci. Fenn. Comment. Phys.-Math. **27**, no.3, 1–8. MR 26#2416.

Nevanlinna,V. (1964): Über die elementaren Beweise der Primzahlsätze und deren äquivalente Fassungen. Ann. Acad. Sci. Fenn. Ser. A.1, **343**, 1–52. MR 29#5800.

Nevanlinna,V. (1973): On constants connected with the prime number theorem for arithmetic progressions. Ann. Acad. Sci. Fenn. Ser. A.1, **539**, 11 pp. MR 49#7219.

Newman,D.J. (1980): Simple analytic proof of the prime number theorem. Amer. Math. Monthly, **87**, 693–696. MR 82h:10056.

Newman,D.J. (1993): A "natural proof" of the nonvanishing of L-series. Contemp.Math., **143**, 495–498. MR 94a:111135.

Niven,I. (1951): Functions which represent prime numbers. Proc. Amer. Math. Soc., **2**, 753–755. MR 13 p.321.

Niven,I.,Powell,B. (1976): Primes in certain arithmetic progressions, Amer. Math. Monthly, **83**, 467–469. MR 54#221.

Niven,I., Zuckerman,H., Montgomery,H.L. (1991): An Introduction to the Theory of Numbers. 5th ed. J.Wiley, New York. MR 91i:11001. [1st ed. 1960, MR 22#5605; 2nd ed. 1966, MR 33#3981; 3rd ed. 1972, MR 49#8921.]

Novák,B. (1975): O elementárním důkazu prvočíslené věty. Čas. Pěst. Mat. **100**, 71–84. MR 56#2937.

Nuriddinov,F.R. (1991): On the remainder term in the asymptotic law of the distribution of prime numbers during its elementary proof. Dokl. Akad. Nauk. Uzb. SSR, no.3, 7–8. (Russian) MR 93m:11089.

Oberhettinger,F. (1956): Note on the Lerch zeta function. Pacific J. Math., **6**, 117–120. MR 17 p.1188.

Obláth,R. (1936): Über Primzahlen in aufeinanderfolgenden Intervallen. Annali di Mat. Pura Appl., (4), **14**, 299–303.

Odlyzko,A.,te Riele,H.J.J. (1985): Disproof of the Mertens conjecture. J. Reine Angew. Math., **357**, 138–160. MR 86m:11070.

Odlyzko,A.M.,Schönhage,A. (1988): Fast algorithms for multiple evaluations of the Riemann zeta function. Trans. Amer. Math. Soc., **309**, 797–809. MR 89j:11083.

Odoni,R.W.K. (1985): On the prime divisors of the sequence $w_{n+1} = 1+w_1w_2\cdots w_n$. J. London Math. Soc., (2), **32**, 1–11. MR 87b:11094.

Ore,O. (1952): On the selection of subsequences. Proc. Amer. Math. Soc., **3**, 706–712. MR 14 p.256.

Ortega-Costa,J. (1950): The explicit formula for the prime number function $\pi(x)$. Revista Mat. Hisp. Amer. (4), **10**, 72–76. MR 12 p. 392. (Spanish)

Osgood,W.F. (1928): Lehrbuch der Funktionentheorie, I. 5th ed. Teubner, Leipzig–Berlin.

Page,A. (1935): On the number of primes in an arithmetic progression. Proc. London Math. Soc., (2), **39**, 116–141.

Pan Chen Biao. (1977): A new proof of the three prime theorem. Acta Math. Sinica, **20**, 206–211. (Chinese) MR 58#27866.

Pan Cheng Dong (1957): On the least prime in an arithmetic progression. Sci. Record, (N.S.), **1**, 311–313. MR 21#4140.

Pan Cheng Dong (1962): On the representation of even numbers as the sum of a prime and a near prime. Sci. Sinica, **11**, 873–888. (Russian) MR 27#1427.

Pan Cheng Dong (1963): On the representation of even numbers as the sum of a prime and a product of not more than 4 primes. Sci. Sinica, **12**, 455–473. MR 28#73.

Pan Cheng Dong (1964): A new application of the Ju.V.Linnik large sieve method. Acta Math. Sinica, **14**, 597–606. MR 30# 3871. (Chinese) [English translation: Chinese Math., Acta **5**, 1964, 642–652. Russian translation: Sci. Sinica, **13**, 1964, 1045–1053.]

Pan Cheng Dong,Ding Xiao Xi,Wang Yuan (1975): On the representation of large even integers as a sum of a prime and an almost prime. Sci. Sinica, **18**, 599–610. MR 57#5897.

Pan Cheng Dong,Pan Chen Biao (1992): Goldbach Conjecture. Science Press, Beijing. MR 95f:11083.

Panteleeva,E.I. (1988): On the Dirichlet divisor problem in number fields. Mat. Zametki, **44**, 494–505. (Russian) MR 91d:11115.

Papadimitriou M. (1973): A simple proof of the formula $\sum_{k=1}^{\infty} k^{-2} = \pi^2/6$. Amer. Math. Monthly, **80**, 424–425. MR 47#2220.

Papadimitriou M. (1975): A recursion formula for the sequence of odd primes. A-mer. Math. Monthly, **82**, p.289. MR 52#246.

Parady,B.K.,Smith,J.F.,Zarantonello,S.E. (1990): Largest known twin primes. Math. Comp., **55**, 381–382. MR 90j:11013.

Patrizio,S. (1976): Generalizzazione del teorema di Wilson alla terne prime. Enseign. Math., (2), **22**, 175–184. MR 55#251.

Patterson,S.J. (1988): An Introduction to the Theory of the Riemann Zeta-function. Cambridge Univ.Press. MR 89d:11072.

Pellegrino,F. (1963): Teorema di Wilson et numeri primi gemelli. Rend. Acc. Naz. dei Lincei, (8), **35**, 258–262. MR 29#2213.

Perelli,A.,Puglisi,G. (1985): On some classical explicit formulae. Boll. Un. Mat. Ital., A, **161** 269–278. MR 87a:11085.

Perelli,A.,Zaccagnini,A. (1995): On the sum of a prime and a kth power. Izv. Ros. Akad. Nauk, ser.mat., **59**, 185–200. MR 96f:11134.

Perott,J. (1881): Sur l'infinité de la suite des nombres premiers. Bull sci. math. et astr., (2), **5**, 183–184.

Perott,J. (1888): Remarque au sujet du théorème d'Euclide sur l'infinité du nombre des nombres premiers. Amer. J. Math, **11**, 99–138; **13**, 1891, 285–308.

Perron,O. (1908): Zur Theorie der Dirichletschen Reihen. J. Reine Angew. Math., **134**, 95–143.

Petersen,J. (1882): On primtal. Tidsskr. for Math., (4), **6**, 138–143.

Phillips,E. (1933): The zeta-function of Riemann; further developments of van der Corput's method. Quart. J.Math. Oxford ser., **4**, 209–225.

Phragmén,E. (1891): Sur le logarithme intégral et la fonction $f(x)$ de Riemann. Öfversigt Kongl. Vet.-Akad. Förh. Stockholm, **48**, 599–616.

Phragmén,E. (1892): Sur la distribution des nombres premiers. Comptes Rendus Acad. Sci. Paris, **114**, 337–340.

Phragmén,E. (1901/02): Sur une loi de symétrie à certaines formules asymptotiques. Öfversigt Kongl. Vet.-Akad. Förh. Stockholm, **58**, 189–202.

Phragmén,E. (1904): Sur une extension d'un théorème de la théorie des fonctions. Acta Math., **28**, 351–368.

Phragmén,E.,Lindelöf,E. (1908): Sur une extension d'un principe classique d'analyse et sur quelques propriétés du fonctions monogènes dans le voisinage d'un point singulier. Acta Math., **31**, 381–406.

Picard,É. (1891): Traité d'analyse. I, 1891; II, 1893; III, 1896. 2nd ed. I,1901; II, 1905; III, 1908. Gauthier-Villars, Paris.

Pillai,S.S. (1927–28): On some empirical theorem of Scherk. J.Indian Math.Soc., **17**, 164–171.

Pillai,S.S. (1944): Bertrand's postulate. Bull. Calcutta Math.Soc., **36**, 97–99. Corr.: *ibidem*, **37** (1945), p.27. MR 6 p.259.

Pilt'jai,G.Z. (1972): The magnitude of the difference between consecutive primes. Studies in Number Theory, **4**, Saratov, 73–79. (Russian) MR 52#13680.

Piltz,A. (1884): Über die Häufigkeit der Primzahlen in arithmetischer Progressionen und über verwandte Gesetze. Habilitationsschrift, Univ. Jena.

Pincherle,S. (1908): Alcune spigolature nel campo delle funzioni determinanti. Atti IV Congr. internat. Mat. **2**, 44–48, Roma.

Pintz,J. (1971): On a certain point in the theory of Dirichlet's L-functions; I,II. (Hungarian) Mat. Lapok, **22**, 143–148, 331–335. MR 48#8411.

Pintz,J. (1973/74): On Siegel's theorem. Acta Arith., **24**, 543–551. MR 49#2595.

Pintz,J. (1976a): Bemerkungen zu der Arbeit "On the sign changes of $(\pi(x) - \text{li } x)$", (Monatsh. Math., 82 (1976), no 2, 163–175) von S.Knapowski und P.Turán. Monatsh. Math., **82**, 199–206. MR 55 #12657c.

Pintz,J. (1976b): Elementary methods in the theory of L-functions,II. On the greatest real zero of a real L-function. Acta Arith., **31**, 273–289. MR 58#5551a.

Pintz,J. (1976/79): Siegel roots and the prime number theorem. Mat. Lapok, **27**, 355–360. (Hungarian) MR 80k:10040.

Pintz,J. (1977a): On the sign changes of $\pi(x) - \text{li } x$. Astérisque, **41/42**, 255–265. MR 56#5466.

Pintz,J. (1977b): On the remainder term of the prime number formula, III. Sign changes of $\pi(x) - \text{li } (x)$. Studia Sci. Math. Hungar., **12**, 345–369. MR 84j:10051c.

Pintz,J. (1977c): Elementary methods in the theory of L-functions. V. The theorem of Landau and Page. Acta Arith., **32**, 163–171. MR 58#5551d.

Pintz,J. (1977d): Elementary methods in the theory of L-functions. VIII. Real zeros of real L-functions. Acta Arith., **33**, 1977, 89-98. MR 58#5551g.

Pintz,J. (1978): On the remainder term of the prime number formula, IV. Sign changes of $\pi(x) - \text{li } (x)$. Studia Sci. Math. Hungar., **13**, 29–42. MR 84j:10051d.

Pintz,J. (1980a): On Legendre's prime number formula. Amer. Math. Monthly, **87**, 733–735. MR 82e:01055.

Pintz,J. (1980b): On the remainder term of the prime number formula, I. On a problem of Littlewood. Acta Arith., **36**, 341–365. MR 84j:10051a.

Pintz,J. (1980c): On the remainder term of the prime number formula, II. On a theorem of Ingham. Acta Arith., **37**, 209–220. MR 84j:10051b.

Pintz,J. (1980d): Oscillatory properties of $M(x) = \sum_{n \leq x} \mu(n)$. Acta Arith., **42**,49–55; II, Studia Sci. Math. Hungar., **15**, 1980, 491–496. MR 84e:10053.

Pintz,J. (1981a): On primes in short intervals, I. Studia Sci. Math. Hungar., **16**, 395–414. MR 84m:10032; II, *ibidem*, **19**, 1984, 89–96. MR 86h:11073.

Pintz,J. (1983a): Oscillatory properties of the remainder term of the prime number formula. Studies in Pure Mathematics, 551–560. Birkhäuser, Basel–Boston. MR 86m:11071.

Pintz,J. (1983b): Oscillatory properties of $M(x) = \sum_{n \leq x} \mu(n)$,III, Acta Arith., **43**, 1983, 105–113. MR 85k:11049.

Pintz,J. (1984): On primes in short intervals, II. Studia Sci. Math. Hungar., **19**, 1984, 89–96. MR 86h:11073.

Pintz,J. (1987): An effective disproof of the Mertens conjecture. Astérisque, **147–148**, 325–333. MR 86m:11070.

Pintz,J. (1988a): Elementary methods in the theory of L-functions,IX. Acta Arith., **49**, 387–394. MR 89m:11081.

Pintz,J. (1988b): A note on the exceptional set in Goldbach's problem. Colloque Théorie Analytique des Nombres "Jean Coquet", 101–115. Orsay 1988. MR 90e:11146.

Pintz,J. (1991): On an assertion of Riemann concerning the distribution of prime numbers. Acta Math. Acad. Sci. Hung., **58**, 383–387. MR 93e:11105.

Pintz,J. (1997): Very large gaps between consecutive primes. J. Number Th., **63**, 286–301. MR 98c:11092.

Plana,G. (1820): Mem. R. Accad. Torino, **25**, 403–418.

Plana,G. (1821/22): Éclaircissements sur la théorie de l'intégrale $\int \frac{dx}{\log x}$, prise depuis $x = 0$. Annales Math. Phys. Appl., **12**, 145–157.

Poincaré,H. (1892): Extension aux nombres premiers complexes des théorèmes de M.Tchebicheff. J. math. pures appl., (4), **8**, 25–68. [Oeuvres, **V**, 441–482. Gauthier-Villars, Paris 1950.]

Poisson,S.D. (1823): Suite de mémoire sur les intégrales définies et sur la sommation des séries. Journal de l'École Royale Polytechnique, **12**, 404–509.

Polignac, A.de (1849a): Recherches nouvelles sur les nombres premiers. Comptes Rendus Acad. Sci. Paris, **29**, 397–401; Rectification: *ibidem*, 738–739.

Polignac, A.de (1849b): Six propositions arithmologiques sur les nombres premiers. Nouv. Ann. Math., **7**, 423–429.

Polignac, A.de (1851): Recherches nouvelles sur les nombres premiers. Bachelier, Paris.

Polignac,A.de (1854): Nouvelles recherches sur les nombres premiers. J. math. pures appl., **19**, 305–333.

Polignac, A.de (1857): Recherches sur les nombres premiers. Comptes Rendus Acad. Sci. Paris, **45**, 406–410, 431–434, 575–580, 882–886.

Polignac,A.de (1859a): Recherches nouvelles sur les nombres premiers. Comptes Rendus Acad. Sci. Paris, **49**, 350–352, 386–387.

Polignac,A.de (1859b): Recherches sur les nombres premiers. Comptes Rendus A-cad. Sci. Paris, **49**, 624–628, 724–729.

Pólya,G. (1918): Zur arithmetischer Untersuchung der Polynome. Math. Zeitschr., **1**, 143–148.

Pólya,G. (1919): Verschiedene Bemerkungen zur Zahlentheorie. Jahresber. Deutsche Math. Ver., **28**, 31–40.

Pólya,G. (1921): Arithmetische Eigenschaften der Reihenentwicklungen rationaler Funktionen. J. Reine Angew. Math., **151**, 1–31.

Pólya,G. (1927): Über die algebraisch-funktionentheoretischen Untersuchungen von J.L.W.V.Jensen. Kgl. Danske Vidensk.Selskab., **7**, No. 17.

Pólya,G. (1930): Über das Vorzeichen des Restgliedes im Primzahlsatz. Nachr. Ges. Wiss. Göttingen, 19–27. [Corrected reprint: Number Theory and Analysis, 233–244. Plenum, New York, 1969. MR 41#8357.]

Pólya,G. (1937): Kombinatorische Anzahlbestimmungen für Gruppen, Graphen und chemische Verbindungen. Acta Math., **68**, 145–254.

Pólya,G.,Szegö,G. (1925): Aufgaben und Lehrsätze aus der Analysis. Springer, Berlin–Heidelberg. [Reprint: Dover, New York, 1945.]

Polyakov,I.V. (1981): On the exceptional set of the sum of a prime and the square of an integer. Izv. Akad. Nauk SSSR, **45**, 1981, 1391–1423. (Russian) MR 83e:10069.

Polyakov,I.V. (1990): Addition of a prime number and the square of an integer. Mat. Zametki, **47**, 90–99. (Russian) MR 91h:11103.

Pomerance,C. (1980): A note on the least prime in an arithmetic progression. J. Number Th., **12**, 218–223. MR 81m:10081.

Popa,V. (1982): Some criteria for a natural number to be a prime. Gaz. Mat., **87**, 6–8. MR 84a:10011.

Popov,O.V. (1994): A derivation of the modern bound for the zeros of the Riemann zeta function by the Hadamard method. Vestnik Mosk. Univ., 42–45. (Russian) MR 95m:11094.

Postnikov,A.G. (1969): A remark on an article by A.G.Postnikov and N.P.Romanov. Usp. Mat. Nauk, **24**, no.5, 263. (Russian)MR 17 p.587.

Postnikov,A.G.,Romanov,N.P. (1955): A simplification of A.Selberg's elementary proof of the asymptotic law of distribution of prime numbers. Usp. Mat. Nauk, **10**, no. 4, 75–87. (Russian)MR 17 p.587.

Potter,H.S.A.,Titchmarsh,E.L. (1935): The zeros of Epstein's zeta function. Proc. London Math. Soc., (2), **39**, 372–384.

Powell,B. (1977): Proof of a special case of Dirichlet's theorem. Fibonacci Quart., **15**, 167–169. MR 55#7903.

Prachar,K. (1954): On integers having many representations as a sum of two primes. J. London Math. Soc., **29**, 347–350. MR 15 p.935.

Prachar,K. (1957): Primzahlverteilung. Springer, Berlin. MR 19 p.393. [2nd ed. Springer 1978. MR 81k:10060.]

Prachar,K. (1961): Über die kleinste Primzahl einer arithmetischer Reihe. J. Reine Angew. Math., **206**, 3–4. MR 23#A2399.

Prachar,K. (1965): Über Zahlen, sie sich als Summe einer Primzahl und einer "kleinen" Potenz darstellen lassen,II. Monatsh. Math., **69**, 62–68. MR 30#4742.

Prachar,K. (1976): Generalisation of a theorem of S.Selberg on primes in progressions. Topics in Number Theory. Proc. J.Bólyai Colloq. **13**, Number Theory, Debrecen 1974, 267-280. North-Holland, Amsterdam. MR 55#7955.

Pringsheim,A. (1890): Zur Theorie der Dirichlet'schen Reihen. Math. Ann., **37**, 38–60.

Pringsheim,A. (1894): Über Functionen, welche in gewissen Punkten endliche Differentialquotienten jeder endlichen Ordnung, aber keine Taylorsche Reihenentwicklung besitzen. Math. Ann., **44**, 41–56.

Prowse,E. (1960): On the analytic continuation of the ζ-function. Canad. Math. Bull., **3**, 289–292. MR 23#A1612.

Putnam,H. (1960): An unsolvable problem in number theory. J.Symb. Logic, **25**, 220–232. MR 28#2048.

Quesada Chaverri,F. (1980): Formulas which express number-theoretic functions without factorization into primes. Cienc. Tech. **4**, 13–20. (Spanish) MR 83j:10009.

Rabinowitsch,G. (1913): Eindeutigkeit der Zerlegung in Primzahlfaktoren in quadratischen Zahlkörpern. J. Reine Angew. Math., **142**, 153–164.

Rademacher,H. (1924): Beiträge zur Viggo Brunschen Methode in der Zahlentheorie. Abh. Math. Sem. Hamburg, **3**, 12–30. [Collected Papers, **1**, 259–277. MIT Press, 1974.]

Rademacher,H. (1930): Ein neuer Beweis für die Funktionalgleichung der ζ-Funktion. Math. Zeitschr., **31**, 12–30. [Collected Papers, **1**, 623–628. MIT Press, 1974.]

Raikov,D.A. (1938): A generalization of the theorem of Ikehara-Landau. Mat. Sb., **3**, 559–568. (Russian)

Ramachandra,K. (1970): A note on numbers with a large prime factor. J. London Math. Soc., (2), **1**, 303–306. MR 40#118.

Ramachandra,K. (1970): A note on numbers with a large prime factor,II. J.Indian Math. Soc., **34**, 39–48. MR 45#8616.

Ramachandra,K. (1974): On the frequency of Titchmarsh's phenomenon for $\zeta(s)$. J. London Math. Soc., (2), **8**, 683–690. MR 50#9815.

Ramachandra,K. (1975a): Some new density estimates for the zeros of the Riemann zeta-function. Ann. Acad. Sci. Fenn. Ser. A.1, **1**, 177–182. MR 52#5588.

Ramachandra,K. (1975b): On a theorem of Siegel. Nachr. Ges. Wiss. Göttingen, 43–47. MR 52#3081.

Ramachandra,K. (1980): One more proof of Siegel's theorem. Hardy- Ramanujan J., **3**, 25-40. MR 81k:10061. Add. and corr. *ibidem*, **4**, 1981, suppl. 12. MR 83a:10073.

Ramachandra,K. (1990): A simple proof of Siegel-Tatuzawa theorem. Bull. Calcutta Math. Soc., **82**, p.222. MR 92h:11071.

Ramachandra,K. (1995): Simplest, quickest and self-contained proof that $\frac{1}{2} \le \theta \le 1$ (θ being the least upper bound of the real part of the zeros of $\zeta(s)$). J.Indian Math.Soc., **61**, 7–12. MR 96b:11123.

Ramachandra,K.,Sankaranarayanan,A. (1993): On some theorems of Littlewood and Selberg,I. J. Number Th., **44**, 281–291. MR 94f:11081.

Ramanujan,S. (1916): Some formulae in the analytic theory of numbers. Messenger of Math., **45**, 81–84. [Collected Papers, 133–135. Cambridge University Press 1927.]

Ramanujan,S. (1919): A proof of Bertrand's postulate. J. Indian Math. Soc., **11**, 181–182. [Collected Papers, 208–209. Cambridge University Press 1927; reprint: Chelsea 1962.]

Ramaré,O. (1995): On Šnirel'man's constant. Ann. Scuola Norm. Sup. Pisa, Cl. Sci., (4), **22**, 645–706. MR 97a:11167.

Ramaré,O.,Rumely,R. (1996): Primes in arithmetic progressions. Math. Comp., **65**, 397–425. MR 97a:11144.

Ramaswami,V. (1934): Notes on Riemann's ζ-function. J. London Math. Soc., **9**, 165–169.

Rankin,R.A. (1938): The difference between consecutive prime numbers. J. London Math. Soc., **13**, 242–244.

Rankin,R.A. (1939): Contributions to the theory of Ramanujan's tau-function and similar arithmetical functions,I. Proc. Cambridge Philos. Soc., **35**, 351–356.

Rankin,R.A. (1940): The difference between consecutive prime numbers, II, Proc. Cambridge Philos. Soc., **36**, 1940, 255–256. MR 1 p.292.

Rankin,R.A. (1947): The difference between consecutive prime numbers, III, J. London Math. Soc., **22**, 226–230. MR 9 p.498.

Rankin,R.A. (1950): The difference between consecutive prime numbers, IV, Proc. Amer. Math. Soc., **1**, 143–150. MR 11 p.644.

Rankin,R.A. (1962/63): The difference between consecutive prime numbers, V. Proc. Edinburgh Math. Soc., (2), **13**, 331–332. MR 28#3978.

Rédei,L. (1958): Über die algebraischzahlentheoretische Verallgemeinerung eines elementarzahlentheoretischen Satzes von Zsigmondy. Acta Sci. Math., **15**, 12–19. MR 20#3845.

Regimbal,S. (1975): An explicit formula for the kth prime number. Math. Mag., **48**, 230–232. MR 51#12676.

Remak,R. (1909): Elementare Verallgemeinerung einer bekannter Eigenschaft der Zahl 30. Archiv Math. Phys., (3), **15**, 186–193.

Rényi,A. (1947): On the representation of an even number as the sum of a single prime and a single almost-prime number. Doklady Akad. Nauk SSSR, **56**, 455–458. (Russian) MR 9 p.136.

Rényi,A. (1948): On the representation of an even number as the sum of a single prime and a single almost-prime number. Izv. Akad. Nauk SSSR, **12**, 57–78. (Russian) MR 9 p.413.

Révesz,Sz.Gy. (1980): On the least prime in an arithmetic progression. Studia Sci. Math. Hungar., **15**, 83–87. MR 84d:10049.

Reyssat,É. (1980): Irrationalité de $\zeta(3)$ selon Apéry. Sém. Delange–Pisot–Poitou, **20** (1978/79), exp.6. MR 81i:10041.

Ribenboim,P. (1988): The Book of Prime Number Records. Springer, Berlin–Heidelberg. MR 89e:11052. [2nd ed. 1989. MR 90g:11127.]

Ribenboim,P. (1991): Little book of big primes. Springer, Berlin–Heidelberg. MR 92i:11008.

Ribenboim,P. (1994): Nombres premiers: mystères et records. Presses Universitaires de France, Paris. 95j:11004.

Ribenboim,P. (1996): The New Book of Prime Number Records. Springer, Berlin–Heidelberg. MR 96k:11112.
Ricci,G. (1933a): Sul teorema di Dirichlet relativo alla progressione aritmetica. Boll. Un. Mat. Ital., **12**, 304–309.
Ricci,G. (1933b): Ricerche aritmetiche sui polinomi. Rend. Circ. Mat. Palermo, **57**, 433–475.
Ricci,G. (1933c): Sui grandi divisori primi delle coppie di interi in posti corrispodenti di due progressioni aritmetiche. Applicazione del metodo di Brun. Annali di Mat. Pura Appl., (4), **11**, 91–110.
Ricci,G. (1934): Sui teoremi di Dirichlet e di Bertrand-Tchebyschef relativi alla progressione aritmetica. Boll. Un. Mat. Ital., **13**, 7–17.
Ricci,G. (1936): Sur la congettura di Goldbach e la costante di Schnirelmann. Boll. Un. Mat. Ital., **15**, 183–187.
Ricci,G. (1937): Sur la congettura di Goldbach e la costante di Schnirelmann,II,III. Ann. Scuola Norm. Sup. Pisa, Cl. Sci., (2), **6**, 71–90, 91–116.
Ricci,G. (1952): La differenza di numeri primi consecutivi. Rendiconti Sem. Mat. Univ. e Politecnico Torino, **11**, 149–200. MR 14 p.727; Corr.: *ibidem*, **12**, p.315. MR 15 p.602.
Ricci,G. (1954): Sull'andamento della differenza di numeri primi consecutivi. Riv. Mat. Univ. Parma, **5**, 3–54. MR 16 p.675.
Richards,I. (1974): On the incompatibility of two conjectures concerning primes; a discussion of the use of computers in attacking a theoretical problem. Bull. Amer. Math. Soc., **80**, 419—438. MR 49#2601.
Richert (1894): Über die Lösung von Gleichungen durch bestimmte Integrale. Jahresbericht der Realgymnasiums zu Stralsund, Ostern 1894, 1–10.
Richert,H.-E. (1949a): Über Zerfällungen in ungleiche Primzahlen. Math. Zeitschr., **52**, 342–343. MR 11 p.502.
Richert,H.-E. (1949b): Über Zerlegungen in Paarweise verschiedene Zahlen. Norsk Mat. Tidsskr., **31**, 120–122. MR 11 p.646.
Richert,H.-E. (1953): Aus der additiven Primzahltheorie. J. Reine Angew. Math., **191**, 179–198. MR 15 p.102.
Richert,H.-E. (1967): Zur Abschätzung der Riemannscher Zetafunktion in der Nähe der Vertikalen $\sigma = 1$. Math. Ann., **169**, 97–101. MR 34#5779.
Richert,H.-E. (1969): Selberg's sieve with weights. Mathematika, **16**, 1–22. MR 40#119.
Riehe,A. (1889): Ein Satz aus der Zahlentheorie. Zeitschr. Math. Phys., **34**, 190–191.
Riele,H.J.J.te (1979a): Computations concerning the conjecture of Mertens. J. Reine Angew. Math., **311/312**, 356–360. MR 81c:10055.
Riele,H.J.J.te (1979b): Tables of the first 15000 zeros of the Riemann zeta function to 28 significant digits. Math. Centrum Amsterdam. MR 80f:10050.
Riele,H.J.J.te (1985): Some historical and other notes about the Mertens conjecture and its recent disproof. Nieuw Arch. Wisk., (4), **3**, 237–243. MR 87d:01022.
Riele,H.J.J.te (1987): On the sign of the difference $\pi(x) - \mathrm{li}\,(x)$. Math. Comp., **48**, 323–328. MR 88a:11135.
Riemann,B. (1851): Grundlagen für eine allgemeine Theorie der Functionen einer veränderlichen complexen Grösse. Dissertation, Göttingen. [Gesammelte mathematische Werke, 3–47, Teubner, Leipzig, 1876; reprint: Dover 1953.]
Riemann,B. (1860): Ueber die Anzahl der Primzahlen unter einer gegebener Größe. Monatsber. Kgl. Preuß. Akad. Wiss. Berlin, 671–680. [English translation in Edwards (1974); Werke, 145–155, Teubner, Leipzig 1892; reprint: Dover 1953.]
Riesel,H. (1985): Prime numbers and computer methods for factorization. Birkhäuser, Basel. MR 88k:11002. [2nd.ed., 1994. MR 95h:11142.]

Riesel,H.,Göhl,G. (1970): Some calculations related to Riemann's prime number formula. Math. Comp., **24**, 969–983. MR 45#3222.

Riesel,H.,Vaughan,R.C. (1983): On sums of primes. Arkiv Mat., **21**, 45–74. MR 84m:10042.

Ripert,L. (1903): L'Intermédiaire des Math., **10**, p.66.

Robbins,N. (1994): On Fibonacci numbers and primes of the form $4k+1$. Fibonacci Quart., **32**, 15–16. MR 94k:11006.

Robin,G. (1983): Estimation de la fonction de Tchebychef θ sur le k-ième nombre premier et grandes valeurs de la fonction $\omega(n)$ nombre de diviseurs premiers de n. Acta Arith., **42**, 367–389. MR 85j:11109.

Robin,G. (1984): Sur la différence $\mathrm{Li}(\theta(x)) - \pi(x)$. Ann. Fac. Sci. Toulouse, Math., (5), **6**, 257–268. MR 86j:11090.

Rodosskij,K.A. (1954): On the least prime in an arithmetic progression. Mat. Sb., (N.S.), **34**, 331–356. (Russian) MR 15 p.935.

Rodosskij,K.A. (1956): On the exceptional zero. Izv. Akad. Nauk SSSR, **20**, 667–672. (Russian) MR 18 p.563.

Rogel,F. (1889): Zur Bestimmung der Anzahl [von] Primzahlen, welche nicht grösser als eine gegebene Zahl sind. Archiv Math. Phys., (2), **7**, 381–388.

Rogel,F. (1890): Zur Bestimmung der Anzahl [von] Primzahlen unter gegebenen Grenzen. Math. Ann., **36**, 304–315.

Rogel,F. (1895): Ueber Primzahlmengen. Sber. der Kgl. Böhmischen Gesellschaft der Wissenschaften, Math.-Nat.Kl., No.22.

Rogel,F. (1897): Recursive Bestimmung der Anzahl [von] Primzahlen unter gegebener Grenzen. Sber. der Kgl. Böhmischen Gesellschaft der Wissenschaften, Math.-Nat.Kl., No.22.

Rogel,F. (1900): Die Bestimmung der Anza[h]l der unter einer gegebenen Grenze liegenden Primza[h]len. Archiv Math. Phys., (2), **17**, 225–237.

Rohrbach,H.,Weis,J. (1964): Zum finiten Fall des Bertrandschen Postulats, J. Reine Angew. Math., **214/215**, 432–440. MR 28#5024.

Romanov,N. [Romanoff,N.] (1934): Über einige Sätze der additiven Zahlentheorie. Math. Ann., **109**, 668–678.

Rooney,P.G. (1994): Another proof of the functional equation for the zeta function. J. Math. Anal. Appl., **185**, 223–228. MR 95g:11083.

Ross,P.M. (1975): On Chen's theorem that each large even number has the form $p_1 + p_2$ or $p_1 + p_2p_3$. J. London Math. Soc., (2), **10**, 500–506. MR 52#10646.

Rosser,B. (1941): Explicit bounds for some functions of prime numbers. Amer. J. Math, **63**, 211–232. MR 2 p.150.

Rosser,J.B. (1949): Real roots of Dirichlet L-series. Bull. Amer. Math. Soc., **55**, 906–913. MR 11 p.332.

Rosser,J.B. (1950): Real roots of real Dirichlet L-series. J.Res. Nat. Bur. Stand., **45**, 505–514. MR 12 p.804.

Rosser,J.B.,Schoenfeld,L. (1962): Approximate formulas for some functions of prime numbers. Illinois J. Math., **6**, 64–94. MR 25#1139.

Rosser,J.B.,Schoenfeld,L. (1975): Sharper bounds for the Chebyshev functions $\theta(x)$ and $\psi(x)$. Math. Comp., **29**, 243–269. MR 56#15581a.

Rosser,J.B.,Yohe,J.M.,Schoenfeld,L. (1969): Rigorous computation and the zeros of the Riemann zeta-function. Information Processing 68, **I**, 70–76. North-Holland, Amsterdam. MR 41#2892.

Rota,G.-C. (1964): On the foundations of combinatorial theory, I. Theory of Möbius functions. Z. Wahrscheinlichkeitstheor. Verw. Geb., **2**, 340–368. MR 30#4688. [Gian-Carlo Rota on Combinatorics, 3–31. Birkhäuser, Basel 1995.]

Rotkiewicz,A. (1960): Elementarny dowód istnienia dzielnika pierwszego pierwotnego liczby $a^n - b^n$. Prace matem., **4**, 21–28. MR 24#A1237.

Rotkiewicz,A. (1961): Démonstration arithmétique de l'existence d'une infinité de nombres premiers de la forme $nk+1$. Enseign. Math., (2), **7**, 277–20. MR 25#38.
Roux,D. (1956): Sui numeri primi delle progressioni aritmetiche. Boll. Un. Mat. Ital., (3), **11**, 55–63. MR 17 p. 1057.
Rubinstein,M. (1993a): A formula and a proof of the infinitude of primes. Amer. Math. Monthly, **100**, 388-392.
Rubinstein,M. (1993b): A simple heuristic proof of Hardy and Littlewood conjecture. Amer. Math. Monthly, **100**, 456–460. MR 94a:11143.
Rubinstein,M.,Sarnak,P. (1994): Chebyshev's bias. Experiment. Math., **3**, 173–197. MR 96d:11099.
Rumely,R. (1993): Numerical computations concerning the ERH. Math. Comp., **61**, 415–440. MR 94b:11085.
Ryavec,C. (1973): The analytic continuation of Euler products with applications to asymptotic formulae. Illinois J. Math., **17**, 608–618. MR 48#259.
Saffari,B. (1970a): Sur les oscillations des fonctions sommatoires des fonctions de Möbius et de Liouville. Comptes Rendus Acad. Sci. Paris, **271**, A578–A580. MR 42#7607.
Saffari,B. (1970b): Sur la fausseté de la conjecture de Mertens. Comptes Rendus Acad. Sci. Paris, **271**, A1097–A1101. MR 43#6167.
Salerno,S.,Vitolo,A.,Zannier,U. (1991): Functional properties for Dirichlet series with periodic coefficients. Ricerche Mat., **40**, 209–222. MR 94a:11139.
Sancery,L. (1875/76): De la répartition des nombres entre les diviseurs de $\varphi(M)$, lorsque M est une puissance d'un nombre premier impair, ou le double d'une telle puissance. Bull. Soc. Math. France, **4**, 17–29.
Sándor,J. (1988): Über die Folge der Primzahlen. Mathematica (Cluj), **30**, 67–74. MR 90f:11007.
Šanin,A.A. (1964): Determination of constants in the method of Brun-Šnirelman. Volzh. Mat. Sb. **2**, 261–265. (Russian) MR 33#2617.
Santos,B.R. (1976): Twelve and its totitives. Math.Mag., **49**, 239–240. MR 54#5098.
Saouter,Y. (1998): Checking the odd Goldbach conjecture up to 10^{20}. Math. Comp., **67**, 863–866. MR 98g:11115.
Sato,D., Straus,E.G. (1966a): Generalized interpolation by analytic functions. Bull. Amer. Math. Soc., **72**, 32–36. MR 36#1670.
Sato,D., Straus,E.G. (1966b): Generalized interpolation by analytic functions. J. Math. Sci., **1**, 53–76. MR 34#2884.
Sato,D., Straus,E.G. (1970): p-adic proof of non-existence of proper prime representing algebraic functions and related problems. J. London Math. Soc., (2), **2**, 45–48. MR 40#2599.
Šatunovskij,S. (1893): Solution of the problem 446. Vestnik Eksper. Fiz. i Elem. Mat., **15**, 276–278. (Russian)
Satyanarayana,U.V. (1963): On the inversion property of the Möbius μ-function, I,II. Math. Gaz., **47**, 38-42. MR 26#3653; **49**, 1965, 171–178. MR 32#1156.
Schaper,H.von (1898): Über die Theorie der Hadamardschen Funktionen und ihre Anwendung auf das Problem der Primzahlen. Dissertation, Göttingen.
Scheffler,H. (1893): Beleuchtung und Beweis eines Satzes aus Legendre's Zahlentheorie. Foerster, Leipzig.
Scheibner,W. (1860a): Ueber die Anzahl der Primzahlen unter einer beliebigen Grenze. Zeitschr. Math. Phys., **5**, 232–252.
Scheibner,W. (1860b): Über unendliche Reihen und deren Convergenz. Hirzel, Leipzig.
Schering,E. (1869): Die Fundamental-Classen der zusammensetzbaren arithmetischer Formen. Abh. Kgl. Ges. Wiss. Göttingen, **14**. [Gesammelte Math. Werke, **1**, 135–148, Mayer & Müller, Berlin 1902.]

Schering,E. (1887): Zahlentheoretische Bemerkung, J. Reine Angew. Math., **100**, 447–448. [Gesammelte Math. Werke, **1**, 103–104, Mayer & Müller, Berlin 1902.]

Schering,E. (1909a): Beweis des Dirichletschen Satzes, dass durch jede eigentlich primitive quadratische Form unendlich viele Primzahlen dargestellt werden. [Gesammelte mathematische Werke, **2**, 357–365, Mayer & Müller, Berlin 1909.]

Schering,E. (1909b): Brief an P.Bachmann von 10.V.1893. [Gesammelte mathematische Werke, **2**, 433, Mayer & Müller, Berlin 1909.]

Scherk,H.F. (1833): Bemerkungen über die Bildung der Primzahlen aus einander. J. Reine Angew. Math., **10**, 201–208.

Schinzel,A, (1961): Remarks on the paper "Sur certaines hypothéses concernant les nombres premiers". Acta Arith., 7, 1961, 1–8. MR 24#A70.

Schinzel,A. (1962): Remark on the paper of K.Prachar "Über die kleinste Primzahl einer arithmetischer Reihe". J. Reine Angew. Math., **210**, 121–122. MR 27#118.

Schinzel,A. (1963): A remark on the paper of Bateman and Horn. Math. Comp., **17**, 445–447. MR 27#3609.

Schinzel,A., Sierpiński,W. (1958): Sur certaines hypothèses concernant les nombres premiers. Acta Arith., **4**, 185–208; Erratum: *ibidem*, **5**, p.259. MR 21#4936.

Schlömilch,O. (1849): Lehrsatz. Archiv Math. Phys., **12**, p.415.

Schlömilch,O. (1858): Über eine Eigenschaft gewisser Reihen. Zeitschr. Math. Phys., **3**, 130–132.

Schlömilch,O. (1878): Ueber die Summen von Potenzen der reciproken natürlichen Zahlen. Zeitschr. Math. Phys., **23**, 135–137.

Schlömilch,O. (1895): Ueber einen zahlentheoretischen Satz von Legendre. Zeitschr. Math. Phys., **40**, 125.

Schmidt,E. (1903): Über die Anzahl der Primzahlen unter gegebener Grenze. Math. Ann., **57**, 195–204.

Schnee,W. (1908): Über irreguläre Potenzreihen und Dirichletsche Reihen. Dissertation, Berlin.

Schnee,W. (1909a): Über Mittelwertsformeln in der Theorie der Dirichletschen Reihen. SBer. Kais. Akad. Wissensch. Wien, **118**, Abt.2a, 1439–1522.

Schnee,W. (1909b): Zum Konvergenzproblem der Dirichleschen Reihen,Math. Ann., 337–349.

Schnee,W. (1910): Über die Koeffizientendarstellungsformel in der Theorie der Dirichletschen Reihen. Nachr. Ges. Wiss. Göttingen, 1–42.

Schnee,W. (1930): Die Funktionalgleichung der Zetafunktion und der Dirichletschen Reihen mit periodischen Koeffizienten. Math. Zeitschr., **31**, 378–390.

Schnirelman,L.G. (1930): On additive properties of numbers. Izv. Donsk. Politehn. Inst., **14**, 3–28. (Russian)

Schnirelman,L.G. (1933): Über additive Eigenschaften von Zahlen. Math. Ann., **107**, 649–690.

Schoenfeld,L. (1969): An improved estimate for the summatory function of the Möbius function. Acta Arith., **15**, 221–233. MR 39#2716.

Schoenfeld,L. (1976): Sharper bounds for the Chebyshev functions $\theta(x)$ and $\psi(x)$, II. Math. Comp., **30**, 337–360. MR 56#15581b; Corr.: *ibidem*, p.900. MR 56#15581c.

Schönemann,T. (1846): Grundzüge einer allgemeinen Theorie der höhern Congruenzen, deren Modul eine Primzahl ist. J. Reine Angew. Math., **31**, 269–325.

Schönhage,A. (1963): Eine Bemerkung zur Konstruktion grosser Primzahldifferenzen. Archiv Math., **14**, 29–30. MR 26#3680.

Schulze,J.C. (1778): Neue und erweiterte Sammlung logarithmischer [...] Tafeln. I, Berlin.

Schuppener,G. (1994): Geschichte der Zeta-Funktion von Oresme bis Poisson. Deutsche Hochschulschriften, **533**. Hänsel–Hohenhausen, Egersbach. MR 95k:01012.

Schur,I. (1912/13): Über die Existenz unendlich vieler Primzahlen in einigen speziellen arithmetischen Progressionen. SBer. Berlin. Math. Ges., **11**, 40–50. [Gesammelte Abhandlungen, **2**, 1–11, Springer, Berlin–Heidelberg 1973.]

Schur,I. (1929): Einige Sätze über Primzahlen mit Anwendung auf Irreduzibilitätsfragen. SBer. Akad. Wiss. Berlin, Phys. Math. Kl., 125–136; II. *ibidem*, 370–391. [Gesammelte Abhandlungen, **3**, 140–151, 152–173, Springer, Berlin–Heidelberg 1973.]

Schwarz,W. (1969): Einführung in Methoden und Ergebnisse der Primzahltheorie. Bibliographisches Institut, Mannheim. MR 41#8350.

Schwarz,W.,Spilker,J. (1994): Arithmetical Functions. Cambridge University Press. MR 96b:11001.

Sedrakian,N.,Steinig,J. (1998): A particular case of Dirichlet's theorem on arithmetic progressions. Enseign. Math., **44**, 3–7. MR 99k:11108.

Segal,S.L. (1969a): A Tauberian relative of the Landau-Ingham Tauberian theorem. Proc. Amer. Math. Soc., **20**, 287–294. MR 38#4854.

Segal,S.L. (1969b): A general Tauberian theorem of Landau-Ingham type. Math. Zeitschr., **111**, 159–167. MR 40#1624.

Selberg,A. (1942): On the zeros of Riemann's zeta-function. Skr. Norske Vid. Akad. Oslo, No.10, 1–59. MR 6 p.58. [Collected Papers, **1**, 85–141. Springer, Berlin–Heidelberg 1989.]

Selberg,A. (1943): On the normal density of primes in small intervals, and the difference between consecutive primes. Arch. Math. Naturv., B, **47**, No.6, 87–105. MR 7 p.48. [Collected Papers, **1**, 160–178. Springer, Berlin–Heidelberg 1989.]

Selberg,A. (1946): Contributions to the theory of the Riemann zeta-function. Arch. Math. Naturvid., **48**, 89–155. MR 8 p.567. [Collected Papers, **1**, 214–280. Springer, Berlin–Heidelberg 1989.]

Selberg,A. (1947): On an elementary method in the theory of primes. Det Kong. Norske Vid.Selsk.Forh., **B19**, 64–67. MR 9 p.271. [Collected Papers, **1**, 363–366. Springer, Berlin–Heidelberg 1989.]

Selberg,A. (1949a): An elementary proof of Dirichlet's theorem about primes in an arithmetic progression. Ann. of Math., (2), **50**, 297–304. MR 10 p.595. [Collected Papers, **1**, 371–378. Springer, Berlin–Heidelberg 1989.]

Selberg,A. (1949b): An elementary proof of the prime-number theorem. Ann. of Math., (2), **50**, 305–313. MR 10 p.595. [Collected Papers, **1**, 379–387. Springer, Berlin–Heidelberg 1989.]

Selberg,A. (1950): An elementary proof of the prime number theorem for arithmetic progressions. Canad. J. Math., **2**, 66–78. MR 11 p.419. [Collected Papers, **1**, 398–410. Springer, Berlin–Heidelberg 1989.]

Selberg,A.,Chowla,S. (1967): On Epstein's zeta function. J. Reine Angew. Math., **227**, 86–110. MR 35#6632. [A.Selberg, Collected Papers, **1**, 521–545. Springer, Berlin–Heidelberg 1989.]

Selmer,E.S. (1942): Über Primzahlen von der Form x^2+1. Norske Vid. Selsk. Forh. **15**, 149–152. MR 8 p.503.

Selmer,E.S. (1943a): Eine neue hypothetische Formel für die Anzahl der Goldbachschen Spaltungen einer geraden Primzahl und eine numerische Kontrolle. Arch. Math. Naturvid., **46**, no.1, 1–18. MR 7 p.48.

Selmer,E.S. (1943b): Eine numerische Untersuchung über die Darstellung der natürlichen Zahlen als Summe einer Primzahl und einer Quadratzahl. Arch. Math. Naturvid., **47**, no.2, 21–39. MR 7 p.49.

Selvaray,S. (1991): A note on Riemann sums and improper integrals related to the prime number theorem. J. Approx. Th., **66**, 106–108. MR 92f:11121.

Serre,J.-P. (1970): Cours d'arithmétique. Presses Universitaires de France, Paris. MR 41#138; 2nd ed. 1977. MR 58#16473. [English translation: Springer, Berlin–Heidelberg 1973. MR 49#8956.]

Serret,J.A. (1850): Mémoire sur le nombre des valeurs que peut prendre une fonction quand on y permute les lettres qu'elle renferme. J. math. pures appl., **15**, 1–44.

Serret,J.A. (1852): Note sur un théorème de la théorie des nombres. J. math. pures appl., **17**, 186–189.

Serret,J.A. (1855): Théorème de Fermat généralisé. Nouv. Ann. Math., **14**, 261–262.

Shah,N.M.,Wilson,B.M. (1919): On an empirical formula connected with Goldbach's theorem. Proc. Cambridge Philos. Soc., **19**, 238–244.

Shanks,D. (1959): Quadratic residues and the distribution of primes. Math. Tables Aids Comput., **13**, 272–284. MR 21#7186.

Shanks,D. (1960): On the conjecture of Hardy and Littlewood concerning the number of primes of the form $n^2 + a$. Math. Comp., **14**, 321–332. MR 22#10960.

Shanks,D. (1961): On numbers of the form $n^4 + 1$. Math. Comp., **15**, 186–189. MR 22#10941.

Shanks,D. (1963): Supplementary data and remarks concerning a Hardy—Littlewood conjecture. Math. Comp., **17**, 188–193. MR 28#3013.

Shanks,D. (1964): On maximal gaps between successive primes. Math. Comp., **18**, 646–651. MR 29#4745.

Shanks,D. (1972): Bateman's constant reconsidered and the distribution of cubic residues. Math. Comp., **26**, 265–285. MR 46#1734.

Shanks,D. (1991): Euclid's primes. Bull. Inst. Comb. Appl., **1**, 33–36. MR 92f:11013.

Shanks,D.,Lal,M. (1972): Bateman's constants reconsidered and the distribution of cubic residues. Math. Comp., **26**, 265–285. MR 46#1734.

Shanks,D.,Wrench,J.W.Jr. (1974): Brun's constant. Math. Comp., **28**, 293–299. Corr.: *ibidem*, p.1183. MR 50#4510.

Shapiro,G. (1949): On the non-vanishing at $s = 1$ of certain Dirichlet series. Amer. J. Math, **71**, 621–626. MR 11 p.13.

Shapiro,H.N. (1949): Some assertions equivalent to the prime number theorem for arithmetic progressions. Comm. Pure Appl. Math., **2**, 293–308. MR 11 p.419.

Shapiro,H.N. (1950a): On the number of primes less or equal x. Proc. Amer. Math. Soc., **1**, 346–348. MR 12 p.80.

Shapiro,H.N. (1950b): On primes in arithmetic progressions, I,II. Ann. of Math., (2), **52**, 217–230, 231–243. MR 12 p.81.

Shapiro,H.N.,Warga,J.(1950): On representation of large integers as sums of primes, I. Comm. Pure Appl. Math., **3**, 153–176. MR 12 p.244.

Sheffer,I.M. (1927): On entire function interpolation. Amer. J. Math, **49**, 329–342.

Shen,M.-K. (1964): On checking the Goldbach conjecture. BIT, **4**, 243–245. MR 30#3051.

Shiu,-P. (1991): The intermediate value theorem and the density of prime numbers. Bull. Inst. Math. Appl. **27**, no. 4, 89–92. MR 92e:11093.

Siebeck,H. (1846): Die recurrenten Reihen, von Standpuncte der Zahlentheorie aus betrachtet. J. Reine Angew. Math., **33**, 71–77.

Siebert,H. (1968): Darstellung als Summen von Primzahlen, Diplomarbeit, Marburg.

Siegel,C.L. (1921): Approximation algebraischer Zahlen. Math. Zeitschr., **10**, 173–213. [Gesammelte Abhandlungen, **1**, 6–46. Springer, Berlin–Heidelberg 1966.]

Siegel,C.L. (1922): Bemerkung zu einem Satz von Hamburger über die Funktionalgleichung der Riemannscher Zeta-funktion. Math. Ann., **86**, 276–279. [Gesammelte Abhandlungen, **1**, 154–156. Springer, Berlin–Heidelberg 1966.]

Siegel,C.L. (1932): Über Riemanns Nachlaß zur analytischen Zahlentheorie. Quellen u. Studien zur Geschichte der Math. Astr. Phys., **2**, 45–80. [Gesammelte Abhandlungen, **1**, 274–310. Springer, Berlin–Heidelberg 1966.]

Siegel,C.L. (1936): Über die Classenzahl quadratischer Körper. Acta Arith., **1**, 83–86. [Gesammelte Abhandlungen, **1**, 406–409. Springer, Berlin–Heidelberg 1966.]

Sierpiński,W. (1952a): Sur une formule donnant tous les nombres premiers. Comptes Rendus Acad. Sci. Paris, **235**, 1078–1079. MR 14 p.355.

Sierpiński,W. (1952b): Sur une propriété des nombres premiers. Bull. Soc. Roy. Sci. Liége, **21**, 537–539. MR 15 p.11.

Sierpiński,W. (1964): Les binômes $x^2 + n$ et le nombres premiers. Bull. Soc. Roy. Sci. Liége, **33**, 259–260. MR 29#4746.

Singh,D. (1977): A simple proof of Bertrand's postulate. Math. Student, **45**, p.84. MR 81j:10009.

Sinisalo,M.K. (1993): Checking the Goldbach conjecture up to $4 \cdot 10^{11}$. Math. Comp., **61**, 931–934. MR 94a:11157.

Skewes,S. (1933): On the difference $\pi(x) - \mathrm{li}(x)$, I. J. London Math. Soc., **8**, 277–283.

Skewes,S. (1955): On the difference $\pi(x) - \mathrm{li}(x)$, II. Proc. London Math. Soc., (3), **5**, 48–69. MR 18 p.112.

Smarandache,F. (1991): Characterization of N prime numbers simultaneously. Libertas Math., **11**, 151—155. MR 92i:11005.

Smith,H.J.S. (1857): On the history of the researches of mathematicians on the subject of the series of prime numbers. Proc. Ashmolean Soc., **3**, No.35, 128–131. [The Collected Mathematical Papers, **1**, 35–37, Clarendon Press, Oxford 1894.]

Smith,R.A. (1981): A note on Dirichlet's theorem. Canad. Math. Bull., **24**, 379–380. MR 83e:10066.

Šnirelman,L.G. See Schnirelman,L.G.

Sobirov,A.Š. (1973): On the question of providing an elementary proof of the asymptotic law of the distribution of primes,III. Izv. Akad. Nauk Tadžik. SSR, No 2. 8–13. (Russian) MR 50#9817.

Sobirov,A.Š. (1974): On the prime number theorem in its elementary method of proof. Izv. Akad. Nauk Uzb. SSR., ser. Fiz.-Mat. Nauk, **18**, No 1, p.64. (Russian) MR 49# 234.

Sobirov,A.Š,Inojatova,K.G.(1972): On the elementary proof of the asymptotic law of the distribution of prime numbers. Izv. Akad. Nauk Uzb. SSR., ser. Fiz.-Mat. Nauk, **16**, No.2, 79–81. (Russian) MR 47#3331.

Sondow,J. (1994): Analytic continuation of Riemann's zetafunction and values at negative integers via Euler's transformation of series. Proc. Amer. Math. Soc., **120**, 421–424. MR 94d:11066.

Spears,N. (1970): A simplified proof of Chebyshev's theorem. Duke Math. J., **37**, 709–713. MR 42#3033.

Specht,W. (1956): Elementare Beweise der Primzahlsätze. Deutscher Verlag der Wissenschaften, Berlin. MR 19 p.250.

Speiser,A. (1939): Die Funktionalgleichung der Dirichletschen L-Funktionen. Monatsh. Math. Phys., **48**, 240–244. MR 1 p.70.

Spira,R. (1969): Calculation of Dirichlet's L-functions. Math. Comp., **23**, 489–497. MR 40#1004.

Srinivasan,B.R. (1961): Formulae for the nth prime. J. Indian Math. Soc., (N.S.), **25**, 33–39. MR 26#1289.

Srinivasan,B.R. (1962): An arithmetic function and an associated formula for the nth prime, I. Norske Vid. Selsk. Forh. (Trondheim), **35**, 68–71. MR 27#101; II. *ibidem*, 72–75. MR 27#102.

Srinivasan,S. (1984): On infinitude of primes. Hardy-Ramanujan J., **7**, 21–26. MR 87b:11010.

Srinivasan,S.,Sampath,A. (1983): A combinatorial proof of the generalized Selberg's identity. J.Indian Math. Soc. (N.S.), **47**, 187–192. MR 89b:11011.

Srinivasan,S.,Sampath,A. (1988): An elementary proof of the prime number theorem with a remainder term. J.Indian Math. Soc. (N.S.), **53**, 1–50. MR 90i:11094.

Srivastava,H.M. (1987): Some infinite series associated with the Riemann zeta function. Yokohama Math.J., **35**, 47–50. MR 89b:11071.

Stäckel,P. (1896): Ueber Goldbachs empirisches Theorem: Jede grade Zahl kann als Summe von zwei Primzahlen dargestellt werden. Nachr. Ges. Wiss. Göttingen, 292–299.

Stäckel,P. (1907/08): Eine vergessene Abhandlung Leonhard Eulers über die Summe der reziproken Quadrate der natürlichen Zahlen. Bibl. mat. **8**, 37–54. [L.Euler, Opera omnia, **I.14**, 156–176, Teubner, Lipsiae et Berolini 1924.]

Stäckel,P. (1916): Die Darstellung der geraden Zahlen als Summen von zwei Primzahlen. Sber. Akad. Wiss. Heidelberg.

Stanievitch,V. (1892): Sur un théorème arithmétique de Poincaré. Comptes Rendus Acad. Sci. Paris, **114**, 109–112.

Stanley,G.K. (1928): On the representation of a number as a sum of squares and primes. Proc. London Math. Soc., **29**, 122–144.

Stark,E.L. (1979): Another way to sum $\sum(1/k^2)$. Math. Gaz., **63**, p.47. MR 80b:10006.

Stark,H.M. (1967a): On the zeros of Epstein's zeta function. Mathematika, **14**, 47–55. MR 35#6633.

Stark,H.M. (1967b): A complete determination of the complex quadratic fields of class-number one. Michigan J. Math., **14**, 1–27. MR 36#5102.

Stark,H.M. (1969): On the "gap" in a theorem of Heegner. J. Number Th., **1**, 16–27. MR 39#2724.

Stark,H.M. (1971): A problem in comparative prime number theory. Acta Arith., **18**, 311–320. MR 44#6642.

Stark,H.M. (1993): Dirichlet's class number formula revisited. Contemp. Math., **143**, 571–577. MR 94a:11133.

Staś,W. (1961): Über die Umkehrung eines Satzes von Ingham. Acta Arith., **6**, 435–446. MR 26#3679.

Staś,W.,Wiertelak,K. (1974): Some estimates connected with the distribution of primes. Topics in Number Theory, Proc. J.Bólyai Colloq. **13**, Debrecen 1974, 373–382. North-Holland, Amsterdam. MR 55#12665.

Stečkin,S.B. (1968): A simple proof of the prime number theorem of Čebyšev. Usp. Mat. Nauk, **23**, no.5, 221–222. (Russian) MR 38#1056.

Stečkin,S.B. (1970a): Certain extremal properties of positive trigonometric polynomials. Mat. Zametki, **7**, 411–422. (Russian) MR 41#8355.

Stečkin,S.B. (1970b): The zeros of the Riemann zeta function. Mat. Zametki, **8**, 419–429. (Russian) MR 43#6168.

Steffensen,J.B. (1912): Analytiske Studier med Anvendelser paa Taltheorien. 1–148. Kopenhagen.

Steffensen,J.B. (1914a): Über eine Klasse von ganzen Funktionen und ihre Anwendung auf die Zahlentheorie. Acta Math., **37**, 75–112.

Steffensen,J.B. (1914b): Über Potenzreihen, im besondern solche, deren Koeffizienten zahlentheoretische Funktionen sind. Rend. Circ. Mat. Palermo, **38**, 376–386.

Stein,M.L.,Stein,D.R. (1965): New experimental results on the Goldbach conjecture. Math. Mag., **38**, 72–78. MR 32#4109.

Steinig,J. (1969): The changes of sign of certain arithmetical error-terms. Comment. Math. Helv., **44**, 385–400. MR 41#1658.

Stephens,P.J. (1976): Prime divisors of second order linear recurrences. I, II. J. Number Th., **8**, 313–332, 333–345. MR 54#5142,5143.

Sterneck,R.D.von (1893): Bemerkungen zur Theorie der zahlentheoretischer Ableitungen. Monatsh. Math. Phys., **4**, 53–56.

Sterneck,R.D.von (1896): Über einige spezielle zahlentheoretische Functionen. Monatsh. Math. Phys., **7**, 37–48.

Sterneck,R.D.von (1898): Bemerkung über die Summierung einiger zahlentheoretischer Funktionen. Monatsh. Math. Phys., **9**, 43–45.

Sterneck,R.D.von (1900): Zur Tschebyschef'schen Primzahlentheorie. SBer. Kais. Akad. Wissensch. Wien, **109**, Abt.2a, 1137–1158.

Sterneck,R.D.von (1901): Empirische Untersuchungen über den Verlauf der zahlentheoretischen Function $\sigma(n) = \sum_{x=1}^{x=n} \mu(x)$ im Intervalle von 150 000 bis 500 000. SBer. Kais. Akad. Wissensch. Wien, **110**, Abt.2a, 1053–1102.

Stieltjes,T.J. (1885a): Sur une fonction uniforme. Comptes Rendus Acad. Sci. Paris, **101**, 153–154. [Oeuvres complètes, **1**, 457–458, P.Noordhoff, Groningen 1914. Reprint: Springer, Berlin–Heidelberg 1993. MR 95g:01033.]

Stieltjes,T.J. (1885b): Sur une loi asymptotique dans la théorie des nombres. Comptes Rendus Acad. Sci. Paris, **101**, 368–370. [Oeuvres complètes, **1**, 459–461, P.Noordhoff, Groningen 1914. Reprint:Springer, Berlin–Heidelberg 1993. MR 95g:01033.]

Stieltjes,T.J. (1887): Note sur la multiplication de deux séries. Nouv. Ann. Math., (3), **6**, 210–215. [Oeuvres complètes, **2**, 95–99, P.Noordhoff, Groningen 1918. Reprint: Springer, Berlin–Heidelberg 1993. MR 95g:01033.]

Stieltjes,T.J. (1889): Sur le développement de log $\Gamma(a)$. J. math. pures appl., (4), **5**, 425–444. [Oeuvres complètes, **2**, 211–230, P.Noordhoff, Groningen 1918. Reprint: Springer, Berlin–Heidelberg 1993. MR 95g:01033.]

Stieltjes,T.J. (1890): Sur la théorie des nombres. Ann. Fac. Sci. Toulouse, **4**, 1–103. [Oeuvres complètes, **2**, 265–377, P.Noordhoff, Groningen 1918. Reprint: Springer, Berlin–Heidelberg 1993. MR 95g:01033.]

Stoica,G. (1987): A recurrence formula in the study of the Riemann zeta function. Stud.Cerc. Mat., **39**, 261–264. (Romanian) MR 88i:11060.

Størmer,C. (1898): Sur une équation indéterminée. Comptes Rendus Acad. Sci. Paris, **127**, 752–754.

Subbarao,M.V. (1966): On relatively prime sequences. Amer. Math. Monthly, **73**, 1099–1102. MR 34#4236.

Subbarao,M.V. (1974): On two congruences for primality. Pacific J. Math., **52**, 261–268. MR 50#2049.

Subbarao,M.V. (1980): On the nonexistence of certain Euler products. Canad. Math. Bull., **23**, 371–372. MR 81k:10064.

Suzuki,T. (1913): A theorem on the series of prime numbers. Tôhoku Math. J., **3**, 83–86.

Sweeney,D.W. (1963): On the computation of Euler's constant. Math. Comp., **17**, 170–178. MR 28#3522.

Sylvester J.J. (1860): On the equation

$$P(m) + E(\frac{m}{m-1})P(m-1) + E(\frac{m}{m-2})P(m-2) + \cdots + E(m) = \frac{m(m+1)}{2}.$$

Quart. J.Math. Oxford ser., **3**, 186–190. [Collected Math. Papers, **2**, 225–228, Cambridge 1908.]

Sylvester J.J. (1871a): On the partition of an even number into two primes. Proc. London Math. Soc., **4**, 4–6. [Collected Math. Papers, **2**, 709–711, Cambridge 1908.]

Sylvester J.J. (1871b): On the theorem that an arithmetical progression which contains more than one, contains an infinite number of prime numbers. Proc. London Math. Soc., **4**, 7–8. [Collected Math. Papers, **2**, 712–713, Cambridge 1908.]

Sylvester J.J. (1881): On Tchebycheff's theory of the totality of the prime numbers comprised within given limits. Amer. J. Math, **4**, 230–247. [Collected Math. Papers, **3**, 530–545, Cambridge 1909.]

Sylvester J.J. (1883): Note sur le théorème de Legendre cité dans une Note insérée dans les Comptes rendus. Comptes Rendus Acad. Sci. Paris, **96**, 463–465. [Collected Math. Papers, **4**, 88–90, Cambridge 1912.]

Sylvester J.J. (1888a): On certain inequalities relating to prime numbers. Nature, **38**, 259–262. [Collected Math. Papers, **4**, 592–603, Cambridge 1912.]

Sylvester J.J. (1888b): Preuve élementaire du théorème de Dirichlet sur les progressions arithmétiques dans le cas oú la raison est 8 ou 12. Comptes Rendus Acad. Sci. Paris, **106**, 1278–1281, 1385–1386. [Collected Math. Papers, **4**, 620–624, Cambridge 1912.]

Sylvester J.J. (1892): On arithmetical series. Messenger of Math., **21**, 1–19, 87–120. [Collected Math. Papers, **4**, 687–731, Cambridge 1912.]

Sylvester J.J. (1896): On the Goldbach–Euler theorem regarding prime numbers. Nature, **55**, 1896–97, 196–197, 269. [Collected Math. Papers, **4**, 734–737, Cambridge 1912.]

Szekeres,G. (1974): On the number of divisors of $x^2 + x + A$. J. Number Th., **6**, 434–442. MR 51#422.

Szydło,B. (1989): Über Vorzeichenwechsel einiger arithmetischer Funktionen,I,II. Math. Ann., **283**, 139–149; 151–163. MR 90a:11105.

Szydło,B. (1991): Über eine Methode, die Anzahl der Vorzeichenwechsel arithmetischer Funktionen von unten abzuschätzen. Archiv Math., **57**, 267–272. MR 92g:11090.

Tate,J. (1950): Fourier analysis in number fields and Hecke's zeta function. Thesis, Princeton Univ. MR 36#121. [Reprinted in Cassels,Fröhlich (1961).]

Tatuzawa,T. (1950): On the zeros of Dirichlet's L-functions. Proc. Japan Acad. Sci., **26**, 1–13. MR 14 p.249.

Tatuzawa,T. (1951): On a theorem of Siegel. Japan J. Math., **21**, 163–178. MR 14 p.452.

Tatuzawa,T. (1962): On Bertrand's problem in an arithmetic progression. Proc. Japan Acad. Sci., **38**, 293–294. MR 25#5042.

Tatuzawa,T. (1985): On a proof of Siegel's theorem. Number Theory and Combinatorics, 383–416. World Scient. Publ., Singapore. MR 87d:11065.

Tauber,A. (1897): Ein Satz aus der Theorie der unendlichen Reihen. Monatsh. Math. Phys., **8**, 273–277.

Tee,G.J. (1972): Simple analytic expressions for primes and for prime pairs. New Zealand Math. Mag., **9**, 32–44. MR 45#8601.

Tee,G.J. (1974): A refinement of Mill's prime-generating function. New Zealand Math. Mag., **11**, 9–11. MR 49#7203.

Teege,H. (1901): Beweis, daß die unendliche Reihe $\sum_{n=1}^{n=\infty} \left(\frac{p}{n}\right) \frac{1}{n}$ einen positiven von Null verschiedenen Wert hat. Mitt. Math. Ges. Hamburg, **4**, 1–11.

Templer,M. (1980): On the primality of $k! + 1$ and $2 \cdot 3 \cdots p + 1$. Math. Comp., **34**, 303–304. MR 80j:10010.

Tenenbaum,G. (1990a): Introduction à la théorie analytique et probabiliste des nombres. Inst. Elie Cartan, Paris. [2nd ed. Soc. Math. France, Paris 1995. MR 97e:11005a; English translation: Cambridge University Press, Cambridge 1995. MR 97e:11005b.]

Tenenbaum,G. (1990b): Sur une question d'Erdős et Schinzel,II. Invent. math., **99**, 215–224. MR 92j:11100.
Tenenbaum,G., Mendès-France,M. (1997): Les nombres premiers. Presses Universitaires de France. MR 98d:11105.
Teuffel,E. (1954): Eine Rekursionsformel für Primzahlen. Jahresber. Deutsche Math. Ver., **57**, 34–36. MR 15 p.685.
Teuffel,E. (1955): Beweise für zwei Sätze von H.F.Scherk über Primzahlen. Jahresber. Deutsche Math. Ver., **58**, Abt.1, 43–44. MR 17 p.587.
Thompson,J.G. (1953): A method for finding primes. Amer. Math. Monthly, **60**, p.175. MR 14 p.621.
Thue,A. (1897): Mindre meddelelser,II; Et bevis for at primtallenes antal er uendeligt. Archiv for Math. og Natur., **19**, No.4, 1–5.
Thurnheer,P. (1980): Une généralisation du théorème de Wiener-Ikehara. Comptes Rendus Acad. Sci. Paris, **290**, p.499. MR 81c:10051.
Titchmarsh,E.C. (1926): A series inversion formula. Proc. London Math. Soc., (2), **26**, 1–12.
Titchmarsh,E.C. (1927): A consequence of the Riemann Hypothesis. J. London Math. Soc., **2**, 247–254.
Titchmarsh,E.C. (1928): On an inequality satisfied by the zeta-function of Riemann. Proc. London Math. Soc., **28**, 683–690.
Titchmarsh,E.C. (1930a): A divisor problem, Rend. Circ. Mat. Palermo, **54**, 414-429. Corr.: *ibidem*, **57**, 1933, 478–479.
Titchmarsh,E.C. (1930b): On the zeros of Riemann's zeta function. Proc. London Math. Soc., (2), **30**, 319–321.
Titchmarsh,E.C. (1930c): The Theory of Riemann Zeta-function. University Press, Cambridge.
Titchmarsh,E.C. (1931): On van der Corput's method and the zeta-function of Riemann, Quart. J.Math. Oxford ser., **2**, 161–173; II, *ibidem*, 313–320.
Titchmarsh,E.C. (1933): On the function $\frac{1}{\zeta(1+it)}$. Quart. J.Math. Oxford ser., **4**, 64–70.
Titchmarsh,E.C. (1934): On van der Corput's method and the zeta-function of Riemann,IV. Quart. J.Math. Oxford ser., **5**, 98–105.
Titchmarsh,E.C. (1935): The zeros of the Riemann zeta-function. Proc. Roy. Soc. London, **A 151**, 234–255; *ibidem*, **A 157**, (1936), 261–263.
Titchmarsh,E.C. (1942): On the order of $\zeta(1/2+it)$. Quart. J.Math. Oxford ser., **13**, 11–17. MR 4 p.131.
Titchmarsh,E.C. (1947): On the zeros of the Riemann zeta function. Quart. J.Math. Oxford ser., **18**, 4–16. MR 8 p.567.
Titchmarsh,E.C. (1951): The Theory of Riemann Zeta-function. Oxford. MR 13 p.741. [2nd ed. 1986.]
Toeplitz.O. (1913): Über eine bei den Dirichletschen Reihen auftretende Aufgabe aus der Theorie der Potenzreihen von unendlich vielen Veränderlichen. Nachr. Ges. Wiss. Göttingen, 417–432.
Torelli,G. (1901): Sulla totalità dei numeri primi fino ad un limite assegnato. Atti R. Accad. Sci. fis. mat. Napoli, (2), **11**, Nr.1, 1–222.
Treiber,D. (1995): Zur Reihe der Primzahlreziproken. Elem. Math., **50**, 164–166. MR 96f:11114.
Trost,E. (1953): Primzahlen. Birkhäuser, Basel-Stuttgart. MR 15 p.401. [2nd ed. 1968. MR 41#160.]
Tsang,K.-M. (1986): Some Ω-theorems for the Riemann zeta-function. Acta Arith., **46**, 369–395. MR 88a:11081.
Tsangaris,P.G.,Jones,J.P. (1992): An old theorem on GCD and its application to primes. Fibonacci Quart., **30**, 194–198. MR 93e:11004.

Tschakaloff, L. (1940): Trigonometrische Polynome mit einer Minimumeigenschaft. Ann. Scuola Norm. Sup. Pisa, Cl. Sci., (2), **9**, 13–26. MR 1 p.225.

Tsumura,H. (1994): On evaluation of the Dirichlet series at positive integers by q-calculation. J. Number Th., **48**, 383–391. MR 96a:11083.

Turán,P. (1936/37): Über die Primzahlen der arithmetischen Progression, I. Acta Litt. Sci. Reg. Univ. Hungar. Franc.-Jos., sect. Math., **8**, 226–235. MR 1 p.135. [Collected Papers, **1**, 64–73. Akadémiai Kiado, Budapest 1990.]

Turán,P. (1938/40): Über die Primzahlen der arithmetischen Progression, II. Acta Litt. Sci. Reg. Univ. Hungar. Franc.-Jos., sect. Math., **9**, 187–192. MR 1 p.135. [Collected Papers, **1**, 144–149. Akadémiai Kiado, Budapest 1990.]

Turán,P. (1950): On the remainder-term of the prime-number formula, I,II. Acta Math. Acad. Sci. Hung., **1**, 48–63. MR 13 p.208; *ibidem*, 155–166. MR 14 p.137. [Collected Papers, **1**, 515–530, 541–551. Akadémiai Kiado, Budapest 1990.]

Turán,P. (1953): Eine neue Methode in der Analysis und deren Anwendungen. Akadémiai Kiadó. Budapest. MR 15 p.688. [English version: J.Wiley 1984. MR 86b:1059.]

Turán,P. (1954): On Lindelöf's conjecture. Acta Math. Acad. Sci. Hung., **5**, 145–163. MR 16 p.999. [Collected Papers, **1**, 765–783. Akadémiai Kiado, Budapest 1990.]

Turán,P. (1959): Nachtrag zu meiner Abhandlung "On some approximative Dirichlet polynomial in the theory of zeta-function of Riemann". Acta Math. Acad. Sci. Hung., **10**, 277–298. MR 22#6774. [Collected Papers, **2**, 1142–1163. Akadémiai Kiado, Budapest 1990.]

Turán,P. (1961): On a density theorem of Yu.V.Linnik. Magyar Tud. Akad. Mat. Kut. Int. Közl., **6**, 165–179. MR 26#3681. [Collected Papers, **2**, 1242–1255. Akadémiai Kiado, Budapest 1990.]

Turán,P. (1965): On a characterization of Dirichlet's L-functions. Ann. Univ. Sci. Budapest, **8**, 65–69. MR 34#1285. [Collected Papers, **2**, 1703–1707. Akadémiai Kiado, Budapest 1990.]

Turing,A.M. (1953): Some calculations of the Riemann zeta-function. Proc. London Math. Soc., (3), **3**, 99–117. MR 14 p.1126.

Uchiyama,S. (1963): On a theorem concerning the distribution of almost primes. J.Fac. Sci. Hokkaido Univ., **1 17**, 152–159. MR 28#2101.

Uchiyama,S. (1967): On the representation of large even integers as sums of a prime and an almost prime, II. Proc. Japan Acad. Sci., **43**, 567–571. MR 37#179.

Uchiyama,S. (1971): On some products involving primes. Proc. Amer. Math. Soc., **28**, 629–630. MR 43#3227.

Vahlen K.Th. (1895): Ueber eine Verallgemeinerung der Euler'schen φ Funktion. Zeitschr. Math. Phys., **40**, 126–127.

Vakhitova,E.V. (1992): Some applications of a one-dimensional sieve with weights. Mat. Zametki, **51**, 139–141. (Russian) MR 93k:11085.

Vallée-Poussin,C.J.de la (1896a): Recherches analytiques sur la théorie des nombres premiers, I–III. Ann. Soc. Sci. Bruxelles, **20**, 183–256, 281–362, 363–397.

Vallée-Poussin,C.J.de la (1896b): Démonstration simplifiée du théorème de Dirichlet sur la progression arithmétique. Mem. couronnés et autres mémoires publiées par l'Acad. Roy. Sci. Bruxelles, **53**, 6–8.

Vallée-Poussin,C.J.de la (1897): Recherches analytiques sur la théorie des nombres premiers, IV,V. Ann. Soc. Sci. Bruxelles, **21**, 251–342, 343–368.

Vallée-Poussin,C.J.de la (1899): Sur la fonction $\zeta(s)$ de Riemann et le nombre des nombres premiers inférieurs à une limite donnée. Mem. Couronnés de l'Acad. Roy. Sci. Bruxelles, **59**. [Reprint: Colloque Th.des Nombres, Bruxelles 1955, 9–66, Liége 1956. MR 16 p.676.]

Vallée-Poussin,C.J.de la (1916): Sur les zéros de $\zeta(s)$ de Riemann. Comptes Rendus Acad. Sci. Paris, **163**, 418–421, 471–473.
Vandeneynden,C. (1972): A proof of Gandhi's formula for the nth prime. Amer. Math. Monthly, **79**, 625. MR 46#3425.
Vandeneynden,C. (1980): Proofs that $\sum 1/p$ diverges. Amer. Math. Monthly, **87**, 394–397. MR 81g:10013.
Vasilkovskaja,E.A. (1977): Mertens' formula for an arithmetic progression. Naučn. Trudy Gos. Univ. Taškent, No. 548, Voprosy Mat. 14–17, 139–140. (Russian) MR 58#27848.
Vaughan,R.C. (1972): On Goldbach's problem. Acta Arith., **22**, 21–48. MR 48#6045.
Vaughan,R.C. (1975): Mean value theorems in prime number theory. J. London Math. Soc., (2), **10**, 153–162. MR 51#12742.
Vaughan,R.C. (1976): A note on Šnirelmans approach to Goldbach's problem. Bull. London Math. Soc., **8**, 245–250. MR 54#12695.
Vaughan,R.C. (1977): On the estimation of Schnirelman's constant. J. Reine Angew. Math., **290**, 93–108. MR 55#10408.
Vaughan,R.C. (1980): An elementary method in prime number theory. Acta Arith., **37**, 111–115. MR 82c:10055.
Vecchi,M. (1913): Un nouve aspetto dato al teorema di Goldbach. Atti della Reale Accademia dei Lincei, Roma. **22**, sem. II, 654–659.
Venugopalan,A. (1983a): Formula for primes, twinprimes, number of primes and number of twinprimes. Proc. Indian Acad. Sci. Math. Sci. **92**, 49–52. MR 85d:11011; Errata: *ibidem*, **93**, 1984, p.66. MR 87i:11016.
Venugopalan,A. (1983b): Primes between $p_n + 1$ and $p_{n+1}^2 - 1$. Hardy Ramanujan J. **6**, 45–48. MR 85f:11002.
Verma,D.P. (1963): Laurent's expansion of Riemann's zeta function. Indian J.Math., **5**, 13–16. MR 28#5046.
Verma,K. (1991): Laurent's expansion of Hurwitz and Riemann's zeta functions about $s = 1$. Ganita, **42**, 65–70. MR 93i:11106.
Vinogradov,A.I. (1957): Application of Riemann's $\zeta(s)$ to the Eratosthenian sieve. Mat. Sb., **41**, 49–80; Corr.: *ibidem*, 415–416. (Russian) MR 20#2836.
Vinogradov,A.I. (1963): On the remainder in Merten's formula. Doklady Akad. Nauk SSSR, **148**, 262–263. (Russian) MR 26#1292.
Vinogradov,A.I. (1965): The density hypothesis for Dirichlet L-series. Izv. Akad. Nauk SSSR, **29**, 903–934. (Russian) MR 33#5579; Corr.: *ibidem*, **30**, 1966, 719–720. MR 33#2607.
Vinogradov,A.I. (1985): On a binary problem of Hardy–Littlewood. Acta Arith., **46**, 33–56. (Russian) MR 87j:11110.
Vinogradov,I.M. (1937): Representation of an odd number as a sum of three prime numbers. Doklady Akad. Nauk SSSR, **15**, 291–294. (Russian)
Vinogradov,I.M. (1940): A certain general property of the distribution of prime numbers. Mat. Sb., **7**, 365–372. (Russian) MR 2 p.40.
Vinogradov,I.M. (1958): A new evaluation of $\zeta(1 + it)$. Izv. Akad. Nauk SSSR, **22**, 161–164. (Russian) MR 21#2624.
Vinogradov,I.M. (1976): Special Variants of the Method of Trigonometric Sums. Nauka, Moskva 1976. (Russian) MR 57#9659.
Vivanti,G. (1893): Sulle serie di potenze. Rivista di mat., **3**, 111–114.
Voronoi,G.F. (1903): Sur un probléme du calcul des fonctions asymptotiques. J. Reine Angew. Math., **126**, 241–282.
Waage,E. (1913): Zur Tschebyscheff'schen Primzahltheorie. SBer. Kais. Akad. Wissensch. Wien, **122**, Abt.2a, 701–719.

Waage,E. (1914): Zur Tschebyscheff'schen Primzahltheorie. SBer. Kais. Akad. Wissensch. Wien, **123**, Abt.2a, 493–510.

Waerden,B.L. van der (1950): Über Landau's Beweis des Primzahlsatzes. Math. Zeitschr., **52**, 649–653. MR 11 p.655.

Wagstaff,S.S.Jr. (1978): The least prime in an arithmetic progression with prime difference. J. Reine Angew. Math., **301**, 114–115. MR 80c:10044.

Wagstaff,S.S.Jr. (1979): Greatest of the least primes in arithmetic progressions. Math. Comp., **33**, 1073–1080. MR 81e:10038;

Wagstaff,S.S.Jr. (1993): Computing Euclid's primes. Bull. Inst. Combin. Appl., **8**, 23–32. MR 94e:11139.

Walfisz,A. (1924): Zur Abschätzung von $\zeta(1/2+it)$. Nachr. Ges. Wiss. Göttingen, 155–158.

Walfisz,A. (1936a): Zur additiven Zahlentheorie. Acta Arith., **1**, 123–160.

Walfisz,A. (1936b): Zur additiven Zahlentheorie,II. Math. Zeitschr., **40**, 592–607.

Walfisz,A. (1958): Übe die Wirksamkeit einiger Abschätzungen trigonometrischer Summen. Acta Arith., **4**, 108–180. MR 21#2623.

Walfisz,A. (1963): Weylsche Exponentialsummen in der neueren Zahlentheorie. Deutscher Verlag der Wissenschaften, Berlin. MR 36#3737.

Wang,Tianze (1995): On the exceptional set for the equation $n = p + k^2$. Acta Math. Sinica, **11**, 156–167. MR 97i:11104.

Wang,Tianze,Chen,J.-R. (1993): On odd Goldbach problem under general Riemann hypothesis. Science in China, A, **36**, 682–691. MR 95a:11090.

Wang Yuan (1957): On some properties of integral-valued polynomials. Advancement in Math., **3**, 416–423. MR 20#4531.

Wang Yuan (1958): On sieve methods and some of their applications,I. Acta Math. Sinica, **8**, 413–429. (Chinese) MR 21#1958. [English translation: Sci. Sinica, **8**, 1959, 357–381. MR 21#4944.]

Wang Yuan (1959): On sieve methods and some of their applications. Acta Math. Sinica, **9**, 87–100. (Chinese) MR 21#5615. [English translation: Sci. Sinica, **11**, 1962, 1607–1624. MR 26#3685.]

Wang Yuan,Xie Sheng-gang,Yu Kun-rui (1965): Remarks on the difference of consecutive primes. Sci. Sinica **14**, 786–788. MR 32#5617.

Wang,Wei (1991): On the least prime in an arithmetic progression. Acta Math. Sinica, **7**, 279–288. MR 93c:11073.

Ward,M. (1930): A generalisation of a familiar theorem concerning prime numbers. J. London Math. Soc., **5**, 106–107.

Ward,M. (1954): Prime divisors of second order recurring sequences. Duke Math. J., **21**, 178–188. MR 16 p.221.

Ward,M. (1961): The prime divisors of Fibonacci numbers. Pacific J. Math., **11**, 379–386. MR 25#2029.

Waring,E. (1770): Meditationes Algebraicae. Cantabrigine. [3 ed. 1782; English translation: American Math. Soc., Providence 1991.]

Waterhouse,W.C. (1970): The sign of the Gaussian sum. J. Number Th., **2**, p.363. MR 42#4480.

Watt,N. (1989): Exponential sums and the Riemann zeta-function,II. J. London Math. Soc., (2), **39**, 385–404. MR 90k:11111.

Watt,N. (1995): Short intervals almost all containing primes. Acta Arith., **72**, 131–167. MR 96f:11115.

Weber,H. (1882): Beweis des Satzes, dass jede eigentlich primitive quadratische Form unendlich viele Primzahlen darzustellen fähig ist. Math. Ann., **20**, 301–329.

Weber,H. (1897): Über Zahlengruppen in algebraischen Körpern,II. Math. Ann., **49**, 83–100.

Weber,H. (1903): Über Abel's Summation endlicher Differenzenreihen. Acta Math., **27**, 225–233.
Ween, S.C.van (1978): Thomas Jan Stieltjes (1856–1894), Nieuw Arch. Wisk., (3), **26**, 84–95. MR 80f:01022.
Weierstrass,K. (1842): Definition analytischer Funktionen einer Veränderlichen vermittelst algebraischer Differentialgleichungen. Mathematische Werke, **1**, 75–84, Mayer & Müller, Berlin 1894.]
Weierstrass,K. (1876): Zur Theorie der eindeutiger analytischen Functionen. Abh. Kgl. Preuß. Akad. Wiss. Berlin, 11–60. [Mathematische Werke, **2**, 77–124, Mayer & Müller, Berlin 1895.]
Weierstrass,K. (1885): Über die analytische Darstellbarkeit sogenannter willkürlicher Funktionen reeller Argumente. SBer. Kgl. Preuß. Akad. Wiss. Berlin, 633–639, 789–805. [Mathematische Werke, **3**, 1–37, Mayer & Müller, Berlin 1903.]
Weil,A. (1983): Number Theory. Birkhäuser, Basel. MR 85c:01004.
Weil,A. (1989): Prehistory of the zeta-function. Number Theory, Trace Formulas and Discrete Groups, 1–9, Academic Press, Boston. MR 90e: 01029.
Weisner,L. (1935): Abstract theory of inversion of finite series. Trans. Amer. Math. Soc., **38**, 474–484.
Wendt,E. (1895): Elementarer Beweis des Satzes, dass in jeder unbegrenzter arithmetischen Progression $my + 1$ unendlich viele Primzahlen vorkommen. J. Reine Angew. Math., **115**, 85–88.
Western,A.E. (1922): Note on the number of primes of the form $n^2 + 1$. Proc. Cambridge Philos. Soc., **21**, 108–109.
Western,A.E. (1934): Note on the magnitude of the difference between successive primes. J. London Math. Soc., **9**, 276–278.
Westzynthius,E. (1931): Über die Verteilung der Zahlen, die zu der n ersten Primzahlen teilerfremd sind. Comm.Phys.Math. Helsingfors, (5), **25**, 1–37.
Weyl,H. (1916): Über die Gleichverteilung von Zahlen mod.Eins. Math. Ann., **77**, 313–352. [Gesammelte Abhandlungen, **1**, 563–599. Springer, Berlin–Heidelberg 1968.]
Weyl,H. (1921): Zur Abschätzung von $\zeta(1+it)$. Math. Zeitschr., **10**, 88-101. [Gesammelte Abhandlungen, **2**, 181–194. Springer, Berlin–Heidelberg 1968.]
Whittaker,E.T.,Watson,G.N. (1902): A Course of Modern Analysis. University Press, Cambridge. [2nd ed. 1915, 3rd ed. 1920.]
Wiegandt,R. (1959): On the general theory of Möbius inversion formulas and Möbius product. Acta Sci. Math., **20**, 164–180. MR 21#7176.
Wiener,N. (1932): Tauberian theorems. Ann. of Math., (2), **33**, 1–100. [Collected Works, **2**, 519–619. MIT, 1979.]
Wiener,N.,Geller,L. (1950): Some prime number consequences of the Ikehara theorem. Acta Sci. Math., **12**B, 25–28. MR 11 p.644. [N.Wiener, Collected Works, **2**, 754–757. MIT, 1979.]
Wiertelak,K. (1971): On the application of Turán's method to the theory of Dirichlet L-functions. Acta Arith., **19**, 249–259. MR 45#1855.
Wiertelak,K. (1972): A note on the remainder term of the prime-number formula for arithmetical progressions. Comment. Math. (Prace Mat.), **16**, 149–157. MR 48#8413.
Wigert,S. (1895): Remarque sur les nombres des nombres premiers inférieurs à une quantité donnée. Ofversigt Kgl. Vetensk. Akad. Förhandl., **5**, pp. 341 seq..
Wigert,S. (1906/07): Sur l'ordre de grandeur du nombre des diviseurs d'un entier. Arkiv f. Math. Astr. Fys., **3**, No.18, 1–9.
Wigert,S. (1909): Sur un théorème de la théorie des fonctions analytiques. Arkiv f. Math. Astr. Fys., **5**, no.8, 1–10.
Wigert,S. (1914): Sur quelques fonctions arithmétiques. Acta Math., **28**, 113–140.

Wigert,S. (1920): Sur la théorie de la fonction $\zeta(s)$ de Riemann. Arkiv f. Math. Astr. Fys., **14**, No.12, 1–17.

Willans,C.P. (1964): On formulae for the nth prime number. Math. Gaz., **48**, 413–415. MR 31#2198.

Williams,G.T. (1953): A new method of evaluating $\zeta(2n)$. Amer. Math. Monthly, **60**, 19–25. MR 14 p.536.

Williams,H.C. (1993): How was F_6 factored. Math. Comp., **61**, 463–474. MR 93k:01046.

Williams,K.S. (1971): On $\sum_{n=1}^{\infty}(1/n^{2k})$. Math.Mag., **44**, 273–276. MR 45#3997.

Williams,K.S. (1974): Mertens' theorem for arithmetic progressions. J. Number Th., **6**, 353–359. MR 51#392.

Wilson,B.M. (1922): Proof of certain formulas enunciated by Ramanujan. Proc. Cambridge Philos. Soc., (2), **21**, 235–255.

Wilton,J.R. (1916): Note on the zeros of Riemann's ζ-function. Messenger of Math., **45**, 180–183.

Wintner,A. (1946): The fundamental lemma in Dirichlet's theory of the arithmetical progressions. Amer. J. Math, **68**, 285–292. MR 7 p.416.

Wirsing,E. (1962): Elementare Beweise des Primzahlsatzes mit Restglied, I. J. Reine Angew. Math., **211**, 205–214. MR 27#119.

Wirsing,E. (1964): Elementare Beweise des Primzahlsatzes mit Restglied, II. J. Reine Angew. Math., **214/215**, 1–18. MR 29#3457.

Wirtinger,W. (1902): Einige Anwendungen der Euler-Maclaurin'schen Summenformel, insbesondere auf eine Aufgabe von Abel. Acta Math., **26**, 255–271.

Wójcik,J. (1966): A refinement of a theorem of Schur on primes in arithmetic progressions. Acta Arith., **11**, 1966, 433–436. MR 34#4227,

Wójcik,J. (1966/67): A refinement of a theorem of Schur on primes in arithmetic progression,II. Acta Arith., **12**, 97–109. MR 34#4228.

Wójcik,J. (1968/69): A refinement of a theorem of Schur on primes in arithmetic progressions, III. Acta Arith., **15**, 193–197. MR 39#161.

Wolfskehl,P. (1900): Nombres entiers N jouissant de cette propriété que les nombres premiers à N et non supérieurs à N soient tous les nombres premiers. L'Intermédiaire des Math., **7**, p.254.

Wolfskehl,P. (1901): Ueber eine Aufgabe der elementaren Arithmetik. Math. Ann., **54**, 503–504.

Wolke,D. (1979): Fast-Primzahlen in kurzen Intervallen. Math. Ann., **225**, 233–242. MR 81i:10055.

Wolke,D. (1980): On the explicit formula of Riemann-von Mangoldt. Studia Sci. Math. Hungar., **15**, 231–239. MR 84b:10059.

Wolke,D. (1983): On the explicit formula of Riemann-von Mangoldt,II. J. London Math. Soc., (2), **28**, 406–416. MR 86b:11056.

Wolke,D. (1985): Eine Bemerkung zum Primzahlsatz. Monatsh. Math., **100**, 337–339. MR 87b:11086.

Wormell,C.P. (1961): Formulae for primes. Math. Gaz., **51**, 36–38.

Wrench,J.W.Jr. (1952): A new calculation of Euler's constant. Math. Tables Aids Comput., **6**, p.255.

Wrench,J.W.Jr. (1961): Evaluations of Artin's constant and the twin constant. Math. Comp., **15**, 396–398. MR 23#A1619.

Wright,E.M. (1951): A prime-representing function. Amer. Math. Monthly, **58**, 616–618. MR 13 p.321.

Wright,E.M. (1952): The elementary proof of the prime number theorem. Proc. Roy. Soc. Edinburgh, sect. A, **63**, 257–267. MR 14 p.137.

Wright,E.M. (1954): A class of representing functions. J. London Math. Soc., **29**, 63–71. MR 15 p.288.

Wu,Dong Hua (1987): An improvement of J.R.Chen's theorem. Shanghai Keji Daxue Xuebao, p.94–99. (Chinese), MR 88g:11072.
Wu,J. (1990): Sur la suite des nombres premiers jumeaux. Acta Arith., **55**, 365–394. MR 91j:11074.
Wu,J. (1992): P_2 dans les petits intervalles. Sém. Th.Nombres, Paris, 1989–90, Progr. Math., **102**, 233–267.
Wunderlich,M. (1965): Another proof of the infinite prime theorem. Amer. Math. Monthly, **72**, p.305.
Yamamoto,K. (1955): Theory of arithmetic linear transformations and its application to an elementary proof of Dirichlet's theorem. J. Math. Soc. Japan, **7**, 424–434. MR 18 p.18.
Yin,Wen-lin (1956): Note on the representation of large integers as sums of primes. Bull. Acad. Pol. Sci., **4**, 793–795. MR 19 p.16.
Young,J.,Potter,A. (1989): First occurence prime gaps. Math. Comp., **52**, 221–224. MR 89f:11019.
Zaccagnini,A. (1992): On the exceptional set for the sum of a prime and a k-th power. Mathematika, **39**, 400–421. MR 94g:11086.
Zagier,D. (1997): Newman's short proof of the Prime Number Theorem, Amer. Math. Monthly, **104**, 705–708. MR 98j:11069.
Zassenhaus,H. (1949): Über die Existenz von Primzahlen in arithmetischen Progressionen. Comment. Math. Helv., **22**, 232–259. MR 10 p.594.
Zeitz,H. (1930): Elementare Betrachtungen über eine zahlentheoretische Behauptung von Legendre. Berlin.
Zeitz,H. (1932): Neue Beweise einer elementaren Abschätzung der n-ten Primzahl. SBer. Berlin. Math. Ges., **31**, 4–8.
Zeuthen,H. (1896): Geschichte der Mathematik im Altertum und Mittelalter. Kopenhagen.
Zhang,M.Y.,Ding,P. (1983): An improvement to the Schnirelman constant. J.China Univ. Sci. Techn., **13**, Math.Issue, 31–53. MR 85d:11090.
Zhang,N.Y. (1981): On the Stieltjes constant of the zeta function. Bejing Daxue Xuebao, no.4, 20–24. (Chinese) MR 84h:10056.
Zhang,N.Y.,Williams,K.S. (1994): Some results on the generalized Stieltjes constants. Analysis, **14**, 147–162. MR 95k:11110.
Zhang Wen-Bin (1987): Chebyshev type estimates for Beurling generalized prime numbers. Proc. Amer. Math. Soc., **101**, 205–212. MR 88m:11081.
Zinoview,D. (1997): On Vinogradov's constant in Goldbach's ternary problem. J. Number Th., **65**, 334–358. MR 98f:11107.
Zsigmondy,K. (1892): Zur Theorie der Potenzreste. Monatsh. Math. Phys., **3**, 265–284.
Zsigmondy,K. (1893): Zur Verallgemeinerung der Function $\varphi(m)$ in der Zahlentheorie. J. Reine Angew. Math., **111**, 344–346.

Author Index

Subject Index

Springer Monographs in Mathematics

Abhyankar, **Resolution of Singularities of Embedded Algebraic Surfaces,** 2nd Edition, 1998
Arnold, **Random Dynamical Systems,** 1998
Aubin, **Some Nonlinear Problems in Riemannian Geometry,** 1998
Baues, **Combinatorial Foundation of Homology and Homotopy,** 1999
Brown, **Buildings,** 1st Edition 1989, Reprint 1996, 1998
Crabb/James, **Fibrewise Homotopy Theory,** 1998
Dineen, **Complex Analysis on Infinite Dimensional Spaces,** 1999
Elstrodt/Grunewald/Mennicke, **Groups Acting on Hyperbolic Space,** 1997
Fedorov/Kozlov, **A Memoir on Integrable Systems,** 1998
Flenner/O'Carroll/Vogel, **Joins and Intersections,** 1999
Griess, **Twelve Sporadic Groups,** 1998
Ivrii, **Microlocal Analysis,** 1998
Kozlov/Maz'ya, **Differential Equations with Operator Coefficients,** 1999
Landsman, **Mathematical Topics Between Classical and Quantum Mechanics,** 1998
Malle/Matzat, **Inverse Galois Theory,** 1999
Mardešić, **Strong Shape and Homology,** 1999
Narkiewicz, **The Development of Prime Number Theory,** 2000
Ranicki, **High-dimensional Knot Theory,** 1998
Ribenboim, **The Theory of Classical Valuations,** 1999
Rudyak, **Thom Spectra, Orientability and Cobordism,** 1998
Sznitman, **Brownian Motion, Obstacles and Random Media,** 1998
Üstünel/Zakai, **Transformation of Measure on Wiener Space,** 1999

Druck: Druckpartner Rübelmann, Hemsbach